W.T. Eckley

Practical Anatomy

Including a Special Section on the Fundamental Principles of Anatomy

W.T. Eckley

Practical Anatomy
Including a Special Section on the Fundamental Principles of Anatomy

ISBN/EAN: 9783337249823

Printed in Europe, USA, Canada, Australia, Japan

Cover: Foto ©berggeist007 / pixelio.de

More available books at **www.hansebooks.com**

INCLUDING A SPECIAL SECTION ON THE FUNDAMENTAL PRINCIPLES OF ANATOMY

EDITED BY

W. T. ECKLEY, M.D.

PROFESSOR OF ANATOMY IN THE COLLEGE OF PHYSICIANS AND SURGEONS, UNIVERSITY OF ILLINOIS; PROFESSOR OF ANATOMY
IN THE NORTHWESTERN UNIVERSITY DENTAL SCHOOL; PROFESSOR OF ANATOMY IN THE CHICAGO CLINICAL
SCHOOL, AND DIRECTOR OF THE CHICAGO SCHOOL OF ANATOMY AND PHYSIOLOGY; MEMBER
OF THE AMERICAN MEDICAL ASSOCIATION, THE CHICAGO PATHOLOGICAL
SOCIETY, THE CHICAGO MEDICAL SOCIETY, THE MEDICO-
LEGAL SOCIETY OF CHICAGO, ETC.

AND

MRS. CORINNE BUFORD ECKLEY

INSTRUCTOR IN ANATOMY IN THE NORTHWESTERN UNIVERSITY DENTAL SCHOOL; PROFESSOR OF ANATOMY IN THE
NORTHWESTERN UNIVERSITY WOMAN'S MEDICAL SCHOOL; PROFESSOR OF ANATOMY IN THE
CHICAGO SCHOOL OF ANATOMY AND PHYSIOLOGY

WITH 347 ILLUSTRATIONS, MANY OF WHICH ARE IN COLORS

PHILADELPHIA

P. BLAKISTON'S SON & CO.

1012 WALNUT STREET

1899

G OF WM. F. FELL & CO.,
1220-24 SANSOM ST.,
PHILADELPHIA.

PREFACE.

We have the honor of submitting this little book as a dissecting-room guide to "Morris' Human Anatomy."

The growing tendency to specialize in the practice of medicine has reacted on the branch of medical science that continges every medical specialty—anatomy ; with the result that our schools are compelled to keep pace in teaching this branch.

Those areas of the body that can not, under existing conditions, be profitably studied in the dissecting-room are properly presented by the lecturer on anatomy. Thus the student has the benefit of the professor's dissections and investigation, and the additional opportunity for minute, detailed dissection and study, which he could not enjoy in the dissecting-room, where gross anatomy is the main topic. The minute and descriptive anatomy are, then, to be studied in Morris.

In this book the gross anatomy only will be considered. For instance, the contents of the orbit,—as the nerves, vessels, muscles,—as found on the cadaver, will be considered, but the anatomy of the eyeball will not be taken up. The gross anatomy of the peritoneum and other abdominal contents will be considered, while the student will be referred to Morris for the special anatomy of each individual organ.

Several pages have been devoted to the sympathetic nerve,—which every student should thoroughly dissect,—for the purpose only of teaching the student to find on the cadaver the relation between sympathetic and their parent somatic nerves.

The method of studying structures in the normal order in which they are exposed in dissection is followed, it is believed, as nearly as it is possible to accomplish this very desirable end.

To aid the memory in fixing salient points seen in dissection, frequent review quizzes are given.

In the introductory chapter are certain rules, principles, and generalizations which underly, to greater or less extent, the science of anatomy. It may be urged that matters of this nature are too elementary for medical students. However this may be, experience teaches that students, even after having studied the subject for months, have a memory well stored with anatomical terms which are mere abstract ideas ; they are too often ignorant of the fundamental principles of our nomenclature.

If anatomy is a nomenclaturic science [Can any department of learning be a science without a nomenclature ?] then certainly the place for the student to learn this nomenclature is in school. The practice of assuming too much knowledge on the part of the student I believe to be wrong in principle. At any rate, we must remember that this book is written, not for those who know anatomy, but for beginners.

To impress on the student the necessity of tracing muscles to their exact origin and insertion, illustrations of bones having such attachments indicated in color have been introduced. These and nearly all other illustrations are from the work to which the text most frequently refers—"Morris' Anatomy." We wish also to express our thanks to both author and publisher for the use of the excellent diagrams from Potter's "Compend of Anatomy," and for the many illustrations of the muscles taken from Gould's (Illustrated) "Medical Dictionary." In addition to these we have contributed about sixty original drawings, for the most part diagrammatic. We wish also to express our thanks to Dr. Theodor Tieken, Dr. F. R. Sherwood, and Dr. D. Loring for valuable suggestions and to Dr. W. A. Mansfield in particular, who read the original copy and made timely corrections.

A book that would fail to give due consideration to those structures of the body which can be easily demonstrated on the cadaver by any careful student would fall short of the objects of this work. A book that would attempt in approximately 400 pages to give the same that Morris devotes 1400 pages to, would, to be charitable, be just 1000 pages too small.

The sequence of structures revealed by dissection, and the great number of illustrations in the book, render an index almost superfluous. In fact, by the aid of the description of the illustration a given muscle or vessel can be found with certainty, even without an index. Still, to make the book as convenient as possible, a brief index is appended.

W. T. ECKLEY.
C. B. ECKLEY.

5816 SOUTH PARK AVE., CHICAGO.

CONTENTS.

LIST OF ILLUSTRATIONS.

PRACTICAL ANATOMY.

FUNDAMENTAL PRINCIPLES OF ANATOMY.

In the following introductory pages will be found some rules and observations on anatomical study which the student must learn, if he would become master of dissecting-room technique. Under the above heading are considered :

1. Anatomical nomenclature.
2. Anatomical tissues applied.
3. Functions of periosteum.
4. Eminences and depressions **of bone.**
5. Anatomical **weak** points.
6 Anatomical visceral roots.
7 Anatomical antagonism in muscles and nerves.
8. Shape of muscle an index to the nerve-supply
9. Hilton's law regarding articular nerves.
10. Non-apposition of anatomical structures.
11. Anatomical sheaths ; rationale of their formation.
12. Dissection defined.
13. Fanciful names **for** muscles.
14. Sphincters and orbiculars distinguished.
15. Geometrical usage in anatomy.
16. Substantive adjectives : use of.
17. Law of projectiles : its application to grooves.
18. Application of law of projectiles **to** reflex pain.
19. Fasciæ, superficial and deep.
20. Aponeuroses : varieties of.
21. Origin and insertion of muscle.
22. Synovial membranes.
23. Serous and mucous membranes.
24. Anatomical fibrous arches.

ANATOMICAL NOMENCLATURE.

This chapter **is** intended to explain in advance some of the features of anatomical nomenclature, a knowledge **of** which will facilitate an understanding of what is seen, **read,** and heard.

Anatomical nomenclature is the technical language of anatomy. Chemistry, music, **law,** all the arts and sciences, have their nomenclature. One of the surest indications of thorough system in any science is the nomenclature of that science, if, indeed, a department can be called scientific without a nomenclature.

Compound words are used to express the collective sense arising from anatomical union, conjunction, and relation. Words of this class are met many times daily, not only in anatomy, but in every branch of the medical sciences. Sterno-hyoid, omo-hyoid, musculo-aponeurotic, temporo-mandibular, express relations in anatomy that are thoroughly understood by the physician.

The principle in compound words is clearly in the nature of a copartnership. In the expression, radio-ulnar articulation, radius and ulna are both concerned in the formation of this joint ; each bone has an anatomical interest in the joint, hence a physiological responsibility. The name of an articulation, then, to represent fully the copartnership, must include all the parties to the contract. The number of partners must be two; it may be any number (1) Sterno-cleido-mastoid ; (2) fronto-malo-spheno-ethmo-lachrymo-maxillo-nasal articulation.

RULE FOR WRITING MINIMAL COMPOUND WORDS.—Place the shorter word first in the ablative, *o ;* the larger word second, in some euphonious adjective ; connect the two words by a hyphen. The rule for writing maximal compound words is the same. Here the shortest word is placed first ; it and all succeeding words end in *o* except the last.

Exceptions to the rule are (1) to indicate direction of motion ; (2) to conform to national or individual ideas of euphony. We may express the direction taken by a missile passing from the radial to the ulnar side thus : radio-ulnarly, or the opposite by ulno-radially. In many cases euphony demands the larger word first. Where exceptions are as important as these, they seem to acquire the dignity of law ; hence remember in writing compound words either word may be placed first.

Anatomical Opposites in Location or Function.—The fundamental idea in each is *antagonism ;* of *location* in the former, of *function* in the latter. An understanding of one opposite implies the *necessity* of the other's presence : anterior surface of the scapula is positive evidence of the existence of a posterior surface of the scapula. The principle is sometimes abused—*e. g., internal* iliac muscle is referred to in some texts ; there is no *external* iliac muscle ; etc. These redundancies, while undesirable, since they are misleading, are nevertheless tolerated in some good works on anatomy by a sort of license that characterizes the unique redundancy in legal and scientific phraseology. Musculus *flexor* carpi radialis and musculus *extensor* carpi radialis are anatomical examples of anatomical structures opposite *in function.* Remember, too, that structures opposite in function are usually also opposites in point of location on the part of the body in which found.

EXAMPLES OF COMMON OPPOSITES.

Superior extremity.	Inferior extremity
Anterior surface.	Posterior surface.
Ventral mesentery.	Dorsal mesentery.
Central organ.	Peripheral organ.
Somatic nerves.	Sympathetic nerves.
Proximal end.	Distal end.
Tendinous origin.	Tendinous insertion
Flexor muscles.	Extensor muscles.
Pronator agents.	Supinator agents.
Greater trochanter.	Lesser trochanter.
Base of lung.	Apex of lung.
Levator muscle.	Depressor muscle.
Gluteus maximus.	Gluteus minimus.
Pectoralis major.	Pectoralis minor.
Peroneus longus.	Peroneus brevis.
Colica dextra.	Colica sinistra.
Centripetal vessels.	Centrifugal vessels.
Compact tissue.	Cancellous tissue.
Longitudinal axis.	Transverse axis.

CLASSICAL WORDS IN COMMON USE IN ANATOMY.

Name.	Gen. Sing.	Plural.	Gen. Plural.	English.
Fascia,	. æ,	. Fasciæ.	Fasciarum.	Sash.
Glandula,	. æ,	. Glandulæ.	Glandularum.	Gland.
Vena,	. æ,	. Venæ.	Venarum.	Vein.
Arteria,	. æ,	. Arteriæ.	Arteriarum.	Artery.
Cartilago,	. inis,	. Cartilagines.	Cartilaginum.	Cartilage.
Chondrum,	. i,	. Chondra.	Chondrium.	Cartilage.
Tendo,	. is,	. Tendines.	Tendinium.	Tendon.
Musculus,	. i,	. Musculi.	Musculorum.	Muscle.
Nervus,	. i,	. Nervi.	Nervorum.	Nerve.
Vas,	. is,	. Vasa.	Vasorum.	Vessel.
Os,	. sis,	. Ossa.	Ossium.	Bone.
Os,	. ris,	. Ores.	Orium.	Mouth.
Aponeurosis,	is,	. Aponeuroses.	Aponeurosium.	Band.
Costa,	æ,	. Costæ.	Costarum.	Rib.
Viscus,	us,	Viscera.		Organ.
Pons,	tis,	. Pontes.	Pontium.	Bridge.
Trochlea,	æ,	. Trochleæ.	Trocharum.	Pulley.
Gaster,				Stomach.
Intestinum,	. i,	. Intestini.	Intestinorum.	Intestine.

Many adjectives are in use, as magnus, longus, brevis, latus, profundus, sublimis, orbicularis; they are declined and used as the nature of the case would indicate.

Rationale of Naming of Certain Muscles.—It is important that the student should, early in his course of dissection, understand why certain muscles are named as they are named in the books. In fact, the student should have a knowledge of the *principle* involved in this part of the nomenclature. To one unfamiliar with this principle the names *tibialis anticus, pronator quadratus, extensor minimi*, may seem perfect and complete designations. The impression, however, is erroneous, as the above terms are incomplete. In other words, the terms above given are only abbreviations, and are properly used when the *manner* by which usage sanctions them is understood. The principle is this: The complete technical name of a muscle contemplates: (1) The structure; (2) the structure in the capacity of an agent; (3) an agent of something; (4) an agent adjectively qualified in such a manner as to distinguish it from others of a synergistic class.

STRUCTURE.	AGENT.	AGENT OF WHAT.	DESCRIPTIVE	ADJECTIVES.
Musculus	flexor	carpi	radialis	
Musculus	flexor	carpi	ulnaris	
Musculus	flexor	cubiti	bicipitalis	
Musculus	extensor	carpi	radialis	longior
Musculus	extensor	carpi	radialis	brevior
Musculus	extensor	digitorum	communis	
Musculus	extensor	digiti	minimi	
Musculus	levator	labii	superioris	

We unconsciously acquire a habit of using the abbreviated name, and this explains in part the lamentable fact that so many students leave school with a memory well stored with words—words numerous, ponderous, classical—words which, by virtue of their abbreviational arrangement, leave on the memory no notion of the *physiological* function of the muscle. At times the habit of thinking of names of muscles almost makes us fail to recognize the full name. The following table will illustrate this point

ANATOMICAL NAME.	TECHNICAL NAME.
Tibialis anticus, ...	Musculus flexor tarsi tibialis anticus.
Tibialis posticus, ...	Musculus extensor tarsi tibialis posticus.
Peroneus longus, ...	Musculus extensor tarsi peroneus longus.

ANATOMICAL NAME.	TECHNICAL NAME.
Peroneus brevis,	Musculus extensor tarsi peroneus brevis.
Peroneus tertius,	Musculus flexor tarsi peroneus tertius.
Triceps,	Musculus extensor cubiti tricipitalis.
Biceps,	Musculus flexor cubiti bicipitalis.
Supinator longus,	Musculus supinator brachio-radialis longior.
Supinator brevis,	Musculus supinator brachio-radialis brevior.
Pronator quadratus,	Musculus pronator radii quadratus.

Grammatical Note.—In constructing technical names of muscles, note (1) the agent, and (2) the descriptive adjectives are in apposition with musculus. The thing or part of which another part is the agent is in the genitive case. The student will note that adjectives are mainly of the first and second declensions, some of the third; sublimis belongs to the third, so do not write it *sublimus.*

TECHNICAL USE OF PREPOSITIONS.

An almost unlimited number of words designative of relation are formed by the judicious use of classical prepositions. It will be noted that these words, while governing case in the language from which they are taken, cease to be prepositional in anatomical nomenclature in the great majority of instances. *Still they may be so used;* and upon the distinction now about to be made depends the proper construction of the new words which come into being by thousands by the simple process of incorporating classical prepositions. In the following cases the words are used prepositionally, and written thus :

    1. Iter e tertio ad quartum ventriculum.
    2. Processus e cerebello ad testes.
    3. Portio inter duram et mollem.
    4. Musculus accessorius ad flexorem, etc.
    5. Musculus accessorius ad musculum, etc.

In the majority of cases the word is absorbed to form a new word, bearing a close relation *in location, and not in structure,* to the word into which it was incorporated. The following examples will illustrate the process of forming new words and their proper writing : Subcostal, intermaxillary, infraorbital, supraorbital, intraabdominal, extrauterine. Remember, then, this rule for writing these words : *Where they are prepositionally used,* the preposition governs the accusative ; *where they are used in a corporate capacity,* the preposition and the qualified word, or its adjective representative, are written as one word.

Inter is used to express relation between anatomical structures having the same name. Not only does it express relation in this sense, but it also names every structure in the locality : Intercostal space, muscle, fasciæ, vessels, and nerves.

Intra—as intrauterine, intrapelvic, intracranial, intrathoracic, intraabdominal, intratympanic, intramural, intraintestinal, intramuscular,—means inside of or in the substance of. Its verbal opposite is *extra.*

Sub implies location under, in the sense of occupying a deeper plane : Subcutaneous, subserous, submucous, subperitoneal, subpleural, subsynovial, subconjunctival. It has no opposite, none being necessary to make descriptions of this nature more definite. Its legitimate field of usefulness in our nomenclature seems to be in expressing relations in the concrete rather than in the abstract. Submucous means a territory under all mucous membrane ; subcutaneous, under the skin everywhere, regardless of location of skin or kind of animal. How such expressions as submuscular, subosseous, or subgastric would shock our conception of the proper use of this word ! Still, such barbarisms as submaxillary, submental, and sublingual are sanctioned by usage—and good usage, too.

Supra is limited in its application to a few areas, for the reason that it does not combine well to form euphonious words, and euphony, you will observe, is a potent factor in determining the volume of our anatomical vocabulary. The chief use of the term is in physical diagnosis, where regions requiring no definite limitations are indicated by *supra* in combination with the name of the locality under consideration : Supraorbital, suprasternal, supraclavicular, suprascapular, suprapubic, supratrochlear, supratrochanteric, supracondylar, supragluteal, suprahyoid, supraacetabular. *Infra* is the opposite, but can not with propriety be used in some cases, as infracondylar ridge, since no such ridge exists.

Juxta has the specific meaning of one above another *with parallelism of parts.* Its use is confined to osteology almost wholly, with no good reason. In the vertebral column we have juxtaposition of bodies, transverse, spinous, and articular processes ; juxtaposition of laminæ, pedicles, and intervertebral notches. Still, for practical purposes, the biceps and brachialis are in juxtaposition. Usage determines the scope of legitimate nomenclature.

Pre, as in preaxial, the opposite of post-axial ; in presystolic ; in prevertebral, in speaking of the three great prevertebral gangliated plexuses of the sympathetic nervous system. The use of this word is limited.

Peri.—The prepositional excellence of anatomical nomenclature seems to culminate in the word now under consideration. For this there is good reason its euphonious nature combines with everything of an anatomical kind, either in substantive or adjective form, and besides it expresses a relation comprehending all other relations—*around.* Every visible anatomical structure possesses a protective coat or sheath, which surrounds and protects the same. Many of these are expressed by *peri* in combination, as the following will show : Periosteum, perimysium, periuterine, pericardium, peritendineum, peridontium, peripulmonum, perichondrium, perineurium, peritoneum, perinæum, periglottis, periorbita, perivenal, periproctic, peristaltic, perivisceral.

In the course of your dissection make a study of each technical word used, and you will soon acquire a habit of thinking in anatomical terms when doing anatomical work. Remember every anatomical region is occupied by structures, and these structures have names and relations. If you learn the principles on which the technical language of anatomy is founded, then you shall have passed the first milestone on your way to a medical education.*

THE ANATOMICAL TISSUES.

In the dissecting-room you meet normal tissues, as muscle, nerve, and skin, in all their various forms and combinations. It is here you should apply practically those things you have learned in such thorough detail in the histological laboratory. It is true you deal here with these tissues, not with the microscope, but with the unaided eye and the sense of touch. In the laboratory you learned to recognize epithelial tissue in general and special, and you classified the same according to the shape and strata of the cell. Here it is enough for you to remember that the *epithelial tissues cover the free surface of the skin and mucous membrane.*

The Bulky Portion of Your Work is Made Up of Muscles.—Of these you learn origin, insertion, relation, nerve-supply, blood-supply, investment, and function. This is gross anatomy. You must also remember histologically that *all muscles belong to the muscular tissues,* that the characteristic of these tissues is

---

* The technical terms used in anatomy are chiefly of Latin derivation ; the grammatical construction is purely so. In view of the importance of being familiar with our nomenclature, I would recommend Robinson's Latin Grammar, both for those who have given but little time to classical study and also for those who wish to review the same.

contractility, and that muscles occur, (1) in the *voluntary* form ; (2) the *involuntary* form. You will be expected to classify each muscle according to the above forms. To aid you in this, the following table is inserted :

1. In the alimentary canal (œsophagus, stomach, intestines, large and small), and in the embryological offspring of the duodenum (the common bile duct, the hepatic ducts, the gall-bladder, and the pancreatic duct) and salivary glands.
2. In the genito-urinary tract of the male and female: the muscular part of the vas, the seminal vesicles, Cowper's glands, the corpora spongiosa and cavernosa, and the prostate in the male ; in the uterus, vagina, Fallopian tubes, round and broad ligaments, and in the erectile tissue of the nipple and external genitals of the female. In the bladder, urethra, and all parts of the urethra *in both male and female.*
3. In the trachea, bronchi, and pleura ; in all arteries, veins, and lymphatics , in the iris, ciliary body, and eyelids ; in the skin of the scrotum ; in the hair follicles and sebaceous glands.
4. You will see later that the territory where you are to locate the involuntary form of muscular tissue coincides with the distribution of the *sympathetic nerve*, or the nerve of organic life.

You Will Find Brain and Nerves.—The structural part of these is purely histological ; still, you will be expected to remember that nerve-tissue consists of cells, fibres, and neuroglia, and a connective-tissue framework. You will find nerves, called *sympathetic*, supplying the viscera and all others ; the *cerebro-spinal*, supplying the skin, joints, muscles, and organs of special sense. *You should remember that all brain and nerve matter belong to the nervous tissues.*

The connective tissues, in their many forms, will be found everywhere. I wish thus, in advance, to teach you where to find and how to classify the same. All the peristructures mentioned in the chapter on nomenclature are forms of connective tissue. This is the most widely distributed of all the tissues. Your task now is to learn to recognize its forms as you meet them.

FORMS OF CONNECTIVE TISSUE YOU WILL MEET IN THE DISSECTING-ROOM.

1. *Mucous Form.*—This is a clear, jelly-like substance surrounding the umbilical cord. Ligate this and see how easily a thread cuts through the same. This is the jelly of Wharton you will hear about. *It is one form of connective tissue.*
2. *Areolar Form.*—Immediately on removing the skin you will notice the areolar form of connective tissue. It is also called *superficial fascia* in this locality. It contains a variable amount of fat. It is called *areolar membrane* in some of the older text-books. On separating one muscle from another, you will see this *areolar tissue.* Its specific name, in this locality, is *intermuscular fascia.*
3. *Cartilage forms* of connective tissue occur in the trachea, larynx, and external ear, where the presence of this tissue gives these structures their strength and elasticity. You will find it completing the space between the ribs and sternum; you will find it covering the articular surfaces of bone in movable joints.
4. *Bone Form.*—This is coextensive with the osseous skeleton. Every bone, be it long, short, flat, irregular, is *connective tissue in this form.* Each bone is covered, at its articular surface, when it moves upon another bone, by

cartilage : this cartilage is one form of connective tissue. Each bone is covered by periosteum, except at the articular surfaces; this periosteum is connective tissue.

5. *Elastic Form.*—In the interlaminar ligaments, or ligamenta subflava, you will find a form of connective tissue, *yellowish in color,* and also dense, elastic, and strong. You will find this as well in the epiglottis. Wherever you find structures possessing (1) *elasticity,* (2) *yellowish color,* as the aorta and structures above mentioned, they owe these qualities to the *elastic form of connective tissue.*

6. *Adipose Form.*—You will find this in the superficial fascia or areolar tissue under the skin; in the great omentum, about the heart, in the iliac region of the pelvis, and in the ischio-rectal fossa. Remember, fat is the adipose form of connective tissue. This tissue represents the great storehouse for potential energy. The student who desires to see the human body in possession of this potential energy will select a fat subject for dissection; while the student who desires to keep his hands clean will continue to clamor for lean material. If you intend to do surgical operations, remember this: The emaciated patient is seldom operated on. Only those possessing fat and vitality are considered good risks. If you dissect a fat cadaver, then you may expect to see man as he is in health. A lean man is man minus fat—not a fair sample.

7. *Supporting Form.*—Dissect any organ, and you will find it covered by a layer of connective tissue, and the parts making up the interior supported by the same. *Boil a muscle,* and you will see the connective tissue interior. Macerate thoroughly a spleen, and by gentle compression in warm water remove the spleen-pulp, and you will have remaining the *connective tissue* of this organ—the *supporting form of connective tissue.* The nerve-cells and nerve-fibres are likewise supported by connective tissue, called *neuroglia.*

8. *Dense Form.*—This form you will find in tendons and fasciæ; in the cornea and sclera, in the dense aponeuroses; in the strong intermuscular septa.

QUERY.—How can a tissue appear in so many forms? How is it possible to recognize in the *dentine of the teeth and in the areolar tissue between muscles* one and the same tissue? How can you classify bone and the jelly of Wharton in the same category?

EXPLANATION.—Each tissue consists of two constituents: (1) *Cells,* (2) *intercellular substances.* The *former* are products of the ovum; the latter of the *cells themselves.* Each constituent has a certain part to perform in the organ in which the tissue is found. Upon the cell depends the life of the tissue; upon the intercellular substance depend the jelly-like condition, the dentine hardness, the fascial strength, the areolæ of intermuscular connective tissue and superficial fascia, the resistance of muscle, the elasticity of the ligamentum subflavum, the cement substance of the epithelial tissues covering the surfaces of the integument and the various mucous membranes.

Homely illustration of how a tissue can appear under so many different forms and still remain connective tissue : (1) Permit *water* to represent the intercellular substance; (2) permit *plaster-of-Paris* to represent the cellular element. As you mix the two, the water will become more and more condensed, until you have finally obtained a substance of bony hardness. The plaster remains plaster, having undergone no changes. In the tissue under consideration the cellular element remains the same: The bone-cell is morphologically the same as the jelly-cell, the tendon-cell, and the fascia-cell. Condensation of the intercellular substance is when the changes have really taken place.

In your dissections, when you find fascia, tendon, bone, cartilage, sclera, cornea, peristructures, supporting structures, dentine, and Wharton's jelly, you are to remember, these structures, while widely varying in form, all belong to the *greater connective tissue family.*

RÉSUMÉ OF ANATOMICAL TISSUES.

1. *Muscular tissue,* for voluntary and involuntary contraction, upon which depend the aggressive and defensive attitudes of the body, as well as the movements of hollow conduits and certain viscera.
2. *Nervous tissue,* upon which depends intellection, reception of impression from an environment, and transmission of brain impulses, whether voluntary or involuntary.
3. *Connective tissue,* on which depends strength, framework, and various investing structures. Common forms specialized are tendon, ligament, capsule, bone, cartilage, dentine, cementum, etc.
4. *Epithelial tissue,* upon which depend external and internal protection, giving as these tissues do a certain resiliency and stability to the areas which they cover, as the skin and mucous membranes.

FIG. 1.—DISPLACEMENT OF COTTON BY GROWING PLANTS.

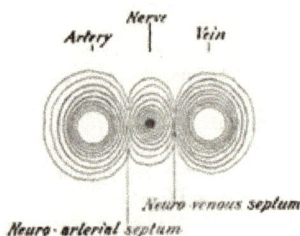

FIG. 2.—CONNECTIVE TISSUE DISPLACED BY THE ARTERY, NERVE, AND VEIN; THE CONNECTIVE TISSUE LEFT BETWEEN THE ARTERY AND NERVE AND BETWEEN THE NERVE AND VEIN ARE TWO SEPTA.

Rationale of the formation of anatomical sheaths, tunics, togas, capsules, and all the peristructures referred to and enumerated in the section on anatomical nomenclature.

In your dissections you will meet the capsules of glands and the sheaths of vessels. Where an artery, vein, and nerve are in the same sheath, they will be separated from each other by a *septum.* In your first dissection you will be disappointed at the frail condition of these sheaths—at their lack of organization. You will expect to find the sheath of an artery resemble a piece of heavy cloth sewed tightly about the trunk of a tree; your chagrin will be manifest when you find it resembles more the investment of straw given tender shrubs to prevent depredations of rodents in our parks. As any attempt to remove the *protective* from the tree would result in the complete destruction of the protective, as you removed straw after straw, so, likewise, when you attempt to disensheath a vessel you arrive at the vessel, but the sheath has disappeared. Remember, then, *sheaths are actual entities,* and their architecture is adapted to the needs and requirements of the vessel ensheathed, just as the cobweb is architecturally suited to the needs of the spider; still, when you molest the latter or attempt to dissect the former, the result is the same—each almost completely disappears, such is the delicacy of their respective material.

For the sake of gaining a conception of the architecture of sheaths and the rationale of their formation by displacement of connective tissue, let the following explanation be studied : Figure 1 represents a pot of earth in which various seeds have been planted ; the pot is covered with a layer of absorbent cotton, and exposed to the sun and rain. In time the cotyledons begin to peep through the cotton ; as each becomes larger in diameter, it displaces more cotton. As the cotton displaced by the growing plant may represent the sheath of the plant, so, in like manner, to aid the memory, let us compare a growing artery, bone, muscle, nerve, or gland surrounded by connective tissue (which represents the cotton in the simile of the growing plant) to the plant. The connective tissue would be displaced as the anatomical structure became thicker, and this *displaced tissue* we call the sheath, capsule, tunic, toga, or peristructure.

Now, by the same process account for periosteum, perimysium, perineurium, and all peristructures and capsules.

PERIOSTEUM AND ITS FUNCTIONS.—Periosteum is the anatomical investment of bone. It covers all the bone except the articular surfaces. It is, as you must demonstrate on your dissection, most intimately adherent to those parts of the bone having the greatest number of irregularities. Its functions are as follows :

1. *Osteogenetic,* for it makes bone grow in thickness.
2. *Protective,* since it hinders progress of contiguous inflammation.
3. *Attachmental,* since it gives attachment to muscles.
4. *Nutritive,* since it feeds the bone with blood.
5. *Ligamentous,* since it forms all capsular ligaments.
6. *Retentive,* since it tends to retain ends of broken bones.

EMINENCES, DEPRESSIONS, AND SURFACES OF BONE are determined in shape by the muscles associated therewith, and take the *name of the muscle.* You are to look on muscular traction as the factor that determines the size of an eminence or the depth of a fossa. In fact, the surfaces of bone are named according to the occupant. In a given bone, as the humerus, name the articular surfaces according to the occupants: (1) Scapular; (2) radial ; (3) ulnar. Name the surfaces according to the muscles. The posterior surface has an upper and a lower tricipital surface, corresponding to the humeral heads of the triceps, etc. Bicipital tuberosity, iliac fossa, gluteal ridge, deltoid impression, are instances. You will observe that this rule is not always observed ; still the value of the rule is no less.

ANATOMICAL WEAK POINTS comply with two conditions : (1) A location in the continuity of the structure where there is a *sudden abrupt change in the direction.* (2) *A junctional area,* as between skin and mucous membrane, as where the small intestine opens into the large. In these localities fractures occur in bone, aneurisms in arteries, morbid growths at muco-cutaneous areas. In your dissection bear these areas in mind. Look for intussusceptions at the weak points in the colon ; for fractures at the necks of bones, etc.

ANATOMICAL ROOTS are places uncovered by serous membrane, in connection with viscera, where the vessels, nerves, and conduits enter to carry on (1) the *functional activity* and (2) the *nutritive activity* of the organ. In your dissection you will give special study to the *root-structures* of the heart, lung, liver, spleen, kidney, intestine, ovary, testicle, etc. The *functional activity* is represented by what the organ does for the body in general ; the *nutritive activity* by what the organ does for its own nutrition. The nutrient artery to organs is accompanied by sympathetic nerves, which regulate the supply of blood each organ shall receive.

ANATOMICAL ANTAGONISM IN MUSCLES necessitates a consequent antagonism in the nerves that supply these muscles, hence you will understand the meaning of flexor, extensor, abductor, and adductor nerves. The very nature of logical antagonism implies a degree of equality in length, weight, strength, and vantage for origin and insertion. At the hip, flexors, extensors, abductors, and adductors are seen. The antagonistic nerves are the anterior crural, the great sciatic, the obturator, and the gluteal.

SHAPE OF MUSCLES AN INDEX TO THE PLACE WHERE THE MUSCLE RECEIVES ITS NERVE-SUPPLY.—I desire you to study carefully the following drawings, and in your dissection you must faithfully determine the general shape of the muscle and trace out the nerve thereto.

1. *Triangular Muscles.*—The pectorales major and minor, the supra- and infra-spinati, the subscapularis, the deltoid, the gluteals, the pyriformis, and the obturators are examples ; take their nerve-supply near the apex.

FIG. 3.—TRIANGULAR.   FIG. 4.—QUADRANGULAR.   FIG. 5.—FUSIFORM.   FIG. 6.—CLUBBED.

2. *Quadrate Muscles.*—The pronator quadratus, the brachialis anticus, and multifidus spinæ are good examples ; take their nerve-supply near the centre.
3. *Fusiform Muscles.*—Take their nerve-supply near the middle. This, as the biceps, is the same as No. 2, compressed at the ends.
4. *Clubbed Muscles.*—This includes a large number, as those taking origin from the outer and inner humeral condyles take their nerve-supply early. In your work notice how very early the median gives off branches to supply muscles of this type.
5. *Serrati muscles* and those arising by more or less pronounced digitations, as the serrati superior and inferior posticus, the diaphragm and the planiform muscles of the abdominal walls, seem to conform to the type of triangular muscles, taking their nerve-supply near the apices of the digitate origins ; still, if the nerves are traced out, they will be found to approach the centre and conform to the quadrate type.

Remember, this tracing out of nerves belongs to macroscopic anatomy. To gain an idea of the manner in which the ultimate tissue is innervated, you must study up *motorial end plates* in physiology and histology. The only object in introducing the subject of shape of a muscle as an index to the nerve-supply entrance is to impress the necessity of tracing out, as far as possible, in your dissection the branches of the nerves ; possibly the scheme introduced may aid the memory.

✓ Hilton's Law : Nerve trunks that supply muscles that move a joint supply the joint acted upon with *articular branches ;* they also supply the skin, covering the fullest insertion of the muscle. Given, then, the number of antagonistic groups of muscles, you are by this rule able to know the source from which the joint is innervated.

Anatomical Apposition Is Nowhere Found.—In dissecting in any region of the body bear this in mind. *Structures are always disjunctively connected by a layer of connective tissue.* Found between muscles, it is called areolar or intermuscular connective ; beneath the pleura, subpleural ; beneath the peritoneum, subperitoneal ; beneath mucous membrane, submucous ; beneath the periosteum, subperiosteal. Notice the mortar between the bricks in a wall ; this corresponds to connective tissue in anatomy.

Dissecting Is the Art of Dividing the Connective Tissue that Intervenes between Adjacent Structures, and the division must be made in such a manner as to do no violence to anatomical compounds,—their nerve-supply, their blood-supply, their excretory ducts, their capsules, and their anatomical relations.

Fanciful Names for Muscles, names founded on some one dominant idea, are frequently met. It is to be regretted that our nomenclature could not have been founded on the idea of *function* throughout. Note here that the substantive, *musculus,* while seldom expressed in our texts, is *always implied.* The following will show you some of the names of muscles founded on fanciful ideas, and also the full name as it should appear if founded on the idea of function

Fanciful Names.	Physiological Names.
Sartorius (tailor),	Musculus extensor tibialis et adductor femoris.
Rhomboideus (geom.),	Musculus levator scapulæ rhomboideus.
Transversalis (direct.),	Musculus compressor abdominis transversalis.
Rectus (geom.),	Musculus flexor thoracis rectus.
Pyramidalis (geom.),	Musculus tensor lineæ albæ pyramidalis.
Vastus Externus,	Musculus extensor vastus externus.
External Oblique,	Musculus compressor abdominis obliquus externus.
Serratus Magnus,	Musculus depressor scapulæ serratus magnus.
Quadratus Lumborum,	Musculus flexor spinæ lateralis quadratus lumborum.
Latissimus Dorsi,	Musculus abductor brachii latissimus dorsi.
Longissimus Dorsi,	Musculus extensor spinæ longissimus dorsi.

Orbicular Muscles and Sphincter Muscles.—You will find the above terms often used interchangeably. The important areas occupied by these muscles make the majority of the same of prime interest to the general surgeon, the genito-urinary specialist, the gynæcologist, the obstetrician, the oculist, the laryngologist, the rectal specialist, the patient. Still, no rational classification of these muscles exists. At the present all orbiculars are sphincters, but not all sphincters are orbiculars, as the following table, founded on usage, will show :

Synonymous Usage.

Orbicularis oris,	Sphincter oris.
Orbicularis ani,	External sphincter ani.
Orbicularis palpebrarum,	Sphincter palpebrarum.

UNIVERSAL USE OF SPHINCTER.

Musculus sphincter vaginæ.
Musculus sphincter urethræ.
Musculus sphincter prostatæ.
Musculus sphincter œsophagei.
Musculus sphincter laryngis.

Musculus sphincter iridis.
Musculus sphincter ilei.
Musculus sphincter gulæ.
Musculus sphincter vesicæ.

As the following comparison will show, there is no good reason for confusion of terms and an annoying interchange between orbicular muscles and sphincter muscles. Such is usage.

Summary of Differences Between Orbiculars and Sphincters:

Orbiculars.

1. The orbiculars are supplied by spinal nerves.
2. The orbiculars are all voluntary.
3. The orbiculars are all dermal muscles.
4. The orbiculars are at muco-cutaneous margins.
5. The orbiculars move rapidly.
6. The orbiculars have parietal blood-supply.
7. The orbiculars are very sensitive to pain.
8. They refer their pain to place of irritation.
9. They report pain foudroyantly.
10. Their nerve-supply is somatic—never sympathetic.
11. When lacerated, they repair slowly.
12. They are synonymous for sphincters in three cases only.

Sphincters.

1. Sphincters are supplied by the sympathetic.
2. They are all involuntary—unstriped fibre.
3. They are not dermals, but viscerals.
4. They are never at muco-cutaneous areas.
5. They always move slowly.
6. They have visceral blood-supply.
7. They are not very sensitive to pain.
8. They refer pain to somatic areas.
9. Their own report of pain is dull and slow.
10. Their nerve-supply is sympathetic.
11. When lacerated, they repair readily.

The reader is referred to reflexes along somatic and sympathetic lines in the dissection of the female pelvis.

The Uterus.

This would seem to be the most fully grown and fully developed structure among the organs having sphincteric openings. Its proper sphincters are known as the *os internum* and *os externum.* It complies in every way to eleven points characteristic of sphincters.

GEOMETRICAL TERMS AND FIGURES.—You will have much descriptive usage along mathematical lines. This always aids the memory, for it appeals to the basis of every one's education.

The axilla has base, apex, angles, and boundaries.
The neck has roof, floor, boundaries, and triangles.
The nasal fossa has roof, floor, walls, and openings.
The orbit has roof, floor, base, apex, angles, and walls.

The mouth has roof, floor, sides, openings, dental arches.
The tympanum has roof, floor, and **four walls.**
Hunter's canal has roof, floor, boundaries, and extremities.
Scarpa's triangle has roof, floor, base, apex, boundaries.
The ischio-rectal fossa has base, apex, walls, sacral and pubic **ends.**
The thorax has apex, base, anterior, posterior, **and** lateral **walls.**
Inguinal canal has rings, roof, floor, and two walls.
Popliteal space has roof, floor, extremities, and boundaries.

SUBSTANTIVE ADJECTIVES.—These are so frequently used that I wish the student early to learn to supply the missing substantive, not only in the case of muscles, as referred to in *nomenclature*, but also in the case of arteries, veins, nerves, and lymphatic glands. In social life, after long and thorough acquaintance with your fellows, you justifiably dub your friends John, James, and the like, the while mindful of the full baptismal name; these liberties of address you would consider improper in addressing comparative strangers. Likewise in anatomy you speak of structures **as follows:**

1. A FRIEND.	2. A STRANGER.
The radial artery or vein,	Arteria radialis or vena radialis.
The ulnar artery or vein,	Arteria ulnaris or vena ulnaris.
The brachial artery or vein,	Arteria brachialis or vena brachialis.
The femoral artery or vein,	Arteria femoralis or vena femoralis.
The hepatic artery or vein,	Arteria hepatica or vena hepatica.
The splenic artery or vein,	Arteria splenica or vena splenica.
The vertebral artery or vein,	Arteria vertebralis or vena vertebralis.
The mesenteric artery or vein,	Arteria mesenterica or vena mesenterica.
The lingual artery or vein,	Arteria lingualis or vena lingualis.
The facial artery or vein,	Arteria facialis or vena facialis.

Lymphatic Glands.

Cervical gland,	Glandula lymphatica cervicalis.
Mesenteric gland,	Glandula lymphatica mesenterica.
Inguinal gland,	Glandula lymphatica inguinalis.
Axillary gland,	Glandula lymphatica axillaria.
Epitrochlear gland,	Glandula lymphatica epitrochlearis.

Nerves.

Radial,	Nervus radialis.
Ulnar,	Nervus ulnaris.
Femoral,	Nervus femoralis.
Lingual,	Nervus lingualis.
Facial,	Nervus facialis.
Circumflex,	Nervus circumflexus.
Great sciatic,	Nervus sciaticus magnus.
Sympathetic,	Nervus sympatheticus.

Remember, arteries, nerves, **veins,** lymphatics, and muscles are spoken of, the country over, in the adjective abbreviated form by those who know what they are talking about; you, however, are not to speak of these structures *in the abbreviated form* until **you are** familiar with the classical name. After such knowledge has been **acquired,** then *always be governed by usage.*

LAW OF PROJECTILES.—In anatomy and physiology, in obstetrics and surgery, in therapeutics and chemistry, we must frequently invoke a reason for the *location* of a nerve or vessel; for the *direction* taken by pus or a bullet; for a *misplaced* foetal head in utero; or for the *tracts pursued* by pain and motion, when these manifestations **are** far removed from their *logical locality.* The above—pain, pus, fœtal, ovoid, bullet, and motion—are all *projectiles*, and act in accordance with this law : A projectile follows—

1. The point of least resistance.
2. The line of greatest traction.
3. The resultant of the two.

Grooves Transmitting Vessels and Nerves, and an Application of the Law of Projectiles.—You may be surprised, in your subsequent dissection work on the cadaver, to find arteries and nerves, as a rule, in grooves. This, however, is the case, and you are to *name the groove according to the name of the artery*, except in those cases where the channel so expands as to be of great surgical importance, in which case the space would seem rather to give its name to the vessel. You are to learn and study these grooves according to *roof, floor, boundaries, and contents*. If you can do this on the cadaver, then you will be competent to do the same thing on a patient. You ask, Why do these structures occupy grooves ? Philosophy answers, *projectiles follow the line of least resistance.*

APPLICATION OF THE LAW OF PROJECTILES IN THE CASE OF PAIN REMOTE FROM PLACE OF INJURY.—*The cranio-spinal nerves* are called somatic. *Nerves that supply the thoracic, abdominal, and pelvic viscera* are called *sympathetic* or *visceral.*

A *burn* on the surface of the body is painful. The pain is violent and quickly reported to the brain. No one, not even the physician, is in doubt as to the *location of the pain.* The individual nerve-fibres of somatic nerves are of *large calibre*, and the course of the same is *not interrupted* by ganglia. These conditions would seem to favor rapid, direct transmission of both pain and motion.

In cancer of the stomach the pain is in the *abdominal walls.* In *enteralgia* the pain is in the region of the *umbilicus.* In *renal colic* the pain is in the end of the *penis.* In *ovaritis* the pain is in the *back, chest, scalp*, and *upper* or *lower extremities.* In each instance above cited the pain is referred to a territory supplied by cerebro-spinal and not by sympathetic nerves. In each instance the pain originates in a territory supplied by sympathetics. In each instance pain is a projectile, and must obey the universal law of projectiles. It is, then, the duty of physician and student alike to trace out the anatomical tracks by which the pain rationally travels. You will note, further, that in each case above cited the pain was referred or reflected to the skin *via* those somatic nerves which had their origin nearest the sympathetic plexuses from which the affected organs drew their sympathetic nerve-supply. The only rational conclusion then is, it would seem, this : Pain is referred to somatic areas because these nerves are so constructed as to offer minimal resistance to the transmission of pain. In other words, projectiles follow the line of least resistance. *In your dissections you will be expected to find the communications between somatic and sympathetic nerves.*

FASCIÆ (PLURAL); FASCIA (SINGULAR).

The word "fascia" will probably occur more frequently in your anatomical reading than any other word. The term means a sack or bundle, but any amount of derivation, any amount of description in books, will give no adequate idea of the application of the word fascia in all its various forms as met in dissecting. The structure, to be understood and appreciated, must be seen and studied on the cadaver. I submit this little outline on fascia, hoping thereby to give the student a working basis for intelligent dissection.

There are two grand divisions of fasciæ :

1. **The superficial fascia,** found under the skin in every region of the body. This is also called in histology the areolar form of connective tissue.

2. **The deep fascia or fascia profunda** has the following characters:

(1) It is dense, heavy, and strong in some places; weak in other places.

(2) It is attached to bone in subcutaneous areas.

(3) It has perforations for cutaneous vessels and nerves.

(4) It is attached to all eminences of bone in the vicinity of joints.

(5) It is a dense form of connective tissue.

(6) It receives many different names in different localities.

You will study the following description, and, if necessary, refer to it frequently in your dissections.

SPECIAL NAMES AND SPECIAL FUNCTIONS OF FASCIA PROFUNDA.—(1) In the *upper extremity;* (2) in the *lower extremity;* (3) in *association with muscles;* (4) in *surgical areas;* (5) forming *intermuscular septa.*

In the upper extremity the deep fascia occurs as:

1. *Anterior annular ligament* lies in front of the carpus. Under this pass the median nerve, flexor sublimis digitorum, flexor profundus digitorum, flexor longus hallucis. This is ¼ of an inch thick, extending from the os trapezium on thumb side to the pisiform bone and unciform process of the unciform bone on ulnar side. It is continuous below with the palmar fascia, and above with the deep fascia covering the muscles. Its function is to bind down the structure under it.

2. *Posterior annular ligament* is behind the carpus. Under it pass the three extensors of the thumb, the three carpal extensors, the extensor indicis, the extensor minimi digiti, the extensor communis digitorum. It is continuous above with the deep fascia covering the muscles, below with the fascia of the back of the hand. It has seven synovial compartments.

3. *Palmar fascia* has an outer part, the thenar, covering the thenar eminence; an inner part, covering the hypothenar eminence; a middle part, covering the great distributing area to the fingers; a neuro-vasal area, in which are the median nerve, the ulnar nerve, and the superficial palmar arch. It is continuous above with the annular ligament, below with the ligamenta vaginales; laterally, with the fascia dorsalis.

4. *Dorsal fascia* is on the dorsum of the hand. It binds the tendons down and holds them together. It is continuous above with the annular ligament, below, with the extensor tendons. It is called fascia dorsalis manus, in contradistinction to the fascia dorsalis pedis.

Ligamenta vaginales form dense sheaths for the flexor tendons of the fingers. They are lined with synovial membrane called theca. The thecæ terminate in thecal culs-de-sac for all the fingers between the thumb and little finger opposite the metatarso-phalangeal articulation. The thecæ for the little finger and thumb terminate above in the general synovial sac under the anterior annular ligament. Very often the thecæ for the little finger and thumb terminate as do the other three digits.

SPECIAL NAMES AND SPECIAL FUNCTIONS FOR THE DEEP FASCIA OF THE LOWER EXTREMITY

Fascia lata, the deep fascia on the front of the thigh. It has an iliac and a pubic part, separated by the saphenous opening. It is continuous above with Poupart's ligament; below, with eminences about the knee. It is very dense and strong.

Ilio-tibial band is in reality the aponeurotic insertion of the tensor vaginæ femoris muscle. The fascia or band is inserted into the eminences about the outer part of knee.

Anterior annular ligament extends from malleolus to malleolus. Under it are found the tibialis anticus, the extensor proprius hallucis, the extensor longus digitorum, the peroneus tertius, the anterior tibial artery and nerve. It is continuous above with the deep fascia ; below, with the dorsal fascia.

External annular ligament extends from outer malleolus to os calcis. Under it are the peroneus longus and peroneus brevis. On it rest the short saphenous vein and nerve.

Internal annular ligament extends from inner malleolus to os calcis. Under it are the tibialis posticus, the flexor longus digitorum, a sheath containing the posterior tibial nerve, artery, and veins ; behind this sheath is the tendon of the flexor longus hallucis.

Dorsal fascia—fascia dorsalis pedis—is on the back of the foot. It covers the extensor muscles and the dorsalis pedis artery. It is continuous above with the annular ligament ; below, with the extensor tendons to the toes.

Plantar fascia is on the sole of the foot. It has three parts,—an outer or hypothenar, an inner or thenar portion, a central portion. These cover three muscles, forming the first layer. Behind, it is attached to the os calcis ; in front, it is continuous with the ligamenta vaginales.

Ligamenta vaginales are strong sheaths for the flexor tendons. They are lined by synovial membrane. These thecæ all terminate in thecal culs-de-sac 1.5 inches above the toe-clefts.

DEEP FASCIA ASSOCIATED WITH MUSCLES.—The location of some muscles is such that their action extraordinarily develops the fascial investment on one surface of the muscle. This is always at the expense of the fascia on the other side of the muscle. The fascia then takes the name of the muscle, as follows :

1. *Masseter muscle*, masseteric fascia.
2. *Temporal muscle*, temporal fascia.
3. *Deltoid muscle*, deltoid fascia.
4. *Supraspinous muscle*, supraspinous fascia.
5. *Infraspinous muscle*, infraspinous fascia.
6. *Obturator internus muscle*, obturator fascia.
7. *Iliac muscle*, iliac fascia.
8. *Transversalis muscle*, transversalis fascia.
9. *Pectoralis major muscle*, pectoral fascia.
10. *Biceps muscle*, bicipital fascia.

DEEP FASCIA IN SURGICAL AREAS.—1. *The deep fascia* helping to form the base of the axilla is called *axillary, suspensory*. It is strong, continuous in front with the pectoral ; externally, with the brachial ; posteriorly, with the fascia of the latissimus dorsi.

2. *Pelvic fascia* has the following names according to muscle, function, etc. : Iliac, iliac muscle ; psoas, psoas muscle ; obturator, obturator internus muscle ; anal fascia, levator ani muscle ; pubo-prostatic, anterior ligament of bladder ; recto-vesical ; pelvic white line ; inner layer of the triangular ligament.

3. *Popliteal fascia* helps to form the roof of the popliteal space. It possesses some transverse muscular fibres, by which the resiliency of the fascia is kept up in forced extension of the leg. It is perforated by the short saphenous vein and nerve.

4. *Cubital fascia* covers the cubital fossa in the retiring angle of the elbow. It covers the median nerve, the tendon of the biceps, the brachial artery and its terminals, the radial and ulnar, and some of their branches. It is also called bicipital fascia, semilunar fascia, and falciform fascia.

DEEP FASCIA FORMING INTERMUSCULAR SEPTA.—1. *The group of muscles* on the front of the thigh is separated from the adductor group internally, and from the flexor group posteriorly ; also, the flexor group is separated from the adductor group.

2. *The peronei muscles* are separated from the flexors behind and the group

on the anterior surface of the tibia and fibula in front. The superficial group behind is separated from the deep group.

3. *The radial group* of the forearm is separated from the flexors in front and the extensors behind. The superficial is separated from the deep group on the anterior part of the forearm.

4. *Groups of muscles*, acting in harmony to discharge some one physiological function, form a musculature. Adjacent musculatures are always separated by septa of deep fascia. Each musculature has its own nerve-supply. Fascial septa are always attached to the bone, being continuous there with the periosteum.

APONEUROSIS.—The term aponeurosis, in the plural, aponeuroses, applies to dense, strong fascia, in localities where the action of the fascia is intimately associated in more than a secondary manner with that of a muscle. You will find the aponeurosis of the diaphragm, of the internal and external oblique and transversalis muscles of the abdomen, the occipito-frontal aponeurosis, the lumbar, pharyngeal, and vertebral aponeuroses. In these places this structure forms either an integral part of the muscle itself, or constitutes the main structure in the region in which it is found. The terms aponeurosis of **investment and insertion** refer to the structure as forming either a cover for a muscle or its insertion. A single aponeurosis may do both of these. For instance, the aponeurosis of investment for the gluteus maximus muscle continues downward, **and the** muscle is *aponeurotically inserted* thereby into the deep fascia of the **thigh.**

ORIGIN AND INSERTION OF MUSCLES.—These are very arbitrary terms. Usually the more fixed of two points is called the origin, the less fixed point, the insertion. Insertion may be by tendon or by aponeuroses. Notice that many muscles take a large part of their origin from the deep fascia investing them. Attachment applies to both origin and insertion.

ANATOMICAL FIBROUS ARCHES.—You will, in the course of your reading, see the above expression. The two heads of muscles are always connected by a fibrous arch. The following will illustrate the point: The gastrocnemius, soleus, biceps, flexor sublimis digitorum, pronator radii teres, flexor longus pollicis, the biceps of the thigh. Vessels and nerves are frequently described as passing under the fibrous arch of a muscle.

THE SUPERFICIAL FASCIA—
1. Is found immediately under the skin ; .
2. It has two strata—an upper and a **lower** ;
3. The **upper stratum** contains fat—the panniculus adiposus ;
4. The lower stratum contains the cutaneous vessels and **nerves** ;
5. It has no fat in the eyelids, penis, **and** scrotum ;
6. It is not attached to bony eminences ;
7. It has some muscles, called dermals ;
8. Its local special names are cribriform and colles ;
9. Its fat in the palms and **soles** is called granular fat.

THE SEROUS MEMBRANES.—You will be able to demonstrate on **the cadaver—**
1. They are all related to the lymphatic system ;
2. They form air-tight cavities, except the female peritoneum ;
3. They are all thin and transparent ;
4. They completely or partially invest organs ;
5. Peritoneum, pleura, and pericardium are proper serous **membranes** ;
6. The linings of blood-vessels and joints are subdivisions.

3

SYNOVIAL MEMBRANES.—In your practical work on the cadaver you will find : (1) **Articular** synovial membrane lining the capsules of movable joints ; (2) vaginal synovial membrane lining the ligamenta vaginales, surrounding the tendons to the fingers and toes, and also in **other** places where a tendon would be exposed to friction ; (3) bursal synovial membrane between an eminence of bone and a muscle playing thereover. The **secretion** of these synovial membranes is, by virtue of its viscidity, fitted to resist friction. Synovial membranes belong to the class of serous membranes.

THE MUCOUS MEMBRANES.—These communicate with **the air.** You will demonstrate them lining the digestive tract, the genito-urinary passages, and the respiratory tract. The margin between skin and mucous membrane is called a muco-cutaneous margin. You must demonstrate this margin in all regions of the body. The specific name of any muco-cutaneous margin is determined by the name of the region, as : (1) The nasal ; (2) ocular ; (3) oral ; (4) anal ; and (5) vaginal muco-cutaneous margin. These are all anatomical weak points, because they are *junctional areas.*

NERVE-ENDINGS.—You should remember that muscles receive both sensory and motor nerves ; that nerve-endings are in tendon, blood-vessel, all the membranes, and skin ; that you are not supposed to dissect out these nerve-endings—this belongs to histological and physiological research. You are to note where the muscle receives its nerve-supply,—usually on the under surface,—and then trace the same as far as possible.

THE HEAD AND NECK.

Any attempt to dissect the general or special regions of the head and neck must be preceded by a thorough review of those osteological parts associated—(1) with the attachment of muscles ; (2) with the limitation of surgical or physiological areas ; (3) with the transmission of vessels and nerves ; (4) with junctional areas, whether the union is fixed, as between the teeth and alveolus, or movable,

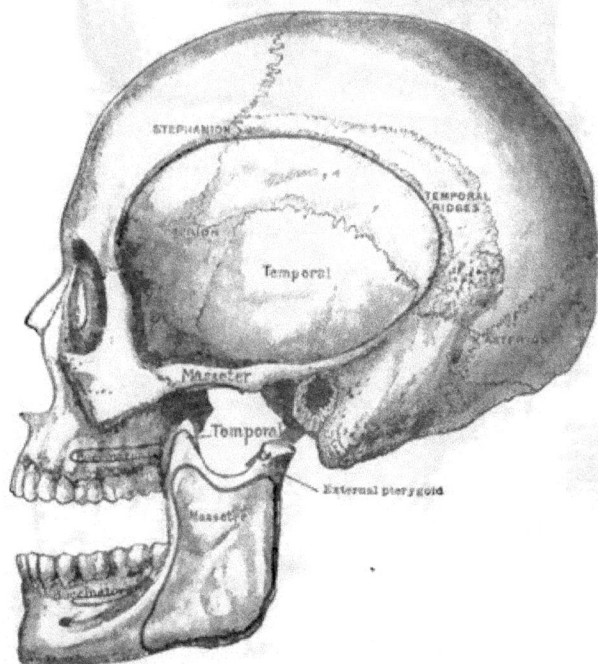

FIG. 7.—THE SKULL. (Norma lateralis.)

as between the mandible and the temporal bone. To accomplish this interesting review let the illustrations in the book, the skull, and the cadaver all be used. (Fig. 7.)

1. **The complete temporal ridge** has—(1) a *frontal* portion ; (2) a *parietal* portion ; (3) a *temporal* portion. From it arises the temporal muscle and its aponeurosis. The greater part of the *temporal fossa* is limited by it. It begins

at the external angular process of the frontal bone and ends in the posterior root
of the zygomatic process of the temporal bone.   The two ends are connected
by the *zygomatic arch.*

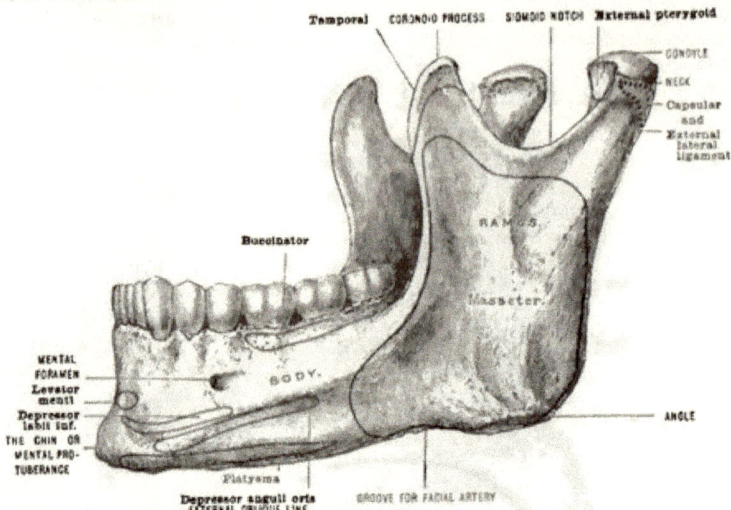

FIG. 8.—THE MANDIBLE.   (Outer view.)

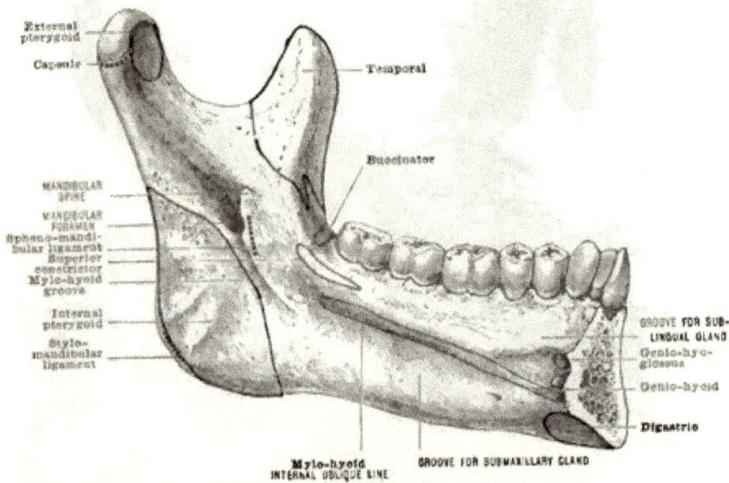

FIG. 9.—THE MANDIBLE.   (Inner view.)

2. The zygomatic arch has (1) a malar and (2) a temporal portion.   It is
formed by the zygomatic process of the malar and the zygomatic process of
the temporal bone.   The arch has an *upper* and a *lower border ;* an *outer* and an
inner surface.

3. **The mandible** has a body and a ramus. On the **body** find: (1) The mental foramen; (2) the external oblique line; (3) the alveolar process; (4) the symphysis menti; (5) the mental tubercles; (6) the incisive fossa; (7) an internal oblique line or mylo-hyoid ridge; (8) a mandibular spine; (9) a mandibular foramen; (10) superior and inferior genial tubercles; (11) a digastric fossa; (12) a submaxillary groove for the submaxillary gland.

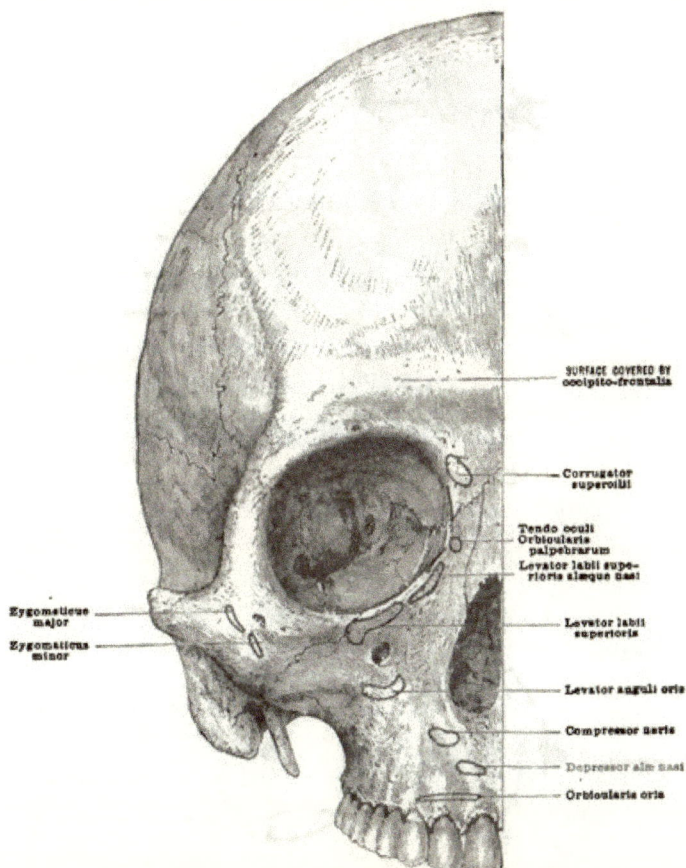

FIG. 10.—THE SKULL. (Norma facialis.)

The ramus has (1) a coronoid process; (2) a sigmoid notch; (3) a condylar process. The condylar process consists of a condyle and a neck.

ON THE FACE LOCATE (Fig. 10): (1) The anterior nares; (2) the nasal bones; (3) the temporal, frontal, orbital, and maxillary processes of the malar bone; (4) the supraorbital and infraorbital arches; (5) the internal and external angular processes of the frontal bone; (6) the floor of the orbit, and in it the infraorbital canal, containing an artery, nerve, and vein of the same name; (7) the infraorbital foramen, the terminus of the infraorbital canal; (8) the supraorbital

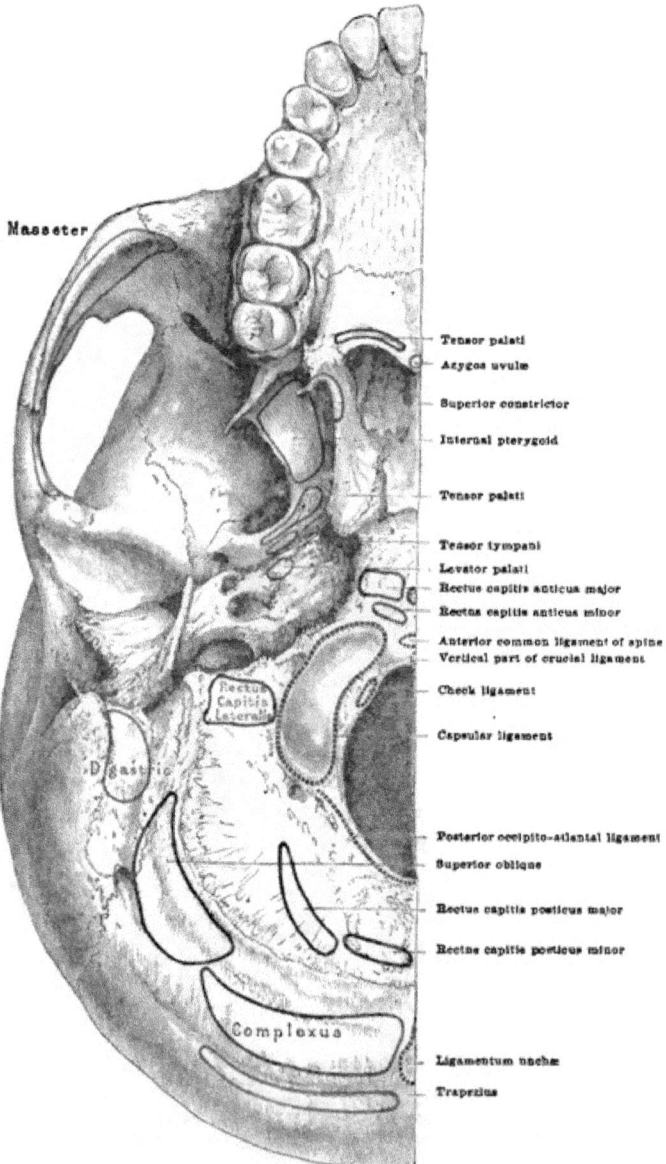

Masseter

Tensor palati
Azygos uvulæ
Superior constrictor
Internal pterygoid

Tensor palati

Tensor tympani
Levator palati
Rectus capitis anticus major
Rectus capitis anticus minor
Anterior common ligament of spine
Vertical part of crucial ligament
Check ligament

Capsular ligament

Rectus Capitis Lateralis

Digastric

Posterior occipito-atlantal ligament
Superior oblique

Rectus capitis posticus major

Rectus capitis posticus minor

Complexus

Ligamentum nuchæ
Trapezius

FIG. 11.—THE SKULL. (Norma basilaris.)

foramen in the frontal bone ; (9) the intermaxillary suture ; (10) the naso-frontal suture ; (11) the superciliary ridge ; (12) the interfrontal or metopic suture or its remains.

ON THE BASE OF THE SKULL (Fig. 11) LOCATE : (1) The foramen magnum ; (2) the occipital condyles ; (3) the digastric groove for the posterior belly of the digastric muscle ; (4) the jugular process for the rectus capitis lateralis ; (5) the styloid process of the temporal bone ; (6) the glenoid fossa with the Glaserian fissure dividing the articular from the non-articular part of the same ; (7) the internal and external pterygoid plates of the pterygoid process of the sphenoid bone ; (8) the pterygoid fossa for the internal pterygoid muscle ; (9) the jugular foramen for the ninth, tenth, eleventh nerves and jugular vein ; (10) the carotid canal in the petrosa for the internal carotid artery and sympathetic nerve ; (11) the foramen spinosum for the great meningeal artery ; (12) the posterior nares ; (13) the hard palate ; (14) the hamular process, around which plays the circumflexus palati.

LOCATE THE ATTACHMENT OF THESE MUSCLES (Figs. 7–10) : (1) Temporal ; (2) masseter ; (3) pterygoid, internal and external ; (4) buccinator ; (5) platysma ; (6) digastric ; (7) orbicularis oris ; (8) depressor labii inferioris ; (9) levator menti ; (10) levator labii superioris ; (11) levator anguli oris ; (12) corrugator supercilii ; (13) zygomaticus, major and minor ; (14) compressor naris ; (15) depressor alæ nasi ; (16) levator labii superioris alæque nasi ; (17) the rectus capitis anticus, major and minor ; (18) the tensor palati ; (19) the azygos uvulæ ; (20) levator palati ; (21) tensor tympani ; (22) depressor anguli oris.

LOCATE THE FOLLOWING FISSURES : (1) The fronto-parietal ; (2) the temporo-parietal ; (3) the interparietal or sagittal ; (4) the occipito-parietal ; (5) the interfrontal ; (6) the fronto-nasal ; (7) the fronto-maxillary ; (8) the malo-frontal ; (9) the malo-zygomatic ; (10) the malo-maxillary ; (11) the naso-maxillary ; (12) the intermaxillary ; (13) the symphysis menti ; (14) the spheno-maxillary ; (15) the petro-occipital ; (16) the palato-maxillary ; (17) temporo-mandibular and occipito-atloid articulations are the only movable joints in this region.

INCISIONS FOR LOCATING THE FOLLOWING STRUCTURES :

1. The supraorbital nerve, artery, and vein (fifth nerve and seventh nerve).
2. The infraorbital nerve, artery, and vein (fifth nerve and seventh nerve).
3. The mental nerve, artery, and vein (fifth nerve and seventh nerve).
4. The facial artery and vein and their branches.
5. The facial nerve and its facial branches and communications.
6. The duct of Stenson and transverse facial artery.

The Mental Nerve.—Turn the skin aside from the intersectional point indicated in figure 12. With the forceps dissect out the *leash of mental nerves*, anp find them continuous with branches of the *facial nerve*, or *seventh cranial*. (Fig. 16.) Find also the *mental branch* of the *inferior dental artery* coming through the same foramen—the mental.

The Supraorbital Nerve.—Turn the skin back, beginning at the intersectional point in the figure. Find with the forceps the leash of the supraorbital nerve (Fig. 16) anastomosing with the *supraorbital* part of the *facial portion of the seventh nerve.*

The Infraorbital Nerve.—Turn the skin back, beginning at the intersectional point in the figure. Here, with the forceps, you will find a large leash of nerves, the terminus of the infraorbital nerve. You will see that its branches are continuous with the infraorbital portion of the seventh cranial nerve. Find also the branches of the infraorbital artery accompanying the nerve.

Stenson's duct is the excretory duct for the parotid gland. Turn the skin back from the intersectional point indicated. Above the duct you will find the transverse facial artery, a small branch of the temporal. Below the duct you will find branches of the facial nerve.

The levator labii superioris covers the infraorbital nerve and artery. This muscle arises from the margin of bone below the orbit. (Fig. 10.) It is inserted into the orbicularis oris.

The levator anguli oris arises from the canine fossa below the infraorbital foramen. It is inserted into the angle of the mouth. It may be seen by pulling the leash of nerves to one side. The leash of nerves lies between the above two muscles. (Fig. 10.)

The Depressor Labii Inferioris.—Remove the skin a little further in the region of the mental nerves, and find the depressor of the lower lip. This muscle is quadrangular in form. Fully one-half of its outer surface lies under

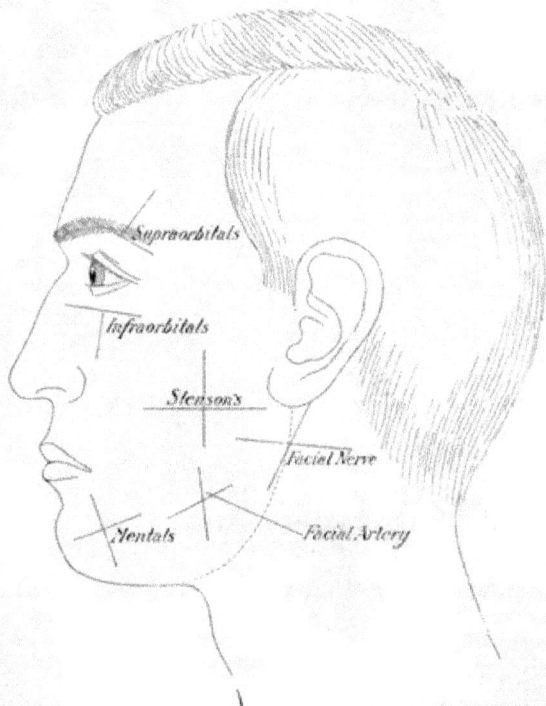

Fig. 12.—Primary Incisions in Dissection of the Face.

the depressor anguli oris (Fig. 13). The *muscle arises* from the upper part of the external oblique line of the mandible, and is *inserted* into the lower lip.

The depressor anguli oris (Fig. 13) will be found arising from the lower part of the external oblique line of the mandible. It is triangular in form. It overlaps the outer one-half of the depressor labii inferioris. Now cut the origin of this muscle, and pull the same aside and expose the whole of the muscle that it overlaps. (Fig. 17.)

CAUTION.—In all dissections about the face care must be taken to cut close to the skin. The branches of the seventh nerve lie beneath the muscles and can not be injured if you take the precaution just given.

The Orbicularis Oris.—Remove the skin around the mouth and expose this muscle. (Fig. 14.) This muscle has a rather strong fascial attachment to the alveolus of the superior maxilla.

The zygomatici major and minor (Fig. 14) arise from the malar bone, and are inserted into the outer part of the upper lip.

The levator labii superioris alæque nasi is inserted into the wing of the nose and the upper lip. It arises from the nasal process of the superior maxilla, on the margin of the orbit.

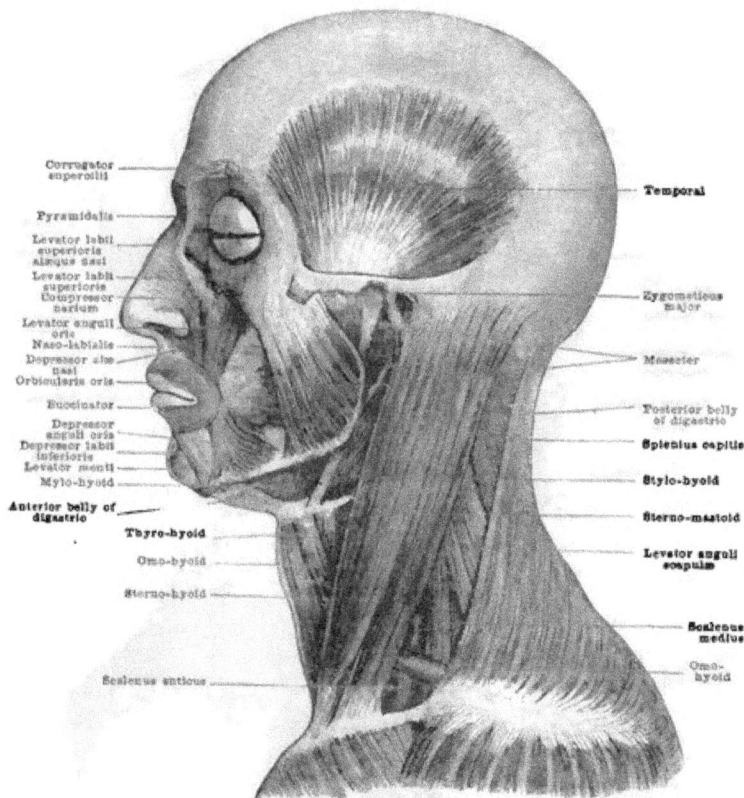

Corrugator
superciili

Pyramidalis

Levator labii
superioris
alæqus nasi

Levator labii
superioris
Compressor
narium

Levator anguli
oris

Naso-labialis

Depressor alæ
nasi

Orbicularis oris

Buccinator

Depressor
anguli oris

Depressor labii
inferioris

Levator menti

Mylo-hyoid

**Anterior belly of
digastric**

Thyro-hyoid

Omo-hyoid

Sterno-hyoid

Scalenus anticus

Temporal

Zygomaticus
major

Masseter

Posterior belly
of digastric

Splenius capitis

Stylo-hyoid

Sterno-mastoid

Levator anguli
scapulæ

Scalenus
medius

Omo-
hyoid

FIG 13—THE DEEPER LAYER OF THE MUSCLES OF THE FACE AND NECK.

The orbicularis palpebrarum surrounds the base of the orbit. It consists of two sets of fibres. One, called the *palpebral*, covers the *palpebræ* or lids ; the other is external to this, and blends above with the occipto-frontalis muscle. It has, internally, a *firm, triple attachment* to the internal angular process of the frontal bone, and to the nasal process of the superior maxilla. Between these two you will find a short, stout tendon that can be mistaken for nothing else— the *tendo oculi* or *tendo palpebrarum.* Figure 10 shows its origin from the nasal process of the superior maxilla in front of the lachrymal groove. Trace the tendon toward the upper and lower lids and observe how it divides.

The **Levator Menti, or Levator Labii Inferioris** (Fig. 13).—This is the muscle by which the lower lip is protruded and elevated at the same time. Cut through the mucous membrane and you will come to the muscular fibres. This muscle arises from the incisive fossa, and is inserted into the integument of the chin, on a plane lower than the origin of the muscle.

The **compressor narium** (Fig. 14), a small muscle, arises from the superior maxilla. (Fig. 10.) It is inserted, by an aponeurosis across the bridge of the nose, into its fellow of the opposite side. This is a small muscle at best, and very hard to demonstrate on prepared material.

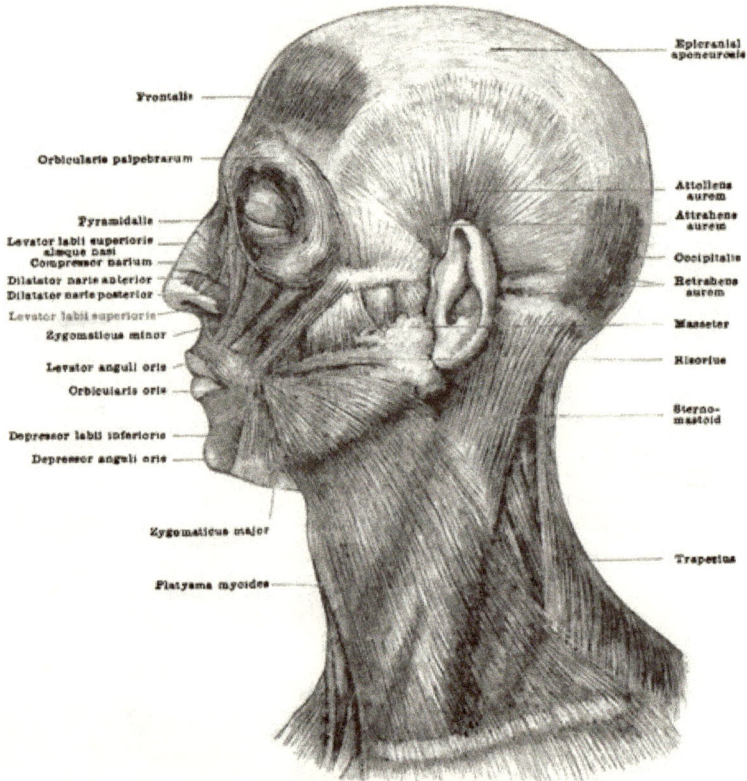

FIG. 14.—THE SUPERFICIAL MUSCLES OF THE HEAD AND NECK.

The **corrugator supercilii** (Fig. 13) is seen by cutting in the mid-line above the nose through the skin and fasciæ. It arises from the superciliary ridge (Fig. 10), and is inserted into the under part of the orbicularis palpebrarum. In action it produces the deep vertical furrows of the forehead.

The **pyramidalis nasi** (Fig. 14) is a part of the occipito-frontalis, continued on to the nose, and *inserted* into the *compressor narium*.

The **dilator naris anterior** arises from the lateral cartilage of the nose, and is *inserted* into the skin near the margin of the nose, well in front.

The dilator naris posterior arises from the superior maxillary bone, from the nasal notch, and is *inserted* into the skin of the margin of the nostril, well back. These two dilator muscles antagonize the *compressor narium*. Remove the skin from the forehead, and expose the anterior belly of the *occipito-frontalis*. We are at present concerned only with the insertion of this muscle. Its anterior fibres are continued on to the nose as the pyramidalis nasi ; its middle and outer blend with the outer portion of the orbicularis palpebrarum.

The muscles of facial expression, those which you have just dissected, are called *dermal muscles*, on account of their *insertion*, being of such a nature as to move the skin, in a most unique manner. These muscles are innervated by the *facial* or *seventh cranial nerve*. (Fig. 16.) They are in the *superficial fascia*— in fact, they occupy the fascia to such a degree that some good authors speak of the *absence* of facial superficial fascia. By their action they confer on the human face facial expression in its broadest sense.

Is not the facial nerve more than an ordinary motor nerve ? In other words,

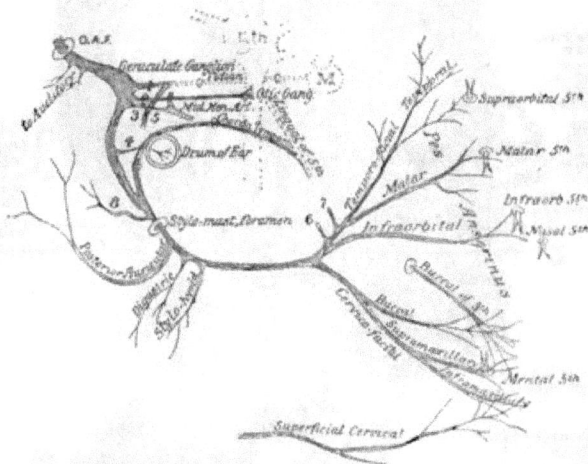

FIG. 15.—SCHEME OF FACIAL NERVE COMMUNICATING WITH THE FIFTH CRANIAL NERVE.

is not this nerve eligible to promotion from the rank and file of ordinary motor nerves to a place among those nerves which are designated *special sense nerves?* For the sake of arousing your interest in the difficult dissection of this nerve, you will pardon a digression for the purpose of answering the above questions.

A slight blow on the ligamentum patellæ produces reflex extension of the leg on the thigh. The steps were (1) conduction of pain, by a *sensory nerve*, to a cortical motor area, and (2) an almost simultaneous contraction of the *extensor quadriceps femoris ;* (3) a record of the blow was recorded as memory. A complete moto-sensory cycle was the result. The nerve that conveyed the sensation of pain to the brain is called a nerve of common sensation ; the one that produced motion in the extensor muscles is called a motor nerve.

The *retina* responds to light, the auditory nerve to sound, the olfactory, to odors. Each nerve records its experience in the brain. Now, see what the *facial nerve* does! It produces motion of a reflex character, as do ordinary motor nerves. It is the only nerve by which vision, audition, olfaction, and gustation manifest themselves in the primitive state. The sight of cruelty, the sound of martial music, the odor of Limburger cheese, the taste of some extra-ordinary viand, are facially expressed, and read off by the world as disapproval, patriotic enthusiasm, repulsion, satisfaction. If, then, auditory, optic, and gustatory nerves are capable of conveying an impression to the brain for record, are they entitled to the designation "special sense" any more than is the facial, which does its own work, and facially expresses to the world the recorded experiences of all the so-called nerves of special sense? Is not this nerve a special sense nerve of facial pantomime?

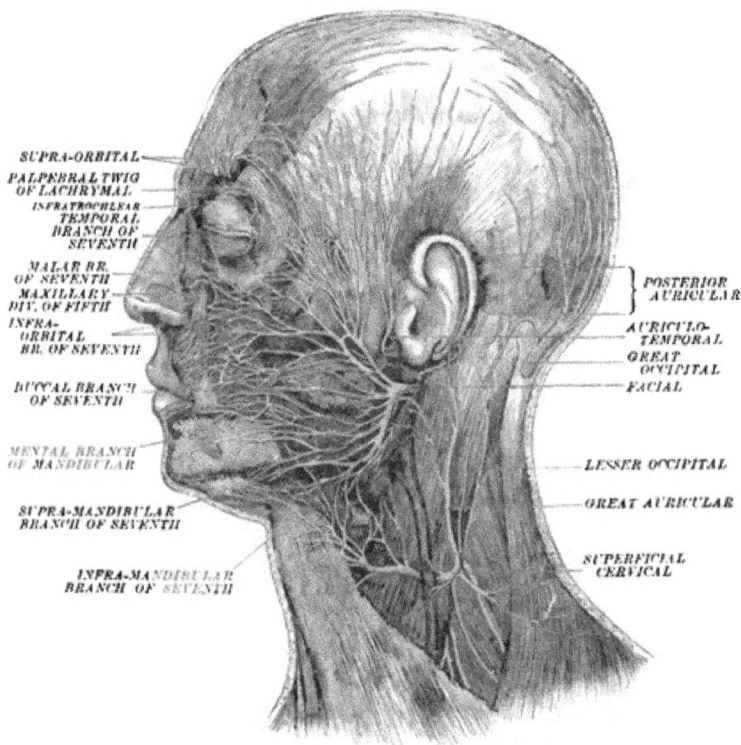

FIG. 16.—SUPERFICIAL DISTRIBUTION OF THE FACIAL AND OTHER NERVES OF THE HEAD.
(After Hirschfeld and Leveillé.)

The Facial Nerve (Fig. 16): **Its Dissection on the Face.**—Remove the skin. Cut down through the parotid gland in the direction of the vertical line in figure 12. It is necessary to go through about one-half of an inch of gland substance. Take the forceps, and find a nerve trunk about the size of a knitting-needle. This will be the *inferior division* of the nerve—the *cervico-facial part.* Follow this branch up a little further, and, very deeply located, you will find the *main trunk* of the nerve. From this you can trace out all the facial branches, in this manner: Put a small piece of tape around the main trunk. Pull on this, and at the same time cut the skin, fascia, and gland substance, in the direction of the branches, with scissors. You will find dissection of this nerve somewhat difficult the first time you make the trial. Any amount of advice, added to what has previously been said, will not help you. Just be careful, and expose one branch at a time.

Communications of the Facial Nerve on the Face with other Nerves (Fig. 15):

1. The seventh nerve + mental branch of the fifth nerve at mental foramen.
2. The seventh nerve + infraorbital branch of the fifth nerve at infraorbital foramen.
3. The seventh nerve + supraorbital branch of the fifth nerve at supraorbital foramen.
4. The seventh nerve + auriculo-temporal branch of fifth on temporal muscle.
5. The seventh nerve + malar branch of temporo-malar of fifth nerve on malar bone.
6. The seventh nerve + temporal branch of temporo-malar of fifth nerve above zygoma.
7. The seventh nerve + great auricular nerve of cervical plexus behind the ear.
8. The seventh nerve + small occipital nerve of cervical plexus behind the ear.
9. The seventh nerve + lachrymal and infratrochlear branches of the fifth nerve.
10. The seventh nerve + great occipital. Posterior division of second cervical nerve.

It will appear from the foregoing that the nerve-supply of the face is a very complicated proposition. Such, however, is not the case, since the whole nerve-distribution can be reduced to:

1. *Motor nerves* to the muscles of expression = seventh nerve.
2. *Sensory nerves* to the skin over the muscles = fifth nerve.
3. *Sympathetic* nerves—the nervi molles—on facial artery
4. *Communicating* branches to adjacent areas.

Divisions and Facial Branches of the Facial Nerve.—The nerve divides into two branches: (1) An upper division that supplies the upper half of the face and the temporal region; (2) a lower division that supplies the lower half of the face and the neck. (Fig. 16.)

The temporo-facial division gives off:
1. Temporal branches to the temple and forehead.
2. Infraorbital branches below the orbit.
3. Malar branches to the zygomatic muscles.

The cervico-facial division gives off
1. A buccal branch to the buccinator muscle.
2. A supramandibular branch above the jaw.
3. An inframandibular branch below the jaw

The name *pes anserinus* is applied to the divergence of these six nerves from the two primary divisions of the seventh nerve.

The **auriculo-temporal nerve** will be found in front of the ear just behind the temporal artery. It is a sensory branch of the fifth, and supplies the side of the scalp and the external in front with sensation.

Find these arteries on the face (Fig. 17):
1. The frontal in the inner angle of the orbit.
2. The nasal above the tendo oculi.
3. The lachrymal in the outer angle to the upper lid.
4. The transverse facial on the malar bone.

5. The supraorbital, with the supraorbital nerve.
6. The infraorbital, with the infraorbital nerve.
7. The mental, with the mental nerve.

8. **The facial artery** (Fig. 17) crosses the mandible in front of the *masseter muscle*. It is surrounded by some very delicate *sympathetic nerves*,

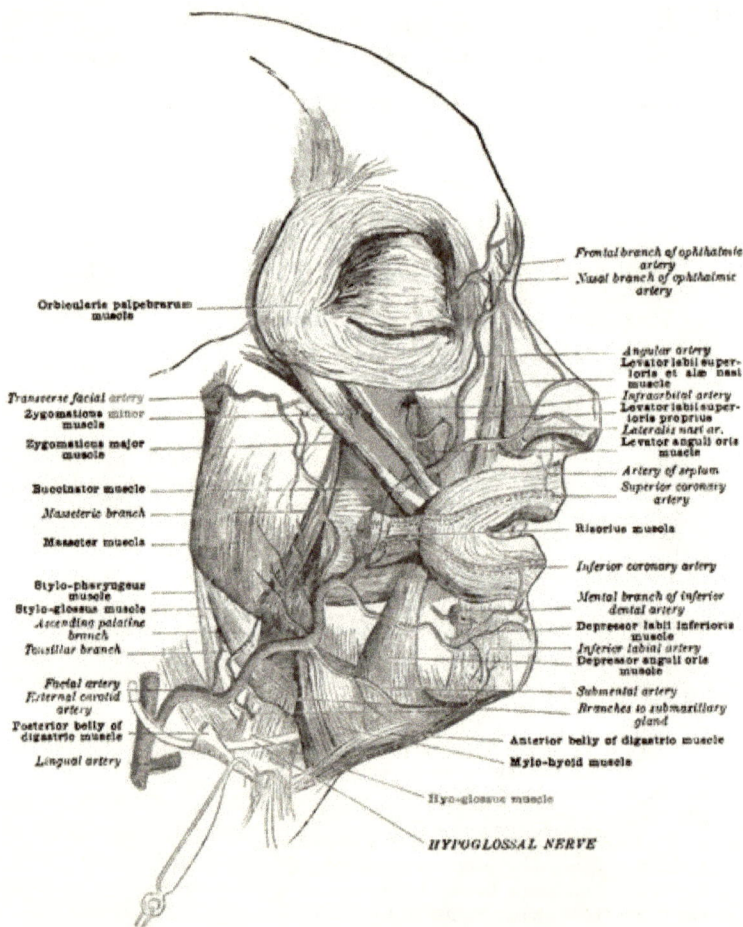

Frontal branch of ophthalmic artery
Nasal branch of ophthalmic artery

Orbicularis palpebrarum muscle

Angular artery
Levator labii superioris et alæ nasi muscle
Infraorbital artery
Levator labii superioris proprius
Lateralis nasi ar.
Levator anguli oris muscle

Transverse facial artery
Zygomaticus minor muscle
Zygomaticus major muscle

Artery of septum
Superior coronary artery

Buccinator muscle
Masseteric branch
Masseter muscle

Risorius muscle

Inferior coronary artery

Stylo-pharyngeus muscle
Stylo-glossus muscle
Ascending palatine branch
Tonsillar branch

Mental branch of inferior dental artery
Depressor labii inferioris muscle
Inferior labial artery
Depressor anguli oris muscle

Facial artery
External carotid artery
Posterior belly of digastric muscle
Lingual artery

Submental artery
Branches to submaxillary gland
Anterior belly of digastric muscle
Mylo-hyoid muscle

Hyo-glossus muscle

HYPOGLOSSAL NERVE

FIG. 17.—SCHEME OF THE FACIAL ARTERY.

which can be demonstrated, if you remove all the fat, with ether, and permit the ether to evaporate. These nerves are the *nervi molles*. They regulate blushing.
Branches of the Facial Artery on the Face :
1. The inferior coronary, between mucous membrane and orbicularis oris.
2. The superior coronary, between mucous membrane and orbicularis oris.
3. The inferior labial, under the muscles of the chin.

4. The lateralis nasi, to the side of the nose.

5. The angular, to the inner angle of the orbit.

6. Muscular branches to the various muscles.

Anastomosis takes place between the facial and all the arteries on the face above enumerated.

The veins of the face (Fig. 18) accompany the arteries, and take the same names, as a rule. (Fig. 17.) They must be dissected with the arteries and

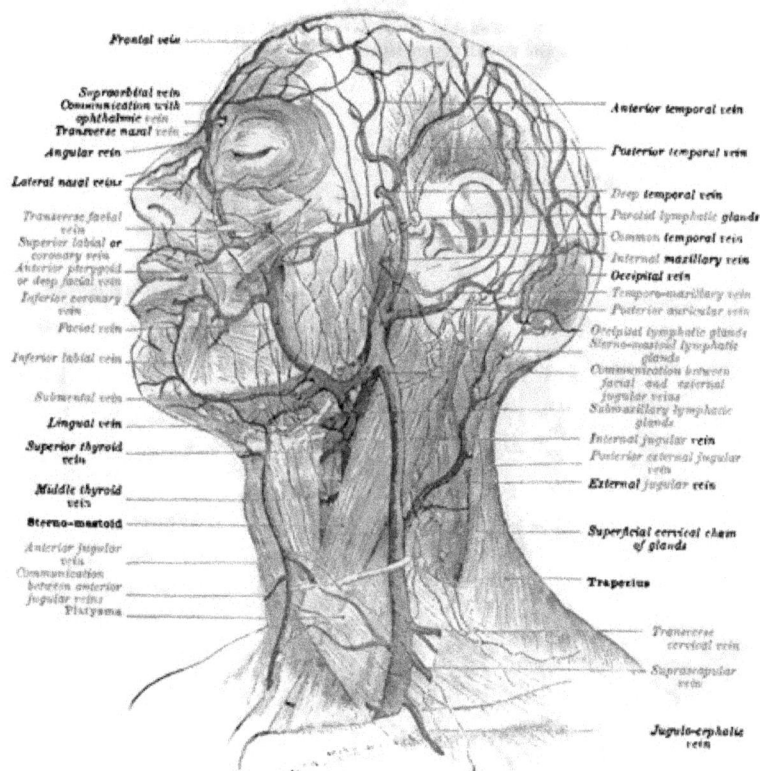

Frontal vein

Supraorbital vein
Communication with
ophthalmic vein
Transverse nasal vein

Angular vein

Lateral nasal veins

Transverse facial
vein
Superior labial or
coronary vein
Anterior pterygoid
or deep facial vein
Inferior coronary
vein

Facial vein

Inferior labial vein

Submental vein

Lingual vein

Superior thyroid
vein

Middle thyroid
vein

Sterno-mastoid

Anterior jugular
vein
Communication
between anterior
jugular veins
Platysma

Anterior temporal vein

Posterior temporal vein

Deep temporal vein
Parotid lymphatic glands
Common temporal vein
Internal maxillary vein
Occipital vein
Temporo-maxillary vein
Posterior auricular vein
Occipital lymphatic glands
Sterno-mastoid lymphatic
glands
Communication between
facial and external
jugular veins
Submaxillary lymphatic
glands
Internal jugular vein
Posterior external jugular
vein
External jugular vein

Superficial cervical chain
of glands

Trapezius

Transverse
cervical vein

Suprascapular
vein

Jugulo-cephalic
vein

FIG. 18.—THE SUPERFICIAL VEINS AND LYMPHATICS OF THE SCALP, FACE, AND NECK.

nerves. These veins must be handled very carefully, not to be injured, as they are very easily ruptured.

Observe the deep temporal vein piercing the temporal fascia, above the zygoma. This returns blood from the *temporal muscle.* Observe the confluence of the internal maxillary vein, in front of the ear, in the substance of the parotid gland, with the temporal vein, and the result of their confluence—the *temporo-maxillary vein.* Observe the communication between the facial vein and the external jugular; also the communication between the angular, supraorbital, and ophthalmic veins. Erysipelas of the face may reach the meninges through this communication. The ophthalmic vein opens into the cavernous sinus.

The parotid gland (Fig. 20) is located by limitation :

1. Superiorly, limited by the zygomatic arch.
2. Inferiorly, limited by the angle of the mandible.
3. Anteriorly, limited by the mid-line of the masseter.
4. Posteriorly, limited by the mastoid process and sterno-mastoid.
5. It sends in prolongations as follows (Fig. 19) :

The inner surface is irregular, sending (1) a large process of gland tissue, in front of the styloid process of the temporal bone, the same also occupying the non-articular part of the glenoid cavity ; (2) a large process behind the styloid and under the mastoid process and sterno-cleido-mastoid muscle.

Parotid fascia is the name given to the deep fascia covering the outer sur-

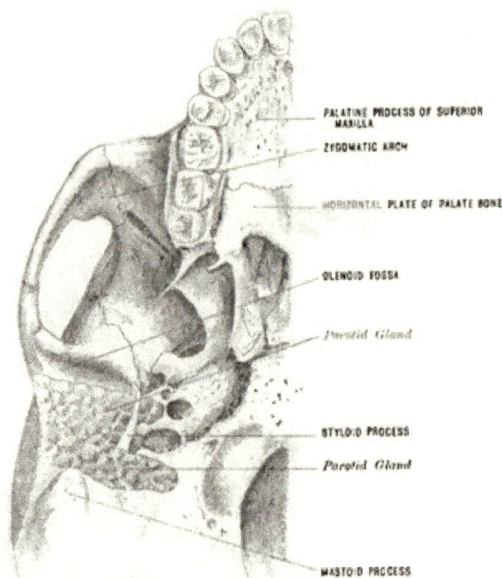

FIG. 19.—SHOWING SCHEMATICALLY THE DEEP PART OF THE PAROTID GLAND EMBRACING THE STYLOID PROCESS.

face of this gland. You have already cut through the gland vertically to find the seventh nerve.

Contents of the Parotid Gland :

1. The *facial* nerve and pes anserinus.
2. The *auriculo-temporal* branch of the fifth nerve.
3. The *external carotid* artery and its three terminals.
4. The *posterior auricular*, temporal, and transverse facial arteries.
5. The *internal maxillary* artery and vein.
6. The *temporo-maxillary* vein.
7. A branch of the great auricular nerve to the seventh.

Excretory Duct of the Parotid (Fig. 20).—Stenson's duct crosses the masseter muscle on a line from the centre of the upper lip to the lobule of the ear. It perforates the buccinator muscle, and opens into the vestibule of the oral cavity, opposite the second upper molar tooth.

1. *How may the dermal muscles be grouped?*

They may be grouped about the four apertures : the orbit, the ear, the nose, and the mouth. Their action seems to be secondary to the special senses of hearing, smell, sight, and taste, since they open and close these openings to a variable extent.

2. *Give the bony attachments of the orbicularis oris.*

This muscle is attached to the upper and also to the lower incisive fossa, and to the alveolar processes.

3. *Has the orbicularis oris any antagonists? If so, name them.*

Yes ; superiorly, levator labii superioris, levator anguli oris, zygomaticus minor, zygomaticus major ; inferiorly, depressor labii inferioris, depressor anguli oris, levator menti, risorius.

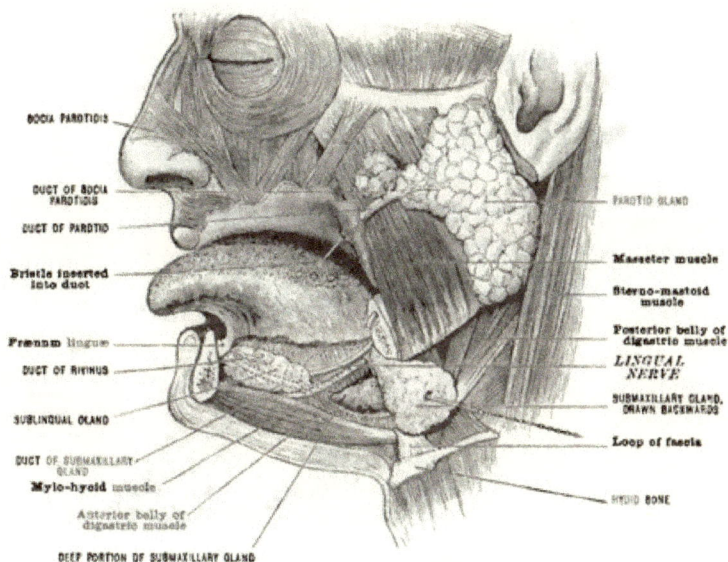

FIG. 20.—THE SALIVARY GLANDS.

4. *Name the foramen between the origins of the levator labii superioris and the levator anguli oris.*

The infraorbital, transmitting the infraorbital vessels and nerves.

5. *Locate on a patient the foramen mentale, and tell what it transmits.*

This foramen is located at the junction of the mento-Meckelian and dentary parts of the mandible, one-half of an inch below the gingiva of the second bicuspid tooth ; it transmits the mental branches of the inferior dental vessels and nerves.

6. *What is the function of the mental nerves and with what do they communicate?*

The function is to supply the skin of the lower lip and chin ; they communicate with mental branches of the facial nerve.

7. *Name the aural group of dermal muscles, and give their function.*

Attrahens aurem, attolens aurem, retrahens aurem. These muscles are mere vestiges in man, often incapable of demonstration. They are best studied on domestic animals, as the dog and rabbit. Their function is, first, to enlarge the

4

auditory canal by rendering tense the cartilaginous parts ; and, second, to direct the external ear toward a given noise. These muscles are supplied by the seventh cranial nerve.

8. *Describe the rationale of a complete frown.*

A proper frown consists of vertical and horizontal corrugations. The verticals are produced by the action of the corrugator supercilii on the anterior belly of the occipito-frontalis muscle, acting on this so as to deflect inward its muscular belly and the skin covering the same. The horizontal ridges are the result of the same force, intensified to some extent by gravity.

9. *Locate the infraorbital plexus and give its formation.*

It is located one-half of an inch external to the wing of the nose. It lies

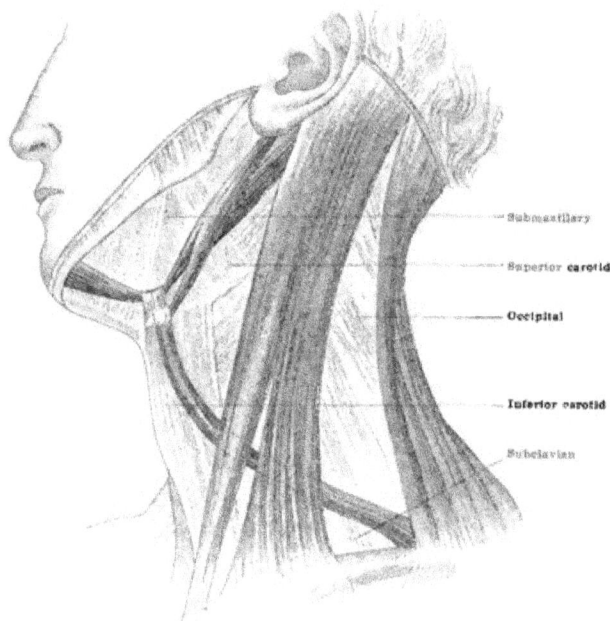

FIG. 21.—CERVICAL TRIANGLES.

under cover of the levator labii superioris, and is formed by a communication between the infraorbital branch of the fifth and the infraorbital branches of the seventh nerve.

10. *Name and locate the great anastomotic blood areas on the face and tell how they are formed.*

The facial branch of the external carotid, the ophthalmic branch of the internal carotid artery, the internal maxillary branch of the external carotid, and the temporal branches of the external carotid are the branches of arterial trunks concerned in the rich blood-supply to the face.

Anastomotic areas : (Fig. 17) (1) The ophthalmic anastomoses with the temporal, forming the supraorbital anastomosis, and with the angular branch of the facial artery at the inner base of the orbit, forming the angular anastomosis. (2) The transverse facial artery anastomoses with the masseteric branch of the facial artery on the outer surface of the masseter muscle, forming the masseteric anastomosis. (3) The infraorbital anastomoses with the facial, below the orbit, to form the infraorbital anastomosis. (4) The mental branch of the inferior dental artery anastomoses with the inferior labial branch of the facial artery, forming the mental anastomosis. (5) *In the facial mid-line* the corresponding arteries of opposite sides of the face anastomose. At the aural, ocular, nasal, and oral muco-cutaneous junctions capillary anastomosis takes place. (6) The arterial anastomoses above referred to are attended by venous anastomoses. The mental, infraorbital, and supraorbital may be compressed at their respective exit foramina. The facial may be compressed on the mandible just in front of the masseter muscle. (7) The important communication on the face is between the ophthalmic and angular, by which route facial erysipelas may reach the meninges, since the ophthalmic vein opens into the cavernous sinus of the dura mater.

11. *Explain briefly the parotid gland.*

The gland is located in a depression having the following limitations above, the zygomatic arch; below, the angle of the mandible and the masto-mandibular line; behind, the external ear and sterno-mastoid muscle; in front, by the masseter muscle; deeply, it embraces the styloid processes and internal carotid artery. The gland contains the seventh nerve and the auriculo-temporal branch of the fifth nerve, both of which furnish it with nerve-filaments. It contains the external carotid artery, which here breaks up into the temporal, internal maxillary, and transverse facial branches, these supply the gland with blood. It contains the temporo-maxillary vein. The fascial covering of the gland is called the parotid fascia. The gland is the largest of the salivary glands, and has an excretory duct called Stenson's.

12. *Trace Stenson's duct.*

It crosses the masseter muscle in a line from the lobule of the ear to the upper lip. It perforates the buccinator muscle, and opens into the vestibule of the cavum oris opposite the second upper molar tooth.

13. *Where does the facial nerve escape from the cranium?*

It escapes through the stylo-mastoid foramen, in the petrosa of the temporal bone. On its escape it gives off the posterior auricular, the digastric, and stylo-hyoid branches. It subsequently forms the *pes anserinus*, from which the muscles of expression are innervated. The facial is a motor nerve.

NOTE.—Many a good dentist has been heard to tell his patient, "All this toothache is caused by the facial nerve." The facial nerve makes you smile when you hear such things; it is the fifth nerve that is concerned in toothache.

. .

THE NECK. SUPERFICIAL DISSECTION.

Locate on the cadaver:

1. The lower border of the mandible and its angle.
2. The mastoid process of the temporal bone, behind the ear.
3. The sterno-clavicular articulation. Does it move?
4. The interclavicular or suprasternal notch. How deep?
5. The acromio-clavicular articulation. Does it move?
6. An imaginary line from the angle of the jaw to the mastoid.
7. The sterno-cleido-mastoid muscle and its double origin.
8. The exact anterior border of sterno-cleido-mastoid.
9. The exact posterior border of sterno-cleido-mastoid.
10. The cervical mid-line—mento-sternal line.

11. The hyoid bone in the mid mento-sternal line.
12. The body of the hyoid bone, one-half of an inch long to touch.
13. The greater horn of the hyoid + lingual artery above same.
14. The thyroid cartilage—Adam's apple—and notch above.
15. The thyro-hyoid space + thyro-hyoid membrane.
16. The thyroid notch + mid-part of thyro-hyoid membrane.
17. The crico-thyroid membrane in crico-thyroid space.
18. The cricoid cartilage and trachea.   (Fig. 27.)

INCISIONS.

1. Through the mid mento-sternal line.
2. From symphysis menti to mastoid process.
3. From sternum to acromion process.

BOUNDARIES OF THE NECK.

1. *Anterior.*—The mid mento-sternal line.
2. *Posterior.*—The anterior border of the trapezius.
3. *Inferior.*—The clavicle and manubrium sterni.
4. *Superior.*—Lower border of mandible and masto-mandibular line.

TRIANGLES OF THE NECK.   (Fig. 21.)

1. *Submaxillary* or digastric.
2. *Superior carotid* or triangle of election.
3. *Inferior carotid*, tracheal, or triangle of necessity.
4. *Occipital*—not suboccipital.
5. *Subclavian* or brachial triangle.

BOUNDARY STRUCTURES OF CERVICAL TRIANGLES.   (Fig. 23.)

1. The *sterno-cleido-mastoid muscle.*
2. The *mid mento-sternal line.*
3. The *lower border of the mandible.*
4. The *masto-mandibular line.*
5. The *omo-hyoid muscle.*
6. The *digastric muscle.*
7. The *stylo-hyoid muscle.*
8. The **manubrium sterni** and *clavicle.*

SURGICAL AND MEDICAL AREAS OF THE NECK.

Larynx—intubation and laryngotomy.
Trachea—tracheotomy, high and low.
Thyroid gland—operations on.
The contents of the carotid sheath.
The *subclavian artery* and its branches.
The *brachial* and *cervical plexuses.*
The *apex of the lung,* one and one-half inches above the first rib.
Tracheal and *bronchial* respiration.
The *superficial lymphatic glands.*   (Fig. 23.)
The *upper set of deep lymphatic glands.*
The *lower set of deep lymphatic glands.*

SPECIFIC BOUNDARIES OF CERVICAL TRIANGLES. (Fig. 23.)

1. *Digastric Triangle.*—Mandible and masto-mandibular line ; anterior belly of digastric muscle ; posterior belly of digastric muscle and stylo-hyoid muscle.

2. *Superior Carotid Triangle.*—Sterno-cleido-mastoid muscle ; anterior belly of omo-hyoid muscle ; stylo-hyoid muscle and posterior belly of digastric muscle.

3. *Inferior Carotid or Tracheal Triangle.*—Mid mento-sternal line ; sterno-cleido-mastoid muscle ; anterior belly of omo-hyoid muscle.

4. *The Occipital Triangle.*—The trapezius muscle ; the posterior belly of the omo-hyoid muscle ; the sterno-mastoid muscle.

5. *The Subclavian Triangle.*—The clavicle, the posterior belly of the omo-hyoid muscle ; the sterno-cleido-mastoid muscle.

Note that the sterno-mastoid divides the neck into two triangles. Of these two the anterior contains three, the posterior two, smaller triangles.

Dissection to show superficial structures in the superficial fascia of the neck, as follows :

1. The platysma myoides muscle—a dermal.
2. The anterior jugular vein is in front of sterno-mastoid.
3. The posterior jugular vein is behind the sterno-mastoid.
4. External jugular vein crosses the sterno-mastoid.
5. The superficial lymphatics—vertical group. (Fig. 23.)
6. The superficial lymphatics—transverse group. (Fig. 23.)
7. The inframandibular branch of the seventh nerve. (Fig. 22.)
8. The superficial cervical nerve, of cervical plexus. (Fig. 22.)
9. The great auricular nerve, of cervical plexus. (Fig. 22.)
10. The small occipital nerve, of cervical plexus. (Fig. 22.)
11. The suprasternal nerve, of cervical plexus. (Fig. 22.)
12. The supraclavicular nerve, of cervical plexus. (Fig. 22.)
13. The supraacromial nerve, of the cervical plexus. (Fig. 22.)

The platysma myoides (Fig. 14) is a dermal muscle. It is allied to the muscles of expression and to the three dermal aural muscles ; they are all remnants in man of the great panniculus carnosus. To expose this muscle it will be necessary to exercise the greatest care in removing the skin. Usually the fibres of origin of this muscle extend two inches below the clavicle. They are continued obliquely upward and forward onto the face. In some persons this muscle is very heavy ; in others it is almost absent.

The Superficial Nerves of the Cervical Plexus and the Spinal Accessory Nerve.—*Dissection.*—Locate the posterior border of the sterno-cleido-mastoid muscle. Very carefully cut the platysma along this posterior border, using the forceps for a director. Next cut through the deep fascia along the posterior border of the muscle. Now look first for a small nerve that parallels the posterior border of the sterno-cleido-mastoid muscle. (Fig. 22.) This is the lesser occipital. Trace it upward, as in the figure.

Three nerves cross the sterno-cleido-mastoid : (1) The mastoid branch, or second small occipital ; (2) the great occipital ; (3) the superficial cervical.

Follow their branches out, taking care to harm no veins. These are the three ascending branches of the plexus.

The spinal accessory nerve is sometimes the subject of surgical procedure. On the living, if you are in doubt as to whether you have the spinal accessory or some other nerve, remember this rule : Pinch the nerve with the forceps, and if it is the spinal accessory you have, the trapezius muscle will elevate the shoulder-blade. The other nerves that this one is sometimes mistaken for are all sensory, and will, if pinched, produce no muscular contraction.

The descending branches of the cervical plexus are (1) the suprasternal, (2) the supraclavicular, (3) the supraacromial.   Trace these out, as in figure 22.

Note that the superficial cervical branch passes behind the external jugular vein : that it is distributed to almost the entire front of the neck ; that it anastomoses above with the inframandibular branch of the seventh nerve and below with the suprasternal.

Note the spinal accessory nerve pierces the tapezius muscle, and has many communications with the other nerves in the occipital triangle.   To this fact is

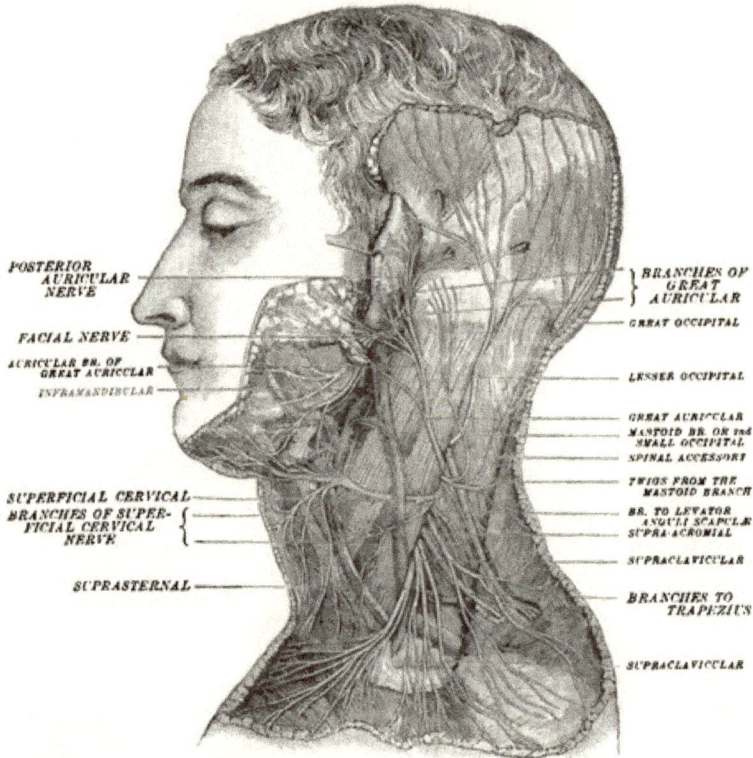

POSTERIOR AURICULAR NERVE

FACIAL NERVE

AURICULAR BR. OF GREAT AURICCLAR

INFRAMANDIBULAR

SUPERFICIAL CERVICAL BRANCHES OF SUPERFICIAL CERVICAL NERVE

SUPRASTERNAL

BRANCHES OF GREAT AURICULAR

GREAT OCCIPITAL

LESSER OCCIPITAL

GREAT AURICCLAR
MASTOID BR. OR 2nd SMALL OCCIPITAL
SPINAL ACCESSORY

TWIGS FROM THE MASTOID BRANCH
BR. TO LEVATOR ANGULI SCAPULÆ
SUPRA-ACROMIAL

SUPRACLAVICULAR

BRANCHES TO TRAPEZIUS

SUPRACLAVICULAR

FIG. 22.—SUPERFICIAL BRANCHES OF THE CERVICAL PLEXUS.   (After Hirschfeld and Leveillé.)

possibly due the twitching of the shoulder in frostbitten ears ; purely a reflex movement.   (Fig. 22.)

The Superficial Lymphatic T.—This will enable you to remember the general distribution of the superficial lymphatic glands in the neck.   (Fig. 23.) In the main the lymphatics follow the veins.   They are readily seen when enlarged by disease ; they are scarcely recognizable in cadavers when not diseased.

The jugular veins (Fig. 18) in the superficial fascia are : (1) The *anterior*, near the mid *mento-sternal line*.   It opens into the *subclavian vein* or into the *external jugular* under the sterno-cleido-mastoid muscle.   (2) The *posterior jugular* vein parallels the posterior border of the sterno-cleido-mastoid muscle. It opens into the external jugular vein. (3) The *external jugular* vein crosses the sterno-cleido-mastoid muscle. It is formed, as you must show, by the confluence of (1) the temporo-maxillary, (2) the posterior auricular and facial communicating branch.

The Deep Cervical Fascia.

I desire the student to have a conception of the deep fascia of the neck, in advance of his work. The following scheme will greatly assist him both in review and during the dissection.

1. The deep fascia ensheathes the contents of the neck in four strata. The

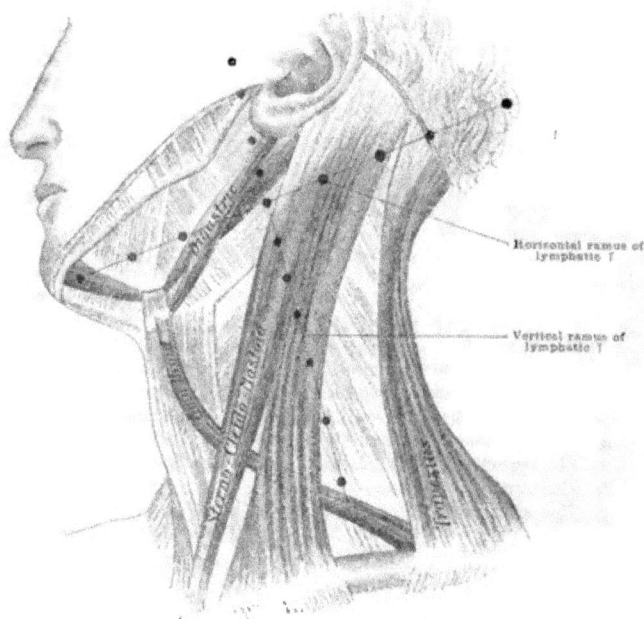

FIG. 23.—SUPERFICIAL LYMPHATICS.

student is expected to learn just what structures are ensheathed by each layer of fascia.

2. The deep fascia has superior attachments, relations, and specializations; it also has inferior attachments and continuations, thoracic and maxillary, which the student must learn if he ever expects to dissect the neck and thorax understandingly.

3. The deep fascia forms septa and sheaths, the rationale of which must be read and studied in the introductory chapter of this book. The deep fascia here, as in every region of the body, belongs to the dense compact variety of connective tissue.

4. The deep fascia determines the course taken by pus and missiles, hence a thorough knowledge of the architecture of the deep fascial envelopes of the neck is just as important as any fascial area in the body.

Figure 24 represents schematically a transverse section of the neck. From this it will be seen there are two spaces: (1) Those formed by delamination of the four layers, including certain structures and their nerve-supply and blood-supply (2) The interfascial spaces containing fat and connective tissue. It will be seen that the first layer delaminates to unsheath the sterno-mastoid muscle. The second layer delaminates to enclose the omo-hyoid, sterno-hyoid, sterno-thyroid, and thyro-hyoid muscles, with their nerve-supply and blood-supply. The third layer delaminates to enclose the larynx and trachea and thyroid, the pharynx and œsophagus, the carotid sheath and its contents. The fourth layer delaminates to ensheath the rectus capitis anticus major and minor and longus colli muscles.

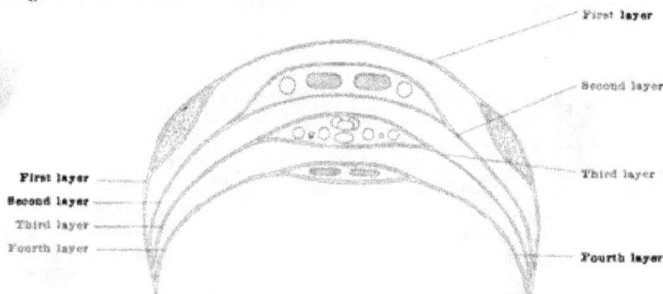

FIG. 24.—DEEP CERVICAL FASCIA—TRANSVERSE SECTION. (Schematic.)

Dissection
1. The temporal muscle and its aponeurosis or fascia.
2. The masseter muscle and its aponeurosis or fascia.
3. The sterno-cleido-mastoid and its fascial covering.
4. The trapezius muscle and its aponeurosis.
5. The zygoma, and its relations.
6. The temporal fossa and its contents.
7. The temporal ridge, its formation of three parts.

Examine your work and answer this quiz:

1. *Name the deep fascia covering the temporal muscle and give all its attachments.*

It is called the temporal fascia. It is attached to the linea suprema of the complete temporal ridge, and to the superior border of the zygomatic arch.

2. *What does the complete temporal ridge consist of?*

It consists of two parallel lines, having a frontal, a parietal, and a temporal portion. The aponeurosis of the muscle arises from the upper, the muscle itself from the lower line.

3. *What structures occupy the superficial fascia, covering the temporal muscle?*

The *superficial temporal arteries* and their veins; the *auriculo-temporal* nerve, a branch of the fifth cranial; some *temporal branches* of the seventh or facial nerve, and the lesser occipital nerve.

4. *What is the masseteric fascia?*

The deep fascia covering the masseter muscle. It is of cervical derivation, being, with the parotid fascia, the upward continuation of the first layer of the deep cervical fascia.

5. *Name the structures on the masseter muscle.*

The parotid gland in part; the pes anserinus of the seventh nerve; Stenson's duct. In front of and under the anterior part of the muscle is some fat, called in the child the sucking pad.

6. *Explain the relations and composition of the zygomatic arch.*

The arch is composed of a malar and a temporal portion. It has a superior border into which is inserted the temporal fascia; an outer surface that is subcutaneous; an inner surface and a lower border from which arises the masseter muscle. Under the arch are found the coracoid process of the temporal bone, into which is inserted the temporal muscle; the sigmoid notch, through which pass the arteries and nerves to the masseter muscle.

7. *What structures lie on, and posterior to, the sterno-cleido-mastoid muscle?*

The muscle is ensheathed by the first layer of the deep cervical fascia. (Fig. 24.) On the muscle lie the auricularis magnus and the superficial cervical nerves. The muscle is crossed by the external jugular vein. The spinal accessory nerve and all the superficial branches of the cervical plexus pierce the deep fascia at the posterior border of the muscle. These nerves are all in the occipital triangle.

8. *Describe the superficial branches of the cervical plexus.* (Fig. 22.)

They are seven in number. The four ascending branches supply the neck and side of the head with sensation. The three descending branches supply the shoulder and upper third of the thorax. Their names are:

1. Lesser occipital parallels the sterno-cleido-mastoid muscle. It supplies the skin over the temporal muscle and the upper posterior part of the ear.

2. The mastoid branch, or second small occipital, supplies the skin over the insertion of the sterno-mastoid muscle into the mastoid process of the temporal bone.

3. The great auricular crosses the sterno-mastoid muscle to the ear, to which and to the skin over the the parotid it is distributed.

4. The superficial cervical nerve crosses horizontally the sterno-mastoid muscle. It supplies with sensation the whole front of the neck. It passes behind the external jugular vein.

5. Suprasternal. This nerve supplies the skin over the origin of the sterno-mastoid and over the manubrium sterni.

6. The supraclavicular supplies the skin over the pectoralis major to the nipple. It probably is this nerve that may account for the very diffuse pain over the head, neck, and shoulders in sore nipples of nursing mothers.

7. The supraacromial supplies the skin over the deltoid and clavicular portion of the trapezius muscles.

The Masseter Muscle (Fig. 25).—Cut this muscle at its origin from the malar process of the maxilla and the inner surface and lower border of the zygoma. As you turn the same down, notice the nerve- and blood-supply coming through the sigmoid notch. A little work with the forceps will remove the fat and develop the insertion of the temporal muscle into the coracoid process. Find the masseter inserted into the outer surface of the ramus of the mandible.

The Temporal Muscle (Fig. 25).—Cut the aponeurosis of this muscle around the whole circumference of the temporal fossa. The aponeurosis will be attached to the zygoma in two layers. Find between these two layers a small branch of the temporo-malar nerve. The nerve will do you no good; but finding the same will be evidence of careful work. Turn the muscle itself down, after having removed the fascia, and see the deep temporal arteries. These supply the muscle with blood. They lie on the bone. They are branches of the internal maxillary. Find the deep temporal vein (Fig. 18) piercing the fascia above the zygoma, to join the superficial temporal.

The sterno-cleido-mastoid muscle (Fig. 25) has two origins : (1) A sternal ;

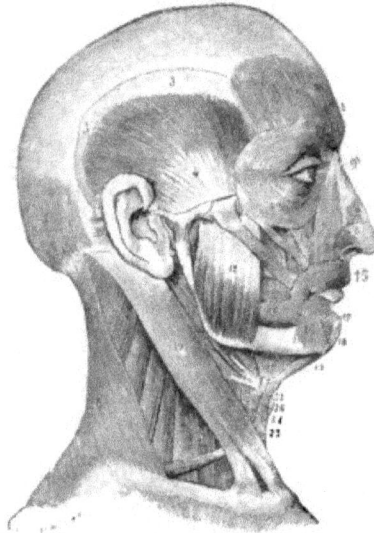

FIG. 25.—MUSCLES OF THE FACE AND NECK.

1. Frontal muscle. 2. Occipital muscle. 3, 3. Epicranial aponeurosis. 4. Temporal muscle. 5. Retrahens aurem. 6. Orbicularis palpebrarum. 7. Levator labii superioris et alæque nasi. 8. Dilator naris. 9. Compressor naris. 9'. Pyramidalis nasi. 10. Zygomatic minor. 11. Zygomatic major. 12. Masseter. 13. Levator anguli oris. 14. Levator labii superioris. 15. Orbicularis oris. 16. Buccinator. 16'. Depressor anguli oris. 17. Depressor labii inferioris. 18. Levator labii inferioris. 19. Sterno-mastoid. 20. Trapezius. 21. Digastric and stylo-hyoid. 22. Anterior belly of digastric. 23. Pulley for tendon of digastric. 24, 24. Omo-hyoid. 25. Sterno-hyoid. 26. Thyrohyoid. 27. Mylo-hyoid. 28. Splenius capitis. 29. Splenius colli. 30. Levator anguli scapulæ. 31. Scalenus posticus. 32. Scalenus anticus.

(2) a clavicular. Its size, location, and extensive nerve-supply make it the most important muscle in the neck. These origins vary in size. In some cases you will find the clavicular part very small, in other cases very large. The sternal origin corresponds to the mastoid insertion ; the clavicular origin corresponds to the occipital insertion. If you will follow the cleft between the two heads, it will lead you to the junction between the occipital and mastoid. (Fig. 23.)

Sterno-mastoid sheath (Fig. 24) is formed by a delamination of the first layer of deep cervical fascia. Cut through this sheath from end to end, just as you would separate two muscles. In fact, as pointed out in the preceding paragraph, there are two muscles here, coalesced to form one.

NERVE-SUPPLY OF THE STERNO-MASTOID.—This muscle, as you may now see, receives numerous twigs from the spinal accessory nerve as this nerve is passing through the muscle. It also receives nerves from the cervical plexus. Lift the muscle very carefully, and demonstrate these nerves on your own work.

The Upper Attachments and Special Structures Formed by the First Layer of Deep Cervical Fascia (Fig. 26).—The same fascia that covers the sterno-mastoid muscle is continued upward and forms special structures. You will now dissect and study these. Cut along the lower margin of the jaw to the symphysis menti. Avoid the facial artery and vein in front of the masseter muscle.

Distribution of the Deep Cervical Fascia Above (Fig. 26).—You now see this fascia extending from the sterno-mastoid muscle to the masseter muscle

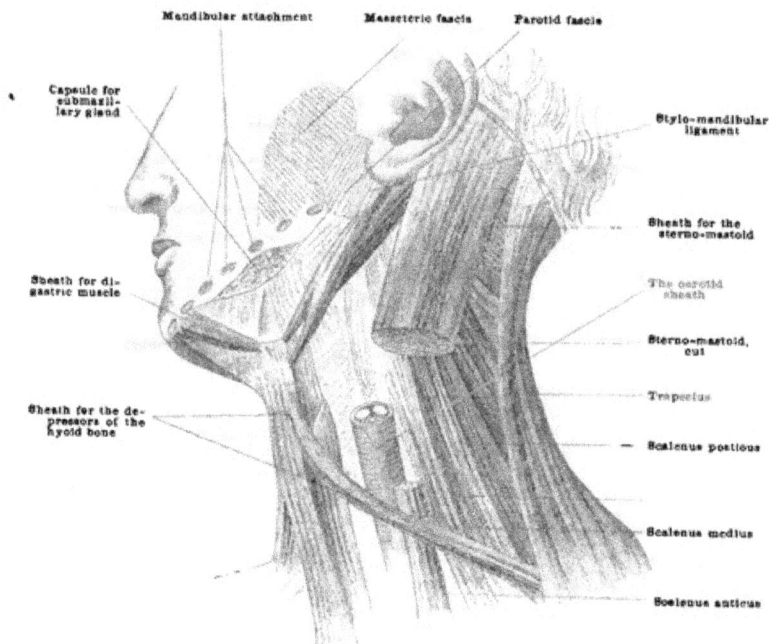

FIG. 26.—SCHEME OF UPPER ATTACHMENTS OF AND SPECIAL NAMES FOR DEEP CERVICAL FASCIA.

and parotid gland, forming their specific fasciæ. You see it extending from the styloid process to the jaw as the stylo-maxillary ligament; you see it forming a capsule for the submaxillary gland, and continuing forward to invest the anterior belly of the digastric muscle.

The submaxillary gland is ensheathed by the first layer of deep cervical fascia. Observe the facial vein passing in front of the gland; and the facial artery, perforating the deep fascia and passing behind the gland, or even through the same.

What nerves accompany the facial artery in its distribution to the face?

Sympathetic branches—the nervi molles. They produce blushing and pallor of the face.

Are any other arteries accompanied by sympathetic nerves, in like manner, in this region?

Yes, branches of the sympathetic accompany every branch of the external carotid artery. Hyperæmia then may occur in the distribution of the temporal, internal maxillary, and lingual arteries. The specific name for this physiological hyperæmia on the face is "blushing."

Relations of the Sterno-cleido-mastoid Muscle.—Review now carefully your dissection, and learn the relations of this muscle thus far exposed. See if your work shows the following points:

1. Did you demonstrate a sternal and clavicular origin and a mastoid and occipital insertion of the muscle?

2. Did you find crossing or lying upon the sheath of the muscle the external jugular vein, the superficial cervical nerve, the great auricular nerve, and the

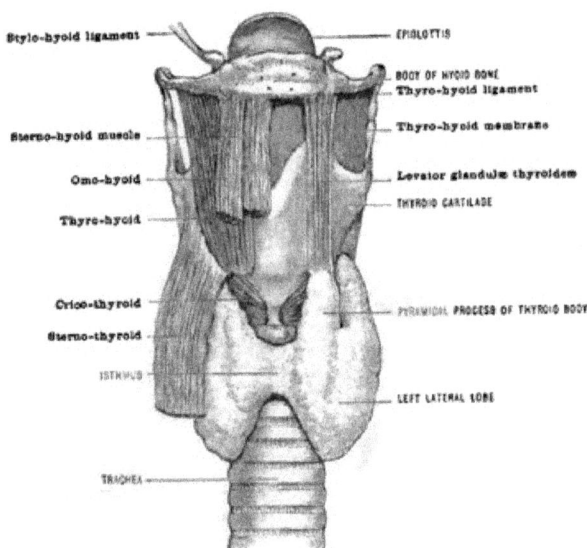

FIG. 27.—THYROID BODY, WITH MIDDLE LOBE AND LEVATOR MUSCLE.

mastoid branch or second small occipital nerve? Did you see emerging through the deep fascia, between the muscle and the trapezius, in the occipital triangle, the spinal accessory nerve and all the superficial branches of the cervical plexus? Did you find the sheath of the muscle attached to the clavicle below? Did you find the same derived from the first layer of the deep cervical fascia and continuous above with the masseteric and parotid fasciæ? with the stylo-maxillary ligament? with the submaxillary and digastric fasciæ?

Having done this, you may lift the muscle, observe its nerve-supply again, from the spinal accessory and cervical plexus, and study the posterior relations.

Posterior relations of the sterno-cleido-mastoid muscle are:

1. The contents of the second layer of the deep cervical fascia.

2. The carotid sheath and its contents. This latter, you will remember, belongs to the third layer of deep cervical fascia.

THE MUSCLES AND NERVES IN THE SECOND LAYER OF DEEP CERVICAL FASCIA.

1. The *sterno-hyoid* and *omo-hyoid*.
2. The *sterno-thyroid* and *thyro-hyoid*.
3. The nerve-supply of these—the *ansa hypoglossal loop*. (Figs. 23 and 29.)

Having turned the sterno-mastoid muscle and the fascia extending from it to the mid-line of the neck aside, you expose the sterno-hyoid and the omo-hyoid of this group. Figure 27 shows you the insertion of these two muscles conjointly into the lower border of the body of the hyoid bone. Cut these two muscles one inch below this insertion, turn the lower part of the same forward carefully, and note the nerve-supply. These muscles must be handled very carefully, in order to preserve them and also their nerves. The omo-hyoid arises from the superior border of the scapula for one inch. The sterno-hyoid arises

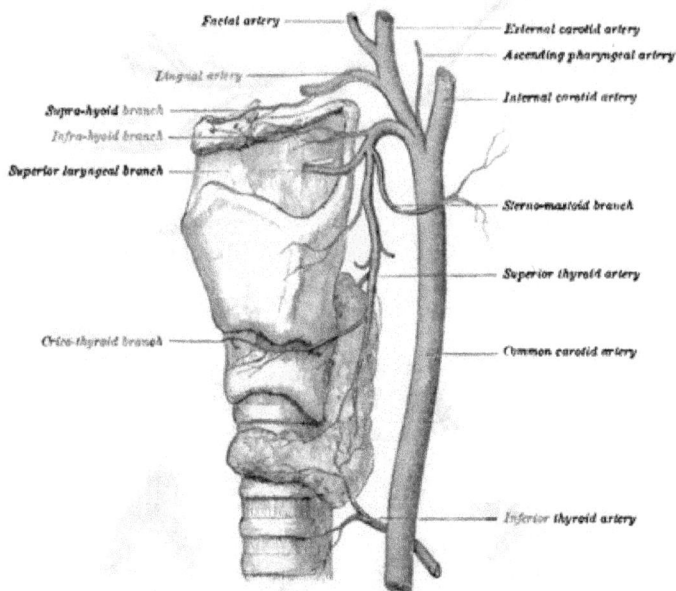

Facial artery — External carotid artery

Ascending pharyngeal artery

Lingual artery — Internal carotid artery

Supra-hyoid branch —

Infra-hyoid branch —

Superior laryngeal branch —

Sterno-mastoid branch

Superior thyroid artery

Crico-thyroid branch —

Common carotid artery

Inferior thyroid artery

FIG. 28.—SCHEME OF SUPERIOR THYROID ARTERY.

from the manubrium sterni. Trace out these origins at a later stage of your work.

The sterno-cleido-thyroid is commonly called sterno-thyroid. This is inserted into the oblique line of the ala of the thyroid cartilage. (Fig. 27.) From here it is continued to the lower border of the outer third body, and the inner half of the greater horn of the hyoid bone as the thyro-hyoid muscle.

The sterno-thyroid arises from the manubrium low down, and also from the clavicle. (Fig. 30.) Detach this muscle from its insertion and turn it down very carefully, and you will see the thyroid **gland** and its isthmus.

The thyro-hyoid removed by detaching it at its origin, you expose the thyro-hyoid membrane, and see entering the same the **superior laryngeal nerve and artery.** (Fig. 28.)

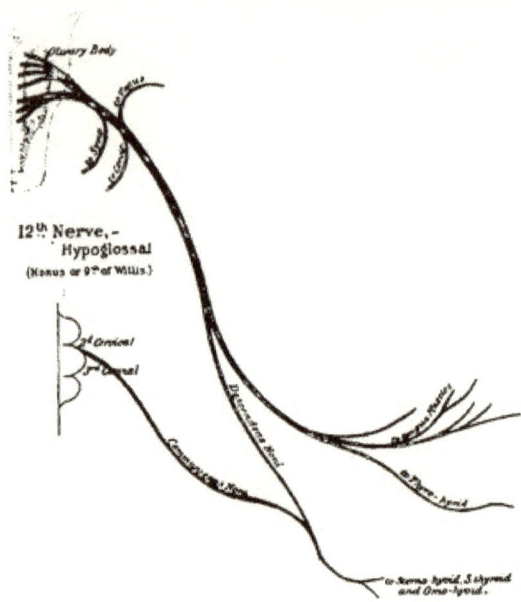

Ciliary Body

12th Nerve,—
Hypoglossal
(Nonus or 9th of Willis.)

1st Cervical

2nd Cervical

Fig. 29.—The Loop formed by Communicating Branches.

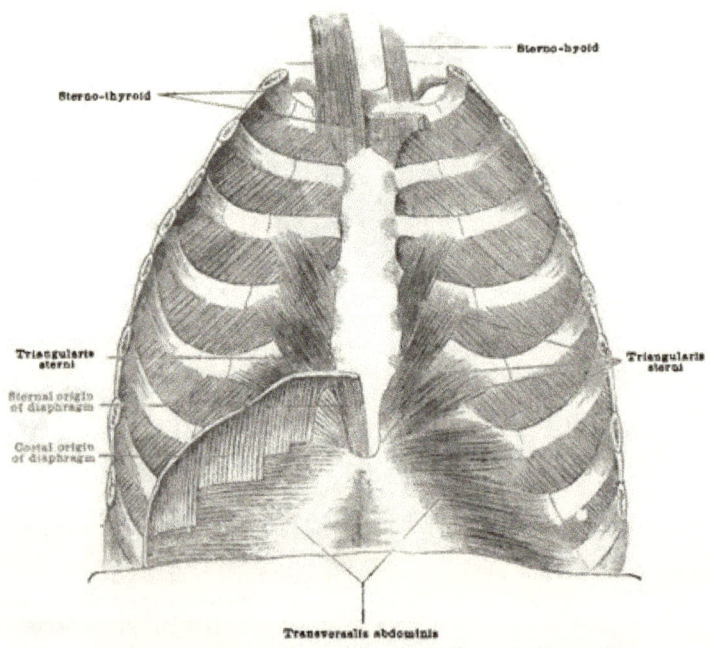

Sterno-hyoid

Sterno-thyroid

Triangularis
sterni

Triangularis
sterni

Sternal origin
of diaphragm

Costal origin
of diaphragm

Transversalis abdominis

Fig. 30.—The Muscles attached to the Back of the Sternum.

54

THE NERVE-SUPPLY TO THE DEPRESSOR MUSCLES OF THE HYOID BONE.—The group you have just dissected comes from a loop called the *ansa hypoglossi*. The expression *cervico-hypoglossal* would be a better term, as it would seem, as it expresses the derivation of the two elements composing the loop. (Fig. 29.)

The loop is formed by the descendens hypoglossi, anastomosing with two communicating branches from the deep part of the cervical plexus. This loop is formed on the front part of the carotid sheath. From it branches are given off to the depressors of the hyoid bone.

---

1. *Give the boundaries of that part of the neck most frequently operated on by the surgeon.*

Anteriorly, the mid- or mento-sternal line. Posteriorly, the anterior border of the trapezius muscle. Superiorly, the mandible and masto-mandibular line. Inferiorly, the clavicle and interclavicular ligament.

2. *Name the cervical triangles and indicate their importance.*

The submaxillary or supra-hyoid triangle is bounded above by the mandible and masto-mandibular line; below, by the digastric and stylo-hyoid muscles; in front, by the mento-sternal line. It contains: (1) The facial vessels and their sympathetic nerves; (2) the submaxillary gland, its blood-vessels, nerves, and capsule; (3) the submental and mylo-hyoid vessels and nerves; (4) the stylo-mandibular ligament; (5) the parotid gland and its contents; (6) the vagus nerve, internal jugular vein, internal and external carotid arteries.

The superior carotid triangle is bounded above by the digastric; below, by the omo-hyoid; behind, by the sterno-mastoid muscle. It contains: (1) The common carotid artery; (2) the external carotid artery; (3) the internal jugular vein; (4) the vagus; (5) spinal accessory, and (6) sympathetic nerves.

3. The inferior carotid or tracheal triangle is bounded above by the omo-hyoid, behind by the sterno-mastoid, in front by the mento-sternal line. It contains: (1) The thyroid gland and its blood-supply; (2) the trachea; (3) the carotid sheath and contents; (4) the inferior laryngeal nerve and inferior thyroid vessels.

The three foregoing triangles lie anterior to the sterno-mastoid muscle. The two following ones lie posterior to the sterno-mastoid muscle.

4. The occipital triangle is bounded in front by the sterno-mastoid, behind by the trapezius, below by the omo-hyoid muscle. It contains: (1) The spinal accessory nerve; (2) the descending branches of the cervical plexus; (3) the transversalis colli vessels; (4) a chain of lymphatic glands.

5. The subclavian triangle is bounded in front by the sterno-mastoid, above by the omo-hyoid, below by the clavicle. It contains: (1) The subclavian artery and vein; (2) the brachial plexus; (3) the external jugular vein; (4) the nerve to the subclavian muscle.

The student will please note that any arbitrary classification of contents is impossible. Structures are found to form partial contents of several triangles. The surgical triangles are to the modern surgeon about what totem poles would be to the city council.

6. *Name the structures in the superficial fascia of the neck.*

(1) The platysma myoides muscle; (2) the anterior, external, and posterior jugular veins; (3) the inframandibular branch of the seventh nerve, and (4) the superficialis colli branch of the cervical plexus.

7. *Name the layers of deep cervical fascia and give the contents of each.*

The first layer surrounds the sterno-mastoid muscle.

The second layer surrounds the hyoid depressors and their vessels and nerves.

The third layer surrounds the larynx, trachea, thyroid, œsophagus, common carotid artery, internal jugular vein, and vagus nerve.

The fourth layer covers the rectus capitis anticus major and minor muscles and the longus colli.

8. *What becomes of the first and second layers below?*

They are attached to the clavicle and sternum.

9. *Give the origin and insertion of the temporal muscle.*

This muscle arises from the complete temporal ridge, from the temporal aponeurosis, from the temporal fossa, and is inserted into the coronoid process of the mandible.

10. *Analyze the coronoid process.*

This has an outer and an inner surface, an anterior and a posterior border, a base, and an apex.

11. *Of what does the complete temporal ridge consist?*

It consists of frontal, parietal, and temporal portions. The ridge has a superior part for the attachment of the temporal fascia; an inferior part for origin of the muscle.

12. *What do you find in grooves on the deep surface of the muscle next the bone?*

The deep temporal arteries—branches of the internal maxillary artery—for the supply of the muscle.

13. *Analyze the zygomatic arch and tell what you find under the same.*

The *arcus zygomaticus* is made up of the zygomatic processes of the malar and temporal bones. It has a superior border, into which is inserted the temporal fascia; an outer surface, which is subcutaneous; an inferior border and an inner surface, which are occupied by the origin of the masseter muscle. Under the arch are the coronoid and condylar processes of the mandible; the sigmoid notch, transmitting the masseteric vessels and nerves; the external pterygoid muscle; and a considerable quantity of fatty connective tissue

14. *Describe the masseter muscle.*

This muscle arises (1) from the inferior border, anterior two-thirds, and (2) from the inner surface and posterior one-third of the lower border of the zygomatic arch. It is inserted into the external surface of the ramus. It is covered on its outer surface by the masseteric fascia, on which lie the pes anserinus, the parotid gland in part, and Stenson's duct.

15. *Describe the sterno-cleido-mastoid muscle.*

This muscle has two origins—a clavicular and a sternal; and two insertions —a mastoid and an occipital. It extends obliquely across the neck, dividing this region into an anterior and a posterior part. Its nerve-supply is from the spinal accessory and deep branches from the cervical plexus. It lies on the carotid sheath in part of its course, and along its posterior margin emerge the superficial branches of the cervical plexus. (Fig. 22.)

16. *Name the superficial branches of the cervical plexus.*

The descending branches are the suprasternal, supraacromial, supra-clavicular; the ascending branches are the great auricular, the small occipital, the lesser occipital, and the superficialis colli.

Quiz on the Structures in the Second Layer of Deep Cervical Fascia.

1. *Name the depressor muscles of the hyoid.*

(1) The sterno-hyoid. (2) The omo-hyoid. (3) The sterno-thyroid. (4) The thyro-hyoid.

2. *How are these muscles ensheathed and from what are they separated in front and behind?*

They are ensheathed by a delamination of the second layer of deep cervical fascia. In front of them is the first layer of deep cervical fascia, which delaminates on the side of the neck to enclose the sterno-cleido-mastoid muscle. Behind these structures are the members making up a group ensheathed by the third layer of deep cervical fascia.

3. *To what did you find the second layer of deep cervical fascia attached superiorly and inferiorly?*

Superiorly it was attached to the hyoid bone; inferiorly to the posterior part of the clavicle and manubrium.

4. *From what source do the depressors of the hyoid bone receive their nerve-supply?*

From the ansa hypoglossal loop, descendens, and communicans noni.

5. *How is the ansa hypoglossal loop formed?*

It is formed by the union of two branches from the deep cervical plexus with one branch from the hypoglossal nerve.

6. *What are these communicating nerves called?*

The one from the hypoglossal nerve is called the *descendens hypoglossi*; the two from the cervical plexus are called the communicantes hypoglossi, and the loop formed by their union is the *hypoglossal loop*.

7. *Have these communicating nerves any synonyms in anatomical literature?*

Yes, in the older texts they are called the descendens noni and communicantes noni. Under the classification of cranial nerves by Willis, there were nine pairs. The hypoglossal belonged to this ninth pair, hence the expressions nervus nonus, nervus descendens noni, and nervi communicantes noni.

8. *Can you think of a compound word that would be more specific, and at the same time a more rational name than ansa hypoglossal?*

Yes, the compound cervico-hypoglossal would express the anatomical parties to the compound, and would harmonize with our rules for writing compound words, by which such relations should always be expressed.

9. *Where did you find this loop?*

On the sheath containing the common carotid artery, pneumogastric nerve, and internal jugular vein. (Fig. 31.)

The third layer of deep cervical fascia contains (Fig. 24):

1. The *thyroid body*—its nerve-supply and vessels.
2. The *larynx* and *trachea*, nerves and vessels.
3. The *common carotid artery*, vagus and internal jugular vein.
4. The *external carotid artery* and its branches in the neck.
5. The *cervical stage* of the internal carotid artery.
6. The *hypoglossal nerve* and its descending branch.
7. The *cervical sympathetic* cord and ganglia.
8. The *arteries from* the transverse aorta and their branches.
9. The *phrenic nerve* and anterior scalenus muscle.
10. The *cervical plexus* and its branches of origin.
11. The *brachial plexus* and its sheath.
12. The *scaleni muscles*—anticus, medius, posticus.

Dissection.—Remove the group of muscles in the second layer of the cervical fascia, and examine the structures in the following order:

1. **Thyroid Gland** (Figs. 27 and 28).—This derives its capsule from the connective tissue in which it is developed. It has two lobes, connected by an isthmus. Its arteries are a superior thyroid, which you will take to its origin, the external carotid. This artery is attended by a vein. Note the anastomosis between the

5

superior thyroid and the inferior thyroid in the substance of the gland. The inferior thyroid artery is a branch of the thyroid axis of the subclavian. Note the crico-thyroid branch on the crico-thyroid membrane. Note also the superior laryngeal artery, the same artery you found when you removed the thyro-hyoid muscle : it accompanies the superior laryngeal nerve. The thyroid gland belongs to the class of ductless structures, as the spleen, thymus, and suprarenal capsules. An enlargement of this gland is called bronchocele ; in which case the isthmus may be divided or one-half of the gland removed. The nerves to the thyroid are of sympathetic derivation, from the middle and inferior cervical ganglia. The inferior thyroid veins you will trace to the left brachio-cephalic vein. The superior and middle veins open into the internal jugular vein. These veins take their origin in a plexus of veins especially large on the posterior part of the gland.

The carotid sheath is situated by the side of the trachea and larynx. In this sheath are the common carotid artery internally, the internal jugular vein externally, and the pneumogastric or vagus nerve between the two.

The internal jugular vein is formed by the confluence of the dural sinuses. These sinuses receive their blood from the brain. The internal jugular veins are made up at the jugular foramina at the base of the skull. The ninth, tenth, and eleventh cranial nerves leave the cranial cavity with the veins. In its course down the neck the jugular vein receives the superior and middle thyroid veins ; and near their termination, they receive the external jugular veins. Behind the clavicle, you will see the internal jugular vein on each side unite with the subclavian vein. The result of this union is a large vein, called the innominate or brachio-cephalic vein. These latter of each side unite with the vena azygos major to form the descending vena cava.

The right common carotid artery begins on the right side, at the bifurcation of the innominate artery. (Fig. 36.) This division occurs opposite the upper margin of the thyroid cartilage. The result of the division gives us an internal and an external carotid artery. Find to the inner side of the common carotid artery the larynx, trachea, thyroid gland, inferior thyroid artery, and recurrent laryngeal nerve.

The left common carotid artery on the left side differs from that on the right as follows : It is given off in the thorax from the highest part of the arch of the aorta. (Fig. 36.) It has the same relations in the neck as the right common carotid. In the thorax the left common carotid artery is behind the manubrium sterni, the origins of the sterno-thyroid and sterno-hyoid muscles, the left brachio-cephalic or innominate vein, and the remains of the thymus gland. Behind the artery are the trachea, œsophagus, and thoracic duct. Internal to the artery are the innominate artery, the inferior thyroid veins, and remains of the thymus gland. External to the artery are the vagus nerve, the left lung, and pleura.

The Glandulæ Concatenatæ.—If the subject be tubercular, you may find the deep cervical lymphatics enlarged. They lie with the internal jugular vein, and may have contracted firm adhesions to the vein or artery or both. This seems to be the tendency of glandular tissue : (1) to penetrate or (2) to embrace. Instance the deep part of the parotid gland, penetrating the glenoid cavity and embracing the styloid process too ; the *anterior* part of the same gland embraces, in a V shape, the posterior border of the ramus of the mandible. Likewise, these lymphatics, when enlarged, embrace vessels, a circumstance which makes their removal often a very dangerous procedure. These glands you will find forming a continuous chain from the base of the skull to the apex of the thorax, where they are continuous with the mediastinal gland-chains.

The pneumogastric nerve, also called the vagus, or par vagum, lies between the common carotid artery and internal jugular vein. Separate the connective tissue between the vessels and find the nerve, deeply located. This nerve gives off in the neck (1) the **superior laryngeal nerve** to the larynx ; (2) the **cervical cardiac branches** ; (3) the **pharyngeal branches** ; (4) the **recurrent laryngeal nerve.**

For the present remember this : The pneumogastric supplies the organs of voice and respiration with motion and sensation ; the organs of circulation and digestion with motion only.

The internal carotid artery on each side begins at the bifurcation of the common carotid. Find this bifurcation and see whether it is opposite the thyroid cartilage or the hyoid bone. The internal carotid has four stages :

1. The *cervical*—to the base of the skull, from the bifurcation.

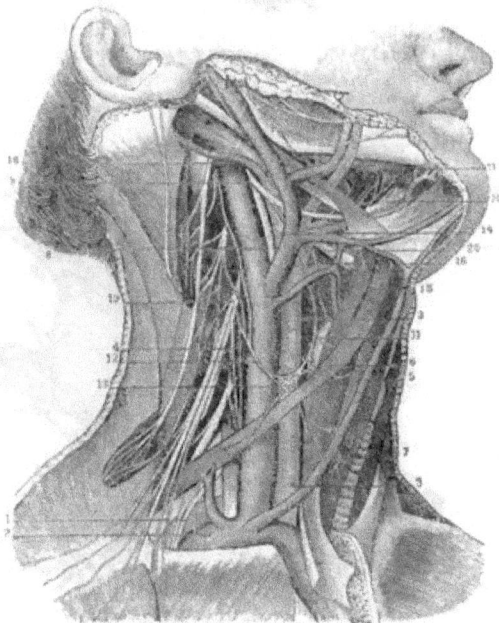

· FIG. 31.—VESSELS AND NERVES OF THE HEAD AND NECK.

1. Subclavian artery. 2. Subclavian vein. 3, 3. Common carotid artery. 4. Internal jugular vein. 5. Anterior jugular vein. 6. Omo-hyoid muscle. 7. Sterno-hyoid muscle. 8. Trunk of pneumogastric nerve. 9. Hypo-glossal nerve. 10. Its terminal portion. 11. Its descending branch. 12. Internal descending branch of cervical plexus. 13. Plexus formed by last two branches. 14. External carotid artery. 15. Superior thyroid artery and vein. 16. Lingual and facial arteries. 17. Facial artery and vein. 18. Occipital artery. 19. Anterior branches of the first four cervical nerves. 20. Superior laryngeal nerve.

2. The *petrosal*—in the carotid canal in the petrosa.

3. The *cavernous*—in the cavernous sinus on the lingula.

4. The *cerebral*—at the base of the brain in the cranium.

The cervical stage gives off no branches in the neck. The function of this artery is to take blood to the brain, eye, and nose. It assists the vertebral in forming the circle of Willis, by which circle the brain receives all its blood. Blood returns from the brain by the internal jugular vein.

External Carotid Artery.—This vessel begins at the bifurcation of the common carotid artery just mentioned. In some tubercular cases you will find a large nest of enlarged glands here at the bifurcation. The following are the branches of the external carotid artery and their attendants :

1. The *superior thyroid artery*, vein, nerve, lymphatic.
2. The *lingual artery*, vein, nerve, lymphatic.
3. The *facial artery*, vein, nerve, lymphatic.
4. The *occipital artery*, vein, nerve, lymphatic.
5. The *posterior auricular artery*, vein, nerve, lymphatic.
6. The *ascending pharyngeal artery*, vein, nerve, lymphatic.
7. The *temporal artery*, vein, nerve, lymphatic.
8. The *internal maxillary artery*, vein, nerve, lymphatic.

NOTE.—It has been pointed out in a foregoing paragraph that sympathetic nerves accompany every branch of the external carotid artery. Morris shows that the deep lymphatics of the head and neck roughly follow the course of the deep arteries, and finally terminate in the glandulæ concatenatæ previously described. Hence in your dissection of the branches of the external carotid,

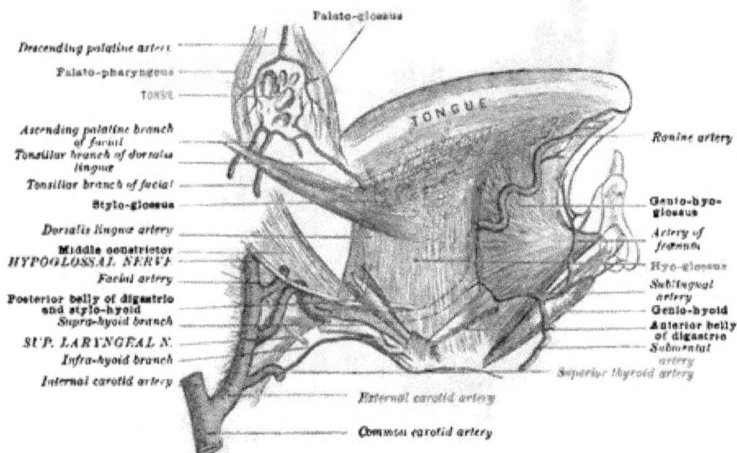

FIG. 32.—SCHEME OF THE LINGUAL ARTERY.

you will remember you can always find the vein corresponding to the artery ; the sympathetics may be found by treating the artery with ether and formaline ; the lymphatic glands can only be demonstrated when enlarged by disease. As to the lymphatic, I would urgently request the student to study carefully the schematic drawings of these vessels by Professor F. R. Sherwood, in Morris' "Anatomy."

• *Caution.*—Never use a cutting instrument in dissecting arteries. The forceps or dissecting hook is all you need to divide the connective tissue. Handle veins with gentle touch ; they are very easily ruptured.

Specific Dissection.—**The lingual artery**: (1) find this vessel above the greater horn of the hyoid bone ; (2) a little below, and running parallel with, the hypoglossal nerve ; (3) passing behind the free margin of the hyo-glossus muscle to the tongue. (Fig. 32.)

The Superior Thyroid Artery.—(1) It is the first branch given off by the external carotid ; (2) it goes to the upper part of the thyroid gland ; (3) it gives off the superior laryngeal branch, which always accompanies the superior laryngeal nerve; (4) these structures, the superior laryngeal nerve and artery, pierce the thyro-hyoid membrane; (5) the anastomosis is feeble with the opposite side; free with the inferior laryngeal artery, and the distribution is to the gland and to the depressor muscles of the hyoid bone that cover the gland. (Fig. 28.)

The facial artery: (1) often given off with the lingual; (2) note its deep course behind the hypoglossal nerve, the stylo-hyoid muscle, the digastric muscle, the parotid gland. It crosses the mandible in front of the masseter muscle. Its specific sympathetic nerves are called nervi molles. (Fig. 17.)

The facial vein (Fig. 18), you will note, takes a superficial course. It lies in

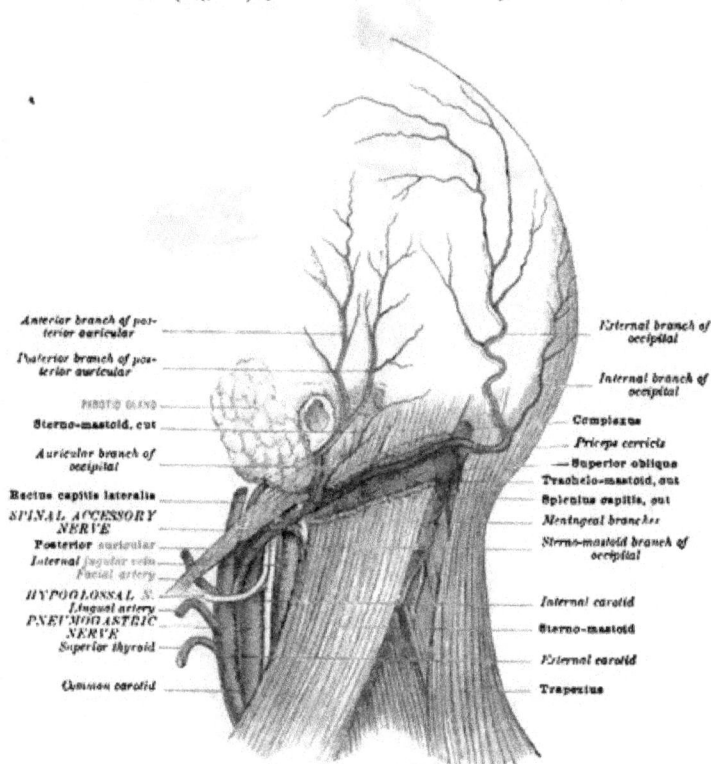

Anterior branch of posterior auricular
Posterior branch of posterior auricular
PAROTID GLAND
Sterno-mastoid, cut
Auricular branch of occipital
Rectus capitis lateralis
SPINAL ACCESSORY NERVE
Posterior auricular
Internal jugular vein
Facial artery
HYPOGLOSSAL N.
Lingual artery
PNEUMOGASTRIC NERVE
Superior thyroid
Common carotid

External branch of occipital
Internal branch of occipital
Complexus
Priceps cervicis
Superior oblique
Trachelo-mastoid, cut
Splenius capitis, cut
Meningeal branches
Sterno-mastoid branch of occipital
Internal carotid
Sterno-mastoid
External carotid
Trapezius

FIG. 33.—SCHEME OF OCCIPITAL AND POSTERIOR AURICULAR ARTERIES.

front of the structures behind which the facial artery passes. It opens into the external jugular vein. In your dissection show all the structures by which the facial artery and vein are separated in their cervical stage.

Branches of the facial artery in the neck are:

1. To the stylo-hyoid, internal pterygoid, masseter, and buccinator.
2. Submaxillary branches four, to the submaxillary gland.
3. Submental artery to structures under the chin.
4. Tonsillar branches to the tonsil and tongue.
5. Ascending palatine branches to the soft palate.

The Occipital Artery (Fig. 33).—(1) Find the hypoglossal nerve passing under and behind and external to it, to gain its place in front of the internal and external carotids ; (2) it passes behind the digastric, sterno-mastoid, splenius capitis ; (3) it gives off the arteria princeps cervicis, which anastomoses with the

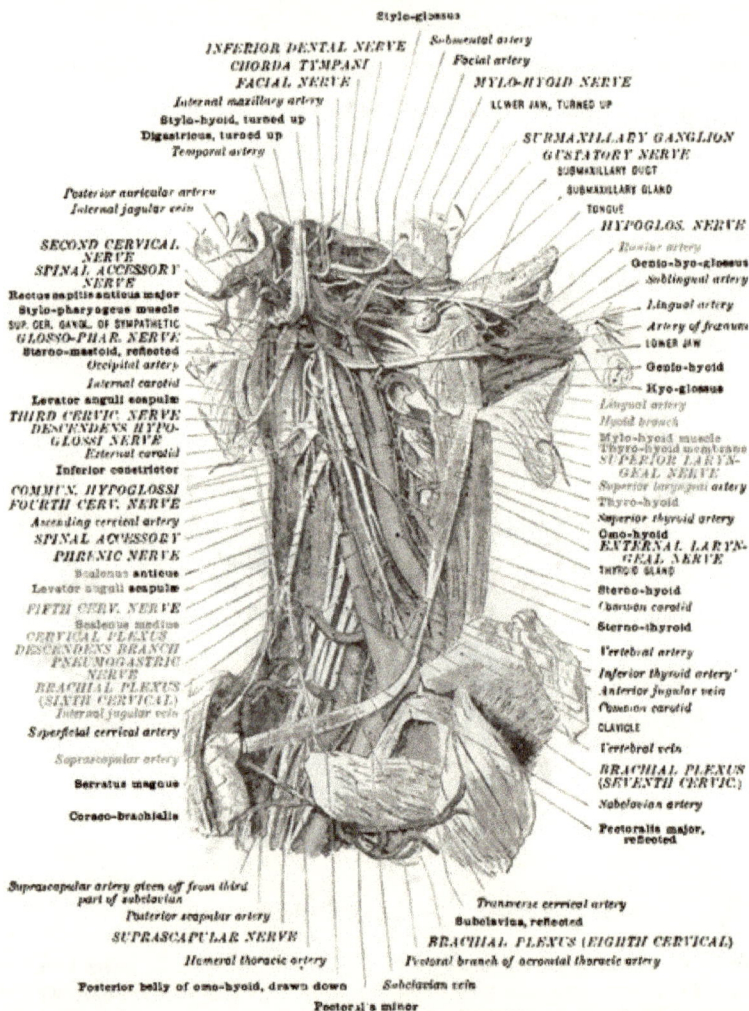

FIG. 34.—THE COMMON CAROTID, THE EXTERNAL AND INTERNAL CAROTID, AND THE SUBCLAVIAN ARTERIES OF THE RIGHT SIDE AND THEIR BRANCHES.
(From a dissection by Dr. Alder Smith in the Museum of St. Bartholomew's Hospital.)

deep cervical, a branch of the superior intercostal, and the vertebral artery. This anastomosis keeps up the collateral circulation, after ligation of the common carotid artery.

The posterior auricular artery arises high. You will find it behind the parotid gland. Trace it very close to the ear; between the ear and the mastoid process of the temporal bone.

The superficial temporal arteries are in the superficial fascia covering the temporal muscle. They are attended by the sensory auriculo-temporal branches of the fifth cranial nerve. Note the deep temporal vein piercing the temporal fascia above the zygoma. (Fig. 18.)

The ascending pharyngeal you will find buried in connective tissue, lying between the internal and external carotid arteries. It arises from the external carotid near the bifurcation. It is the smallest branch of the external carotid. It is distributed to the pharynx and meninges.

The internal maxillary will be dissected with the muscles of mastication and the fifth cranial nerve. (Fig. 51.)

The Hypoglossal Nerve (Fig. 31).—Find this nerve crossing the internal and external carotid arteries a little above the bifurcation of the common carotid artery. Note that this nerve lies a little distance above the lingual artery; that the artery crosses behind the hyo-glossus muscle, the nerve in front of the muscle. The nerve gives off the following branches:

1. To the thyro-hyoid muscle.
2. To the stylo-glossus muscle.
3. To the hyo-glossus muscle.
4. To the genio-hyoid muscle.
5. To the genio-hyo-glossus muscle.
6. To the sterno-hyoid muscle.
7. To the omo-hyoid muscle.
8. To the sterno-thyroid muscle.
9. To the meninges; recurrent branches.

All these branches, except the meningeals, you can readily trace out. This nerve must not be mistaken for the superior laryngeal branch of the pneumogastric. Now compare the two nerves in their relation to the carotid arteries on your dissection. The result of section of this nerve will be considered when you dissect the tongue and outline its complete nerve-supply.

Dissect the following muscles (Fig. 35):

1. The digastric and its intermediary tendon.
2. The stylo-hyoid muscle.
3. The mylo-hyoid muscle.
4. The hyo-glossus muscle.

1. Find the insertion of the stylo-hyoid muscle into the lower border of the body of the hyoid bone at the junction of the greater cornu and body of the hyoid bone. Study the relation of the intermediary tendon of the digastric to the aponeurosis of the stylo-hyoid muscle. Trace the origin of the stylo-hyoid muscle to the base and outer surface of the styloid process of the temporal bone.

The Digastric has an *anterior belly*, that lies on the mylo-hyoid muscle, a *posterior belly*, that lies under the stylo-hyoid muscle just found, an *intermediary tendon*, and a suprahyoid aponeurosis. Carefully detach the anterior belly of the digastric muscle from the digastric fossa of the mandible, and as you pull this detached belly down, divide the connective tissue between this and the mylo-hyoid muscle. Notice the nerve-supply to the anterior belly of the digastric—the mylo-hyoid branch of the inferior dental. The mylo-hyoid muscle is now in full view since you removed the anterior belly of the digastric muscle. It forms the floor of the mouth. It arises from the internal oblique line or mylo-hyoid ridge on the inner surface of the body of the inferior maxilla. It is inserted into the mid-line of the neck into its fellow of the opposite side. This muscle will be studied when you dissect the mouth and tongue.

The Hyo-glossus Muscle (Fig. 32).—This is the muscle that separates the hypoglossal nerve from the lingual artery. It arises from the body and greater and lesser cornua of the hyoid bone. It passes to the tongue. It is the most deeply located muscle in this region.

The great branches from the transverse part of the arch of the aorta may be reviewed. In fact, you can not study them and their relations too much. The *innominate* has been seen dividing into the right subclavian and right common carotid. You may now dissect the branches of the subclavian artery and vein.

The Subclavian Artery and Vein.—The subclavian artery has three stages :

1. From its beginning to the inner border of the scalenus anticus.
2. The artery behind the scalenus anticus. (Fig. 36.)
3. From the scalenus anticus to the lower border of first rib.

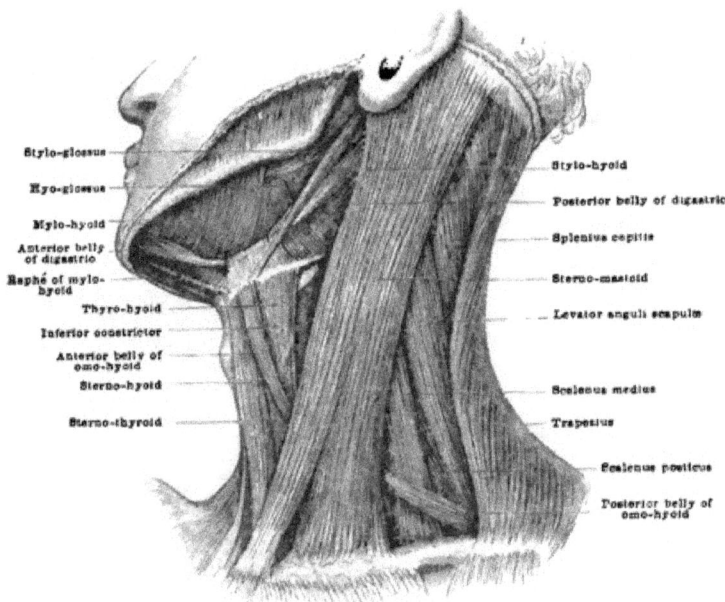

FIG. 35.—ANTERIOR AND LATERAL CERVICAL MUSCLES.

Find the scalenus anticus muscle inserted into the scalene tubercle of the first rib. In a groove in front of this tubercle find the *subclavian vein ;* behind the tubercle find the *subclavian artery.* Behind the artery find the *scalenus medius muscle* inserted into the first rib. Above the artery observe the anterior primary divisions of the fifth, sixth, seventh, and eighth cervical nerves and first dorsal nerve between the scalenus anticus and scalenus medius forming the brachial plexus.

The subclavian arteries differ on the two sides ; the stages and branches, however, are the same in each as to name. The first stage of the left subclavian is longer than the first stage of the right.

The right subclavian artery begins behind the right sterno-clavicular articulation, at the bifurcation of the innominate artery. The artery lies very deeply. In front of it are the skin, **superficial** fascia containing the platysma myoides, the clavicular head of the sterno-cleido-mastoid muscle, the sterno-hyoid and sterno-thyroid muscles. The phrenic nerve, the cardiac branches of the sympathetic nerve, the vagus nerve, the vertebral and internal jugular veins, **cross this artery**. Behind the artery is the neck of the first rib and the longus colli muscle and the recurrent laryngeal nerve; below the artery is the pleura, against wounding which the greatest precautions should be taken in operations in this locality.

The left subclavian artery is longer than the right; it does not arch outward like the artery of the opposite side, but ascends vertically to the inner border of the scalenus anticus muscle and is situated more deeply in the thorax. In front

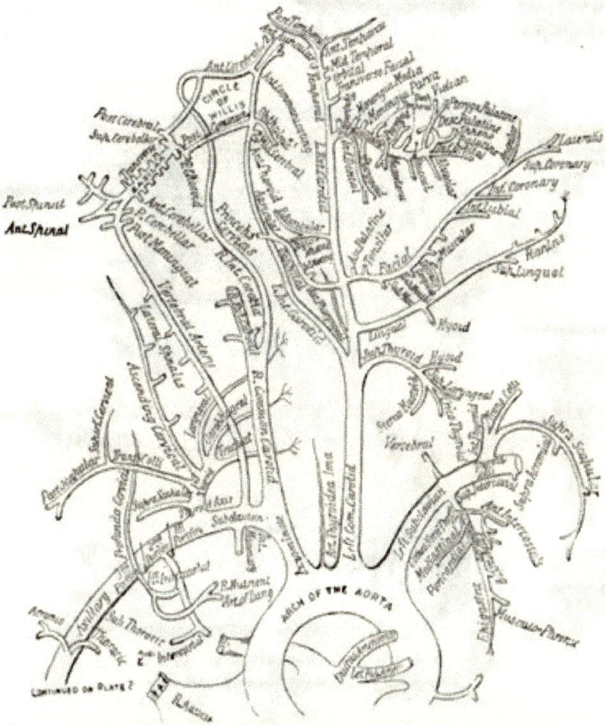

FIG. 36.—SCHEME FOR HEAD AND UPPER EXTREMITY.

of the artery are the lung and pleura, the vagus, phrenic, and cardiac nerves, the left common carotid artery, internal jugular and vertebral veins, and left brachio-cephalic vein, the sterno-hyoid, sterno-thyroid, and sterno-cleido-mastoid muscles. Behind the artery are the œsophagus, thoracic duct, inferior cervical ganglion, longus colli muscle, and vertebral column. To the outer side of the artery is the pleura; to the inner side are the trachea, thoracic duct, and œsophagus.

Branches of the subclavian artery:

The *internal mammary* to anterior thoracic and abdominal walls.

The *vertebral* to the brain and spinal cord and vertebræ.

The *superior intercostal* to the first and second intercostal spaces.

The *suprascapular* to the shoulder.

The *inferior thyroid* to the thyroid gland.

The *transversalis colli.*

1. **The internal mammary** (Fig. 36) is given off from the under part of the first stage of the subclavian artery. Its course, branches, anastomoses, and important relations are in the *anterior thoracic walls*, where you will find its description when you dissect that part.

2. **The suprascapular artery** (Fig. 37) is attended by a vein and nerve of the same name. It crosses the trunks of the brachial plexus, and is distributed to the supra- and infraspinati muscles.

3. **The Inferior Thyroid Artery.**—Trace this artery behind the common carotid artery and sympathetic nerve. Very often the middle cervical ganglion of the sympathetic nerve rests on the inferior thyroid artery. Usually the artery

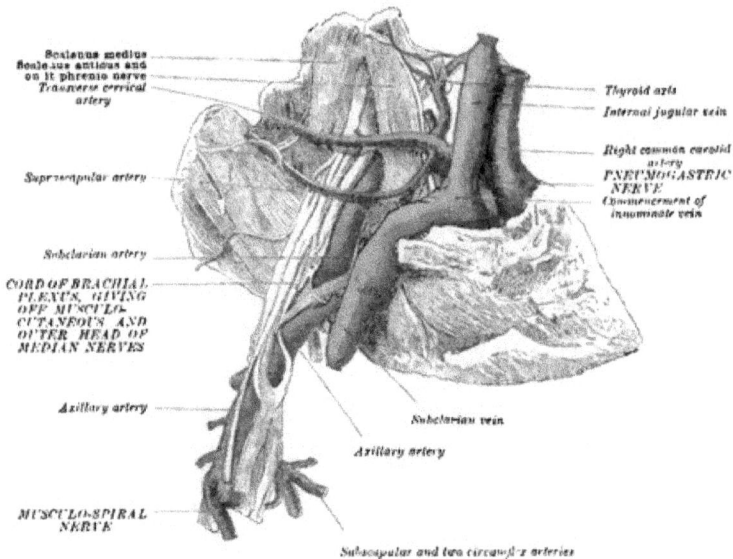

Scalenus medius
Scalenus anticus and
on it phrenic nerve
*Transverse cervical
artery*

Suprascapular artery

Subclavian artery

*CORD OF BRACHIAL
PLEXUS, GIVING
OFF MUSCULO-
CUTANEOUS AND
OUTER HEAD OF
MEDIAN NERVES*

Axillary artery

*MUSCULO-SPIRAL
NERVE*

Thyroid axis

Internal jugular vein

*Right common carotid
artery*
*PNEUMOGASTRIC
NERVE*
*Commencement of
innominate vein*

Subclavian vein

Axillary artery

Subscapular and two circumflex arteries

FIG. 37.—THE SUBCLAVIAN VESSELS.
(From a dissection in the Hunterian Museum.)

lies in front of the recurrent laryngeal nerve; sometimes you will find the reverse is true. In operations on the thyroid gland isolate (1) the recurrent laryngeal nerve, (2) the sympathetic.

The branches of the inferior thyroid artery:

1. The *inferior laryngeal* to the larynx.

2. *Tracheal branches* to the trachea.

3. *Œsophageal* branches to the œsophagus.

4. *Branches to the depressor* muscles of the hyoid bone.

5. *Ascending cervical branches* to the muscles of the neck.

The Superior Intercostal Artery (Fig. 39).—This is given off from the second stage of the subclavian artery. It gives off a communicating branch to the first aortic intercostal, which you will find high in the thorax, when you remove the pleura. The artery gives off its deep cervical branch, which inosculates with the *arteria princeps cervicis*, a branch of the occipital. The anastomosis takes place between the complexus and semispinalis colli muscles. By

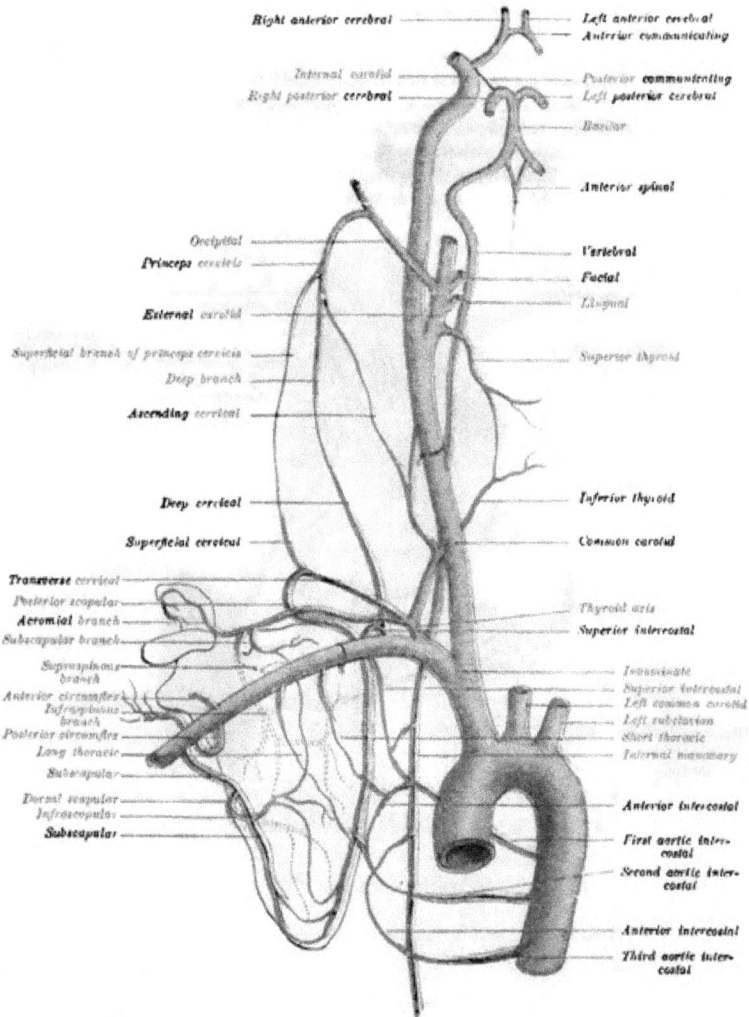

FIG. 38.—THE COLLATERAL CIRCULATION AFTER LIGATURE OF THE COMMON CAROTID AND SUBCLAVIAN ARTERIES.

(A ligature is placed on the common carotid and on the third portion of the subclavian artery.)

this anastomosis the collateral circulation is carried on after ligature of the common carotid artery. (Figs. 38, 39, and 40.)

The vertebral artery (Fig. 40) is given off from the second stage of the subclavian. It passes through the foramina in the transverse processes of the cervical vertebræ, except the lower two, and through the foramen magnum, in the occipital bone, to assist the internal carotid artery in forming, by anastomosis, the *circle of Willis.* The artery is attended by a sympathetic plexus of nerves from the inferior cervical ganglion. The artery is found lying in the vertebral groove, at the junction of the lateral mass and posterior arch of the atlas in the *suboccipital triangle,* at the base of the skull. This triangle is bounded by the superior and inferior oblique and rectus capitis posticus major muscles. The roof of the triangle is the complexus muscle.

Branches of the vertebral artery are as follows:

1. *Muscular* branches to the deep muscles of the neck.

2. *Lateral spinal* arteries to the spinal cord and meninges; the bodies of the vertebræ also receive blood from these branches.

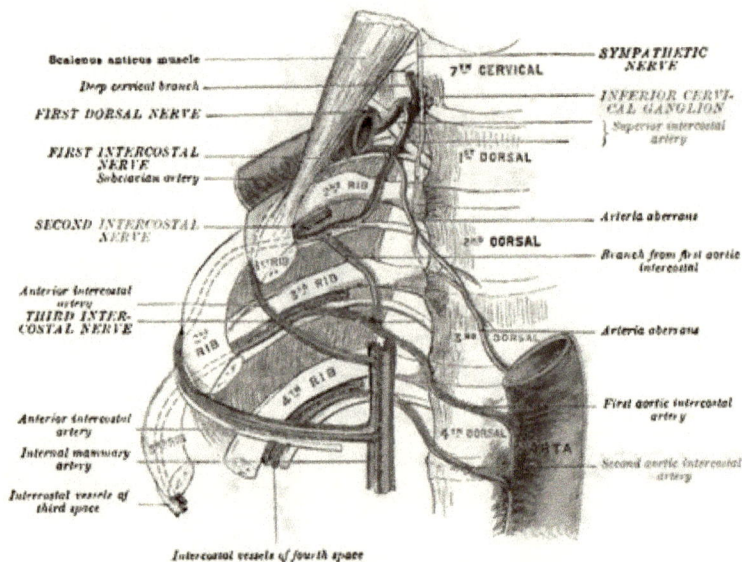

FIG. 39.—SCHEME OF THE SUPERIOR INTERCOSTAL ARTERY.

3. *Anterior* and *posterior* spinal branches that may be well seen when you dissect the cord.

4. *Posterior inferior cerebellar* arteries. These are the largest arteries given off from the vertebral. They will be studied when we consider the structures seen on removing the brain. The two vertebral arteries unite, within the cranium, to form the basilar artery. This artery will be studied when you remove the brain.

The transversalis colli is a branch of the thyroid axis. It gives off the posterior scapular artery, by which collateral circulation is established with the subscapular branch of the axillary artery in cases of ligation of the third stage of the subclavian artery. (Figs. 38, 39, and 40.)

The scaleni muscles are three in number. You have already studied their insertion into the first and second ribs. You will now review the relation between the subclavian vein and artery on the outer surface of the first rib. You will again see the nerves that are to form the brachial plexus coming through between the scalenus anticus and scalenus medius muscles above the subclavian artery. You will find the *phrenic nerve* lying on the anterior surface of the scalenus anticus muscle. You will in time trace it between the subclavian vein and artery into the thorax, down in front of the root of the lung, to the diaphragm, and all the serous membranes continging the diaphragm, for distribution.

Origin of the Scaleni Muscles.—To aid the memory, remember the numerals 3, 6, 3. The scalenus anticus arises from the anterior tubercles of the lower three transverse processes; the scalenus medius from the lower six; the scalenus posticus from the lower three. The last two arise from posterior tubercles. Remember, scarcely any two dissections will show exactly the same origin;

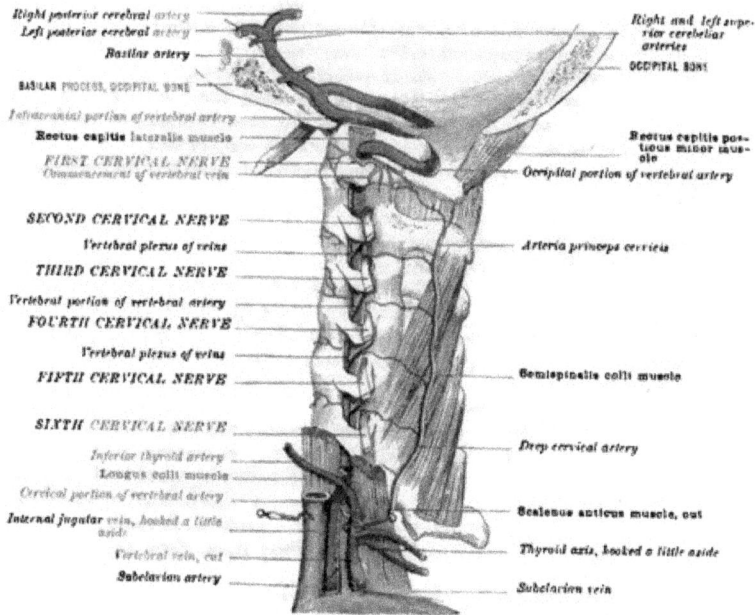

FIG. 40.—SCHEME OF THE VERTEBRAL ARTERY.
The internal jugular and vertebral veins are hooked aside to expose the artery.

hence it is, scarcely any two authors will give the origin of these muscles alike. The foregoing is a general average and easily remembered—3, 6, 3.

The Levator Anguli Scapulæ Muscle.—It is necessary to have exact knowledge of this muscle, to understand your dissection of the plexuses. This muscle arises from the posterior tubercles of the transverse processes of the upper three. You will find the tendons of origin of this muscle becoming fleshy. They will unite to form a flat muscle, two inches broad. This muscle will be inserted into the middle lip of the vertebral border of the scapula, from the superior angle to the vertebral end of the scapular spine. Find its nerve-supply coming from the cervical plexus.

1. *Locate the thyroid gland.*

It is in the third layer of deep cervical fascia. It embraces the cricoid cartilage of the larynx and the upper part of the trachea.

2. *From what source does the thyroid gland receive its blood?*

From the superior thyroid artery, a branch of the external carotid, and from the inferior thyroid, a branch of the thyroid axis of the subclavian artery.

3. *Does anastomosis occur between the right and left thyroid arteries?*

Yes ; this occurs, but very scantily, in the isthmus.

4. *Does the thyroid ever receive blood from any other source?*

Yes ; it may receive an artery either from the arch of the aorta or from the innominate artery, called the *thyroidea ima.*

5. *From what source does the thyroid gland derive its nerve-supply?*

From the middle cervical ganglion of the sympathetic.

6. *By what structures is the thyroid gland covered?*

By the skin, superficial fascia, first and second layers of deep fascia, the latter containing the depressor muscles of the hyoid bone.

7 *Give contents of the carotid sheath and locate the same.*

It contains the common carotid artery, the pneumogastric or vagus nerve, and the internal jugular vein. On the sheath lies the ansa-hypoglossal loop, from which the depressor muscles of the hyoid bone are innervated. The sheath is in the third layer of deep cervical fascia, and crossed near its middle third by the omo-hyoid muscle.

8. *Explain the common carotid artery.*

It lies to the inner side of the sheath, very near the larynx and trachea. To its outer side is the internal jugular vein. It is crossed by the ansa-hypoglossal loop and omo-hyoid muscle. On the right side it begins at the bifurcation of the innominate artery, behind the sterno-clavicular joint ; on the left side it begins at the arch of the aorta. The common carotid arteries end near the hyoid bone by dividing into the external and internal carotids, having given off no branches in their course.

9. *Describe the internal carotid artery.*

It begins at the bifurcation of the common carotid, and has four stages : (1) The cervical stage to the base of the skull, where it enters the temporal bone ; (2) the petrosal stage, where it passes through the carotid canal in the petrosal part of the temporal bone ; (3) the cavernous stage, where it lies in the cavernous sinus of the dura mater, by the side of the body of the sphenoid bone ; (4) the cerebral stage, where it comes through the dura mater to give off its terminal branches to the brain.

10. *Name the branches given off from the intracranial portion of the internal carotid.*

(1) It contributes the anterior cerebral, the middle cerebral, and the posterior communicating to the circle of Willis. (2) It gives off the ophthalmic artery for the supply of the orbit and its contents, the ethmoidal cells and the inner and outer nose in part. (3) It gives off the anterior meningeal arteries to the dura of the anterior fossa of the base of the skull. (4) It supplies the Gasserian ganglion.

11. *What branches are given off from the petrosal stage?*

(1) The Vidian, which is said to anastomose with the Vidian branch of the internal maxillary. (2) The tympanic, which goes to the middle ear and anastomoses with the tympanic branches of the internal maxillary and stylo-mastoid arteries.

12. *Name all the structures between the skin and the internal carotid one-half of an inch below the base of the skull.*

(1) The parotid gland ; (2) posterior belly of the digastric muscle ; (3) stylo-

hyoid muscle; (4) hypoglossal nerve; (5) posterior auricular and occipital arteries; (6) external carotid artery; (7) stylo-glossus and stylo-pharyngeus muscles; (8) stylo-hyoid ligament and pharyngeal branch of the pneumogastric nerve.

13. *Name the branches of the external carotid artery and indicate the territory its branches supply.*

By the posterior auricular, occipital, and temporal branches it supplies the scalp; by the superior thyroid it supplies the thyroid gland and the structures covering the same; by the facial it supplies the dermal muscles of expression and the skin covering them; by the lingual it supplies the tongue; by the internal maxillary it supplies the teeth, the muscles of mastication, the palate, the antrum of Highmore, the nasal fossæ, and the dura mater; by its ascending pharyngeal branch it supplies the pharynx.

14. *Where may you compress the facial artery?*

In front of the masseter muscle on the body of the mandible.

15. *Where may you find the lingual artery for ligation?*

Above the greater cornu of the hyoid bone.

16. *Give names of four important structures in front of the facial artery.*

(1) Hypoglossal nerve; (2) digastric muscle; (3) stylo-hyoid muscle; (4) submaxillary gland. The fascia and skin lie in front of these four structures.

17. *Name an important branch of the superior thyroid artery found in your dissection.*

The superior laryngeal, which, in company with the superior laryngeal branch of the pneumogastric nerve, piercing the thyro-hyoid membrane to supply the larynx.

18. *Where is the internal jugular vein made up?*

In the jugular foramen, by the confluence directly and indirectly of the dural sinuses. The ninth, tenth, and eleventh nerves also pass out through this foramen.

19. *Does the internal jugular vein receive any tributaries in its course?*

Yes; the lingual, facial, superior and middle thyroid, and pharyngeal. This vein unites with the subclavian to form the innominate or brachio-cephalic vein.

20. *What branches does the pneumogastric nerve give off in the neck?*

(1) The pharyngeal, to help form the pharyngeal plexus; (2) the superior and recurrent laryngeal for the supply of the larynx; (3) the cervical cardiac branches to assist in forming the cardiac plexus.

21. *Where did you find the superior laryngeal nerve, and in company with what was it?*

Piercing and lying on the thyro-hyoid membrane with the superior laryngeal branch of the superior thyroid artery.

22. *What does the superior laryngeal nerve supply?*

The mucous membrane of the larynx and trachea by its internal branch; the crico-thyroid muscle by its external branch.

23. *Where did you find the recurrent given off and distributed?*

It was given off from the vagus on the right side, just after this nerve crossed the subclavian artery. The nerve then passed upward and inward, behind the subclavian artery, gained the space next the trachea, and followed the same to the larynx. It supplies branches to all the muscles of the larynx except the crico-thyroid; it supplies, also, branches to the trachea and œsophagus.

24. *What are the glandulæ concatenatæ and what is their surgical importance?*

The deep cervical glands accompanying the internal jugular vein; they extend from the base of the skull to the root of the neck. Their importance surgically is the operation for their removal.

25. *Name the structures a surgeon should avoid injuring in this operation.*

(1) The carotid artery ; (2) the internal jugular vein ; (3) the vagus nerve ; (4) the cervical sympathetic, behind the carotid sheath ; (5) the recurrent laryngeal nerve ; (6) the apex of the parietal pleura.

26. *Name the branches of the hypoglossal nerve.*

(1) A communicating branch to the cervical plexus to form the ansa hypoglossal loop. (2) Muscular branches to the sterno-hyoid, omo-hyoid, sternothyroid, thyro-hyoid, stylo-glossus, hyo-glossus, and genio-hyo-glossus.

27. *Explain the digastric muscle.*

It has an anterior belly, a posterior belly, an intermediary tendon. The attachments are the digastric groove of the temporal bone and the digastric fossa of the mandible, the body of the hyoid bone. The action of the muscle depends on its fixed point. When the depressor muscles of the hyoid bone fix this bone, then the digastric depresses the mandible. In the act of swallowing, the hyoid bone is elevated by the digastric and its probably dismembered synergist, the stylo-hyoid. The seventh nerve supplies the posterior belly ; the mylo-hyoid branch of the inferior dental of the fifth nerve, the anterior belly of the digastric.

28. *What important nerve did you find on the outer part of the hyo-glossus muscle?*

The hypoglossal, the motor nerve of the tongue. The lingual artery passes behind the muscle.

29. *Explain fully the hyo-glossus muscle.*

The action of the muscle is to make the back of the tongue convex, and to retract the tongue. The origin is from the outer third of the anterior part of the body of the hyoid bone and from both the greater and lesser cornua. The muscle is inserted into the side of the tongue.

30. *Name the arteries you found arising from the transverse part of the aortic arch.*

The innominate artery on the right side ; the left common carotid and left subclavian on the left side.

31. *Name the stages of the subclavian artery.*

The first stage is internal to the scalenus anticus muscle ; the second stage is behind the muscle ; the third stage is from the outer margin of the muscle to the lower border of the first rib.

32. *Name the branches of the subclavian artery.*

(1) The internal mammary ; (2) the vertebral ; (3) the suprascapular ; (4) the superior intercostal ; (5) the transversalis colli.

33. *Where would you find for ligation the internal mammary artery?*

A finger's breadth to the right or left of the sternum, in the third or fourth intercostal space.

34. *How does the vertebral artery get to the base of the brain?*

It passes through foramina in the transverse processes of the cervical vertebræ, and through the foramen magnum.

35. *If the common carotid artery were ligated, by what two collateral channels would the blood circulate?*

First, through the thyroid arch. Second, through the occipito-intercostal arch.

36. *How is the thyroid arch formed?*

By anastomosis between the superior thyroid branch of the external carotid artery and the inferior thyroid branch of the subclavian artery. (Fig. 38.)

37. *How is the occipito-intercostal arch formed?*

By an anastomosis between the princeps cervicis of the occipital, and the deep cervical branch of the superior intercostal artery.

38. *What, if any, changes occur in these collateral arches?*

They become much larger, since growth is the correlative of function.

39. *In dissection, where do you find the occipito-intercostal arch?*
Between the semi-spinalis colli and complexus muscles. (Fig. 40.)

40. *Define the word scalene and name the scalene muscles.*
A triangular figure of unequal sides. The scalenus anticus, medius, and posticus.

41. *Where are these muscles inserted?*
The anticus is inserted into the scalene tubercle of the first rib; the medius into the first rib; the posticus into the second rib. The two are separated by the subclavian groove for the subclavian artery.

42. *Give the origin of the scalene muscles according to 3, 6, 3.*
The scalenus anticus arises from the anterior tubercles of the transverse processes of the lower three (3); the medius from the posterior tubercles of the lower six (6); the posticus from the posterior tubercles of the lower three (3) cervical vertebræ.

The **cervical plexus**, to be of technical as well as practical value to the student, should be studied in the following analytical manner, by question and answer, before dissection:

1. *Why is this plexus called cervical?*
On account of its derivation from cervical nerves, its location and major distribution in the neck.

2. *Where is this plexus situated?*
It is situated opposite the four upper cervical vertebræ. It is covered by the sterno-cleido-mastoid muscle. It lies on or is supported by two muscles whose origin we have just seen—the scalenus medius and levator anguli scapulæ.

3. *Is this plexus deeply located?*
No, it is the most superficially located of all the somatic plexuses. It emerges from under the posterior border of the sterno-cleido-mastoid muscle midway between the origin and insertion of this muscle, regardless of length of neck, into the occipital triangle.

4. *What seems to be the predominating function or physiological importance of the cervical plexus?*
To aid primarily and secondarily in respiration, as follows: It furnishes the phrenic nerve. This nerve supplies the diaphragm with motor influence; it supplies the serous membranes investing the diaphragm with sensation; through its relation to the sympathetic it reaches all the abdominal organs that continge the diaphragm. It communicates with the nerve to the subclavius muscle, and, according to Hilton's law, must send articular branches to the sterno-clavicular articulation. (See Hilton's law.) It communicates with the intercostals, by which the ribs are moved in respiration. It communicates with the ansa hypoglossal loop, by which the depressor muscles of the hyoid bone are innervated.

The cervical plexus furnishes communicating branches to the formation of the hypoglossal loop. The hypoglossal nerve is the motor nerve to the tongue. This nerve unites, by its descendens hypoglossi, with the two communicantes hypoglossi, previously referred to, and in this way harmony is established between the nerve that moves the tongue and the nerve that moves the muscles that depress the hyoid bone, on which the tongue rests.

The cervical plexus sends **motor nerves** to the following muscles: sterno-cleido-mastoid, scaleni, the trapezius, the levator anguli scapulæ; through its descending cutaneous branches it supplies the skin over the insertion of these muscles. These muscles all are accessory to respiration in asthma and other diseased conditions requiring more than ordinary tranquil respiration.

The cervical plexus communicates with the pneumogastric, and this nerve supplies the organs of voice and respiration with motion and sensation. In fine, the function of the cervical plexus very clearly is to innervate the muscles of

6

respiration, and to supply the serous membranes associated with the respiratory movement.

5. *How is the cervical plexus formed?*

It is formed by the union of the anterior primary divisions of the first, second, third, and fourth cervical nerves. The communications of the upper two nerves of this plexus are with nerves whose function has to do with respiration, with expenditure of air, or with movement of the tongue—the sympathetic, the pneumogastric, the hypoglossal. The communications of the lower two nerves are with the brachial plexus.

6. *Explain the distribution of the branches of the cervical plexus.*

This is rational, and must be studied in a philosophical light to be remembered and appreciated. Remember, it is one thing to commit to memory a statistical table of the branches of distribution of this plexus, as is often done ; and quite another thing to understand the rationale of a distribution. I desire you to learn the latter first.

1. This plexus has muscular branches which supply muscles that, without a single exception, have two well-defined actions.

2. This plexus has branches that supply the skin or serous membrane, as the case may be, covering these muscles. This is in accordance with Hilton's law, and applies to the articulations, and also to the skin covering the most extensive origin or insertion of the muscle.

3. This plexus has communicating branches, like any other plexus. One set of communicating branches is to those muscles which act synergistically to produce the muscular movements of respiration directly. Another set is to communicate with nerves that supply parts dependent on respiration.

The *muscles* supplied are as follows :

1. The platysma myoides acts toward clavicle or toward mandible.
2. The sterno-cleido-mastoid acts on the head or the thorax.
3. The scaleni act on the ribs or on the neck.
4. The levator anguli scapulæ acts on scapula or neck.
5. The recti capitis antici major and minor act on head or neck.
6. The trapezius acts on head or shoulder girdle.
7. The diaphragm acts on ribs and on cervical fascia.
8. Sterno-hyoid acts toward sternum or toward hyoid bone.
9. Omo-hyoid acts toward scapula or to the hyoid bone.
10. Thyro-hyoid and sterno-thyroid toward sternum or hyoid.
11. Genio-hyoid acts toward mandible or toward hyoid.
12. Genio-hyo-glossus acts forward and backward.
13. Stylo-glossus acts to or from the tongue.
14. Hyo-glossus acts to or from the hyoid.

The *serous membranes* and *integument* supplied in these areas :

1. The pleuræ, pericardium, peritoneum.
2. Skin over the pectoralis major and acromion and clavicle.
3. Skin over the depressors of the hyoid bone.
4. Skin over the genial muscles.
5. Skin over the upper ends of sterno-mastoid, trapezius, and scaleni.

The phrenic supplies all the serous membranes. The descending branches of the plexus supply the skin over the anterior part of the upper thorax and over the shoulder. The lesser and second occipital supply the skin over the sterno-mastoid and trapezius. (Fig. 43.) The superficial cervical supplies the whole front of the neck and the region under the chin.

The *ascending* and *descending branches*, then, of the cutaneous, sensory, or superficial nerves of the plexus are thus accounted for. The supraclavicular, suprasternal, and supraacromial are the descending ; the superficial cervical is a transverse branch; the great auricular, small occipital, and lesser or second occipital are ascending. These branches had to undergo this very distribution to carry out the scheme of distribution of the cervical nerves—a scheme we everywhere invoke when we would account for sensory nerve distribution.

The *communicating branches* are:

1. Extrinsic, as your dissection should show, with the auriculo-temporal, the facial, the intercostals, the circumflex, and the great occipital.

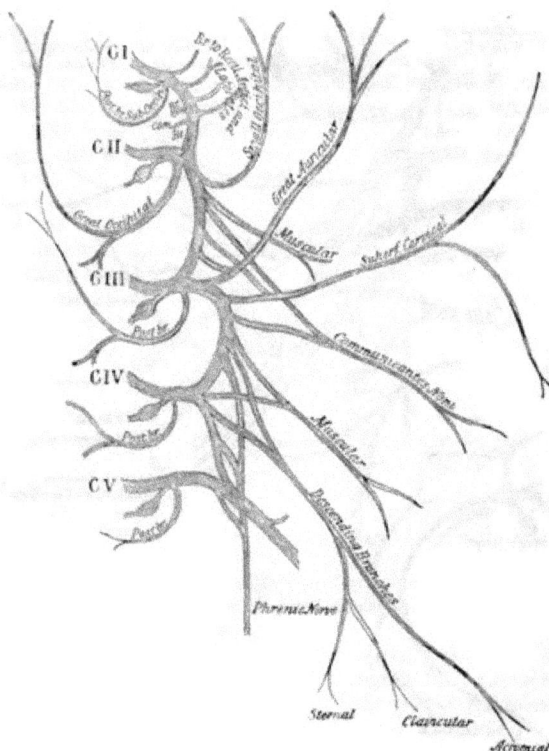

FIG. 41.—SCHEME OF THE CERVICAL PLEXUS.
Note the loops by which one nerve communicates with the next.

2. Intrinsic: with the hypoglossal, sympathetic, pneumogastric, spinal accessory, and brachial plexus, for reasons set forth in a foregoing paragraph.

External and Internal Series of Muscular Branches of the Cervical Plexus.—The scalenus medius, trapezius, levator anguli scapulæ, and sterno-cleido-mastoid are, according to the time-honored classification, given as the external series, while all the remaining muscles would naturally fall to the internal series. It may be said, with all deference, that this classification is categorical only. It rests on no basis of physiology, philosophy, rationalism, or even sense; it is categorical only. It belongs to that part of our nomenclature that had its origin in fanciful creations, and in location only, like sphenoid, ethmoid, and external iliac. It is tolerated in anatomy and revered on account of its antiquity.

The branches of the cervical plexus, tabulated according to the prevailing custom, but carried out to their distribution to show that these nerves are physiologically associated, directly or indirectly, with respiration :

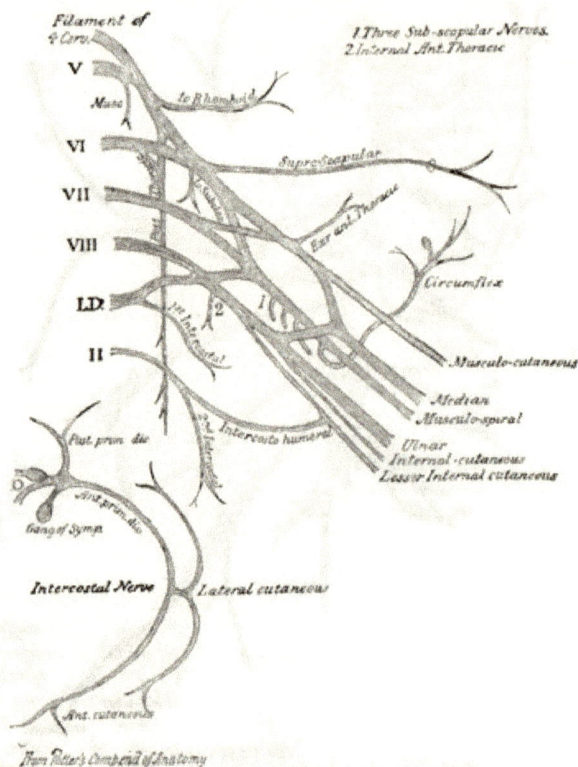

FIG. 42.—SCHEME OF THE BRACHIAL PLEXUS.
Note the acute angles characterizing the union of nerves.

SUPERFICIAL BRANCHES. (FIG. 43.)

ASCENDING.—(1) *Second occipital* supplies the skin over the mastoid insertion of the sterno-cleido-mastoid. (2) *Lesser occipital* supplies the skin of the scalp over the occipital origin of the trapezius, and over the greater insertion of the sterno-mastoid. (3) *Great auricular* supplies the skin over the facial part of the platysma and parotid gland, and over the insertions of the three dermal muscles inserted into the external ear. (4) *Superficial cervical* supplies the skin over the origin of the muscles arising from the genial tubercles, and over the muscles inserted into the hyoid bone.

DESCENDING.—(1) *Suprasternal* supplies the skin over the origins of the sterno-mastoid, sterno-hyoid, sterno-thyroid. (2) *Supraclavicular* supplies the skin over the thoracic part of the platysma, and over the costal attachments of the scaleni. (3) *Supraacromial* supplies the skin over the omo-hyoid and over the clavicular and scapular insertions of the trapezius.

Deep Branches.

INTERNAL.—(1) *Communicating* branches to the pneumogastric, hypoglossal, sympathetic. (2) *Muscular* branches to the rectus capitis anticus major and minor, and rectus capitis lateralis, and longus colli. (3) *Communicans hypoglossi,* two cervical branches that unite with the descendens hypoglossi to form the ansa hypoglossal loop, for the supply of the depressor muscles of the hyoid bone. (4) *The phrenic*—a mixed nerve that supplies the diaphragm and the diaphragmatic parts of the pericardium, pleuræ, and peritoneum.

EXTERNAL.—(1) *Communicating* branches to the spinal accessory nerve, in the trapezius and under the sterno-cleido-mastoid muscle. (2) *Muscular* to the trapezius, levator anguli scapulæ, scalenus medius, and sterno-cleido-mastoid.

The Brachial or Axillary Plexus.—A comprehensive knowledge of this plexus, for dissecting-room purposes, must include the following points:

1. The *formation*—by the union of the anterior primary divisions of the fifth, sixth, seventh, and eighth cervical nerves, and the greater part of the first dorsal or thoracic nerve.

2. The *location of emergence of the above*—between the scalenus anticus and scalenus medius muscles.

3. The *communications*—which are with the fourth cervical; with the second and third thoracic nerves; with the phrenic nerve; with the inferior cervical ganglion of the sympathetic.

4. The *relations*—with the subclavian artery below; with the clavicle and subclavius muscle in front; with the serratus magnus and subscapular muscles behind.

5. The *sheath*—formed by the axillary prolongation of the third layer of deep cervical fascia, investing the plexus and axillary vessels, and finally forming the suspensory or axillary fascia proper.

6. The *formation of three brachial trunks:* upper, middle, and lower. The upper by the fusion of the fifth and sixth; the middle by the seventh; and the lower trunk by the eighth cervical and the greater part of the anterior primary division of the first dorsal nerve.

7. The *splitting of the brachial trunks* into anterior and posterior brachial divisions.

8. The *formation of the posterior cord* by the union of the three posterior brachial divisions.

9. The *formation of the outer cord* by the union of the anterior brachial divisions of the upper and middle trunks.

10. The *formation of the inner cord* by the anterior brachial division of the lower trunk.

11. The *branches of distribution* to muscles; to the skin covering these muscles in general; to the skin covering the insertions of these muscles in particular; to articulations where there is motion produced by muscles supplied by the brachial plexus.

12. The *inosculation* between sensory nerves and motor nerves peripherally.

13. *Branches given off above* the clavicle; branches thrown off below the clavicle, or from the plexus proper.

The branches you will find given off above the clavicle :

1. Muscular branches to the subclavius, scaleni, longus colli, rhomboidei, as indicated in figure 42.

2. A communicating branch to the phrenic nerve. Follow the phrenic nerve up, on the scalenus anticus muscle, and you will find a delicate nerve coming through the muscle.

3. The posterior thoracic nerve, also called the long thoracic, the external

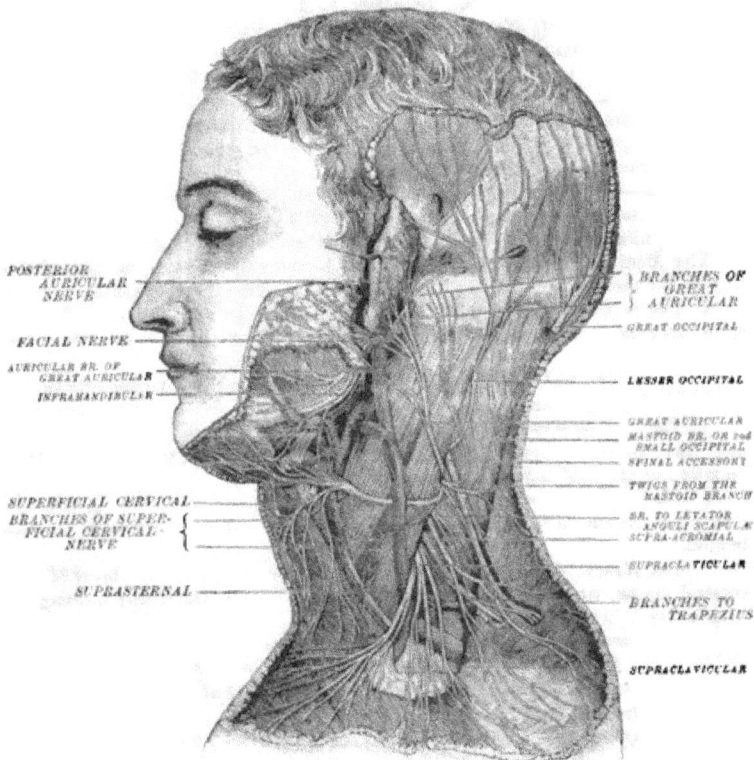

POSTERIOR AURICULAR NERVE

FACIAL NERVE

AURICULAR BR. OF GREAT AURICULAR

INFRAMANDIBULAR

SUPERFICIAL CERVICAL BRANCHES OF SUPERFICIAL CERVICAL NERVE

SUPRASTERNAL

BRANCHES OF GREAT AURICULAR

GREAT OCCIPITAL

LESSER OCCIPITAL

GREAT AURICULAR
MASTOID BR, OR 2nd SMALL OCCIPITAL
SPINAL ACCESSORY

TWIGS FROM THE MASTOID BRANCH

BR. TO LEVATOR ANGULI SCAPULAE
SUPRA-ACROMIAL

SUPRACLAVICULAR

BRANCHES TO TRAPEZIUS

SUPRACLAVICULAR

FIG. 43.—SUPERFICIAL BRANCHES OF THE CERVICAL PLEXUS. (After Hirschfeld and Leveillé.)

phrenic, and external respiratory nerve of Bell, supplies the serratus magnus muscle. You will find this nerve in the axillary space.

4. The suprascapular nerve (Fig. 42) arises from the upper trunk. It passes through the suprascapular foramen, in the superior costa of the scapula and supplies the supra- and infraspinati muscles.

Branches from the Brachial Plexus Proper.—Outer cord: External anterior thoracic nerve to pectoralis major; outer head of the median nerve fuses with inner head from the inner cord of the plexus; musculo-cutaneous nerve to flexors of the forearm and coraco-brachialis.

Inner cord: Internal anterior thoracic to pectoralis minor; inner head of the median nerve fuses with the outer head from the outer cord ; internal cutaneous nerve, to inner forearm ; lesser internal cutaneous, to inner arm ; ulnar nerve, to forearm and hand.

Posterior cord : Three subscapular nerves to posterior wall of axilla ; musculo-spiral to the muscles of the posterior part of arm and forearm ; circumflex to the deltoid and teres minor.

THE MOUTH.

The structures **seen** in the mouth of a patient are :

1. The tongue-tip, **dorsum, and sides.**
2. The dental **arches and the teeth.**
3. The vestibule—the space outside the dental **arches.**
4. The frænum linguæ.
5. **The** uvula in mid-line of the soft palate.
6. Tonsil in the tonsillar recess. (Fig. 47.)
7. The palato-glossal fold of mucous membrane. (Fig. 47.)
8. The palato-pharyngeal fold of mucous membrane. (Fig. 47.)
9. The posterior pharyngeal wall.
10. The orifice of Stenson's duct, opposite second upper molar.
11. The general oral **mucous** membrane.
12. The soft palate **and its** subdivisions.

The fold **of mucous membrane, the** palato-glossal fold, is supported by a muscle—the palato-glossal **muscle.** Also a muscle, the palato-pharyngeal, is **under** the palato-pharyngeal **fold of** mucous membrane. These two muscles **form** the anterior and posterior boundaries of the tonsillar recess, in which is found the tonsil. Having **made** yourself familiar with the physical appearances of the above structures on **the** living, find on the **cadaver** the following

1. The genio-hyoid muscle. (Fig. 47.)
2. **The genio-hyo-glossus muscle.** (Fig. 47.)
3. **The** stylo-hyoid muscle. **(Fig.** 35.)
4. The stylo-glossus muscle.
5. The stylo-pharyngeus **muscle.**
6. The Eustachian orifices and **pharyngeal** tonsil.
7. The lingualis muscle. (Fig. 46.)
8. The tonsil in its recess ; between what **two muscles ?**
9. The palato-pharyngeal fold and muscle.
10. The palato-glossal fold and muscle.
11. The hypoglossal nerve (twelfth cranial). (Fig. 31.)
12. The gustatory nerve (lingual of fifth cranial nerve).
13. The glossal branch of glosso-pharyngeal nerve. (Fig. 34.)
14. The chorda tympani—of **the** seventh **cranial** nerve.
15. The lingual artery and vein. (Figs. 31 and 47.)
16. The salivary apparatus and blood-supply. (Fig. 45.)
17. The mylo-hyoid muscle—the floor of the mouth. (Fig. 35.)
18. The isthmus of **the fauces.**
19. **The anterior** pillar = palato-glossus muscle. (Fig. 47.)
20. The posterior pillar = palato-pharyngeus muscle. (Fig. 47.)
21. The azygos uvulæ muscle.
22. The levator palati muscle.
23. The circumflexus palati (tensor palati).

How to Dissect the Tongue.—The dissection of the neck, you will recall by examining your work, has exposed the anterior belly of the digastric muscle, the hyo-glossus muscle, the mylo-hyoid muscle, the branches of the external carotid artery, the hypoglossal nerve, and, to some extent, the muscles and ligaments attached to the styloid process of the temporal bone. To dissect the tongue and mouth you proceed as follows: Retract the chin, saw through the symphysis menti; then, with a sharp knife, cut through the center of the tongue from tip to hyoid bone. You will then see, separated by connective tissue areas, the mylo-hyoid, the genio-hyoid, and the genio-hyo-glossus muscles as in figure 44.

24. The boundary between the mouth and the pharynx. Examine a patient and you will see the soft palate hanging down between the mouth and pharynx.

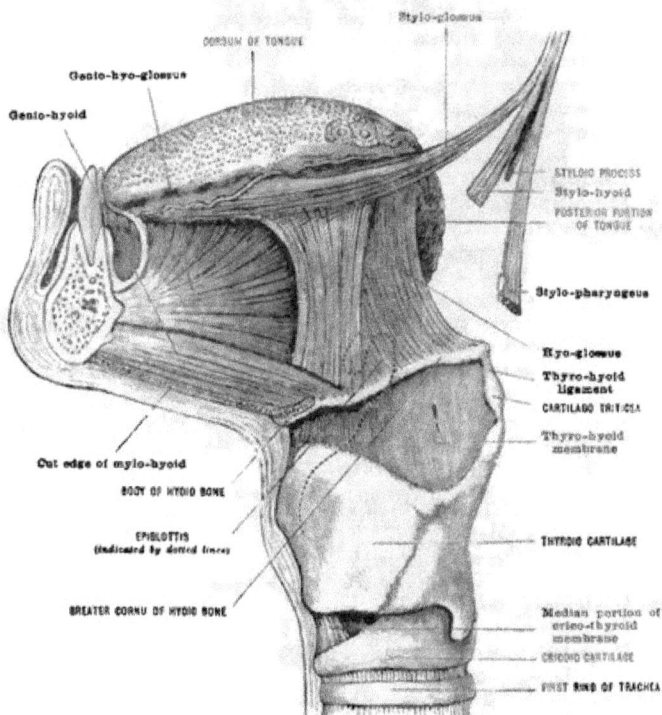

FIG. 44.—SIDE VIEW OF THE TONGUE, WITH ITS MUSCLES.

The passage between the two cavities, bounded by the tongue, anterior pillar of fauces, and uvula is called the isthmus of the fauces. The student should early become familiar with the soft palate, as its function is complex and of a very interesting and practical nature.

Study the movements of the **hyoid bone**:

1. *Elevated*—by the stylo-hyoid and digastric behind and the genio-hyoid and mylo-hyoid in front. These muscles are clearly antagonistic. The former are supplied by the seventh, the latter by the twelfth and fifth nerves.

2. *Forward.*—The genio-hyoid, when not antagonized by the stylo-hyoid, pulls the tongue forward.

3. *Backward.*—The stylo-hyoid and digastric, when not antagonized by the mylo-hyoid and genio-hyoid, pull the hyoid bone backward.

4. *Downward.*—The depressor muscles of the hyoid bone, when not antagonized, pull the hyoid bone down, and with it the tongue. The depressors of the hyoid bone are the sterno-hyoid, omo-hyoid, sterno-thyroid, and thyro-hyoid muscles. Their nerve-supply is from the ansa hypoglossal loop.

Study the simple movements of the **tongue** and the muscles that produce them.

1. *Protrusion beyond the teeth* = deep part of genio-hyo-glossus.
2. *Retraction* = superficial part of genio-hyo-glossus.
3. *Convex dorsum* = hyo-glossi.
4. *Concave dorsum* = inferior part of lingualis and stylo-glossus.
5. *Shorten tongue* = the lingualis.

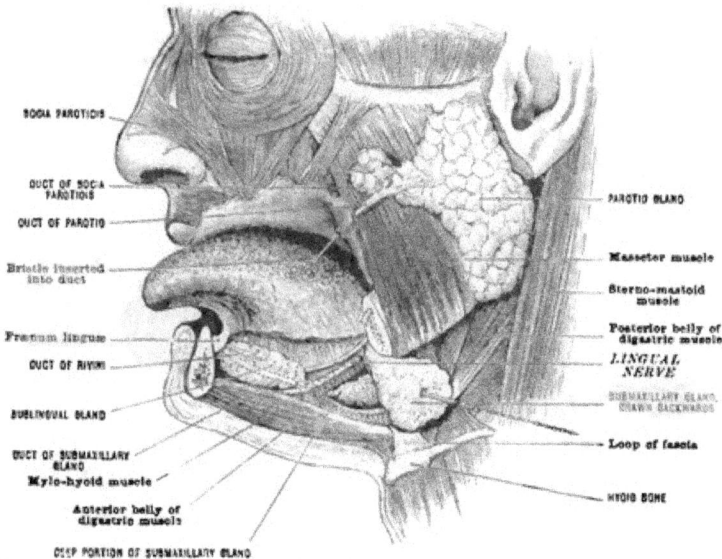

Fig. 45.—Side of the Face and Mouth Cavity, showing the Three Salivary Glands.

There may be a great number of compound movements. Thus, the tongue may be concave-dorsum, in protrusion or retraction; convex-dorsum, the same. In any of the foregoing positions the whole organ may be turned to either side, or the base may be elevated.

Divide the connective tissue between the genio-hyoid and genio-hyo-glossus. (Fig. 44.) The genio-hyoid arises from the inferior genial tubercle of the mandible. (Fig. 9.) It is inserted into the body of the hyoid bone.

The genio-hyo-glossus arises from the **superior genial tubercle.** (Fig. 9.) It has a superficial insertion into the anterior part of the tongue; a deep insertion into the body of the hyoid; above the preceding, of course. This muscle produces the rapid protrusion and retraction of the tongue.

The mylo-hyoid muscle forms the floor of the mouth. Divide the connective tissue between it and the genio-hyoid. (Fig. 44.) This muscle arises from the oblique line of the mandible—the mylo-hyoid ridge. (Fig. 9.) A small part of the posterior part of the muscle is inserted into the body of the hyoid bone; the greater part of the muscle is inserted into the median line with its fellow of the opposite side. (Fig. 35.)

Relations of the Mylo-hyoid Muscle.—Cut through the mucous membrane, between the border of the tongue and the alveolar process of the mandible, and you will see the superior surface of the mylo-hyoid muscle. On this surface find:

1. The gustatory nerve, the sensory (a branch of the fifth cranial), anastomosing, in the substance of the tongue, with the hypoglossal, the motor nerve of the tongue. (Fig. 45.)

2. Find the sublingual salivary gland, communicating, around the posterior margin of the mylo-hyoid, with the submaxillary salivary gland. (Fig. 45.)

3. Find the anterior belly of the digastric muscle (Fig. 35) under—resting upon—the under surface of the mylo-hyoid muscle. Note a branch of the mylo-hyoid nerve to this muscle. This muscle arises from the digastric fossa of mandible. (Fig. 9.)

4. The hyo-glossus muscle arises from the body and both horns of the

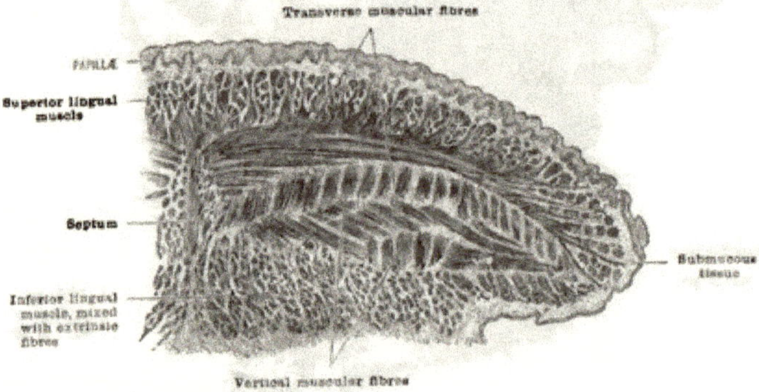

FIG. 46.—TRANSVERSE SECTION THROUGH THE LEFT HALF OF THE TONGUE.
(Magnified.)

(From a preparation by Mr. J. Pollard, Middlesex Hospital Museum.)

hyoid bone. (Fig. 48.) It is inserted into the tongue. On its outer surface find the hypoglossal nerve. (Fig. 34.) On its inner surface see the lingual artery (Fig. 47) dividing in (1) the dorsalis linguæ; (2) the ranine artery. (Fig. 36.)

The Stylo-glossus and Stylo-hyoid.—Trace the former to the side of the tongue, the latter to the side of the body of the hyoid bone. Their origins will be seen at a later stage of the dissection. (Fig. 48.)

The musculi linguales, or proper intrinsic muscles of the tongue, may be studied in the longitudinal section (Fig. 46) to a limited extent on your work.

Deep Dissection of the Pterygo-styloid Region.—In this region you will find the following:

1. **The stylo-maxillary or stylo-mandibular ligament,** a derivative of the first layer of cervical fascia, extending from the styloid process of the temporal bone to the ramus and angle of the mandible. It is between the masseter muscle and the internal pterygoid muscle.

2. **The temporo-mandibular articulation,** with its interarticular fibro-cartilage dividing the glenoid cavity into an anterior and a posterior synovial cavity.

3. **The external pterygoid muscle,** with one *insertion* into the depression in front of the condyle, and another into the interarticular fibro-cartilage.

4. **The internal pterygoid** muscle, inserted into the inner surface of the ramus. See this muscle coming down from the pterygoid fossa.

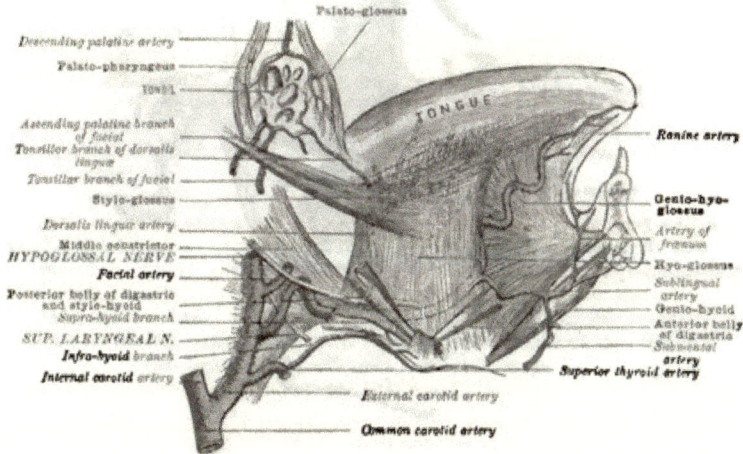

FIG. 47.—SCHEME OF THE RIGHT LINGUAL ARTERY. (Walsham.)

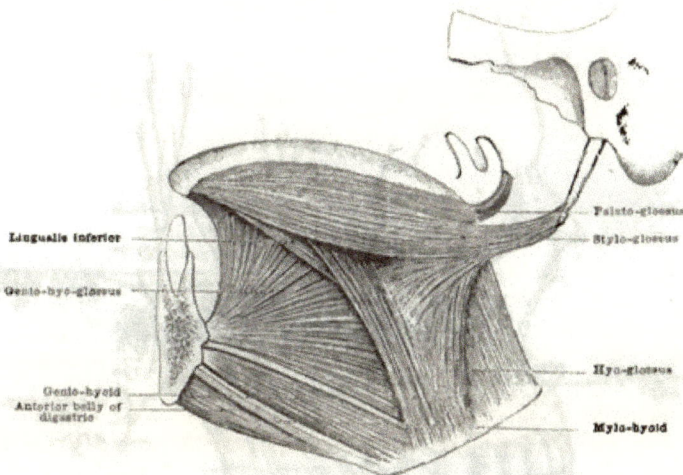

FIG. 48.—SIDE VIEW OF THE MUSCLES OF THE TONGUE.

5. **The inferior dental nerve and artery,** between the neck of the mandible and the internal lateral ligament of the joint. This ligament you will see extending from the spine of the sphenoid to the lingula of the mandible ; under it are the above structures, the inferior dental nerve and vessels.

6. **The internal maxillary artery** (Fig. 51), between the neck of the jaw

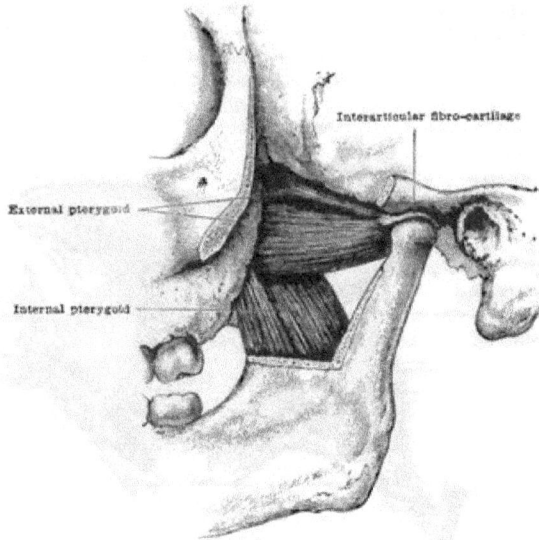

Interarticular fibro-cartilage

External pterygoid

Internal pterygoid

FIG. 49.—THE PTERYGOID MUSCLES.

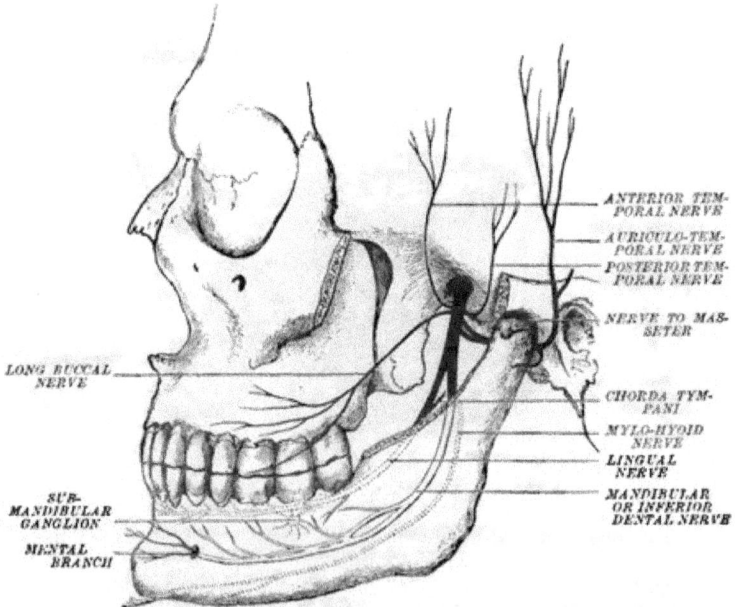

ANTERIOR TEM-PORAL NERVE

AURICULO-TEM-PORAL NERVE

POSTERIOR TEM-PORAL NERVE

NERVE TO MAS-SETER

CHORDA TYM-PANI

MYLO-HYOID NERVE

LINGUAL NERVE

MANDIBULAR OR INFERIOR DENTAL NERVE

LONG BUCCAL NERVE

SUB-MANDIBULAR GANGLION

MENTAL BRANCH

FIG. 50.—DISTRIBUTION OF THE MANDIBULAR DIVISION OF THE TRIGEMINAL NERVE. (Henle.)

and the spheno-mandibular ligament. You will see it on the outer surface of the external pterygoid muscle.

7. **The styloid process,** muscles and ligaments attached.
   1. The stylo-hyoid muscle.
   2. The stylo-pharyngeus muscle.
   3. The stylo-glossus muscle.
   4. The stylo-mandibular ligament.
   5. The stylo-hyoid ligament.

Trace each from its styloid origin to its specific insertion.

Dissection.—The above deep structures will be readily found on dividing the temporo-mandibular ligament. As you do this, notice the interarticular fibro-cartilage. Between the neck of the jaw and the internal lateral ligament find the first stage of the internal maxillary artery.

The external pterygoid muscle (Fig. 49) originates by two heads : (1) From the outer surface of the external pterygoid plate. (2) From the pterygoid ridge on the **outer surface of** the greater wing of the sphenoid bone. It has two insertions : (1) Into the interarticular fibro-cartilage of the temporo-mandibular articulation ; (2) into a depression in front of the condyle of **the jaw.**

The inferior dental nerve (Fig. 51) is a branch of the inferior maxillary division of the fifth cranial. It passes between the ramus of **the jaw and** the spheno-mandibular ligament, in company with an artery of **the** same name, and enters the inferior dental canal. (Fig. 50.) It throws off **in** its course dental branches to the teeth ; a large cutaneous branch, the mental, to the skin of the chin. The mental branch comes to the surface through the mental foramen. (Fig. 8.) The inferior dental artery takes the same course and has **a** similar distribution. (Fig. 51.)

The mylo-hyoid nerve is given off from the inferior dental as the latter is entering the mandibular canal in the mandible. It lies in a groove on the inner **surface** of the lower jaw, and is covered by the periosteum. It supplies **the** mylo-hyoid muscle and the anterior belly of the digastric muscle **with motion.**

---

1. *What is the buccal orifice?*

A transverse slit, bounded by the lips, **and** terminating laterally in the angles ; by it the mouth communicates with the external world.

2. *What is the isthmus of the fauces?*

It is the buccal opening of the pharynx. It is bounded by the **tongue below,** soft palate **above,** and pillars of the fauces laterally, and by it the **mouth** communicates with the pharynx.

3. *How is the soft palate made up, and to what is it attached?*

Its **special** parts are : (1) The uvula ; (2) the anterior and posterior pillars of the fauces ; (3) the tonsillar recess between the pillars. This **should be studied** on the living subject.

4. *What is the vestibule?*

The part of the oral cavity bounded by the cheeks and lips externally and the dental arches internally.

5. *What muscle lies under the anterior pillar of the fauces and helps form the same?*

The palato-glossus muscle, very readily seen.

6. *What is the action of this muscle?*

In the act of swallowing it draws the tongue upward and the anterior pillars downward, hood-like, over the back of the tongue ; this closes the buccal orifice of the pharynx and prevents regurgitation of ingesta into the mouth, thereby enabling the constrictors of the pharynx to grasp the bolus to be swallowed.

7. *What muscle lies under the fold of mucous membrane of the posterior pillar of the fauces?*

The palato-pharyngeus. This muscle is inserted into the superior cornu of the thyroid cartilage, and also into the fibrous part of the pharynx. The action of this muscle is to assist deglutition by elevating the pharynx—by stretching the pharynx over the bolus, as it were.

The act of swallowing may be compared to a most delicately acting cylinder and piston: The tongue is the piston; the isthmus of the fauces, the cylinder in which the piston plays; the junction between the pharynx and the œsophagus

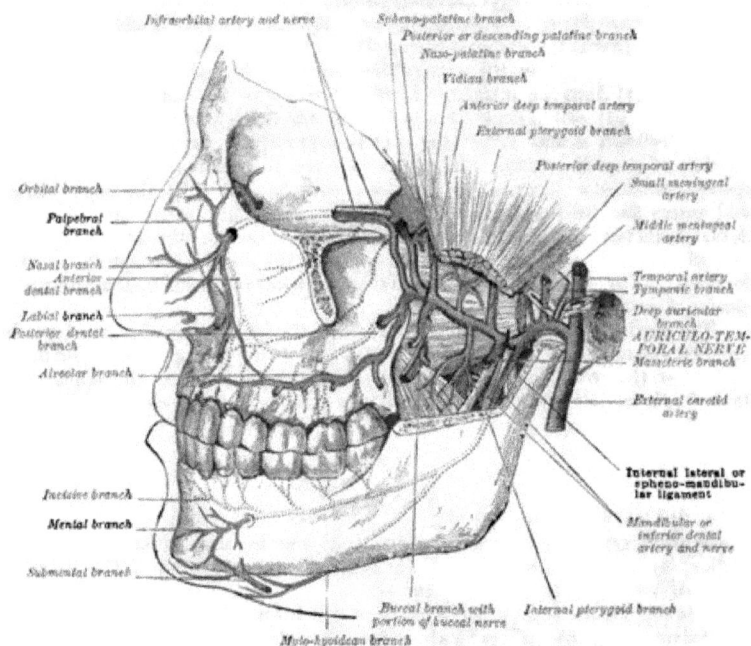

FIG. 51.—SCHEME OF INTERNAL MAXILLARY ARTERY.

is the escape. The digastric and stylo-hyoid muscles elevate the tongue; at the same time the palato-glossus draws the tongue backward, piston-like, through the isthmus, thereby preventing regurgitation and at the same time pressing onward the food. At this juncture the palato-pharyngeus muscle depresses the soft palate, thereby aiding the preceding muscle, and at the same time pulls up the pharynx, just as one pulls a stocking on.

8. *What would be the result were the palato-pharyngeus muscle paralyzed, and do such circumstances arise in the practice of medicine?*

The result would be loss of the delicate part of the act of swallowing. Solids would find their way to the stomach, but fluids, more difficult to grasp, by the constriction of the pharynx would regurgitate through the nose, under these conditions, this being the line of least resistance. Such cases of regurgitation are met as sequels of diphtheria.

9. *Locate the tonsil.*

It is between the anterior and posterior pillars of the fauces; it corresponds in position to the angle of the jaw; the superior constrictor of the pharynx is external to it, while the internal carotid artery is fully one inch from it. It is the ascending pharyngeal artery, between the tonsil and the superior constrictor, and not the internal carotid artery that is in danger of being wounded in tonsillotomy.

10. *Explain the mylo-hyoid nerve.*

It is a branch of the inferior dental, being given off while this nerve is between the ramus and spheno-mandibular ligament. It pierces the spheno-mandibular ligament, passes along the mylo-hyoid groove, and supplies the anterior belly of the digastric and the mylo-hyoid muscle. This nerve is a part of the motor root of the fifth, ensconsed in the sheath of the (sensory) inferior dental part of the fifth, given off from the Gasserian ganglion. (Fig. 53.)

11. *Name and explain fully the structures attached to the styloid process of the temporal bone.* (Fig. 44.)

There are three muscles and two ligaments: (1) The stylo-hyoid muscle. This muscle arises from the back and outer surface of the styloid process, near the base. It lies above the posterior belly of the digastric muscle. Near its insertion into the body of the hyoid, at the junction of the body and greater horn, it bridges the intermediary tendon of the digastric. In its action it is synergistic with the posterior belly of the digastric, the two being supplied with motion by the seventh nerve. (2) The stylo-glossus muscle arises from the front and tip of the styloid process, also from the upper part of the stylo-mandibular ligament. It is inserted into the side and under surface of the tongue. Its action is to draw the sides of the tongue up to make the dorsum of this organ concave from side to side. (3) The stylo-pharyngeus arises from the base of the styloid process, opposite the stylo-hyoid. It is inserted, with the palato-pharyngeus, into the thyroid cartilage, and also into and with the constrictors of the pharynx. (4) The stylo-maxillary or stylo-mandibular ligament will be seen in your dissection as a process of deep cervical fascia, a derivative of the third layer, between the masseter and internal pterygoid muscles. It separates the parotid from the submaxillary gland. It gives origin to one head of the stylo-glossus muscle. (5) The stylo-hyoid ligament extends, as a small elastic fibrous cord, from the tip of the styloid process of the temporal bone to the lesser cornu of the hyoid bone. It frequently becomes ossified. Occasionally it is of enormous size.

12. *Describe the external pterygoid muscle.* (Fig. 49.)

This muscle has two origins and two insertions. The upper head arises from the under surface of the greater wing of the sphenoid bone, limited internally by the foramen ovale, transmitting the third division of the fifth nerve, and the foramen spinosum, transmitting the great meningeal artery; externally, the origin of this head is limited by the pterygoid ridge. The lower head arises from the greater part of the outer surface of the external pterygoid plate. The upper head is inserted into the interarticular fibro-cartilage; the lower into the front of the condyle. The action of the muscle is to draw the condyle well forward, and with it also the interarticular cartilage. It also moves the jaw to the opposite side, and assists to some extent in opening the mouth.

13. *Describe the internal pterygoid muscle.* (Fig. 49.)

It arises from the inner surface of the external pterygoid plate, and from the tuberosity of the palate bone. It is inserted into the inner surface of the ramus of the mandible, limited above by the mylo-hyoid ridge and inferior dental foramen. The action is to close the mouth and draw the jaw forward and to the mid-line.

The nerves to the pterygoid muscles are from the third division of the fifth nerve. (Fig. 53.) The blood-supply comes from the internal maxillary artery, second stage. (Fig. 51.)

Describe the *internal maxillary artery.* (Fig. 51.)
First Stage.—Behind the neck of the mandible. In this stage it gives off:
1. The inferior dental artery to the inferior teeth.
2. The great meningeal artery to the dura and calvarium.
3. The small meningeal artery to the dura mater.
4. The deep auricular branch to external canal.
5. Tympanic, to the membrana tympani.

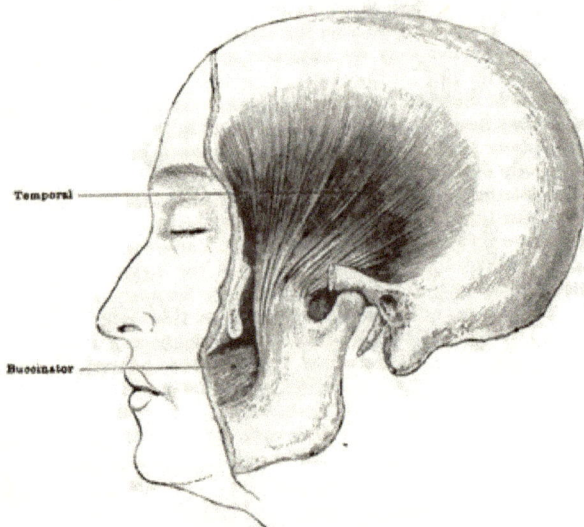

Fig. 52.—The Temporal Muscle.

The *second stage* of the artery gives off branches to muscles of mastication:
1. Deep temporal branches to temporal muscles.
2. Buccal branch to the buccinator muscle.
3. Internal pterygoid branch to internal pterygoid muscle.
4. External pterygoid branch to external pterygoid muscle.
5. Masseteric branch, to the masseter.
The second or pterygoid stage lies on the outer surface of the external pterygoid muscle, under cover of the temporal muscle and the ramus of the mandible.
The *third stage* of the artery gives off these branches:
1. The posterior superior dental artery. This is confusing, and requires an explanation. The dental branches supply the teeth. They enter the posterior superior dental canals, through foramina on the posterior surface of the body of the superior maxilla. The alveolar branches supply the gums. Some branches supply the mucous membrane of the antrum; these might be called antral branches.

2.. The infraorbital artery enters the infraorbital canal in the floor of the orbit. It appears on the face as the infraorbital. It gives branches to the *anterior teeth.* These latter pass through the anterior superior dental canals in the anterior wall of the antrum.

3. Descending palatine passes through the posterior palatine canal, along the roof of the mouth, to the foramen of Stenson, where it anastomoses with the naso-palatine artery. It accompanies the anterior branch of Meckel's ganglion.

4. The spheno-palatine passes through the spheno-palatine foramen into the cavity of the nose. It divides into the *naso-palatine* branch to the septum,

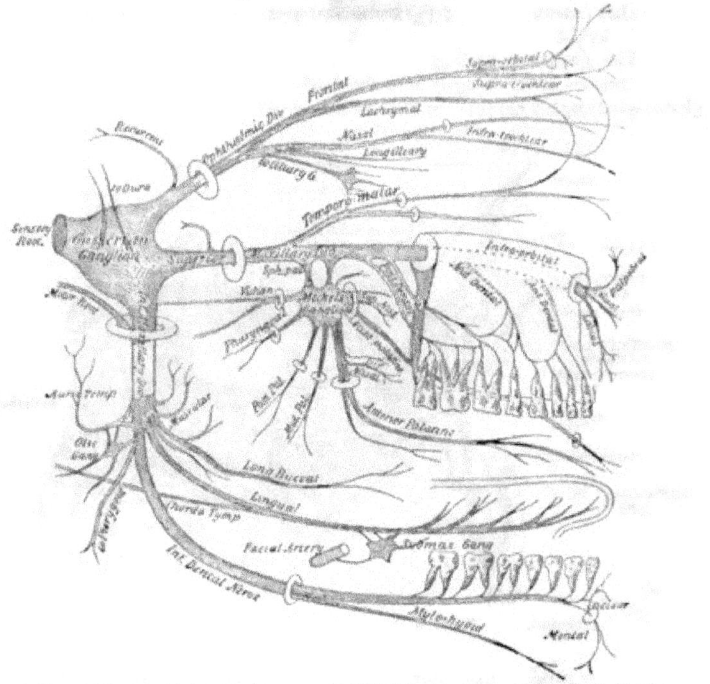

FIG. 53.—SCHEME OF THE DISTRIBUTION OF THE FIFTH CRANIAL OR TRIGEMINAL NERVE. Notice especially: (1) The recurrent branch from the Gasserian ganglion to the dura of the middle fossa and tentorium. (2) Filaments from the carotid plexus. (3) The motor root of the fifth nerve behind the ganglion, but independent thereof. (4) The formation of the otic, submaxillary, Meckel's, and the ciliary ganglion.

which anastomoses with the descending palatine artery, and some external branches to the lateral walls of the nose.

5. The Vidian nerve passes through the Vidian canal with the Vidian nerve. It is distributed to the pharynx and Eustachian tube.

6. The pterygo-palatine passes through the pterygo-palatine canal. It is distributed to the upper part of the pharynx and Eustachian tube.

Describe the inferior maxillary division of the fifth cranial nerve.

It passes through the foramen ovale with the small meningeal artery. It gives off the following branches :

1. Muscular branches to all the muscles of mastication. These accompany

7

muscular branches of the internal maxillary artery and take the name of the muscles.

2. The auriculo-temporal nerve. (Fig. 53.) The middle meningeal artery passes between the two roots of this nerve. This nerve gives an articular branch to the temporo-mandibular articulation. It does this in accordance with Hilton's law. It gives branches to the external auditory meatus and the membrana tympani, and is often responsible for ear-ache in children. It sends branches to the scalp, with the temporal arteries, and may be the terminus of visceral reflexes in the scalp.

3. The inferior dental nerve to (1) the teeth of the mandible; (2) to the skin of the chin (mental nerve); (3) to the anterior belly of the digastric; and (4) to the mylo-hyoid muscle, through the mylo-hyoid nerve.

4. The gustatory or lingual nerve, the great sensory nerve to the tongue. It anastomoses in the tongue with the hypoglossal and the glossal branch of the glosso-pharyngeal nerve. Special notice must be given this nerve, on account

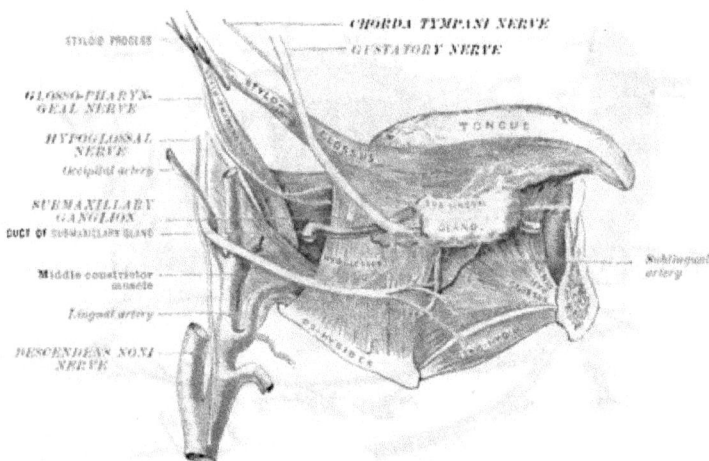

FIG. 54.—MUSCLES, VESSELS, AND NERVES OF THE TONGUE.

of (1) the chorda tympani nerve; (2) the submaxillary ganglion; (3) the otic ganglion.

The chorda tympani nerve (Fig. 54) is a branch of the seventh cranial nerve. It passes through the tympanum. It leaves this cavity by the canal of Huguier in the petrous portion of the temporal bone. It passes between the two pterygoid muscles. It meets the lingual nerve, and accompanies this to the submaxillary gland. This nerve gives branches to the sublingual gland, the lingualis muscle, and the submaxillary ganglion.

The Submaxillary Ganglion. (Fig. 54.)
Give location of the submaxillary ganglion.
It is on the outer surface of the hyo-glossus muscle, attached to the gustatory or lingual nerve. The ganglion is about the size of a pin's head.
Name the roots of the ganglion and give their source.
1. *Sensory root* is from the gustatory nerve, a branch of the fifth nerve.
2. *Motor root* is from the chorda tympani, a branch of the seventh nerve.
3. *Sympathetic root* is from the nervi molles—the facial sympathetics.

Its branches are distributed to the mucous membrane of the floor of the mouth and to the submaxillary gland and its duct.

The otic ganglion (Fig. 53) is on the inferior maxillary division of the fifth nerve. Its roots are :

1. *Motor,* from the inferior maxillary division of the fifth nerve.
2. *Sensory,* from the auriculo-temporal branch of the fifth nerve.
3. *Sympathetic,* from a plexus on the middle meningeal artery.

Give the location and relations of the otic ganglion.

(1) External to it is the inferior maxillary nerve ; (2) Internal to it is the circumflexus (tensor) palati muscle ; (3) posterior to it is the middle meningeal artery.

Aside from the three root communications previously given, has the otic ganglion any other communications ?

Yes ; it communicates with the seventh or facial nerve and with the glossopharyngeal by the lesser petrosal nerve.

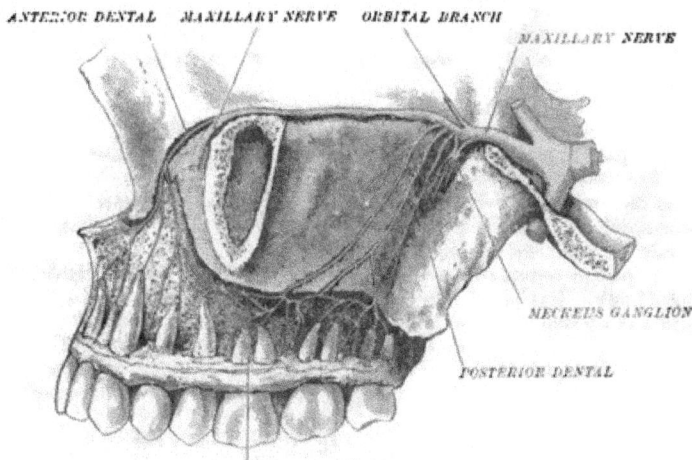

ANTERIOR DENTAL   MAXILLARY NERVE   ORBITAL BRANCH   MAXILLARY NERVE
MECKEL'S GANGLION
POSTERIOR DENTAL
LOOP FORMED BY MIDDLE AND ANTERIOR DENTAL NERVES

FIG. 55.—THE MAXILLARY NERVE SEEN FROM WITHOUT. (Beaunis.)

To what muscles are the branches of the otic ganglion distributed ?

To the tensor palati and tensor tympani.

Describe the superior maxillary division of the fifth cranial nerve. (Fig. 55.)

This nerve, a branch of the Gasserian ganglion, leaves the cranium by the foramen rotundum in the greater wing of the sphenoid bone. The nerve crosses the spheno-maxillary fossa, enters the orbit by the spheno-maxillary fissure, traverses the infraorbital canal in the floor of the orbit, appears on the face at the infraorbital foramen, and here communicates with the infraorbital branch of the seventh nerve to form the infraorbital plexus.

Its branches are both numerous and important :

1. The nasal, labial, and palpebral on the face. (Fig. 53.)
2. Anterior superior dentals to incisor teeth. (Fig. 53.)
3. Posterior superior dentals to molar teeth. (Fig. 53.)
4. Spheno-palatine branches to Meckel's ganglion. (Fig. 53.)
5. Temporo-malar are cutaneous to the cheek and temple.
6. A recurrent meningeal branch to the dura. (Fig. 53.)

Two Cases of Reflex Pain.

A Patient has Pain in the Auditory Canal.—On examination you find no local objective symptoms. Where may the real seat of the trouble be located? The auditory canal is supplied by the auriculo-temporal branch of the fifth cranial nerve. The irritation may be sought for anywhere in the distribution of this nerve, but, logically, those parts of the fifth nerve which experience has proved are most liable to disease and injury should be interrogated first. We may venture the assertion that the teeth, nose, eye, tongue, and antrum of Highmore represent about the order, in point of frequency. Here, then, is the specialist invoked, since a carious tooth, an irritated gum, a septal spur, eye strain, a lingual ulcer, or pus in the antrum of Highmore may be the immediate exciting cause of the pain in the auditory canal. Conversely, a patient may have a violent toothache, and examination may reveal neither exposed pulp nor irritated gum. Conservative dentistry in this case would refer a patient to other specialists, whose field of usefulness lies in the distribution of the fifth pair of cranial nerves, primarily or secondarily, since reflex pain may be primary and secondary.

What is meant by primary and secondary reflex pain?

The expression is a coinage, explained as follows: Where a carious tooth manifests pain in the ear, eye, nose,—anywhere in organs supplied by the same nerve direct,—this is called primary; where, however, a carious tooth manifests pain in distant regions, not supplied by the fifth nerve, but connected therewith by communicating branches, this is called secondary. Let the following cases illustrate the point and apologize for the coinage:

1. A patient consulted a prominent dentist in this city for intractable pain in a molar tooth. Examination revealed no objective symptoms of disease. The patient was advised to consult other specialists in the primary radius of the fifth pair of nerves—viz., the eye, nose, ear, or throat man. The patient disregarded the specific character of the advice, and consulted a specialist in the secondary radius of the fifth nerve, the gynæcologist, who removed an ovarian tumor. The pain in the molar did not return.

2. Hilton reports a case from Romberg of cough and vomiting which ceased only after a pruritus of the external auditory meatus was cured. The nerves concerned in this case were the vagus and the trigeminus.

The general practitioner of medicine who facetiously disclaims confidence in reports of "toothache in the ear, and earache in an ovary," as he is wont to call such cases, is placing himself on record with those who a few years ago hooted at asepsis. Pain is a projectile, and specialists in medicine to-day are doing more than any one else to harmonize physiological speculations and anatomical nerve-distributions and nerve-communications, by furnishing clinical evidence of that line of least resistance which pain, as a projectile, must follow. The student of practical anatomy is urged to give special attention to those parts of nerve-trunks recorded in the text-books as communicating branches, for upon these depends the rationale of reflexes, on which so much is said and written.

The Glosso-pharyngeal Nerve (Fig. 54).—You will find this nerve between the jugular vein and internal carotid artery. It lies above the superior laryngeal nerve, on the stylo-pharyngeus muscle and superior constrictor of the pharynx. It is distributed to the back of the tongue and to the pharynx. This nerve is deeply located. It can not be mistaken. It is the ninth cranial nerve. It leaves the base of the cranium by the jugular foramen with the tenth and eleventh nerves and the jugular vein.

THE MUSCLES OF MASTICATION.

The muscles of mastication are the temporal, buccinator, the internal and external pterygoids, and the masseter. (Figs. 56 and 57.) These are the powerful muscles that act on the mandible, by whose action the food is cut by the incisors, torn by the canines, and triturated by the molars. It would seem that the muscles that depress the jaw belong to the same category; hence we will consider them incidentally here, in detail in another section.

The temporal muscle (Fig. 52) arises from the complete temporal ridge, from the temporal fossa, and from the under surface of the temporal fascia. It is inserted into the coronoid process of the mandible, as far forward as the last molar tooth. The muscle is triangular; hence, according to rule, it will be found taking its nerve-supply near the apex. It derives its nerve-supply from **the** third division of the trigeminus. (Fig. 53.) In your dissection **you** find **this** muscle related superficially to the arcus zygomaticus, the **temporal fascia, the** temporal branches of the seventh nerve, the superficial temporal arteries **and** veins, **and** the auriculo-temporal branch of the fifth nerve. (Figs. 15-17.) Deeply the muscle is related to the temporal fossa and the external pterygoid muscle.

The buccinator muscle (Fig. 62) arises from the alveolar processes above the upper and below the lower molar teeth of the upper and lower jaws respectively; and from the pterygo-maxillary ligament. It is inserted into the outer part of the orbicularis oris. Its function is (1) to draw the corners of the mouth outward and backward against the teeth; (2) to antagonize the tongue by permitting no food to be pushed by this organ beyond the dental arches, into the vestibule during mastication.

The internal pterygoid muscle (Fig. 57) arises from the inner surface of the **external pterygoid** plate, from the tuberosity of the palate bone. It is inserted **into the inner surface** of the **ramus** of the mandible, as high as the inferior dental **foramen and mylo-hyoid** groove. The muscle is quadrangular; hence, according to rule, it takes **its** nerve-supply in **the** centre. The action of the muscle is to close the mouth and draw the jaw forward and to the mid-line.

Name the agents that depress the lower jaw.

The **digastric** muscle, the platysma myoides, and **gravity**.

Which muscles act on the mandible, to protrude the lower jaw?

The external pterygoid muscle and the superficial **part** of the masseter muscle.

How is the mandible returned to its original position in the glenoid?

It is drawn back by the posterior fibres of the temporal muscle, and by the deep fibres of the masseter muscle.

What are the muscles of trituration?

The external pterygoids. They act alternately, drawing the mandible forward and to the opposite side; this movement produces trituration.

What is the function of the temporal, masseter, and internal pterygoid muscles?

They raise the mandible against the upper jaw with great force.

Has the temporal muscle an independent, specific action?

Yes, its specific action is to accentuate incision. The quick snapping movement of the jaw is done by this muscle.

What is meant by superficial and deep part of the masseter muscle?

Your knowledge of the action of the masseter muscle on the mandible is imperfect without an understanding of these two parts: In reality each masseter is two muscles. Each part of the masseter muscle has a separate origin, separate function, separate insertion. One—the superficial—is strong, tendinous, and active in assisting the external pterygoid muscle, in drawing the lower molars across the upper molars, in trituration of food ; the other—the deep portion— is weak, broad, muscular, and almost passive in assisting the posterior segment of the temporal muscle in bringing the mandible back to its position, preparatory to another movement of trituration.

The specific origins and insertions of these separate parts of the masseter muscle make possible the separate physiological functions, as above indicated. Study them separately, as follows :

The superficial part of the masseter (Fig. 56) muscle arises : (1) From

Corrugator supercilii
Pyramidalis
Levator labii superioris alæque nasi
Levator labii superioris
Compressor narium
Levator anguli oris
Naso-labialis
Depressor alæ nasi
Orbicularis oris
Buccinator
Depressor anguli oris
Depressor labii inferioris
Levator menti
Mylo-hyoid
Anterior belly of digastric
Thyro-hyoid
Omo-hyoid
Sterno-hyoid
Scalenus anticus

Temporal
Zygomaticus major
Masseter
Posterior belly of digastric
Splenius capitis
Stylo-hyoid
Sterno-mastoid
Levator anguli scapulæ
Scalenus medius
Omo-hyoid

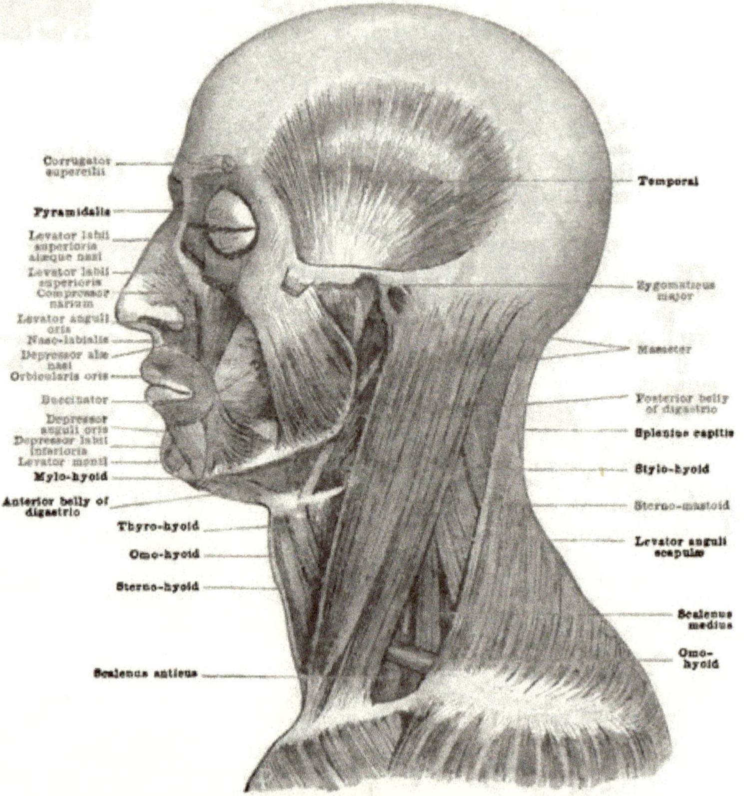

FIG. 56.--THE DEEPER LAYER OF THE MUSCLES OF THE FACE AND NECK.

the malar process of the superior maxilla by a strong tendon ; (2) from the anterior two-thirds of the lower border of the zygomatic arch. From this strong origin the fibres pass backward and downward to their insertion into the lower half and angle of the outer surface of the ramus. (Fig. 58.) The action is to draw the mandible forward. The synergist of this segment of the masseter is the external pterygoid.

The deep portion of the masseter (Fig. 56) muscle arises : (1) From the inner surface of the zygomatic arch ; (2) from the lower border of the posterior third of the zygoma. This portion is inserted into the upper half of the outer surface of the ramus. (Fig. 58.) Its action is to assist the posterior segment of the temporal muscle in drawing the mandible backward.

The external pterygoid muscle (Fig. 57) *originates where and is how inserted ?*

The function of this muscle is second to none, hence its bony and cartilaginous attachments merit special notice. Here, as in the preceding case, there are in reality two muscles. They are, however, described as upper and lower heads of one muscle.

The upper head arises from the pterygoid ridge on the outer and under surface of the greater ala of the sphenoid bone. Its specific insertion is, according to my experience, confined to the interarticular fibro-cartilage of the temporo-mandibular articulation. The function would seem to be to regulate the movements of this

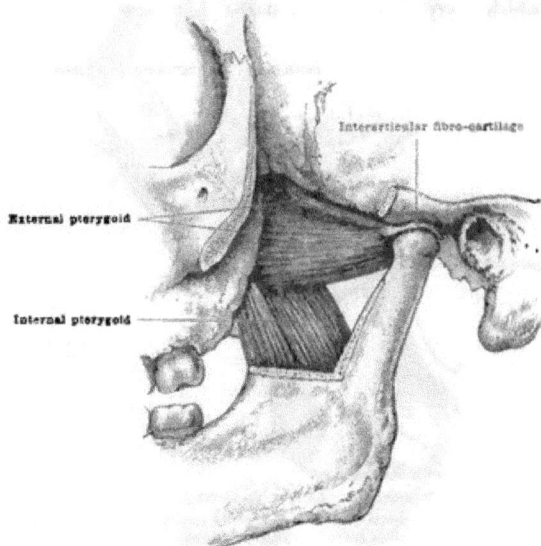

FIG. 57.—THE PTERYGOID MUSCLES.

cartilage to the sliding of the condyle forward on to the eminentia articularis when the mouth is opened. The only apparent use of the interarticular cartilage is, when the mouth is opened, to furnish a cup concavity for the condyle while this latter is gliding over the convex eminentia articularis. (Fig. 58.)

A cracking noise on opening the mouth is frequently both a subjective and an objective symptom. This noise is beyond doubt produced by temporary paralysis of this part of the external pterygoid, the condyle overriding the margin of cartilage that normally should precede the same. The rational treatment would seem to be fixation of the cartilage in the position it occupies when the mouth is open.

Conclusion : The function of the upper head of the external pterygoid is to make tense and carry forward the interarticular fibro-cartilage. The function of the lower head is to most powerfully draw the jaw forward.

The lower head of the external pterygoid muscle arises from the outer surface of the external pterygoid plate, except a small strip at its lower and front part It is inserted into the depression in front of the neck of the condyle. (Fig. 58.) The action of this head is to draw the condyle forward; at the same time the upper head draws the cartilage forward.

Concerning the noise patients frequently hear on opening the mouth, I would say I can not now recall ever having read or heard a rational explanation of its occurrence. The theory I advance above is more for the purpose of getting the student aroused to the necessity of learning the origins and insertions of the pterygoids than for mere theorizing. If I succeed even in this, the object of this paragraph will have been accomplished. I found a cadaver in which the cracking noise attended every time the mouth was passively opened. On dissection I found the cartilage had contracted inflammatory adhesions to the dome of the glenoid cavity. It was on this single case I founded the theory above advanced, in lieu of an explanation for those noises in the temporo-mandibular articulation concerning which every physician and dentist has been consulted by solicitous

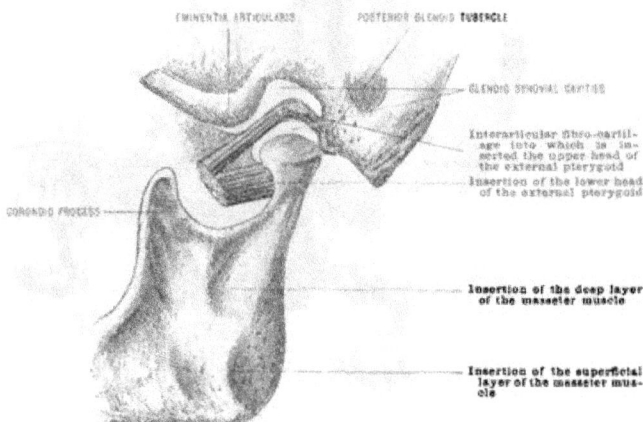

FIG. 58.—SCHEMATIC. TEMPORO-MANDIBULAR ARTICULATION.
This is assumed to represent the normal relation of condyle of mandible, and interarticular cartilage to the eminentia articularis when the mouth is closed.

patients. A single case, however, is a mere bagatelle; numerous cases by different observers will be necessary to give the theory scientific sanction.

In dislocation of the condyle of the mandible, what muscles must be overcome?

Practically all. Theoretically, the posterior part of the temporal and the deep portion of the masseter. As a matter of philosophy, however, malposition of the head of a bone immediately sets up tonic contraction in all the muscles of the group, since pain in the joint is reported to the brain and contraction follows in all the muscles in the articular nerve circuit. See Hilton's law in the introductory chapter.

The temporal muscle has practically two portions: an anterior and a posterior. The latter pulls the jaw backward; the former, acting alone, would pull the jaw forward.

What are the propositions in the nerve-supply to the muscles of mastication?
They are:

1. Nerve-supply, sensory, both to the pulp of the teeth and to the gum

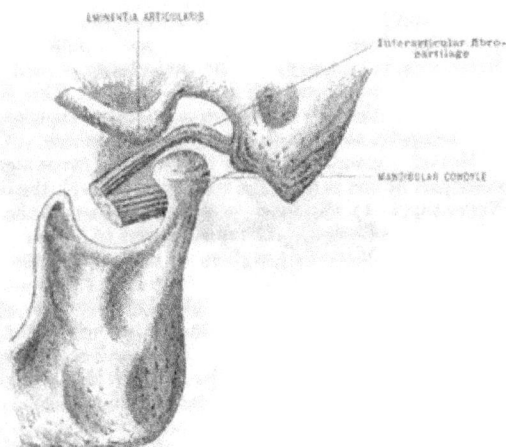

FIG. 59.—SCHEMATIC. TEMPORO-MANDIBULAR ARTICULATION.

This is assumed to represent relation of condyle and cartilage to eminentia articularis when the mouth is open. Here both condyle and cartilage slide forward on to the eminentia articularis; the former by action of the lower, the latter by action of the upper, head.

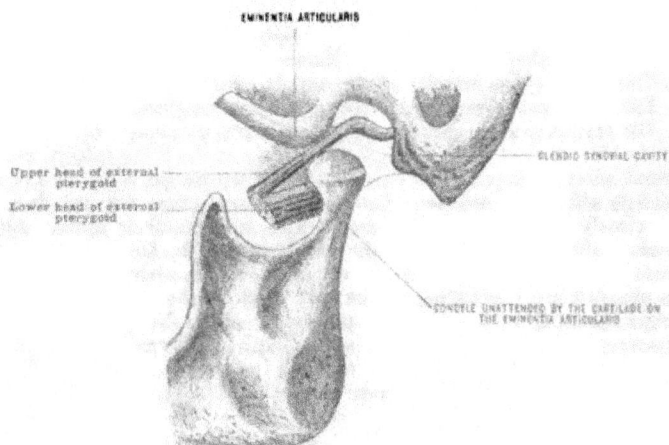

FIG. 60.—SCHEMATIC. TEMPORO-MANDIBULAR ARTICULATION.

This is assumed to represent abnormal relation of condyle and cartilage to eminentia articularis in those cases where the nerve-twig that supplies the upper head of the external pterygoid muscle is paralyzed. A cracking noise on opening the mouth is a symptom. Here the cartilage remains back in the glenoid, there being nothing to pull it forward. The condyle slips over the thick margin and produces the noise.

covering the same. This comes from the fifth cranial nerve, through dental and gingival branches respectively.

2. Nerve-supply, motor, to the muscles that elevate the jaw, and sensory, to the skin covering these muscles. This is from the fifth cranial nerve.

3. Nerve-supply, sensory, to the articulation moved by the muscles of mastication—the temporo-mandibular articulation. This is from the auriculo-temporal branch of the fifth cranial nerve. This is explained by Hilton's law.

4 Nerve-supply, sensory, to the roof of the mouth. This is from the fifth through Meckel's ganglion. Also the sensory nerve-supply to the tongue. The greater part of this is from the fifth nerve through the lingual branch.

5. Nerve-supply to the inner nose, or that part of the system that selects food by the sense of smell. The sensory part of this is in great part from the fifth nerve, through Meckel's ganglion. The special sense of smell is from the olfactory or first cranial nerve.

6. Nerve-supply concerned in deglutition. Here a separate system, the digestive, begins, as is evidenced by a radical change in the nerve-supply. The muscles of the soft palate and pharynx are supplied by the pharyngeal plexus, the principal factors in which are the sympathetic and vagus nerves; these nerves supply all the other organs of the digestive tract.

What is the salivary system ?

This system consists of the salivary glands, the parotid, the submaxillary, the sublingual, and their capsules, their nerve-supply, principally sympathetic, and their blood-supply. The latter is derived from the main artery of the region in which the glands are found.

THE SOFT PALATE.

1. *Prevents regurgitation* of food into the mouth. (Fig. 64.)
2. *Prevents passage* of ingesta into the posterior nares.

The structures concerned are the following :

1. The palato-glossus muscle. Nerve-supply, pharyngeal plexus.
2. The palato-pharyngeus muscle. Nerve-supply, pharyngeal plexus.
3. The levator palati muscle. Nerve-supply, pharyngeal plexus.
4. The tensor palati muscle. Nerve-supply, otic ganglion.
5. The azygos uvulæ. Nerve-supply, pharyngeal plexus.

To see the action of the palato-glossus muscle on the living subject, make a prolonged effort to depress the tongue. When the patient is on the verge of nausea, you will see the anterior pillars of the fauces spread out and draw the tongue closely under the palatine arch. This is, then, their action : during swallowing, after the bolus passes the initial part of the faucial passage, the tongue is forced under this arch to prevent regurgitation, while the constrictors of the pharynx are contracting on the food. During the same act the palato-pharyngei spread in such a way as to protect the upper pharynx and posterior nares, acting with the levator palati ; the circumflexus tightens the soft palate.

The levator palati (Fig. 65) arises from the cartilage of the Eustachian tube, and from the under surface of the petrous portion of the temporal bone. It is inserted into the soft palate near the mid-line.

The circumflexus (Fig. 65), or tensor palati, arises from the scaphoid fossa of the sphenoid bone. (Fig. 11.) The tendon passes around the hamular process of the internal pterygoid plate and is inserted into the median raphe of the soft palate. Some of its fibres are also inserted into the posterior margin of the hard palate.

The azygos uvulæ (Fig. 65) consists of some muscular fibres extending from the posterior nasal spine to the end of the azygos uvulæ. (Fig. 11.)

How is the pharyngeal plexus formed and what does it supply?

This plexus is formed by the union of branches from—

1. The superior cervical ganglion of the sympathetic.
2. The glosso-pharyngeal nerve, the ninth cranial nerve.
3. The pneumogastric or vagus nerve, the tenth cranial nerve.
4. The superior laryngeal nerve, a branch of the vagus.

This plexus supplies the mucous membrane and muscles of the pharynx and all the muscles of the soft palate except the tensor palati, which receives its nerve-supply from the otic ganglion.

The pharyngeal plexus may be seen on your dissection on the outer surface of the middle and inferior constrictors of the pharynx.

What functions may be rationally attributed to the soft palate as a whole?

Its principal function seems to be to direct food to the stomach and protect the other cavities communicating with the pharynx from invasion by the food. (1) The anterior pillars of the fauces and the tongue prevent regurgitation into the mouth. (2) The posterior pillars of the fauces prevent regurgitation into the nose through the posterior nares. (3) The epiglottis protects the larynx. (4) The levator palati forms a temporary roof over the pharynx during deglutition, and thereby protects the Eustachian tube. (5) The tensor palati opens the Eustachian tube during deglutition.

THE PHARYNX.

This cavity is situated behind the nose, mouth, and larynx. It connects the mouth with the œsophagus. It is in communication with:

1. The larynx, which is guarded by the epiglottis.
2. The œsophagus, its downward continuation to the stomach.
3. The tympanum, through the Eustachian tube.
4. The nose, by the posterior nares.
5. The mouth, being partially shut off by the fauces. The proper muscles of the pharynx are: (1) The superior constrictor pharyngis. (2) The middle constrictor pharyngis. (3) The inferior constrictor pharyngis. (4) The stylo-pharyngis muscle. (5) The palato-pharyngeus muscle.

The inferior constrictor you will find arising from (1) the cricoid cartilage; (2) the oblique line of the thyroid; (3) the inferior cornu of the thyroid. It is inserted into the fibrous raphe. (Fig. 62.)

The middle constrictor arises from the cornua of the hyoid bone, and from the stylo-hyoid ligament. (Fig. 62.) It is inserted into the median raphe, the fibres being disposed as follows (1) The inferior fibres extend downward and are overlapped by the inferior constrictor; (2) the superior fibres extend upward and overlap the superior constrictor in part; (3) the middle fibres extend horizontally. (Fig. 62.)

The superior constrictor muscle of the pharynx (Fig. 62) is pale and thin in comparison with the inferior constrictor. The reason of this will be appreciated when you understand the mechanism of deglutition. Growth is the correlative of function: the function of the inferior constrictor is vigorous contraction, hence its roborous fibres; on the other hand, the function of the superior constrictor is a retaining bag, hence its predominance of connective over muscular tissue.

This muscle has the following origins, which are well shown in figure 62: (1) From the internal pterygoid plate, lower third, and its hamular process; (2)

from the pterygo-mandibular ligament; (3) from the posterior fifth of the mylo-
hyoid ridge (Fig. 9); (4) from the side of the tongue. The muscle is inserted
into the pharyngeal spine of the occipital bone and into the median raphe.

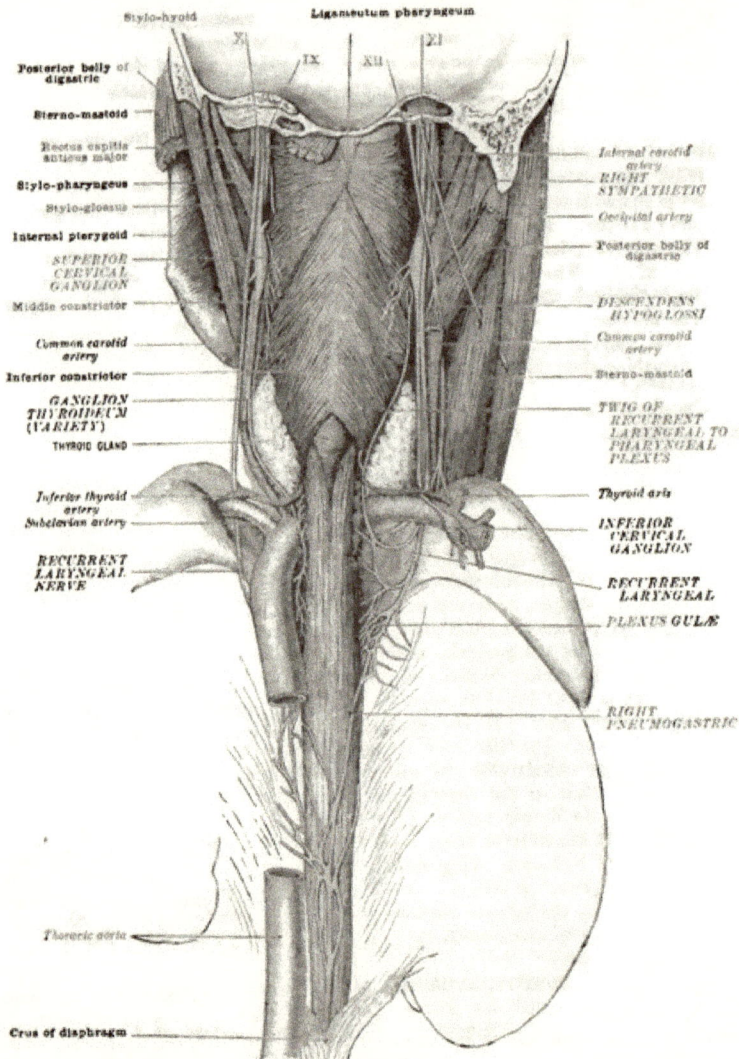

FIG. 61.—DISTRIBUTION OF THE PNEUMOGASTRIC NERVE, VIEWED FROM BEHIND.   (Krause.)

What can you say of the structure of the pharynx?
    It has three coats : a *mucous*, which is continuous with the mucous membrane
of the cavities with which the pharynx communicates ; a muscular coat, special-

ized as constrictors and elevators of the pharynx to facilitate deglutition ; a fibrous coat, the pharyngeal aponeurosis, situated between the mucous and muscular coats, for strength and support. This is also called the pharyngeal aponeurosis.

Between what two points does the pharynx extend?

From the base of the skull to the cricoid cartilage.

What is the pharyngeal tonsil, and where is it located?

It consists of a mass of lymphoid tissue similar to that found in the tonsil. It is located on the back part of the pharynx, between the Eustachian orifices.

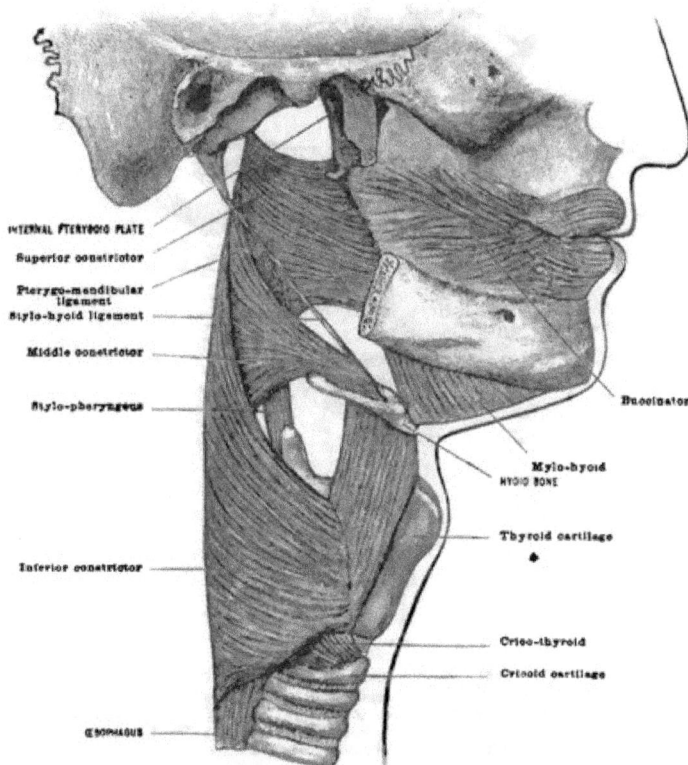

INTERNAL PTERYGOID PLATE
Superior constrictor
Pterygo-mandibular ligament
Stylo-hyoid ligament
Middle constrictor
Stylo-pharyngeus
Inferior constrictor
ŒSOPHAGUS

Buccinator
Mylo-hyoid
HYOID BONE
Thyroid cartilage
Crico-thyroid
Cricoid cartilage

FIG. 62.—THE MUSCLES OF THE PHARYNX.

Distinguish between pharyngeal fascia and pharyngeal aponeurosis.

The *fascia* is behind the pharynx. It is the fourth layer of deep cervical fascia continued upward. The *aponeurosis* is one of the proper coats of the pharynx, located between the mucous and the muscular coats.

What important structures do you find between the superior and middle constrictors of the pharynx?

The glosso-pharyngeal nerve and the stylo-pharyngeal muscle. (Fig. 63.)

What do you find between the middle and inferior constrictors of the pharynx?

The superior laryngeal branch of the pneumogastric nerve, and the superior laryngeal branch of the superior thyroid artery. (Fig. 63.)

Describe the Eustachian tube.

It conveys air from the pharynx to the middle ear. You may see the pharyngeal orifice on the posterior wall of the pharynx, in line with the inferior turbinated bone. You will find the tympanic orifice in the anterior wall of the tympanum, forming the lower of the two compartments of the canalis musculo-tubarius. This you will understand when you dissect the middle ear and the petrous stage of the seventh nerve. The tube is an inch and a half long from the posterior wall of the pharynx to the anterior wall of the tympanum. The tube

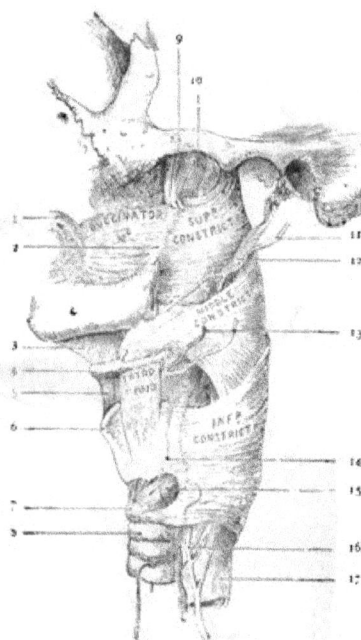

FIG. 63.—MUSCLES OF THE PHARYNX.

1. Orbicularis oris. 2. Pterygo-maxillary ligament. 3. Mylo-hyoideus. 4. Os hyoides. 5. Thyro-hyoid ligament. 6. Pomum Adami. 7. Cricoid cartilage. 8. Trachea. 9. Tensor palati. 10. Levator palati. 11. Glosso-pharyngeal nerve. 12. Stylo-pharyngeus. 13. Superior laryngeal nerve and artery. 14. External laryngeal nerve. 15. Crico-thyroideus. 16. Inferior laryngeal nerve. 17. Esophagus.

has a bony part in the petrous portion of the temporal bone one-half of an inch long; it has a fibro-cartilaginous part one inch long.

Has the cartilaginous tube any practical importance as a guide in dissection?

Yes; it projects between the origins of the levator palati and tensor palati.

Give the origins of the levator palati and the tensor palati, and show their relation to the Eustachian tube.

(1) The levator palati (Fig. 11) arises from the petrous part of the temporal bone and from the lower margin of the Eustachian tube. This muscle you find, then, behind the tube in dissecting, to the inner side. The tensor palati (Fig. 11) arises from the spine of the greater ala of the sphenoid, from the scaphoid fossa, and from the cartilage of the Eustachian tube, and is in relation

with the external surface of the internal pterygoid plate internally, and with the internal pterygoid muscle externally; the tendon turns inward around the hamular process, and is inserted into the aponeurosis of the soft palate and into the under surface of the horizontal plate of the palate bone. This muscle you will find in front of and to the outer side of the cartilaginous part of the Eustachian tube.

What is the salpingo-pharyngeus muscle?

A name by which those fibres of the palato-pharyngeus muscle are designated that arise from the cartilaginous part of the Eustachian tube.

Name all the muscles attached to the Eustachian tube

(1) The levator palati; (2) the tensor palati; (3) the salpingo-pharyngeus, or the Eustachian part of the palato-pharyngeus; (4) the tensor tympani.

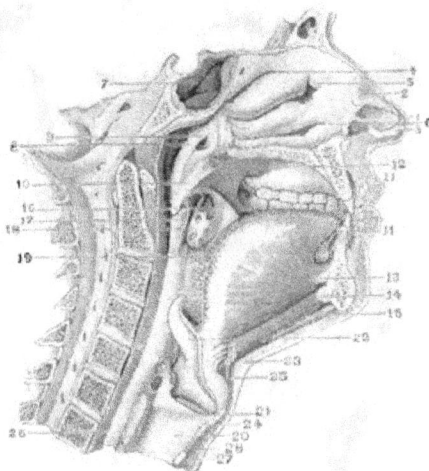

Fig. 64.—Median Section of Mouth, Pharynx, and Larynx.

1, Left nostril. 2. Upper lateral cartilage. 3. Inner portion of lower cartilage. 4. Superior turbinated bone and meatus. 5. Middle turbinated bone and meatus. 6. Inferior turbinated bone and meatus. 7. Sphenoid sinus. 8. Posterior nasal fossa. 9. Internal orifice or pavilion of Eustachian tube. 10. Velum palati. 11, 11. Vestibule of mouth. 12. Palatine vault. 13. Genio-glossus muscle. 14. Genio-hyoid. 15. Mylo-hyoid. 16. Anterior pillar of velum palati. 17. Posterior pillar. 18. Tonsil. 19. Circumvallate papillæ of tongue. 20. Cavity of larynx. 21. Ventricle. 22. Epiglottis. 23. Hyoid bone. 24. Thyroid cartilage. 25. Thyro-hyoid membrane. 26. Posterior portion of cricoid cartilage. 27. Anterior portion. 28. Crico-thyroid membrane.

How is the pharynx innervated?

From the pharyngeal plexus, located on the outer surface of the middle and inferior constrictor muscles and formed by the vagus, glosso-pharyngeal, and by branches from the superior cervical ganglion of the sympathetic.

Give the attachments of the pharynx.

The sphenoid (Fig. 11), the pharyngeal spine of the occipital, the petrous portion of the temporal, the Eustachian tube, internal pterygoid plate, posterior nares, mouth and larynx, prevertebral fascia, and œsophagus.

The rectus capitis anticus major arises from the anterior tubercles of the transverse processes of the third, fourth, fifth, and sixth cervical vertebræ. It is inserted into the under surface of the basi-occipital part of the occipital bone.

The muscle flexes the head. Its nerve-supply is from the first and second cervical.

The rectus capitis anticus minor, a small muscle, arises from the upper part of the lateral mass in front of the articular process of the atlas. It is

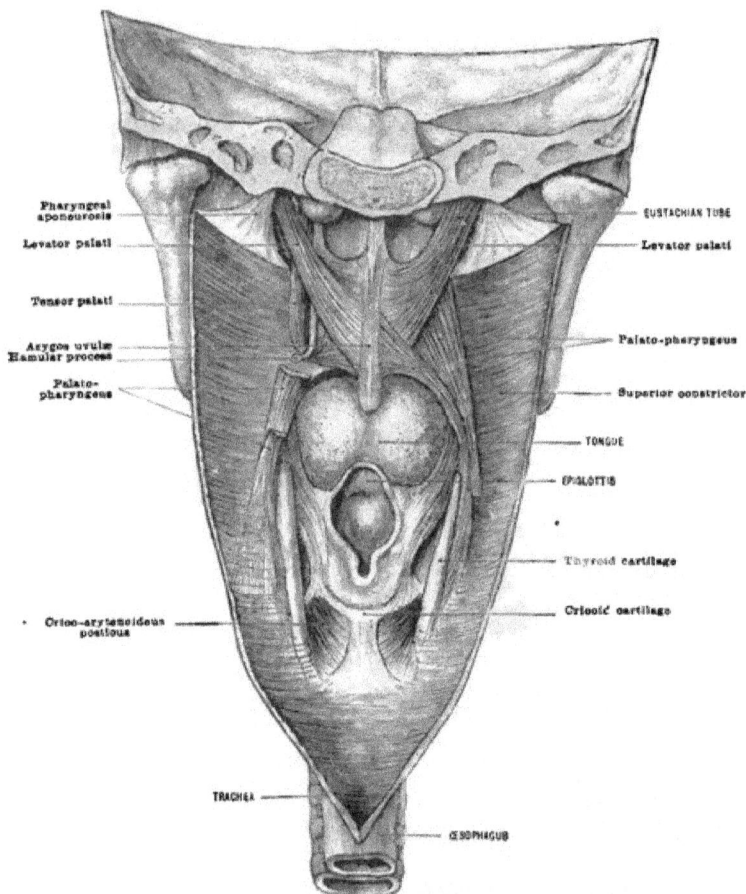

FIG. 65.—VIEW OF MUSCLES OF SOFT PALATE, AS SEEN FROM WITHIN THE PHARYNX.
(Modified from Bourgery.)

inserted into the basilar process of the occipital bone, posterior to the preceding muscle. The muscle acts synergistically with the rectus capitis anticus major.

The rectus capitis lateralis arises from the lateral mass of the atlas. It is inserted into the jugular process of the occipital bone. Its action is to flex the head laterally. This and the preceding muscle are supplied by the first cervical nerve.

The longus colli.

STRAIGHT.

Origin.—1. From bodies of sixth and seventh cervical and first, second, and third dorsal.

Insertion.—2. Into the bodies of the second, third, and fourth cervical.

UPPER OBLIQUE.

Origin.—1. Anterior tubercles of transverse processes of third, fourth, and fifth cervical.

Insertion.—2. Tubercle on anterior arch of the atlas.

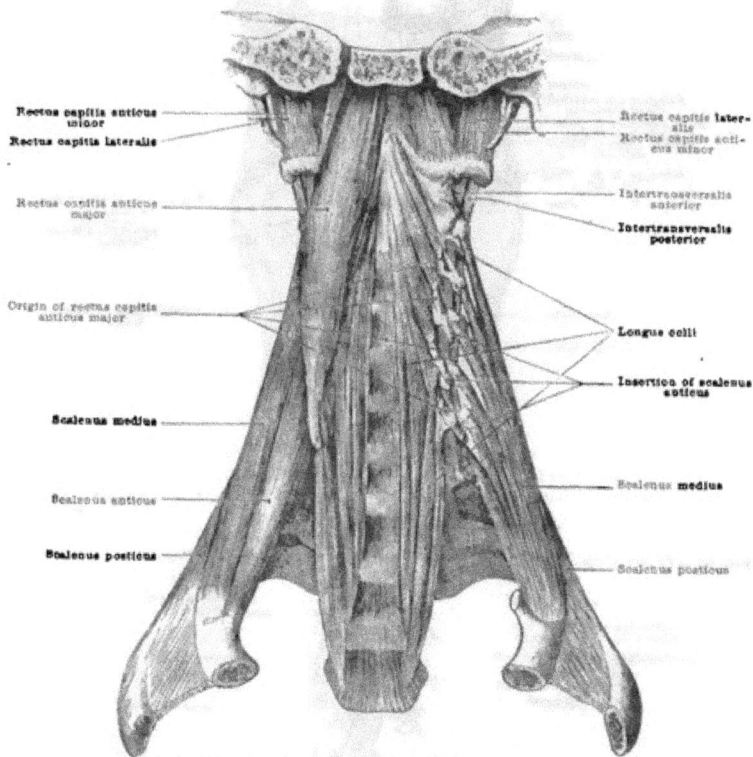

Fig. 66.—The Muscles of the Front of the Neck.

LOWER OBLIQUE.

Origin.—1. Bodies of first, second, and third dorsal vertebræ.

Insertion.—2. Anterior tubercles of transverse processes of fifth and sixth cervical.

These muscles are supplied by the anterior divisions of the cervical nerves. The action is evident : flexion direct, lateral, or combined in rotation.

THE LARYNX.

The upper part of the respiratory tract is called the larynx. Its downward continuation into the neck and thorax is the trachea. The essential organ of voice is a vocal cord on each side of a chink called the rima glottidis. The greater part of the larynx is set aside as a protective for the delicate vocal cords.

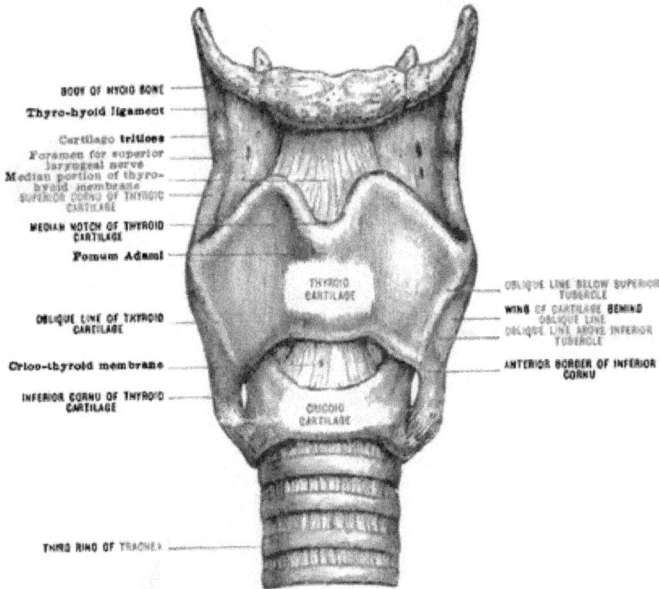

FIG. 67.—FRONT VIEW OF THE CARTILAGES OF THE LARYNX. (Modified from Bourgery and Jacob.)

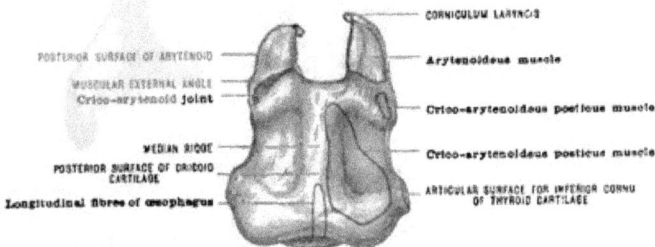

FIG. 68.—BACK VIEW OF THE CRICOID AND ARYTENOID CARTILAGES. (Modified from Bourgery and Jacob.)

A minor part of the larynx is set aside to move the vocal cords. To gain a comprehensive idea of the larynx in your dissection, you must find on your dissection the following structures:

    1. The thyroid cartilage and hyoid bone. (Fig. 67.)
    2. The thyro-hyoid membrane and median notch of thyroid cartilage.

3. The cricoid cartilage and crico-thyroid membrane. (Fig. 67.)
4. The arytenoid cartilages surmounting the cricoid cartilages.
5. The epiglottis in the retiring angle of the thyroid cartilage. (Fig. 69.)
6. The mucous membrane of the larynx.
7. The superior laryngeal nerve of the vagus nerve. (Figs. 63 and 71.)
8. The inferior laryngeal nerve (the recurrent branch—figure 71) of the vagus.

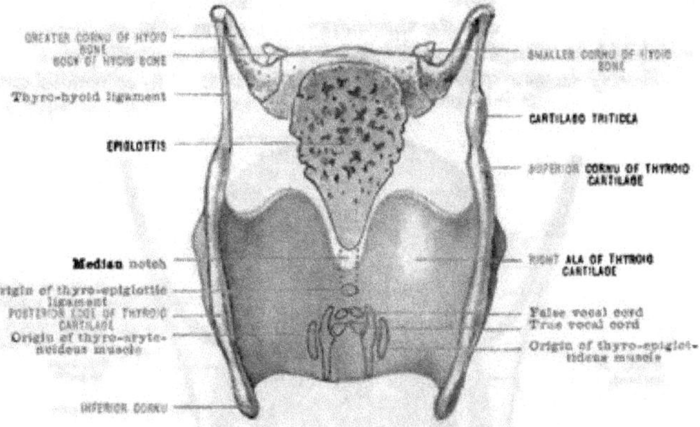

FIG. 69.—POSTERIOR VIEW OF THYROID CARTILAGE WITH EPIGLOTTIS.

9. The superior laryngeal artery, a branch of the superior thyroid artery (Fig. 28.)
10. The sympathetic nerve with the laryngeal arteries.
11. The inferior laryngeal artery, a branch of the inferior thyroid. (Fig. 28.)
12. The vocal cords—true and false.
13. The ventricle, the space between true and false cords.

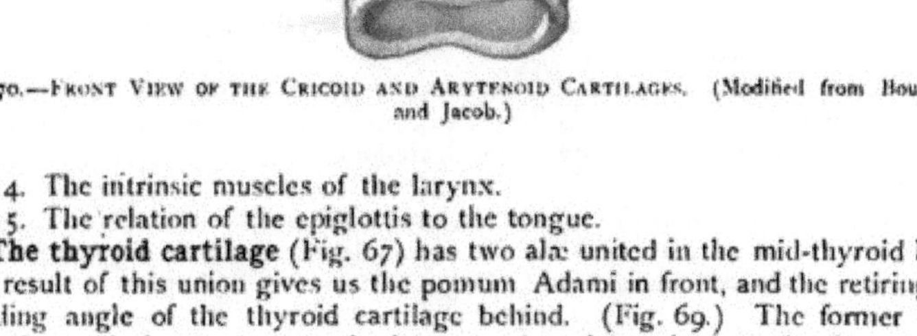

FIG. 70.—FRONT VIEW OF THE CRICOID AND ARYTENOID CARTILAGES. (Modified from Bourgery and Jacob.)

14. The intrinsic muscles of the larynx.
15. The relation of the epiglottis to the tongue.

The thyroid cartilage (Fig. 67.) has two alæ united in the mid-thyroid line. The result of this union gives us the pomum Adami in front, and the retiring or receding angle of the thyroid cartilage behind. (Fig. 69.) The former is a guide in surgical operations on the larynx and trachea ; the latter is of great importance in learning the anatomy and physiology of the vocal cords. Along the superior border of the thyroid cartilage, in the mid-line, you will see the thyroid median notch. Each thyroid ala has a superior cornu and an inferior cornu. The inferior cornu articulates with the cricoid cartilage, forming the crico-thyroid articulation. (Fig. 67.) The superior cornu gives attachment to the lateral thyro-hyoid ligament. The outer surface of the ala has an oblique line, limited above and below by a superior and inferior tubercle respectively. (Fig. 67.)

The **retiring angle** of the thyroid cartilage (Fig. 69) gives origin to : (1) The epiglottis ; (2) the false vocal cords ; (3) the true vocal cords ; (4) the thyro-arytenoid muscle ; (5) the thyro-epiglottideus muscle.

The cricoid cartilage (Figs. 68 and 70) forms, as you will see, a complete ring. It is very narrow in front ; quite wide behind. The arytenoid cartilages

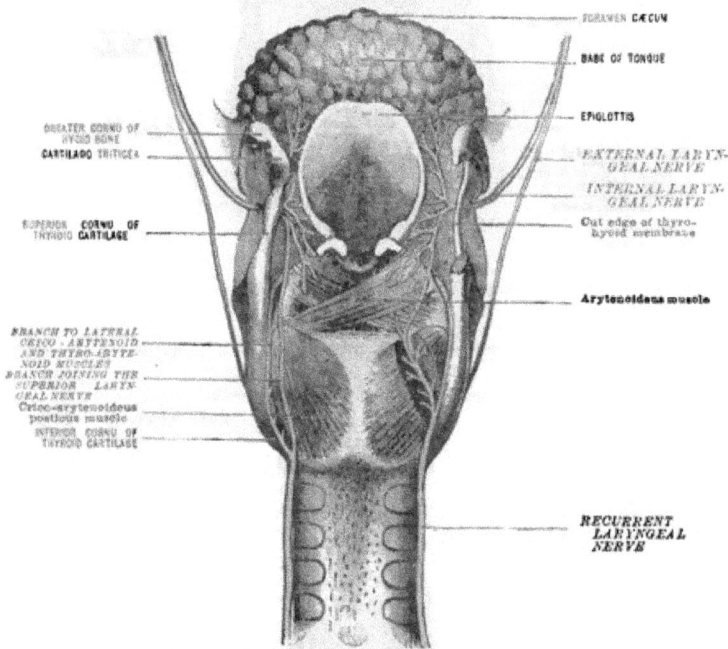

FIG. 71.—NERVES OF THE LARYNX. (Posterior view.)

rest upon the superior border of the cricoid, behind, forming a movable articulation, called the crico-arytenoid.

The arytenoids are small. You must learn the following geometrical description of the arytenoids before you can dissect the larynx understandingly, much less understand the action of the arytenoids :

Apex, obtruncate and points backward and inward.

Base, articulates with the cricoid cartilage. (Fig. 68.)

Posterior surface is occupied by the arytenoideus muscle. (Fig. 68.)

Anterior surface for attachment of false cords and thyro-arytenoideus. (Fig. 70.)

Internal surface is covered by mucous membrane.

Anterior angle, processus vocalis for insertion of true vocal cord. (Fig. 68.)

External angle for insertion of posterior and lateral crico-arytenoid muscles. *The intrinsic muscles* are as follows :

1. The thyro-arytenoideus. (Fig. 70.)
2. The crico-arytenoideus lateralis. (Fig. 70.)
3. The crico-arytenoideus posticus. (Fig. 69.)
4. The arytenoideus. (Fig. 69.)
5. The crico-thyroid. (Fig. 27.)

The intrinsic muscles and the mucous membrane are supplied by (1) the *superior laryngeal* branch of the pneumogastric ; (2) the *inferior laryngeal* branch of the pneumogastric ; (3) the *sympathetic* nerves accompanying the superior and inferior laryngeal arteries ; they take the name of the arteries.

The **superior laryngeal nerve** (Fig. 71) supplies the mucous membrane and the crico-thyroid muscle.

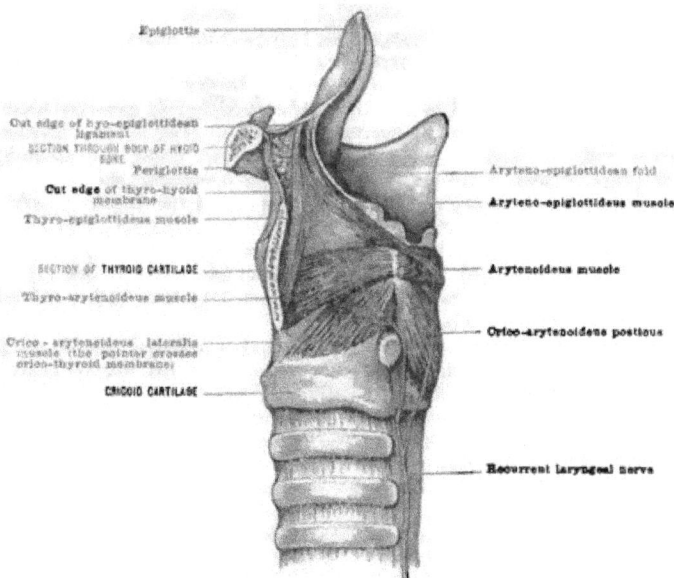

Epiglottis

Cut edge of hyo-epiglottidean ligament
SECTION THROUGH BODY OF HYOID BONE
Periglottis
Cut edge of thyro-hyoid membrane
Thyro-epiglottideus muscle

SECTION OF THYROID CARTILAGE
Thyro-arytenoideus muscle

Crico-arytenoideus lateralis muscle (the pointer crosses crico-thyroid membrane)

CRICOID CARTILAGE

Aryteno-epiglottidean fold
Aryteno-epiglottideus muscle

Arytenoideus muscle

Crico-arytenoideus posticus

Recurrent laryngeal nerve

FIG. 72.—SIDE VIEW OF THE MUSCLES AND LIGAMENTS OF THE LARYNX.

The **inferior laryngeal** supplies the remaining muscles.

The **false vocal cord** extends from the retiring angle of the thyroid cartilage (Fig. 69) to the anterior surface of the arytenoid cartilage. (Fig. 70.)

The **true vocal cord** extends from the retiring angle of the thyroid cartilage (Fig. 69) to the anterior angle of the arytenoid cartilage. (Fig. 70.)

The chink, or rima glottidis, is the space between the two true vocal cords. This chink consists of two parts—a respiratory or posterior and a vocalizing or anterior.

The **thyro-arytenoid muscle** (Fig. 69) extends from the retiring angle of the thyroid cartilage to the anterior surface of the arytenoid cartilage. (Fig. 70.) The action of this muscle is to pull the arytenoid cartilage forward and thereby relax the vocal cords.

The **arytenoid muscle** (Fig. 71) extends from the posterior surface of one to that of the other arytenoid **cartilage**. **It** approximates the vocal cords by drawing the arytenoids together.

The **crico-arytenoideus lateralis** (Fig. 72) arises from the upper border of **the** side of the cricoid cartilage. It is inserted into the external angle of the **base** of the arytenoid cartilage. The action is to draw the vocal cords together **and** close the glottis.

The **crico-arytenoideus posticus** arises on the posterior part of the cricoid. It is inserted into the outer angle of the base of the arytenoid cartilage. This muscle draws the vocal cords apart, and thereby opens the glottis.

The **crico-thyroid muscle** arises from the side of the cricoid cartilage. It is inserted into the **lower corner of the thyroid cartilage**. It tightens the vocal cords.

Describe the superior laryngeal nerve. (Fig. 71.)

It is a branch of the ganglion of the **trunk of the** pneumogastric nerve. It passes behind the internal and external carotid arteries. It is joined by communicating branches from the sympathetic nerve and the pharyngeal plexus. It divides into an internal and an external branch. The external branch pierces the inferior constrictor of the pharynx near the lower border of the thyroid cartilage (Fig. 63), and is distributed principally to the crico-thyroid muscle; some few filaments are distributed to the mucous membrane. The internal division, in company with the superior laryngeal artery, passes under the thyro-hyoid muscle (Figs. 27 and 71), pierces the thyro-hyoid membrane, and is distributed to the mucous membrane, communicating here with the recurrent laryngeal nerve. (Fig. 71.)

Describe the course of the recurrent laryngeal nerves. (Fig. 71.)

The nerve of the right side is given off from the pneumogastric in front of the first stage of the subclavian artery. It passes upward, under, and behind the subclavian artery, then behind the common carotid to the side of the trachea. It lies in a fatty **groove** between the trachea and œsophagus, and, in company with the inferior **laryngeal** artery, passes upward, under the inferior constrictor of the pharynx, **to the** larynx. It is distributed to all the intrinsic muscles of the larynx except the crico-thyroid. The recurrent laryngeal **nerve** of the left side is given off from **the** pneumogastric in front **of the** transverse part **of the** aortic arch, passes under this vessel behind the obliterated ductus arteriosus, **and** gains the fatty space between the trachea and œsophagus, and has the same **subsequent course** and distribution as the nerve on the right side. In their **course between the** œsophagus **and** trachea these nerves give off branches **to these conduits.**

Describe the blood-supply of the larynx. (Fig. 28.)

The superior laryngeal artery is a branch of the superior thyroid. It accompanies the superior laryngeal nerve, piercing with this nerve the thyro-hyoid membrane. The inferior laryngeal artery is a branch of the inferior thyroid. It passes behind the inferior constrictor of the pharynx with its accompanying nerves.

Describe the trachea.

It is an air conduit, extending from the larynx to the tracheal bifurcation, where the bronchial tubes begin. Compared to the vertebral column, in front of which the trachea lies, it extends from the fifth cervical to the fifth thoracic vertebra. It is one inch in width and five inches long.

The structure of the trachea fits it admirably for its specific double function (1) of transmitting air inward and (2) of extruding both air and accumulated **mucus**, bearing crude inhaled impurities. Imperfect rings of cartilage, held together by fibro-elastic membrane possessing some muscular fibres, a rich nerve- and blood-supply, and a lining of mucous membrane, are the anatomical tissues that specially qualify the trachea for its important duty. The front part of the trachea is round, the back part flat. The space between the ends of the imperfect rings of cartilage behind is filled by the musculus trachealis. This consists of transverse and longitudinal unstriped fibres. The arrangement of these is such as to set up a vermicular movement in the trachea when expectoration occurs.

The nerve-supply of the trachea comes from the vagus and also from the recurrent laryngeal, a branch of the vagus, as this latter nerve lies in the fatty groove between the trachea and œsophagus; and from the sympathetic, which latter nerves accompany the tracheal arteries, and take the same name.

The blood-supply of the trachea comes from the inferior thyroid artery, a branch of the thyroid axis of the subclavian artery. (Fig. 36.) The veins join the thyroid plexus of veins. Of the thyroid veins, you will remember, the right opens into the right innominate, the left into the left innominate vein. Each of these has valves where it becomes tributary to the larger vessel. These thyroid veins communicate on the trachea. These are the troublesome vessels in low tracheotomy.

Relations of the Trachea.—If you will review your dissection, you will find in front of the cervical part of the trachea the isthmus of the thyroid gland (Fig. 27), the inferior thyroid veins, the sterno-hyoid and sterno-thyroid muscles (Fig 27), the deep cervical fascia, superficial fascia, and skin. Behind, the trachea is in relation with the œsophagus. On each side are the common carotid arteries, the inferior thyroid artery, the recurrent laryngeal nerve, and the lobes of the thyroid gland. (Fig. 31.) The thoracic part of the trachea, as you will see at a later stage of your work, is covered by the manubrium sterni, the remains of the thymus gland, the left brachio-cephalic vein, the aortic arch, the innominate and left common carotid arteries, and the deep cardiac plexus. Posteriorly is the œsophagus, and laterally the vagus nerves are on each side.

THE NASAL FOSSÆ—INTERNAL NOSE.

Dissection.—With a sharp saw cut through (1) the symphysis of the mandible. Then take a sharp knife and cut through the tongue from base to tip in the midline, as far forward and downward as the hyoid bone. (2) Let an assistant hold the divided halves of the mandible apart, while you saw through the mid-line of the remainder of the face. (Fig. 64.) Precaution: Before you make the latter cut, observe to which side the septum nasi is deflected; cut on the opposite side, so as to have the septum intact for study. The cuts have placed before you for study the mouth and nasal fossæ. Very little dissection in addition to that already done is necessary.

The nasal fossæ have the following geometrical parts:

1. *Roof*, formed by the nasal bones, the nasal spine of the frontal bone, cribriform plate of the ethmoid, the under surface of the body of the sphenoid.

2. A *floor* (Fig. 74), formed by the hard and soft palate, covered by mucous membrane. The hard palate is formed by the palatine process of the superior maxilla and the horizontal part of the palate bone

3. An *inner wall* (Fig. 74), formed by the septum nasi, covered by mucous membrane. The septum nasi is composed of (1) the vomer; (2) the vertical plate of the ethmoid; (3) the cartilaginous septum—quadrangular in some, triangular in other cases.

The *outer surface* is formed by the nasal bones, the nasal process of the superior maxilla, the lachrymal, ethmoid, inner surface of superior maxilla, inferior turbinated, vertical plate of palate bones, and inner surface of pterygoid

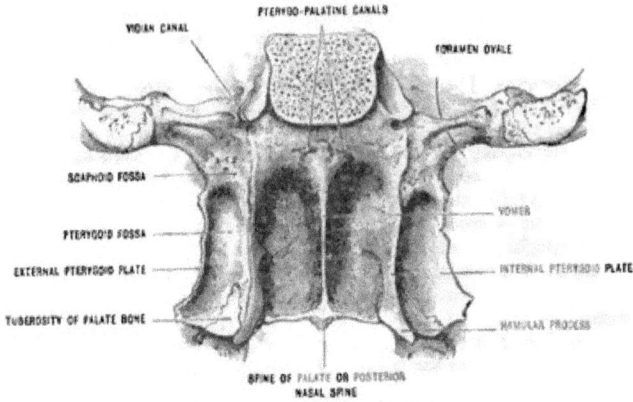

FIG. 73.—THE POSTERIOR NARES.

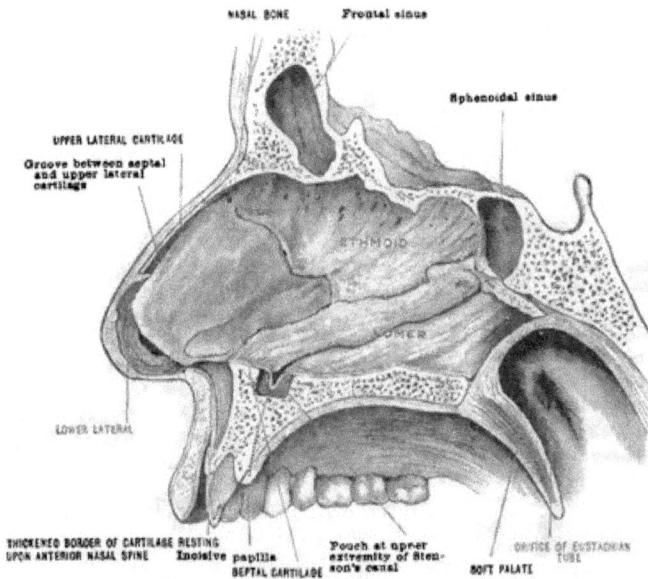

FIG. 74.—SECTION SHOWING BONY AND CARTILAGINOUS SEPTUM.
The dotted line indicates the course of the anterior palatine canal.

process of the sphenoid bone. This surface is covered by mucous membrane. This must be studied on the dry skull.

The anterior nares is the inlet to the nose from the external world. The posterior nares is the opening into the pharynx. (Fig. 73.) Notice : On the outer wall, and even forming part of the same, are the three turbinated bones.

Study figure 74. Remove the mucous membrane from the septum, and find (1) suture between the vomer and ethmoid ; (2) junction between cartilage and ethmoid ; (3) *the naso-palatine groove* on the vomer, containing a nerve and an artery of the same name. The nerve, a branch of Meckel's ganglion, one on each side, passes through the foramina of Scarpa, in the anterior palatine canal (Fig. 74), and anastomoses with the anterior palatine nerves, also branches of Meckel's ganglion. (Figs. 53 and 75.)

The turbinated bones (Fig. 77) are three in number. They occupy the outer wall, as follows fractionally : The superior turbinal, the posterior third ; the middle turbinated, the posterior two-thirds ; the inferior turbinated extends the whole length of the wall.

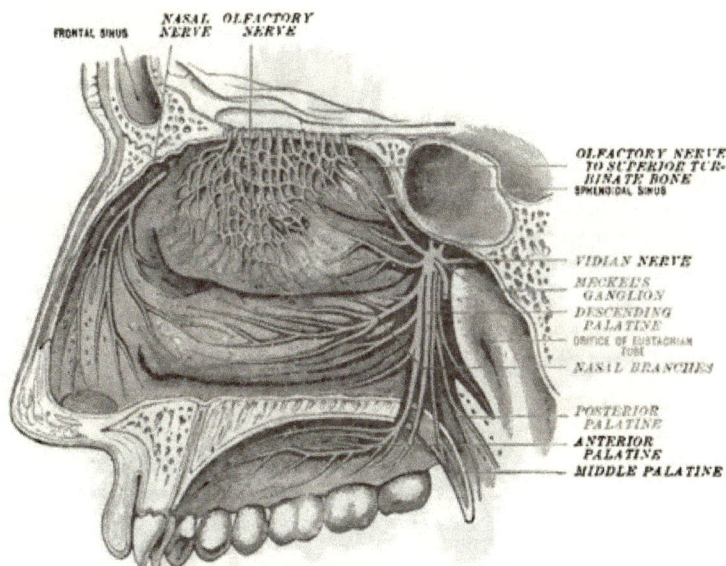

FIG. 75.—NERVES OF THE NASAL CAVITY.

The Ethmo-turbinals (Fig. 77).—This name is given to the superior and middle turbinals. They belong to the lateral mass of the ethmoid bone.

The meatuses are three irregular cavities on the outer wall of the nasal fossæ. Into them open the nasal duct and the frontal, ethmoidal, maxillary, and sphenoidal sinuses :

Inferior, receives the nasal duct conveying the tears.

Middle, receives the openings—the antrum, frontal, and anterior ethmoidal cells.

Superior, the opening of the sphenoidal and posterior ethmoidal cells.

Describe the nasal mucous membrane.

It is called pituitary membrane, from an erroneous idea entertained by the ancients that the nasal deflections had their origin in the pituitary body. It is also called the Schneiderian mucous membrane. In the regions of the turbinals and septum it is thick, vascular, and loosely attached to the bone ; in the bottom of the meatuses and in the intramural sinuses communicating with the nasal fossæ it is thin. It is continuous with the mucous membrane of all the cavities with the nasal fossæ communicante.

Name the arteries that supply the nasal fossæ.

1. The anterior and posterior ethmoidal supply the roof, outer wall, and upper half of the septum. These are branches of the ophthalmic, and leave the orbital cavity by the anterior and posterior ethmoidal foramina. The companion veins have like name, and are tributary to the ophthalmic vein. Infection in this region, then, might extend along the ophthalmic vein to the cavernous sinus, since this vein has no valves.

2. The spheno-palatine supplies the lower half of the septum, the turbinals, the meatuses, the frontal sinuses, and the antra. The companion veins of these arteries are confluent to the pterygoid plexus on the inner side of the internal pterygoid muscle ; the veins take the same name as the arteries. The pterygoid

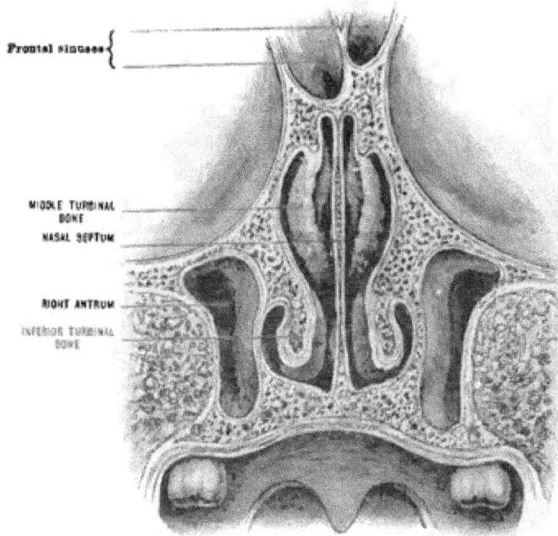

FIG. 76.—TRANSVERSE SECTION PASSING THROUGH THE NASAL FOSSÆ AND ANTRA AT THE POSTERIOR EXTREMITY OF THE MIDDLE TURBINAL BONE. (Seen from the front.)

plexus communicates with the cavernous sinus by the Vesalian vein. Infection from this region may reach the sinus, since the Vesalian has no valves.

3. The descending palatine and Vidian arteries, branches of the internal maxillary, supply the nasal fossæ.

What can you say of the nasal mucous membrane in relation to nose bleeding, and in what way may a violent hemorrhage from the nose abort cerebral apoplexy ?

The venous blood in the mucous membrane of the nose communicates with the superior longitudinal sinus, in the falx cerebri, by an emissary vein which passes through the foramen cæcum in front of the crista galli. The nasal mucous membrane is very vascular and quite loosely attached, a condition favoring easy rupture of its vessels. In cerebral congestion of the face the ocular and nasal mucous membranes become engorged, on account of their communication with the sinuses. On account of its lax attachment the vessels of the nasal mucous membrane frequently give way, and hemorrhage results. This occurs more frequently in children than in adults.

Describe the nerve-supply to the mucous membrane of the nose.

1. The olfactory nerves are distributed to the Schneiderian membrane of the upper third of the septum, and to the superior and middle turbinals.

2. The nasal nerve, a branch of the ophthalmic division of the fifth nerve, supplies the anterior half of the roof, outer wall, and inner wall of the nasal fossæ with common sensation.

3. The upper branches of Meckel's ganglion and the Vidian nerve supply the posterior half of the roof, outer wall and inner wall, and the superior turbinal.

4. The anterior palatine branch of Meckel's ganglion on its way down to the roof of the mouth supplies the middle and inferior turbinals.

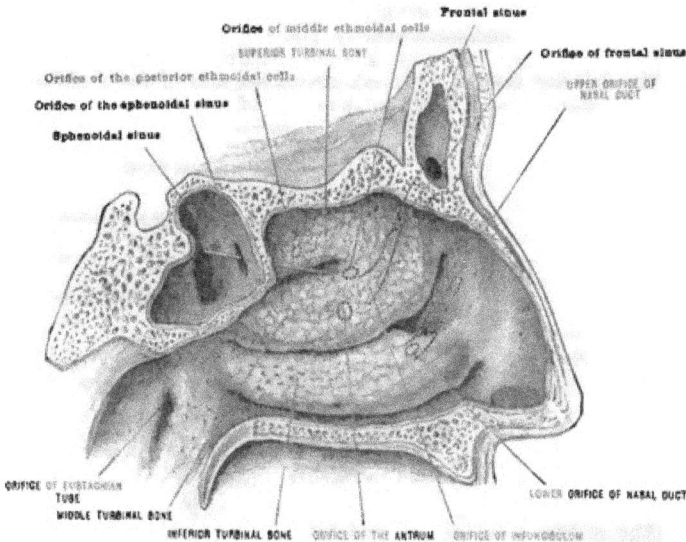

FIG. 77.—SECTION OF THE NOSE, SHOWING THE TURBINAL BONES AND MEATUSES, WITH THE OPENINGS IN DOTTED OUTLINE.

5. The naso-palatine branch of Meckel's ganglion supplies the lower and posterior part of the septum. The anterior superior dental nerves supply the inferior turbinated bone, hence a disease of this bone may cause pain in the ear or in any of the teeth.

The Intramural Sinuses.—This is a convenient term by which to distinguish the air-containing cells, in connection with some of the bones of the skull and face, from the dural sinuses in the dura mater, whose use is to contain and convey blood away from the brain. These sinuses are the ethmoid, the sphenoidal, the mastoid, the maxillary or antrum of Highmore, the frontal, mastoid antrum, and mastoid cells. As they all communicate directly or indirectly with the nasal fossæ, we will consider them in this place. The mastoid cells communicate with the middle ear, and this with the pharynx by the Eustachian tube. The others communicate with the nasal fossæ as indicated in figure 77.

Formation.—The intramural cells are formed by absorption of the middle or diploic plate of bone. They are lined by mucous membrane. Infection is very liable to occur in these sinuses. The difference between the mucous membrane of the sinuses and that of the mouth is this : the latter is more vascular, more hardy, more resisting, both to infection and traumatic causes, on account of its location being such that exposes it to friction. Structure being the correlative of function, an increased blood-supply would logically account for the superior hardihood and less vulnerable character of the mucous membrane of the mouth, and all other regions where friction is a consideration inseparable from the environment.

Take special notice that—

1. The orifice of the Eustachian tube is on a direct line with the inferior meatus.

2. A fine straw or bristle can be passed through the infundibulum (Fig. 77) of the frontal sinus, into the middle meatus.

3. A straw can be passed through the nasal duct into the inferior meatus ; by turning the inferior turbinated bone up, you can easily see the orifice for the nasal duct.

4. **The inferior** meatus is located between the floor of the nasal fossæ and the inferior turbinated bone, and receives the end of the nasal duct. (Fig. 77.)

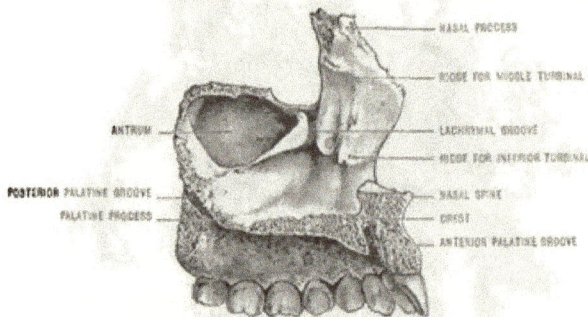

FIG. 78.—THE LEFT MAXILLA. (Inner view.)

5. **The middle meatus** is situated between the inferior and middle turbinated bones. It receives the opening for the antrum, frontal sinus, and anterior ethmoidal cells. (Fig. 77.)

6. **The superior meatus** is located between the middle and superior turbinated bones, and receives the openings for the sphenoidal sinus and for the posterior ethmoidal cells. (Fig. 77.)

The antrum of Highmore (*maxillary sinus*) (Figs. 68 and 76) occupies the interior of the body of the superior maxilla. Its medical and surgical importance entitle it to the following analytical consideration :

1. It has a *roof*, a thin plate of bone that forms the floor of the orbit. In this roof is the infraorbital nerve and vessels. (Fig. 51.)

2. A *floor* formed by the alveolar process of the upper jaw. The fangs of the teeth may produce irregularities of this floor. (Fig. 76.) The surgeon may gain access to the cavity by extracting a first or second molar tooth.

3. *Inner wall.*—The importance of this wall is the presence of the normal opening, by which the antrum communicates with the nasal fossæ. (Fig. 77.) The surgeon may gain access to the antrum for drainage by this wall.

4. *Anterior wall.*—On the inner surface of this wall are the anterior superior

dental canals, containing vessels and nerves for the teeth. (Fig. 51.) On the outer surface of this wall is the infraorbital plexus of nerves, formed by the infraorbital branches of the fifth and seventh nerves. Entrance to the antrum for purpose of drainage is gained by going through this wall too.

5. *Posterior wall.*—On this you will find the posterior dental canals for nerves and vessels to the posterior teeth of the superior maxilla. (Fig. 51.) The apex of the antrum corresponds to the prominence of the cheek.

What can you say of the size of the antrum of Highmore, or maxillary sinus?

It will contain about one ounce of fluid. It is variable in size in different persons. It is longer in the male than in the female. A large bone may have a very small antrum, and vice versâ.

What can you say of the floor of the antrum?

The floor is the strongest wall. It is uneven on account of the roots of the teeth, the first and second molars being usually those producing the unevenness. These tooth-roots may even penetrate the floor of the antrum, and be the exciting cause of disease of the antral mucous membrane. The antrum may be divided into several compartments, or pockets, as they are called, by bony partitions of variable thickness and height, a circumstance which must always enter

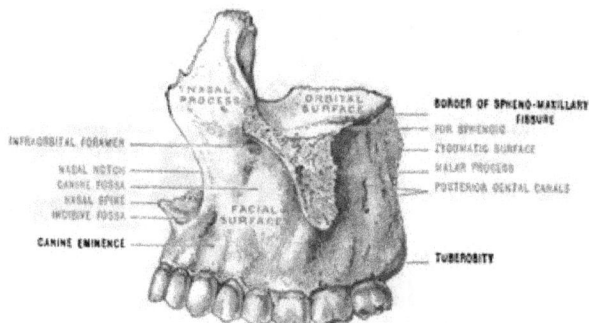

Fig. 79.—The Left Maxilla. (Outer view.)

as a prognostic factor in operations to establish drainage on the antrum. The surgeon should ask himself after each operation, "Have I drained the antrum or only a pocket of the antrum?" A merchant had a wine-cellar flooded with water. He employed a company to remove the water. A hose was thrust through a dark window and the pump started. To the dismay of the merchant, the experts had thrust the hose into a hogshead of wine, and drained the pocket, instead of the cellar.

What can you say of the nerve-supply of the antrum?

The nerve-supply of the mucous membrane of the antrum comes from the anterior and posterior superior dental branches of the fifth nerve. You will remember this: The dental nerves supply the teeth; the skin of the lips and cheeks covering these teeth outside; the gums surrounding these teeth, through small branches called the *nervuli gingivales;* and, lastly, the mucous membrane of the antrum.

Describe the blood-supply of the antrum?

The internal maxillary artery gives off branches not only to the teeth (dental branches), but also branches to the gums (gingival branches), and antral branches to the mucous membrane of the antrum. These antral branches are accompanied by sympathetic nerves.

To what is the nasal nerve distributed?

To the septum, to the outer part of the nose, and to the two lower turbinals. This is a sensory nerve, a branch of the ophthalmic division of the fifth. It is found in the cranial, orbital, and nasal cavities. It traverses the anterior ethmoidal foramen and the nasal slit.

Where is the olfactory nerve distributed?

To the roof of the fossa; to the superior and middle turbinals; to the upper half of the septum nasi. (Fig. 75.)

Where is the spheno-palatine or Meckel's ganglion?

In the spheno-palatine fossa, under the second division of the fifth nerve. It is the source from which the nose and mouth and palate derive their nerve-supply in great part. (Fig. 75.)

The sensory root of Meckel's comes from the second division of the fifth, called the spheno-palatine nerves. (Fig. 53.)

The motor root comes from the seventh or facial nerve, through the large

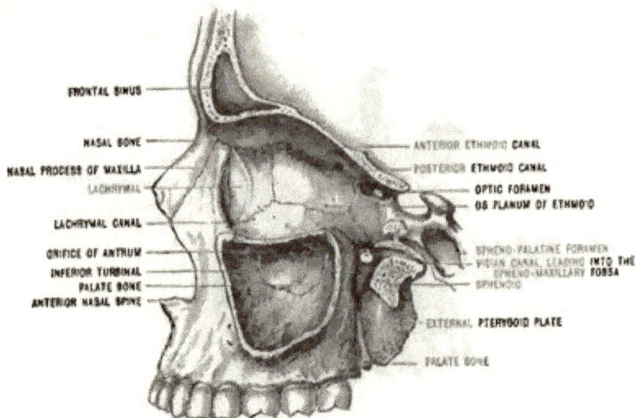

FIG. 80.—A SECTION OF THE SKULL, SHOWING THE INNER WALL OF THE ORBIT, THE BASE OF THE ANTRUM, AND THE SPHENO-MAXILLARY FOSSA.

superficial petrosal. This you can see in a groove under the dura, on the anterior surface of the petrous portion of the temporal bone. The nerve passes behind the Gasserian ganglion, down through the sphenotic foramen. It joins here, at the base of the skull, the large, deep petrosal branch of the carotid plexus. The two now pass through the Vidian canal. The result of the union of the large superficial petrosal of the seventh nerve and the large, deep petrosal of the carotid plexus is the Vidian nerve. (Fig. 53.) The last is the sympathetic root.

Name the branches given off by Meckel's ganglion. (Fig. 53.)

(1) Branches to the orbital periosteum; (2) branches to the mucous membrane of the nose; (3) anterior palatine—to the roof of the mouth (Fig. 55); (4) naso-palatine in groove on nasal septum; (5) branches to the upper pharynx behind Eustachian tube.

I introduce here, by consent of the author, some conclusions on the opening of the infundibulum in relation to the opening for the antrum of Highmore. I have given no special attention to the subject in the dissecting-room; still, incidentally, I have verified in seven cases the views of the author of the following :

CONTINUED STUDY OF THE RELATIONS OF THE FRONTAL SINUS TO THE ANTRUM.

BY THOMAS FILLEBROWN, M.D., D.D.S., BOSTON, MASS.

PROFESSOR OF OPERATIVE DENTISTRY AND ORAL SURGERY, DENTAL SCHOOL OF HARVARD UNIVERSITY.

(Read before the American Dental Association, August 5, 1897, and reprinted from the "Dental Cosmos" for December, 1897.)

" Last year I made a report of some observations I had made on the formation of the infundibulum, showing that in many cases it continued directly to, and terminated in the foramen of, the antrum, and that a fold of mucous membrane mentioned by Merke, in 1834, extended above the foramen, forming a pocket, from the bottom of which the opening into the antrum is situated, thus directing any discharge coming down the infundibulum into the antrum, so that, under ordinary circumstances, no abnormal discharge from the frontal sinus would escape into the nasal passage until the antrum was filled so as to cause a backward overflow.

" I mentioned the fact that Tilleaux, about 1840, noticed that of fluid injected into the frontal sinus, a great part flowed into the antrum, and that Dr. Cryer, in the same year, mentioned the same circumstance. He also showed that a probe could be passed from the antrum into the frontal sinus. I also noticed that Byran had mentioned the fact of occasional communications between the cavities, but considered them anomalies, and that Professor Harrison Allen had discussed the proliferation of empyema of the frontal sinus into the antrum. These observations were of isolated cases, and were not proved or considered indicative of the normal anatomy of the parts.

" I reported that the examination of eight different subjects showed that the infundibulum continued as a deep groove, or tube, open on one side down to the foramen of the antrum, and terminated in it, in every one of the eight cases, and that the pocket described was present in seven of the eight. This seemed to imply that the continuation of the infundibulum to the antral foramen and the presence of the pocket membrane was the normal formation. During the past winter I had opportunity to examine fifteen heads in the Harvard dissecting-room, and found the infundibulum continuing to the foramen of the antrum in every case. The membranous fold was present in every case, except on the left side of one subject. In this case the process was broad and flattened toward the meatus, and, though the mucous fold was absent, the widened process served the same purpose, as it formed a cup-shaped cavity quite as capacious as the pocket on the other side.

" In another case the mucous fold was thickened and had considerable muscular tissue intermingled in its substance.

" On the right side of this subject the infundibulum was very large, and in place of the ordinary foramen there were two openings, both quite large, fully one-fourth of an inch in diameter.

" In another case the pockets were large, irregular in form, and deep, the mucous fold completely covering the infundibulum from the foramen of the frontal sinus to the antrum.

" In one subject the mucous fold was considerably calcified. This condition had obtained to a degree throughout the whole system. This subject was advanced in years. This makes a total of twenty-three cases ; a number, I think, sufficient to establish the fact of the normality of the anatomy of the parts.

" The very few variations only prove the rule. I hoped to secure the co-operation of others in making further examinations during the winter, but the reprints of my paper were too long delayed, and the dissecting season was passed before I could communicate with anatomists and make the necessary arrangements and have the benefit of the results of the examinations of other subjects."

THE COVERINGS OF THE BRAIN.

Those structures intervening between the brain and the outside are brain protectives, and may be called coverings. In number they are three; in a collective sense, outer, middle, inner. The outer is called the scalp; the middle, the calvarium; the inner, the meninges. As above intimated, each of these three coverings is a collective noun, and **must** be reduced, analytically, to its simplest terms of individual structures.

The scalp consists of the following layers:

1. **The skin,** covered by a thick growth of hair in primitive man.
2. **The superficial fascia,** the distribution area where you are to find all the arteries, veins, nerves, and dermal muscles of the scalp.
3. **The occipito-frontalis muscle,** consisting of an anterior and a posterior belly, connected by a broad aponeurosis that takes the name of the muscle.
4. **A subaponeurotic layer of connective tissue,** which you can account for by consulting a paragraph in the introductory chapter on the non-apposition of anatomical structures.

There are, then, four layers. The scalp moves freely. The fibrous covering of the bone is periosteum, and its subperiosteal connective tissue does not differ from like tissue found underlying periosteum everywhere. These two structures do not form integral parts of the scalp because they do not move.

Dissection.—Remove the skin—having shaved the head—by making cross-incisions on the mid-top of the head, thus +.

In the superficial fascia of the scalp you are to find the following structures:

1. The supraorbital nerve and vessels. (Figs. 16 and 18.)
2. The temporal branches of the seventh nerve. (Fig. 16.)
3. The auriculo-temporal branch of the fifth nerve. (Fig. 16.)
4. The temporal arteries—superficial—and veins. (Fig. 18.)
5. The great occipital nerve—second cervical, posterior division. **(Fig. 22.)**
6. The small occipital nerve—cervical plexus. (Fig. 22.)
7. The great auricular nerve—cervical plexus. (Fig. 22.)
8. The posterior auricular artery—branch of the external carotid.
9. The posterior auricular nerve—branch of seventh cranial. (Fig. 16.)

The **above** structures you have already found in your dissection of the face and neck.

10. **The** dermal muscles are as follows (Fig. 14):
    1. **Musculus** attrahens aurem—draws the ear forward.
    2. Musculus attollens aurem—draws the ear upward.
    3. Musculus retrahens aurem—draws the ear backward.

They have origin and insertion as indicated in the figure. They are insignificant, and can only be demonstrated in some cases. The function of these muscles is seen in some of the domestic animals—the dog, horse, and mule.

The occipito-frontalis (Fig. 14) has a posterior attachment to the outer two-thirds of the superior curved line of the occipital bone and to the mastoid process of the temporal bone. This muscular belly terminates anteriorly in a broad epicranial aponeurosis. Follow this forward to the anterior belly. The anterior belly was described with the face.

The calvarium (Fig. 81) or second covering of the brain has three layers called tables. They are named and distinguished as follows : (1) an outer, tough and resisting ; (2) a middle table which is porous or spongy, called *diploë*. This is very porous and contains many diploic veins that bleed rather freely during trephining. The characteristic of this table of some of the bones of the cranium is to become absorbed. The intramural sinuses (see p. 113), as the frontal, ethmoid, sphenoid, mastoid, may be considered as owing their formation to this process of absorption. (3) The inner or vitreous table is hard and brittle. The inner surface is grooved for the accommodation of the meningeal arteries (Fig. 81), the nutrient arteries to the dura mater and calvarium.

The meninges form the third or innermost of the three grand protectives of the brain. This covering consists of a dura mater, an arachnoid, and a pia

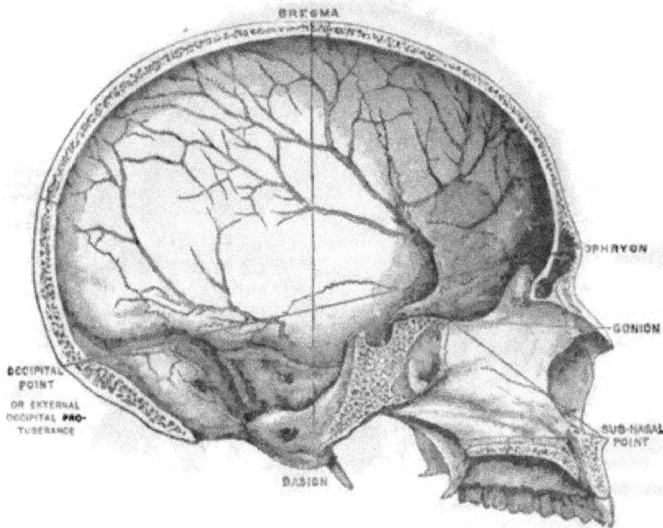

FIG. 81.—THE SKULL IN SAGITTAL SECTION.

mater. The arrangement they bear to each other makes two physiological lymph spaces : (1) a subdural ; (2) a subarachnoid. These terms are self-explanatory, and conform to the rule of subs in the introduction.

The dura mater is adherent to the bony walls of the calvarium and base of the skull. Its attachments are firmer in those localities where (1) great numbers of nerves are leaving ; (2) where there are numerous foramina ; (3) where there are many sutures. The base of the skull possesses these three characters, hence here the dura is most intimately adhered.

The dura mater has four double processes, called : (1) Falx cerebri ; (2) falx cerebelli ; (3) tentorium cerebelli ; (4) diaphragma sellæ. Each process possesses (1) a free margin ; (2) an attached margin ; (3) sinuses for the reception of venous blood from the brain.

The diaphragma sellæ is an inconsiderable dural process attached to the margin of the sella turcica, or pituitary fossa, on the superior surface of the body

9

of the sphenoid bone. In its attached margin is the circular sinus. Under this process of dura is the pituitary body. This body is connected to the tuber cinerium on the floor of the third ventricle of the brain by the infundibulum, which latter passes through an aperture in the diaphragma sellæ.

The falx cerebri (Fig. 82) is between the cerebral hemispheres. It is attached in front to the *crista galli* of the ethmoid bone ; behind, to the *tentorium cerebelli ;* above, to the inner surface of the *calvarium,* in the mid-line, and extends from the tentorium to the crista galli. It contains in its attached margin the *superior longitudinal sinus ;* in its free margin its *inferior longitudinal sinus.* The superior sinus contains the *chordæ Willisii,* mechanical devices for strengthening and holding its walls together. Blood in this sinus flows from before backward ; the sinus is fed by veins which open into it from behind forward.

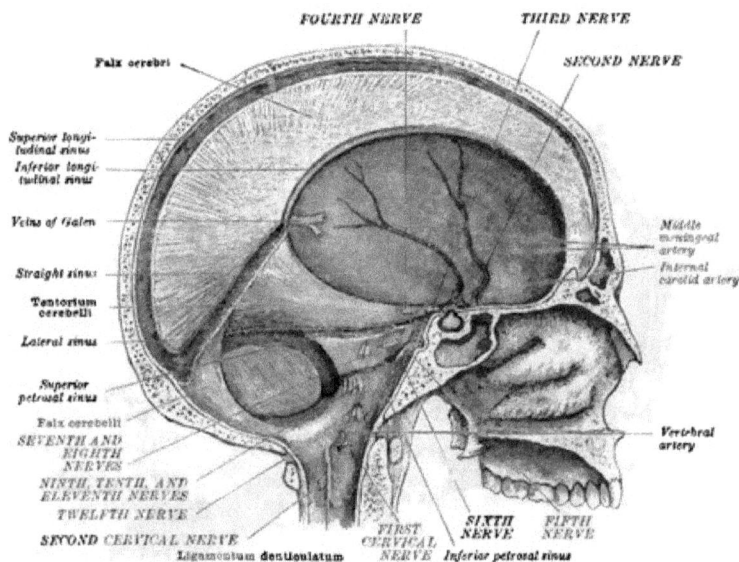

FIG. 82.—THE CRANIUM OPENED TO SHOW THE FALX CEREBRI, THE TENTORIUM CEREBELLI, AND THE PLACES WHERE THE CRANIAL NERVES PIERCE THE DURA MATER. (Sappey.)

The falx cerebelli is the smallest of the three large dural processes of the dura. It is between the cerebellar hemispheres. Its attached margin corresponds to the internal occipital crest of the occipital bone. It contains the *occipital sinus.*

The tentorium cerebelli is midway, both in location and size, between the two preceding processes. It is between the occipital lobes of the cerebrum and the cerebellum. It protects the latter from the weight of the former. It is attached to the horizontal part of the occipital crucial ridges posteriorly ; anteriorly and laterally, to the superior border of the petrous portion of the temporal bone and to the clinoid processes of the sphenoid bone. In the occipital attachment are found the *lateral sinuses ;* in the temporal, the *superior petrosal sinuses.*

The dura mater has sinuses located as follows (Figs. 82 and 83) :

1. Superior longitudinal, in the attached margin of the falx cerebri.
2. Inferior longitudinal, in the free margin of the falx cerebri.
3. Lateral, in the occipital attached margin of the tentorium cerebelli.

4. Superior petrosal, in the petrosal margin of the tentorium cerebelli.
5. Inferior petrosal, at the inner posterior border of the petrosa.
6. Occipital, in the attached margin of the falx cerebelli.
7. Cavernous, external to body of the sphenoid bone.
8. Circular, in diaphragma sellæ, around the pituitary body.
9. Sigmoid, on the inner part of the mastoid process.
10. Transverse, connecting the inferior petrosal; also called basilar.
11. Straight, at junction of falx cerebri and tentorium cerebelli.

The arteries of the dura mater are the great, small, anterior, and posterior meningeal; the anterior and posterior ethmoidal; and branches of the occipital and ascending pharyngeal. **The nerves** to the dura mater are the sympathetic, ophthalmic, hypoglossal, the pathetic, and the Gasserian ganglion.

As you have seen in your dissection, the **dural sinuses** are between the two layers of the dura mater. They are for the reception and transmission of venous

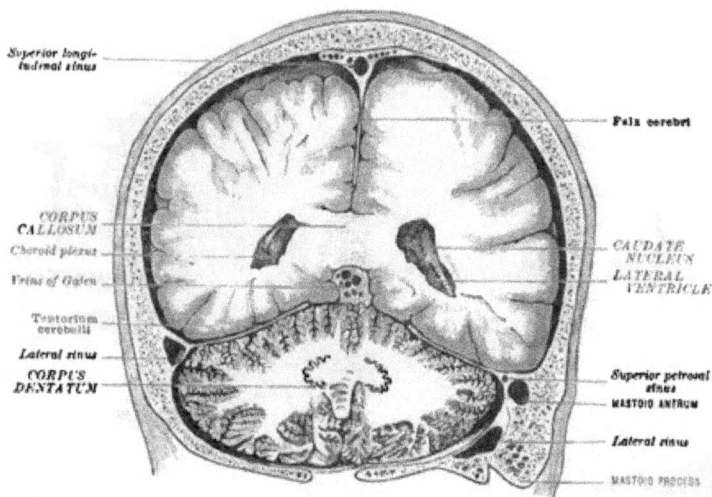

FIG. 83.—CORONAL SECTION OF THE HEAD PASSING THROUGH THE MASTOID PROCESS.
(From a mounted specimen in the Anatomical Department of Trinity College, Dublin.)

blood from the brain. The sinuses are lined by endothelial cells, continuous with the lining of the veins.

The arachnoid membrane you will know by the glistening appearance it gives the brain. Externally it is in relation with the *subdural space;* internally with the *subarachnoid space.*

Where is the subdural space located and what does it contain?

It is located between the dura mater and the arachnoid membrane, and contains a small amount of fluid. Call this fluid subdural.

The subarachnoid space is where located and contains what?

It is located between the arachnoid membrane and the pia mater. It contains the greater part of the cerebro-spinal fluid. Call this fluid subarachnoid.

Does the subdural fluid communicate with the subarachnoid fluid?

No; the two are separated from each other by the arachnoid membrane.

Is the pia mater also intact, so as not to permit communication between the sub-arachnoid fluid and the ventricular fluid in the ventricular cavities of the brain?

No; the pia mater has many slits, by which it permits the fluid in the sub-arachnoid space to communicate with other lymph spaces. The main ones are in the pia mater covering the fourth ventricle. They are three in number: Key, Retzius, Magendie. (Fig. 84.)

The arachnoid is covered externally with endothelial cells. It is joined to the pia by fibrous trabeculæ, called the subarachnoid tissue. At the base of the brain it is separated from the pia mater by the anterior and posterior subarachnoid spaces. The anterior is between the temporo-sphenoidal lobes of the cerebrum and in front of the pons Varolii; the posterior subarachnoid space is between the medulla and cerebellum. These two spaces contain the greater

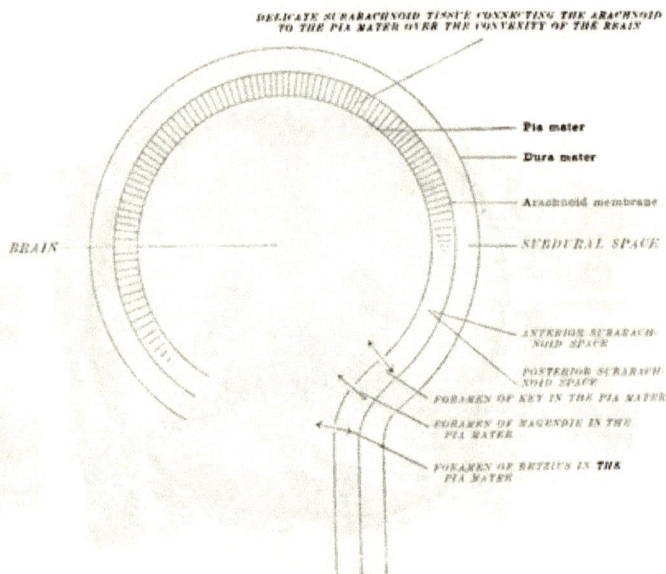

FIG. 84.—SCHEMATIC.

To show the location of the foramina of Key and Retzius in the pia mater, whereby the subarachnoid fluid in the subarachnoid space communicates with the fluid in the fourth ventricle.

part of the subarachnoid fluid. The posterior space communicates with the fourth ventricle of the brain through the pia mater by the foramina of Key and Retzius. (Fig. 84.)

Where is the cerebro-spinal fluid found and what are its uses?

The subarachnoid or cerebro-spinal fluid is found in the subarachnoid spaces of the brain and cord and in the ventricular cavities of the brain. Its use is to mechanically protect the nerve centres from shock, and to fill up space, as fat does in some other parts of the body. Its amount is estimated at less than two ounces. The cerebro-spinal fluid of the brain and cord communicate.

Explain the special features of the subarachnoid space.

As shown in figure 84, the greater part of the space between the arachnoid membrane and the pia mater is occupied by delicate connective tissue. It is impossible to separate the two membranes whenever this is the case. At the base of the brain you will see some well-defined spaces on removing the brain.

The Pacchionian glands will be seen as little stubby hairs or even delicate enlargements on the arachnoid membrane, in the neighborhood of the superior longitudinal sinus. They are enlarged *villi* of this membrane. They often perforate the dura mater. Occasionally they will be found to perforate the bones of the skull. Their use is not positively known; but they seem to transmit subarachnoid fluid into the superior longitudinal sinus.

Of what is the pia mater composed?

It is composed of vessels and connective tissue.

Where is the pia mater found in practical dissection on the cadaver?

The pia mater is found: (1) Covering the surface of the brain; (2) in the interior of the brain, in the third and lateral ventricles, as velum interpositum and choroid plexus; (3) forming the tela choroidea inferior along the roof of the fourth ventricle. This latter is perforated by three foramina, by which fluid in the ventricles communicates with fluid in the subarachnoid space. (Fig. 84.)

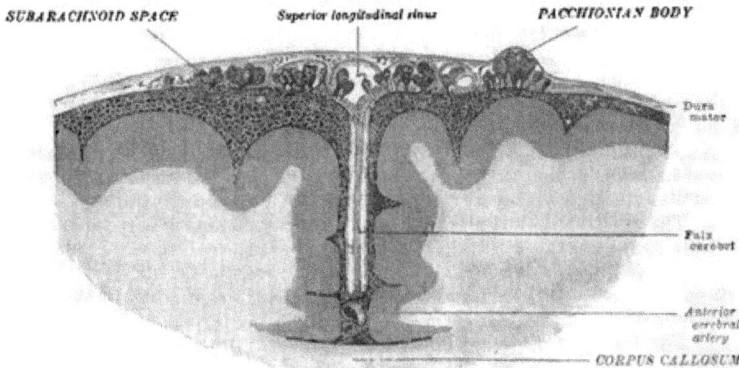

FIG. 85.—CORONAL SECTION THROUGH THE GREAT LONGITUDINAL FISSURE, SHOWING THE MENINGES. (Key and Retzius.)

Distinguish between the dura mater surrounding the brain and the dura surrounding the spinal cord as follows:

1. Attachment to bone, firmly and laxly—periosteal function.
2. Dural processes, for supportive purposes.
3. Dural sinuses, for venous blood.
4. Pacchionian bodies; Pacchionian villi.

These four conditions are found in the cranial dura mater; the dura of the cord has neither attachments, processes, sinuses, nor Pacchionian bodies.

To remove a brain take the following steps: If the case is a post-mortem, then it is necessary to conduct your work so as to leave the subject in a presentable form. In this case make (1) an incision from ear to ear through the scalp, turn the front flap down over the face, the back one down onto the neck; (2) cut through the origin of the temporal muscle, along the complete temporal ridge. Turn this muscle and its heavy fascia down over the zygoma.

To Mark the Calvarium.—Make a line from one-half of an inch above the supraorbital arch on each side, around the head, passing through the external occipital protuberance. Caution: Avoid getting too low in front, lest your incision pass under the roof of the orbit. If you fail to bring your incision to the level of the external occipital protuberance you will have trouble in removing the cerebellum.

To Cut the Calvarium.—Saw in the line as indicated in the preceding paragraph. Avoid cutting too deeply in the temporal region. After you have completed the incision with the saw, take a chisel and hammer and test your work. If any places remain uncut, you can easily finish loosening the calvarium with the chisel. The calvarium must be made loose without cutting through the dura mater. Grasp the cut edge of bone above the orbits, and, while an assistant presses down on the chin, pull the calvarium backward; as it comes off you will hear a noise, produced by the separation of the dura from the bone. On a previous page you have learned that the dura has attachments to the interior of the cranium.

Now make the following observations:

1. Note on the inner surface of the calvarium the **meningeal grooves** for the meningeal arteries. These arteries are between the dura and bone. They furnish blood to both dura and bone. They have nothing to do with the blood-supply to the brain proper.

2. The **Pacchionian bodies and villi** (Fig. 82) will be seen on the outer surface of the dura mater, in some cases where they have perforated this structure. They produce, by pressure atrophy, the Pacchionian depressions in the inner table of the calvarium. They are a villous product of the arachnoid. Their function is not perfectly understood. They are found in great abundance in the region of the superior longitudinal sinus, and even in the sinus itself.

3. You will note the **great or middle meningeal artery**. It consists of several branches, which all spring from one parent stem in the middle fossa of the skull. It lies in the connective tissue that held the dura to the bone.

4. **The superior longitudinal sinus** (Figs. 82 and 83) lies in the mid-line. Cut this open from end to end. Take note of the large amount of clotted blood it contains in its posterior half. Make note also of the chordæ Willisii in the bottom of the sinus. Clear out the clotted blood, and these cords of Willis will come into view.

5. The **Cerebral Veins.**—Cut through the dura, with scissors, one inch external to the sinus just opened, from before backward, parallel with the sinus. Next make a lateral incision from the first incision to the ear. Do this on both sides. Turn the flaps outward. Turn the dura up over the median line, and see the veins opening into the sinus from behind forward.

Origin of the Cerebral Veins.—These veins take blood from the part of the brain supplied by the cortical system given off from the circle of Willis. They discharge the blood into the several dural sinuses.

6. **The subdural space** is exposed now. It is between the dura and arachnoid. It contains a small amount of fluid for lubrication—fluid derived from a source explained in a previous paragraph. This space contains (1) the Pacchionian bodies; (2) the veins to the sinuses; (3) the cranial nerves prior to their emergence.

7. **The falx cerebri** (Fig. 82) you will see occupying the great longitudinal fissure between the cerebral hemispheres. Cut its attachment to the crista galli in front, and as you pull it gently backward, cut the remaining veins opening into the sinus. In the free margin of the falx cerebri is the *inferior longitudinal sinus.*

8. **The arachnoid** membrane is the smooth, glistening, transparent structure that you now see covering the brain. It is transparent, for you see the vessels of the pia mater through it. It is smooth on its outer surface, for it is lubricated by the subdural lymph. It is so very delicate you would scarcely suspect its presence. Under it is the subarachnoid space, also containing fluid.

9. **The Pia Mater.**—As you have just observed, you can not remove the arachnoid from the pia; still, you can see perfectly the vessels of the pia mater through the arachnoid. The presence of the subarachnoid tissue prevents separation of these two structures. (Fig. 84.) The pia mater is composed, as you can see on your dissection, of (1) arteries from the circle of Willis, taking blood to the brain; (2) veins attending these arteries, and finally opening into the dural sinuses; (3) connective tissue holding these arteries and veins, as the warp of a carpet holds the woof.

Remember, the individuality of the pia mater depends not on arteries alone, not on veins alone, not on connective tissue alone; but on all three woven together, the connective tissue being the warp, the vessels the woof.

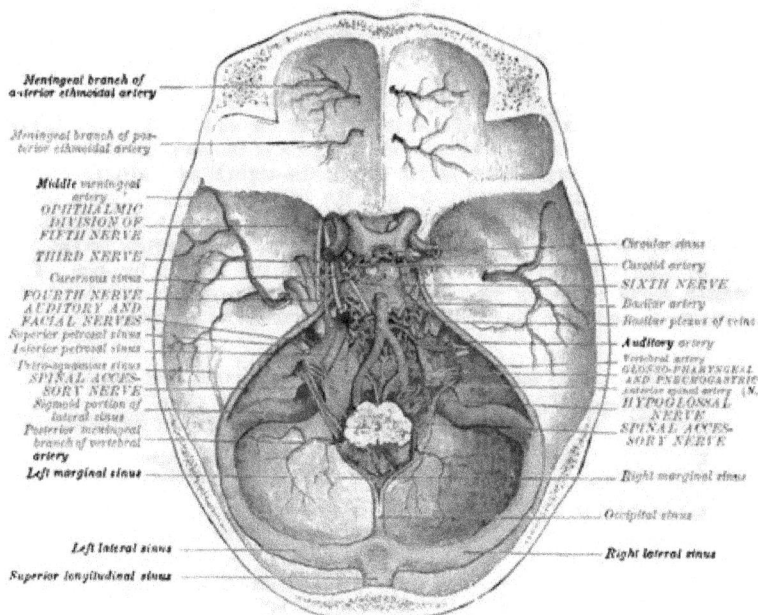

FIG. 86.—THE VENOUS SINUSES.
(From a dissection by W. J. Walsham in St. Bartholomew's Hospital Museum.)

Structures You See on the Base.—Place the cadaver so the head will hang over the end of the table. This position will permit the brain to gravitate to a slight degree out of its bed. Gently shake the brain by rocking the head from side to side. Notice now the following structures:

1. The frontal lobe of the cerebrum in the anterior fossa.
2. The olfactory lobe, on the under surface of the frontal lobe.
3. The optic chiasm and optic nerves.
4. The internal carotid artery—its cerebral stage.
5. The third cranial nerve—motor oculi.
6. The tentorium cerebelli attached to the petrosa.

Hints on Dissection of the Foregoing Structures.—Gently retract the frontal lobes and you will see the optic nerves covered by their prolongation of arachnoid. Remove this with one brainward sweep of the forceps, and expose the optic chiasm. Behind this see the infundibulum, attached above to the brain and below to the pituitary body. See it passing through a hole in the diaphragma sellæ.

The internal carotid artery will be seen to the outer side of the optic nerve. Do not call this large artery the ophthalmic. Cut the optics behind the chiasm. Cut the internal carotid artery close to the anterior clinoid process. You will now expose the third nerve by breaking the prolongation of arachnoid about it.

The fourth cranial nerve lies under the free margin of the tentorium. You will see the tentorium attached to the superior border of the petrosa in front. Cut along this border on both sides. Now gently retract the brain. The fourth nerve will pass to the outer side of the posterior clinoid process and disappear under the dura into the cavernous sinus.

The fifth nerve will now be seen crossing the superior border of the petrosa, near the apex. It also disappears under the dura through the trigeminal notch.

The sixth nerve will be seen going through the dura one inch below the dorsum sellæ. It passes beneath the dura to gain the cavernous sinus with the third and fourth nerves, and the ophthalmic branch of the fifth nerve.

Find now the seventh or facial nerve, the eighth or auditory nerve, and the auditory artery passing through the internal auditory meatus on the posterior surface of the petrosa. Note the funnel of arachnoid that accompanies these structures to the foramen. The ninth, tenth, and eleventh nerves will now be seen leaving the cranial cavity through the jugular foramen. The twelfth nerve leaves as two twigs: after passing through the dura they unite as one nerve, and leave the skull by the anterior condyloid foramen.

FORAMINA AT THE BASE OF THE SKULL AND THE STRUCTURES THEY TRANSMIT.

Foramen Cæcum,	Transmits an emissary vein.
Nasal Slit,	Transmits the nasal branch of the ophthalmic nerve.
Olfactory Foramina,	Twenty in number, transmit the olfactory nerves.
Optic Foramina,	Transmit the optic nerves and the ophthalmic artery.
Sphenoidal Fissure,	Transmits third, fourth, sixth, and first divisions of the fifth nerve, the ophthalmic vein, and sympathetic nerves.
Foramen Rotundum,	Transmits the second or superior maxillary division of the fifth nerve.
Foramen Ovale,	Transmits the third division of fifth nerve and small meningeal artery.
Foramen Spinosum,	Transmits the great or middle meningeal artery.
Foramen, Sphenotic,	Transmits internal carotid and the facial and carotid petrosals for Vidian nerve.
Hiatus Fallopii,	Transmits petrosal branch of the Vidian nerve.
Petrosal Foramen,	Transmits the smaller petrosal nerve.
Carotid Canal,	Transmits internal carotid artery and sympathetic nerves.
Internal Auditory Meatus,	Transmits seventh and eighth nerves and auditory artery.
Jugular Foramen,	Transmits jugular vein, ninth, tenth, and eleventh nerves.
Anterior Condyloid Foramen,	Transmits the hypo-glossal nerve—the twelfth nerve.
Foramen Magnum,	Transmits (1) spinal cord and its meninges; (2) the vertebral arteries and their sympathetic nerves; (3) the spinal accessory nerve—the eleventh nerve. Note the eleventh nerve coming up through the foramen, passing forward, and leaving the cranium with the ninth and tenth nerves by the jugular foramen.

The Cavernous Sinus. — The sinus is located between the apex of the petrous portion of the temporal bone and the sphenoidal fissure. Note that all the nerves you have just seen in the table passing through the sphenoidal fissure also pass through the cavernous sinus.

The sinus contains the following structures:

1. The venous blood common to dural sinuses.
2. The cavernous stage of the internal carotid artery.

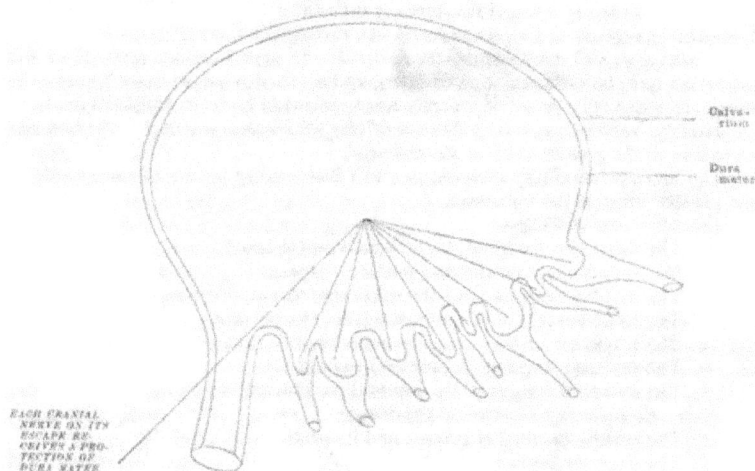

EACH CRANIAL
NERVE ON ITS
ESCAPE RE-
CEIVES A PRO-
TECTION OF
DURA MATER

Calva-
rium

Dura
mater

FIG. 87. — SCHEMATIC.
To show the dura mater forming sheaths for the nerves as they leave the cranium.

3. The sympathetic plexus of nerves on the common carotid artery.
4. The ophthalmic division of the fifth cranial nerve.
5. The motor oculi, the third cranial, nerve.
6. The patheticus, the fourth cranial, nerve.
7. The abducens, the sixth cranial, nerve.

Dissection. — Remove the dura mater. Take hold of it in the region of the frontal bone, and pull it back. It will come off quite readily. Locate the

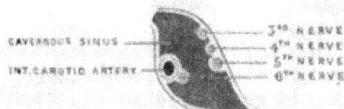

CAVERNOUS SINUS
INT. CAROTID ARTERY

3RD NERVE
4TH NERVE
5TH NERVE
6TH NERVE

FIG. 88. — RELATION OF THE VARIOUS STRUCTURES PASSING THROUGH THE CAVERNOUS SINUS.

internal carotid artery next to the sphenoid wall of the sinus. The small nerve on the artery is the sixth cranial. You will find the others on the outer wall, from above downward: third, fourth, and fifth.

The Gasserian Ganglion. — Remove in the same way the dura mater from the middle fossa as you did from the anterior, and you will expose the Gasserian ganglion in a depression on the anterior surface of the petrous portion of the temporal bone near the apex. The ganglion lies between two layers of subdural connective tissue, called Meckel's fascia. The relation of the ganglion to the internal carotid is of great importance in surgical operations on the ganglion.

In figure 86 the artery will be seen behind the Gasserian ganglion, the third, fourth, sixth, and ophthalmic division of the fifth nerve, and Meckel's fascia. In other words, turn all these nerves aside and you will see the Gasserian ganglion between two layers of subdural connective tissue called Meckel's fascia. The space between these two layers of fascia is called Meckel's space. The space through which the fifth nerve passes, between the tentorium and the superior border of the petrosa, near the apex of the bone, is called the trigeminal notch. Turn the Gasserian ganglion out of its bed and cut through the lower layer of Meckel's fascia, and you will find the internal carotid. In operations for removal of the Gasserian ganglion, this relation of the ganglion to the artery must be borne in mind, since the only protection to the artery is a thin layer of Meckel's fascia.

Find the superior maxillary division of the fifth nerve leaving by the foramen rotundum in the greater wing of the sphenoid.

The inferior maxillary division you will find leaving by the foramen ovale in the greater wing of the sphenoid.

Dissection and study of—

1. The Gasserian ganglion, its branches and relations.
2. Meckel's fascia, anterior and posterior layers.
3. The middle, or great, and the other meningeal arteries.
4. The facial nerve, the seventh cranial in the petrosa.
5. The tympanic cavity, the tympanum or middle ear.
6. The petrosal stage of the internal carotid artery.
7. The cavernous stage of the internal carotid artery.

Locate the following anatomical structures :

1. The middle meningeal groove and its artery.
2. The superior border of the petrosa.
3. The inferior border of the petrosa.
4. The posterior surface of the petrosa.
5. The anterior surface of the petrosa.
6. The internal auditory meatus with the seventh and eighth nerves.
7. The hiatus Fallopii and great petrosal nerve.
8. The tegmen tympani or roof of the tympanum.
9. The foramen lacerum medium. Sphenotic fissure.
10. The foramen spinosum in the alar spine of sphenoid bone.
11. The sphenoidal fissure in the sphenoid bone.
12. The foramen ovale for third division of the fifth nerve.
13. The foramen rotundum for the second division of the fifth nerve.
14. The anterior, middle, and posterior clinoid processes.

Describe the middle meningeal artery.

It is a branch of the first stage of the internal maxillary artery. External to the cranium it lies behind the internal pterygoid muscle. It passes between the two roots of origin of the auriculo-temporal nerve. It enters the cranium by the foramen spinosum, in the great wing of the sphenoid bone, and ramifies between the dura mater and the bone, both of which it supplies with blood. Its two accompanying veins are tributary to the internal maxillary vein. The middle meningeal artery divides in anterior and posterior branches which supply the greater part of the bones of the cranium. (Fig. 81.)

This vessel gives off the following minor branches :

1. Small branches to the Gasserian ganglion and its adjacent dura mater.
2. A branch through the hiatus Fallopii to the seventh nerve and tympanum.
3. Temporal branches perforating the bone to the temporal fossa.
4. Orbital branches passing through the sphenoid fissure to the orbit.

From what other sources does the dura receive its blood-supply?

The anterior meningeal arteries are in the anterior fossa of the skull. They are derived from the ethmoidal branches of the ophthalmic artery, and from the cavernous stage of the internal carotid. The small meningeal is a branch of the first stage of the internal maxillary. It enters the cranium by the foramen ovale with the third division of the fifth nerve. The posterior meningeal arteries come from the vertebral, occipital, and ascending pharyngeal.

Describe the cavernous stage of the internal carotid artery.

It is in the cavernous sinus, but covered by the endothelial lining of the sinus. (Fig. 86.) From the inner end of its petrosal stage the artery ascends to the posterior clinoid process; it then lies by the side of the body of the sphenoid, and then gently curves upward between the middle and anterior clinoid processes, and, lastly, curves backward and perforates the roof of the sinus. (Fig. 86.) On the outer side of the artery is the sixth nerve. It is surrounded by sympathetic nerves.

In this stage the vessel gives off the following branches :

1. Branches to the walls of the cavernous sinus.
2. Branches to the Gasserian ganglion.
3. Anterior meningeal branches to the dura.
4. Branches to the pituitary body in the sella turcica.

Describe the petrosal stage of the internal carotid artery.

It has in this stage two parts : An ascending part, in front of the tympanum and internal ear ; a horizontal part, which you will see on turning the Gasserian ganglion backward. On the artery you will see numerous sympathetic nerves,—the carotid sympathetic plexus,—which are ascending branches of the superior cervical ganglion. The carotid, Vidian, and tympanic are its branches.

What are Meckel's fascia, Meckel's space, Meckel's cave, and the trigeminal notch in the literature of the trigeminal nerve?

Meckel's fascia is the subdural connective tissue on and under the Gasserian ganglion. Meckel's space is between the two layers of dura, external to the cavernous sinus. Meckel's cave is a depression on the anterior surface of the petrosa for the Gasserian ganglion. The trigeminal notch is an osteological term ; it is a depression in the upper border of the petrosa, near the apex, in which the fifth nerve rests before expansion into the Gasserian ganglion. This notch is converted into the trigeminal foramen by the tentorium cerebelli.

Describe the Gasserian ganglion.

This ganglion lies in Meckel's cave, on the anterior surface of the petrosa. It is simply an enlargement on the anterior or sensory root, as it is called, of the fifth cranial nerve. Its posterior relations are the foramen lacerum medium, the great petrosal nerve, and the horizontal part of the petrosal portion of the internal carotid artery. Anterior to it is the dura mater. The ganglion gives off or receives the following branches :

1. Filaments from the carotid plexus of the sympathetic.
2. Nerves to the dura of the middle fossa and tentorium.
3. The ophthalmic nerve to the orbit and nose.
4. The superior maxillary nerve to the upper jaw.
5. The inferior maxillary nerve to the mandibular region.

What is the function of the ganglion and the branches given off therefrom?

It confers the quality of common sensation on all parts to which it is distributed. Behind the ganglion lies a nerve that supplies the muscles of mastication. This is the so-called motor root of the fifth nerve. This nerve has nothing to do with the ganglion. It leaves the base of the skull through the foramen ovale with the inferior maxillary sensory part of the ganglion. (Fig. 53.)

The tympanum or middle ear may be very satisfactorily dissected by the student. Its small size is no argument against careful study of its medical and surgical importance, its contents, and its relations, since it is of easy access for dissection, and resolvable into the form of a geometrical cube for purposes of aiding the memory and facilitating a comprehension of the relative position of those anatomical structures that make the middle ear of such great importance. The student should not undertake a dissection of this part until he has a clear-cut idea of the position, relation, and importance of each wall of the cube. If he begins his dissection knowing that the drum of the ear forms the outer wall of the tympanum proper, that the Eustachian tube and mastoid antrum communicate with the tympanum by the anterior and posterior walls respectively, then he will surely find these structures and think of them in relation to the walls of a cube. To say nothing of the part played by the tympanum in hearing, this region is of interest and importance for the following reasons :

1. The temporo-sphenoidal lobe of the brain lies on its roof.
2. The floor of the tympanum lies on the jugular fossa.
3. The drum of the ear and chorda tympani are on the external wall.
4. The mastoid antrum and cells are behind it.
5. The internal carotid artery is in front of it.
6. The facial nerve skirts two sides of it.
7. The internal ear is in close relation with it.
8. The drum of the ear may become ruptured.
9. The interossicular joints may become ankylosed.
10. Pus may form in the tympanum.

Size and Subdivisions of the Tympanum.—The tympanum is about one-half of an inch in height and one-sixth of an inch in width. It is prolonged forward as the Eustachian tube and backward as the mastoid antrum, so its length is difficult to determine : for practical purposes one-half of an inch is near enough. The subdivisions are the attic and tympanic cavity proper. The latter is quite narrow, and has the membrani tympani or drum as its outer wall ; the former is broader, contains most of the bones of hearing, and has a part of the temporal bone as its outer wall. Remember, the tympanum is not horizontal, but its anterior end slopes downward, forward, and inward to the Eustachian tube ; its posterior end slopes upward, backward, and outward to the *mastoid antrum.*

Study of the tympanum in which this cavity is compared to a box. (Fig. 91.)

1. *The roof of the tympanum* is a thin plate of bone, the tegmen tympani, separating the tympanum from the middle fossa at the base of the skull. It is perforated by foramina that transmit the petrosal branches of the seventh nerve.

2. *The floor* separates the tympanum from the jugular fossa, in which fossa are the internal jugular vein, and the ninth, tenth, and eleventh cranial nerves. In this floor is an aperture, through which Jacobson's nerve, the tympanic branch of the glosso-pharyngeal, passes to form the tympanic plexus.

3. *The outer wall* of the tympanum is formed by the drum below and the squamosa above, these two structures corresponding to the tympanum proper, and the attic respectively. This outer wall is pierced by the following openings : (1) *The iter chordæ posterius,* by which the chorda tympani nerve, a branch of the seventh cranial nerve, enters the tympanum ; (2) *the iter chordæ anterius,* by which the chorda tympani leaves the tympanum ; (3) *the Glaserian fissure,* through which pass the tympanic branch of the internal maxillary artery, the processus gracilis of the malleus, and the laxator tympani muscle.

4. *The inner wall,* you will remember, is called the fourth surface of the petrosa, in Morris' analysis of this part of the temporal bone. This inner wall contains : (1) A ridge of bone covering the seventh nerve in its passage through the tympanum, as heretofore explained ; (2) the fenestra ovalis, leading into the vestibule, and to which is attached the base of the stapes, through

the medium of a periosteal membrane ; (3) the promontory, formed by a turn of the cochlea, and covered by the tympanic plexus ; the promontory is below and in front of the fenestra ovalis ; (4) the fenestra rotunda, covered by the membrana secundaria, and communicating with the scala tympani of the cochlea ; (5) the pyramid, from whose summit emerges the tendon of the stapedius. A special branch of the seventh nerve pierces the pyramid for the supply of the stapedius muscle.

5. *The posterior wall* communicates with the mastoid antrum. As indicated in a previous paragraph, the mastoid antrum is the backward continuation of the upper part of the tympanic cavity, which is called the attic.

6. *The anterior wall of the tympanum* presents but one thing for examination— the canalis musculo-tubarius. This canal is divided into an upper and a lower compartment by a horizontal lamina of bone, called the processus cochleariformis. The upper compartment lodges the tensor tympani muscle ; the lower

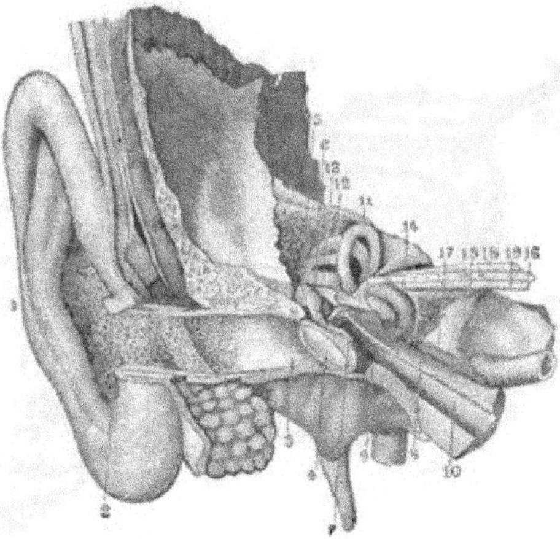

FIG. 89.—EAR AND TYMPANUM.

1. Pinna, or auricle. 2. Concha. 3. External auditory canal. 4. Membrana tympani. 5. Incus. 6. Malleus. 7. Manubrium mallei. 8. Tensor tympani. 9. Tympanic cavity. 10. Eustachian tube. 11. Superior semicircular canal. 12. Posterior semicircular canal. 13. External semicircular canal. 14. Cochlea. 15. Internal auditory canal. 16. Facial nerve. 17. Large petrosal nerve. 18. Vestibular branch of auditory nerve. 19. Cochlear branch.

is the osseous part of the Eustachian tube. These structures can be readily seen in dissection. A common broom straw or hairpin may be thrust through the lower part of the canalis musculo-tubarius as a guide.

Before you attempt to dissect the **seventh nerve** in its tortuous canal through the petrous portion of the temporal bone, study well the following branches this nerve gives off between the internal auditory meatus and the stylo-mastoid foramen. In the auditory canal a short communicating branch, rather large and fatty, extends from the seventh to the eighth nerve. This is called the *portio inter duram et mollem.* The bony canal occupied by the seventh nerve, from the time it leaves the eighth nerve until it emerges from the stylo-mastoid foramen, is called the *aqueductus Fallopii,* or the *facial canal.* In this canal the following branches are given off, as you may see by figure 90.

1. From the geniculate ganglion, the *great superficial petrosal nerve ;* this passes through the cartilage in the foramen lacerum medium, and is joined by

some filaments from the carotid sympathetic plexus, called the great deep petrosal nerve. The union of these two makes a compound nerve called the Vidian
nerve. This Vidian, accompanied by vessels of like name, passes through the
Vidian canal and joins the ganglion of Meckel, conferring on the same motion
and sympathetic qualities. You will recall the fact of osteology that the Vidian
canal is at a junctional area formed by the greater ala of the sphenoid bone, its
lingula and pterygoid process. Meckel's ganglion derives its sensory root from
the spheno-palatine branches of the superior maxillary division of the fifth
cranial nerve.

2. From the geniculate ganglion, the *small superficial petrosal nerve*. This
nerve receives a communicating branch from the glosso-pharyngeal nerve and
leaves the cranium by the canalis innominatus in the sphenoid bone, between the
foramen ovale and foramen spinosum, and joins the otic ganglion on the inner
surface of the third division of the fifth nerve, near the foramen ovale.

3. From the geniculate ganglion, the *external petrosal.* This nerve joins the

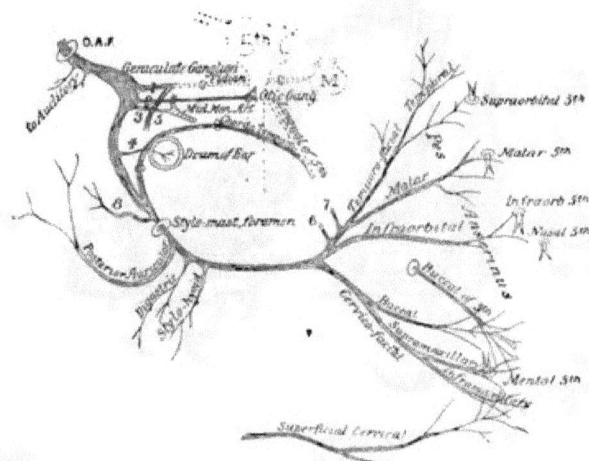

sympathetic plexus on the great or middle meningeal artery, near the foramen
spinosum.

4. From the nerve-trunk, below the geniculate ganglion, the *tympanic branch*
to the stapedius—the smallest muscle in the body. This nerve pierces the pyramid, on the inner surface of the tympanum.

5. A branch from the carotid sympathetic plexus joins the *great superficial
petrosal* just outside the cranium, to form the Vidian nerve.

6. From the main trunk of the nerve (8 in Fig. 90) a branch is given off, just
above the stylo-mastoid foramen, to communicate with the auricular branch
of the vagus nerve.

7. From the main nerve-trunk, just above the stylo-mastoid foramen, the *chorda
tympani nerve* is given off. This nerve passes upward through a bony canal, the
iter chordæ posterius, to the tympanum. It passes forward under the mucous
membrane of the drum of the ear internal to the manubrium of the malleus. The
nerve leaves the tympanum by the iter chordæ anterius, or canal of Huguier.
It communicates with the otic ganglion, passes internal to the external pterygoid

muscle, and joins the gustatory nerve. It is distributed to the submaxillary and sublingual glands and to the dorsum of the tongue.

Dissection of the Seventh Nerve in the Petrosa and of the Tympanum.—Locate the following anatomical structures:

1. The superior border of the petrosa.
2. Anterior and posterior surfaces of the petrosa.
3. The trigeminal notch near apex of petrosa.
4. The cave of Meckel on anterior surface of petrosa.
5. The horizontal part of the carotid canal and its artery.
6. The great petrosal nerve in the hiatus Fallopii.
7. The small petrosal nerve in its small canal.
8. The middle meningeal artery and foramen spinosum.
9. The tegmen tympani or tympanic roof.
10. The internal auditory meatus on posterior surface of the petrosa.

With a small chisel and mallet gently remove the tegmen in such a manner as not to destroy the delicate petrosal nerves on the anterior surface of the petrosa, and then you can look down into the tympanum, or middle ear, the roof now having been removed (tegmen). Now draw a line from the internal auditory meatus to where the great petrosal nerve is seen emerging. This line will locate the course of the seventh nerve in its canal to the tympanum. Now begin at the internal end of the line just drawn and chip off the bone until you have completely exposed the seventh nerve to the tympanum, where you will see the geniculate ganglion giving off petrosal nerves.

The tympanum may be represented as a box with an inner and an outer wall, a roof and a floor, an anterior and a posterior end. The seventh nerve approaches the tympanum, at a point corresponding to the junction between the roof and inner wall, near the front end of the box. The seventh nerve having gained this point, now passes backward in the angle between the roof and inner wall to the corner of the box between the inner wall and posterior end of the box. The nerve here makes a gentle turn and passes down this corner to the floor of the box, and subsequently emerges from the base of the skull through the stylo-mastoid foramen. You may see the geniculate ganglion of the nerve just where the nerve is making the bend to enter the tympanum. Now locate the nerves as described previously, and consult figure 90.

BLOOD-SUPPLY TO THE BRAIN.

Examine the base of the brain. You notice very distinctly the arachnoid membrane quite appreciably separate from the pia. In other words, here you see the subarachnoid space. Now remove carefully the arachnoid and dissect the circle of Willis. You will note the following:

1. The *vertebral arteries* unite to form the basilar artery.
2. *The basilar* divides into the two posterior cerebral arteries.
3. *The internal carotid artery* gives off three large arteries (Fig. 91): (1) The *anterior cerebral,* in the great longitudinal fissure; (2) the *middle cerebral,* in the fissure of Sylvius on the central lobe; (3) the *posterior communicating,* to the posterior cerebral.

The arachnoid membrane, it will be observed, stretches across from frontal to temporo-sphenoidal lobe, so as almost to obscure the fissure of Sylvius. Cut this membrane with the scissors, gently part the lobes, and you will see the island of Reil, on which is the middle cerebral artery with its vein.

The anterior communicating artery is very short, one-eighth of an inch being an average length in a number of cases. You will find it extending from one anterior cerebral artery to the other

The ganglionic or terminal end arteries will be seen in very great numbers on lifting the middle, posterior, and anterior cerebral arteries. They are very minute. Look closely, and you will see them passing through minute perforations to the interior of the brain through the perforated spaces. (Fig. 92.)

The interpeduncular space has (1) boundaries; (2) contents. The contents form the floor of the third ventricle. The boundaries are important structures on the base of the brain which I wish you to study carefully. (Fig. 92.)

The Boundaries of the Interpeduncular Space.—(1) *Antero-laterally,* optic chiasm and optic tracts; (2) *postero-laterally,* pons Varolii and crura cerebri.

Contents of Interpeduncular Space.—(1) The tuber cinereum; (2) infundibulum, pituitary body; (3) the corpora albicantia, part of the fornix; (4) posterior

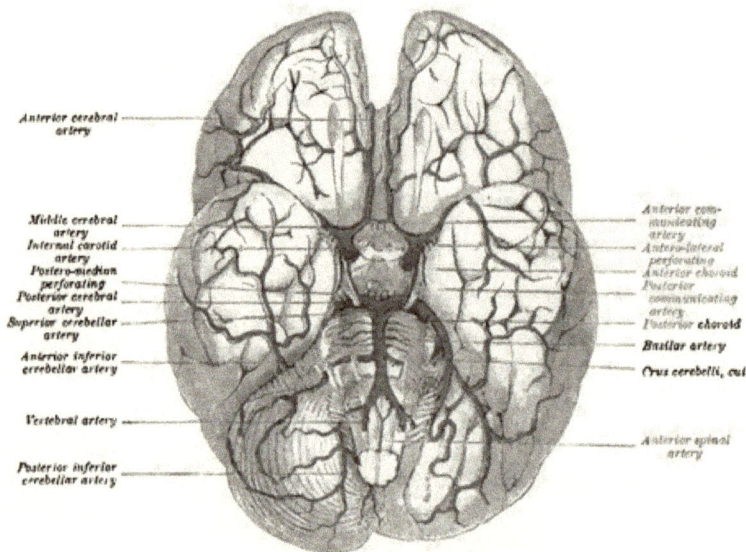

FIG. 91.—THE ARTERIES OF THE BRAIN.
(From a preparation in the Museum of St. Bartholomew's Hospital.)

perforate space, for terminal end arteries; (5) the third cranial nerve, motor oculi nerve.

Notice in particular:

1. The optic tracts crossing the crura cerebri.
2. The third nerve, motor oculi, between the crura cerebri.
3. The terminal end arteries from the circle of Willis. (Fig. 91.)
4. The basilar artery on the body of the pons Varolii. (Fig. 91.)
5. The fourth nerve, coming out between the pons and crus cerebri.
6. The fifth nerve, coming from the side of the pons Varolii.
7. The sixth nerve, coming through between the pons and medulla.
8. The seventh and eighth nerves, between the olivary and restiform bodies.
9. The ninth, tenth, and eleventh nerves, between the olivary and restiform bodies.
10. The twelfth nerve, between the pyramid and olive.

Two systems of circulation are given off from the circle of Willis: (1) *The ganglionic ;* (2) *the cortical.* You have just seen the ganglionic or end arteries entering the brain by the perforated spaces. The cortical arteries and their veins form the woof of the pia mater, the warp of which is connective tissue.

1. *From what source does the brain receive its blood?*
The brain is supplied with blood by the circle of Willis.

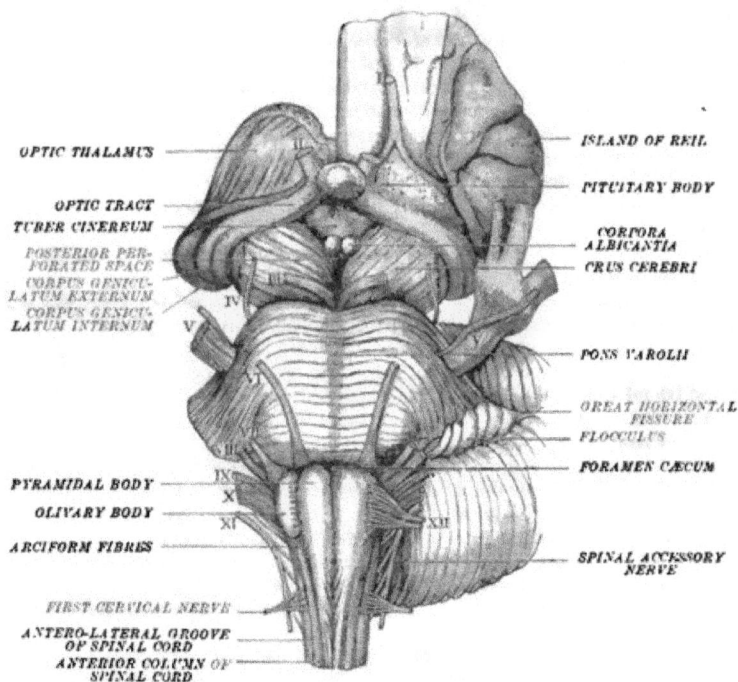

OPTIC THALAMUS

OPTIC TRACT
TUBER CINEREUM

POSTERIOR PER-
FORATED SPACE
CORPUS GENICU-
LATUM EXTERNUM
CORPUS GENICU-
LATUM INTERNUM

PYRAMIDAL BODY
OLIVARY BODY
ARCIFORM FIBRES

FIRST CERVICAL NERVE
ANTERO-LATERAL GROOVE
OF SPINAL CORD
ANTERIOR COLUMN OF
SPINAL CORD

ISLAND OF REIL

PITUITARY BODY

CORPORA
ALBICANTIA
CRUS CEREBRI

PONS VAROLII

GREAT HORIZONTAL
FISSURE
FLOCCULUS
FORAMEN CÆCUM

SPINAL ACCESSORY
NERVE

FIG. 92.—SURFACE ORIGIN OF THE CRANIAL NERVES. (After Allen Thomson.—Quain.)

2. *How is this circle formed?*
It is formed by an anastomosis between branches of the internal carotid and vertebral arteries.

3. *How does the internal carotid reach the brain?*
It comes through the carotid canal and sphenotic foramen, and gives off: (1) The anterior cerebral; (2) the middle cerebral; (3) the posterior communicating.

4. *How does the vertebral artery reach the brain?*
It passes through the foramina in the transverse processes of the cervical vertebræ and the foramen magnum. It divides into the two posterior cerebral arteries.

5. *How is the blood returned from the brain?*
The blood from the ganglionic system is returned to the straight sinus by

the veins of Galen. (Fig. 82.) The blood from the cortical system is returned by various veins to the several dural sinuses.

6. *What becomes of the blood in the sinuses?*

The sinuses converge to form the internal jugular vein. This vein is made up at the jugular foramen, and leaves it in company with the ninth, tenth, and eleventh cranial nerves.

7. *Do the branches given off from the ganglionic system communicate with one another or with branches of the cortical system?*

No.

8. *Where do the branches forming the ganglionic system take their origin, and how are they named?*

They originate from the circle of Willis and from its primary branches for an inch beyond the circle. They are named

Antero-median, from the anterior communicating artery.

Antero-lateral, from the middle cerebral artery.

Postero-median, from the posterior cerebral artery.

Postero-lateral, from the posterior cerebral artery.

9. *What is the most important function of this system of circulation?*

To supply blood to the basal ganglia, the most important part of the brain.

10. *Name the most important branch of the ganglionic system, and give its surgical importance*

The lenticulo-striate artery. It is the largest, and most frequently the seat of embolism or hemorrhage. It passes through the internal capsule.

11. *Do all the branches of the ganglionic system supply basal ganglia?*

No ; the anterior and posterior choroid branches supply the choroid plexuses of the lateral ventricles and some minor adjacent parts.

12. *Has any part of the ganglionic system anything to do with the formation of the pia mater?*

As mentioned in a previous paragraph, the anterior and posterior choroid arteries are branches of the ganglionic system, but they do not supply basal ganglia ; they supply the choroid plexuses, in the lateral ventricles, and these plexuses, with the velum interpositum, are extensions from the general pia mater on the surface of the brain to equally superficial parts of the brain, apparently, but not really, on the interior of the brain.

13. *Are the terms pia mater and cortical system of circulation, for practical purposes, synonymous?*

Yes ; the pia mater consists of vessels, arteries and veins, and connective tissue ; the arteries, for the most part, originate beyond the confines of the ganglionic system. The pia mater is applied to every part of the outer surface of the brain ; it reaches into the fissures and sulci. It forms the velum interpositum and the choroid plexuses.

14. *How are the cerebral veins classified?*

They may be grouped as ganglionic, basilar, and cortical. (1) The central or ganglionic veins come together to form the veins of Galen, having gathered in their course the veins of the choroid plexuses and velum interpositum. The veins of Galen, from the two hemispheres, unite to form the common vein of Galen, a trunk about one-half of an inch in length ; this, as seen previously, opens into the straight sinus. (2) The basilar veins collect the blood from the under surface of the cerebrum, and feed the petrosal, cavernous, and lateral sinuses. (3) The cortical veins gather the blood from the outer and inner surfaces of the hemispheres and feed the superior and inferior longitudinal sinuses in the attached and free margins of the falx cerebri respectively.

Find the following fissures and their contents:

1. *The great longitudinal fissure* contains: (1) The falx cerebri with its superior and inferior longitudinal sinuses; (2) the corpus callosum, on which you may see the anastomosis between the anterior and posterior cerebral arteries.

2. *The great transverse fissure*, between the occipital cerebral lobes and the cerebellum. It contains the tentorium cerebelli.

3. *The fissure of Sylvius*, bounded below by the temporo-sphenoidal lobe; above by the frontal and parietal lobes. It contains: (1) The island of Reil, or central lobe of the cerebrum; (2) the middle cerebral artery and vein and their numerous branches.

4. *The fissure of Rolando* is between the ascending convolutions of the frontal and parietal lobes. Its surgical importance is due to the fact of the location of the cortical motor areas about this fissure.

5. *The occipito-parietal fissure* is very imperfect. As its name implies, it is between the parietal and occipital lobes.

Dissection of the brain does not belong to dissecting-room anatomy, hence the reader is referred to Morris for this very interesting and instructive part of anatomy.

THE ORBIT AND ITS CONTENTS.

For dissecting-room purposes, the following points seem of practical or rather demonstrable importance:

1. **Geometry of the Orbit.**—Roof, floor, outer wall, inner wall, base, apex, angles, and cavity

The *roof* is of importance, since it separates the cavity of the orbit from the frontal lobes of the brain. A missile may reach the brain in this way, since this roof is often very thin.

The *floor* is of importance (1) because it forms a partition between the orbit and the antrum of Highmore, or maxillary sinus; (2) because it is traversed by the infraorbital canal, in which are the infraorbital nerve and vessels.

The *inner wall* is in close relation with the nasal fossæ, and the ethmoid cells, the os planum of the ethmoid, and the lachrymal bones form a large part of the partition between these cavities.

The *outer wall* is formed by the orbital plates of the sphenoid and malar bones, but is of no special importance.

The *apex* of the orbit transmits the optic nerve and the ophthalmic branch of the internal carotid artery. The apex is occupied by the *optic foramen*, between the two pedicles of the lesser ala of the sphenoid bone.

The *base* of the orbit is of great importance on account (1) of its architectural construction, combining beauty, strength, and protection; (2) of the large number of anatomical structures in this region.

Note about the base of the orbit:

1. The *lachrymal gland* (Fig. 104), under the external angular process of the frontal bone, in a fossa called the *lachrymal fossa*. This is called the orbital part of the gland or the superior lachrymal gland; the part of the gland that protrudes below this is called the palpebral portion or the inferior lachrymal gland.

2. The *supraorbital foramen* (it may be double) transmits the supraorbital nerve and vessels. (Figs. 98, 103.) The vessels are branches of the ophthalmic.

The nerve is a branch of the frontal part of the ophthalmic, itself a branch of the fifth cranial nerve.

3. **The trochlea** is found on the inner wall, near the base of the orbit. Around this trochlea plays the tendon of the superior oblique muscle of the eyeball. (Fig. 100.) Here, too, occurs the anastomosis between the supratrochlear branch of the frontal and the infratrochlear branch of the nasal nerve.

The nasal duct has its beginning in the internal inferior angle of the orbital base. (Fig. 104.) By this duct excessive tears reach the nasal fossæ.

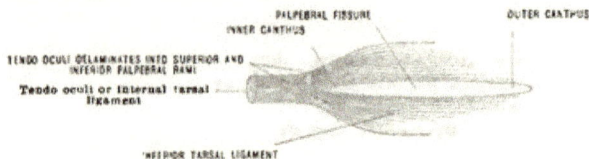

FIG. 93.—TENDO OCULI AND TARSAL CARTILAGES.

The infraorbital nerve (Fig. 51), emerges from the orbit by the infraorbital foramen just below the base of the orbit. This sends off anterior superior branches to the teeth just prior to its emergence.

The tendo oculi crosses the lachrymal sac in front (Fig. 94), giving off an aponeurosis to the same. Observe the caruncula lachrymalis between the two diverging rami of the tendo oculi.

The tendo oculi, or internal tarsal ligament, has its origin from the nasal process of the superior maxilla. (Fig. 10.) The superior and inferior tarsal

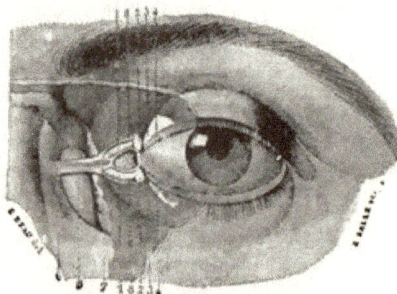

FIG. 94.—RELATIONS OF THE EYE AND THE LACHRYMAL EXCRETORY APPARATUS.

1, 1. Canaliculi. 2, 2. Puncta lachrymalia. 3, 3. Inner extremity of tarsal cartilage. 4, 4. Free borders of lids. 5. Lachrymal sac. 6. Attachment to maxillary bone of superior tendon. 7. Bifurcation of lachrymal sac. 8, 8. Two branches.

ligaments, of periosteal derivation (Fig. 21), are attached to the margin of the base of the orbit both above and below. The external tarsal ligament is not divided.

The Angles of the Orbit.

1. The *superior internal.* This contains the *anterior* and *posterior ethmoidal foramina.* The former transmits the anterior ethmoidal artery and nasal nerve; the latter, the posterior ethmoidal artery.

2. The *superior external* angle. This contains in front the lachrymal fossa already described, and posteriorly the sphenoidal fissure. This fissure transmits

the third, fourth, sixth, and the three divisions of the ophthalmic branch of the fifth nerve, the ophthalmic vein, and some minor structures.

3. The *inferior external* angle. In this is the spheno-maxillary fissure. This fissure transmits the superior maxillary nerve and its orbital branches, the ascending branches from Meckel's ganglion, the infraorbital branch of the internal maxillary artery and its vein.

4. The *inferior internal* angle contains the beginning of the osseous part of the lachrymal apparatus that conveys tears to the nasal fossæ. This apparatus will be considered under the head of orbital contents. (Fig. 104.)

The periosteum **of the** orbit (Fig. 95) lines the walls of the cavity, delaminates at the orbital margin of the base, one layer uniting with the periosteum externally, the other layer forming the tarsal ligaments and tarsal cartilages above and below. The orbital fascia is to the contents of the orbit, other things equal, what the deep fascia of the thigh is to the structures of the thigh. Keep this in mind, for here, as in the thigh, you will find sheaths, capsules, and septa derived from this fascia.

In your dissection you will find the following structures—

  1. The optic nerve—the special nerve of the sense of sight. (Fig. 100.)

  2. The motor oculi nerve, or third nerve. (Fig. 98.)

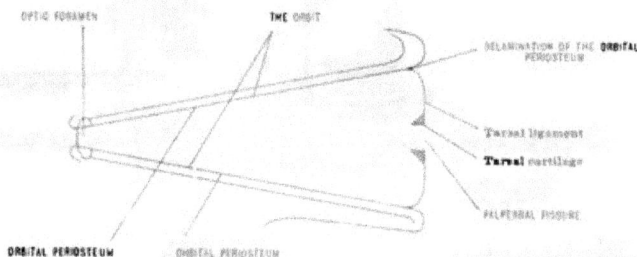

FIG. 95.—SCHEMATIC REPRESENTATION TO SHOW RELATION OF PERIOSTEUM TO TARSAL CARTILAGES AND TARSAL LIGAMENTS.

  3. The patheticus, or fourth nerve. (Fig. 98.)

  4. The abducens, or sixth nerve. (Fig. 98.)

  5. Ophthalmic branch of the fifth nerve. (Fig. 98.)

  6. The lachrymal gland and its capsule. (Fig. 103.)

  7. The ophthalmic artery and its branches. (Fig. 103.)

  8. The ophthalmic vein and its branches. (Fig. 103.)

  9. The ciliary ganglion and its branches. (Fig. 98.)

 10. The orbital connective tissue and fat.

 11. The levator palpebræ muscle. (Fig. 100.)

 12. The superior rectus muscle. (Fig. 100.)

 13. The external rectus muscle. (Fig. 99.)

 14. The inferior rectus muscle. (Fig. 99.)

 15. The internal rectus muscle. (Fig. 99.)

 16. The superior oblique muscle. (Fig. 100.)

 17. The inferior oblique muscle. (Fig. 100.)

 18. The capsule of Tenon.

 19. The eyeball with sclerotica and iris. (Fig. 99.)

Dissection of the Orbital Contents—Steps:

  1. Remove the orbital plate and supraorbital arch by cutting through the

internal and external angular processes of the frontal bone. (Fig. 96.) Carefully remove the bone without injuring the periosteum of the orbit.

2. Cut through the orbital periosteum and find, immediately under the periosteum, the frontal nerve. The frontal nerve lies on the levator palpebræ with the supraorbital artery and vein. It is a branch of the ophthalmic part of the fifth

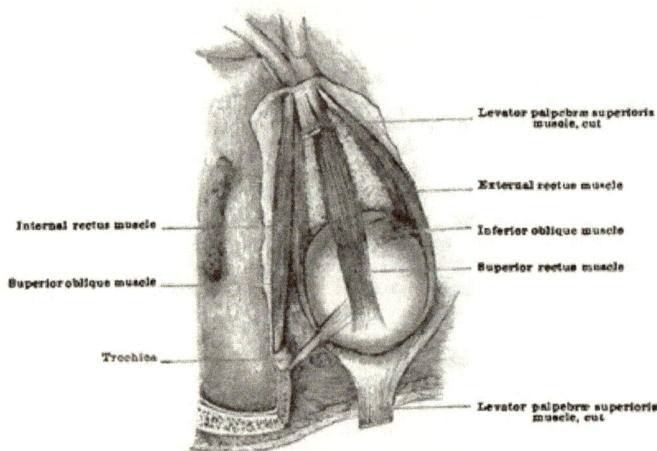

FIG. 96.—VIEW OF LEFT ORBIT FROM ABOVE, SHOWING THE OCULAR MUSCLES
(From Hirschfeld and Leveillé.)

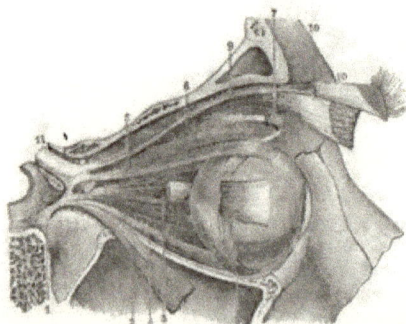

FIG. 97.—MUSCLES OF THE EYE.

1. Tendon of Zinn. 2. External rectus divided. 3. Internal rectus. 4. Inferior rectus. 5. Superior rectus. 6. Superior oblique. 7. Pulley for superior oblique. 8. Inferior oblique. 9. Levator palpebræ superioris. 10, 10. Its anterior expansion. 11. Optic nerve.

nerve. You will see it divide into two branches: (1) the supratrochlear and (2) the supraorbital. The supratrochlear inosculates with the infratrochlear branch of the nasal nerve. The supraorbital comes through the supraorbital foramen, and is distributed to the forehead and inner part of the upper eyelid.

The levator palpebræ superioris muscle (Fig. 97) lies under the preceding frontal nerve. It is inserted into the tarsal ligament of the upper lid. It lies on the superior rectus muscle. Cut the muscle in the middle, turn the ends out of the way, and expose the superior rectus muscle. (Fig. 96.)

The Superior Rectus Muscle (Fig. 96).—This muscle lies under the levator palpebræ. It is inserted into the sclerotic coat of the eyeball. Cut it, and turn the ends aside. You will see under this muscle a bed of fat. In the centre of the bed of fat and connective tissue is the optic nerve. (Fig. 103.)

The Optic Nerve.—To the outer side of this nerve you will see the external rectus muscle. Between the nerve and this muscle you will see the ciliary ganglion.

Study the ciliary ganglion according to this outline (Fig. 98):

1. Location between the optic nerve and external rectus muscle.
2. Size, about as large as a pin's head, but spider-like.
3. Roots: (1) A motor, from the branch of the third nerve to the inferior oblique; (2) a sensory, from the nasal nerve; (3) a sympathetic, from the cavernous plexus.
4. Distribution and function. Branches are given off from this ganglion, called the short ciliary nerves. They go with the optic nerve surrounding the

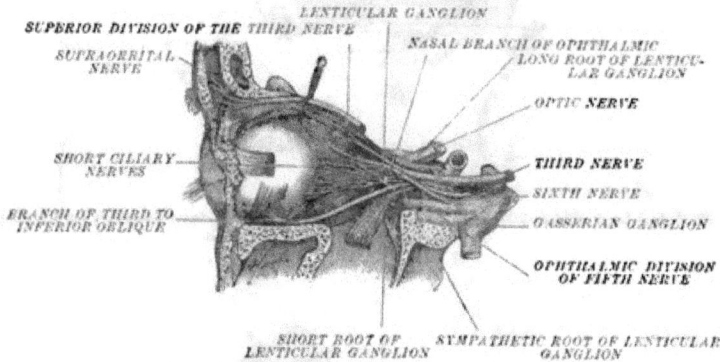

Fig. 98.—Nerves of the Orbit, from the Outer Side.
(From Sappey, after Hirschfeld and Leveillé.)

same. On piercing the sclerotic coat they are joined by the long ciliary branches from the nasal nerve. (Fig. 98.) These branches are distributed to the iris and ciliary muscle.

From what source does the ciliary ganglion derive its sympathetic influence, and how does the same enter the orbit?

It derives it from the superior cervical ganglion. The ascending fibres enter the cranium on the internal carotid artery. Here some fibres accompany the branches of this artery to the brain. Other fibres leave the artery in its cavernous stage, pass through the sphenoidal fissure to the orbit, and supply the ciliary or lenticular ganglion.

Give the influence of paralysis of the third nerve on the iris.

Since the iris derives its motor influence, through the ciliary ganglion, from the third nerve, loss of power of contraction of the pupil would follow.

The optic nerve, you will find, is surrounded by very small nerves and arteries. (Figs. 98 and 103.) The nerves are: (1) Short ciliary, from the ciliary ganglion; (2) long ciliary, from the nasal nerve. The arteries are ciliary, derived from the ophthalmic or some of its branches. The optic nerve derives its sheath from the dura mater.

The **external rectus muscle** lies to the outer side of the optic nerve. It is attended by the **lachrymal nerve**, a branch of the ophthalmic nerve, and the **lachrymal artery**, a branch of the ophthalmic artery. You will dissect the three together and associate them in your memory. The lachrymal structures are on their way to the lachrymal gland. Turn the external rectus muscle

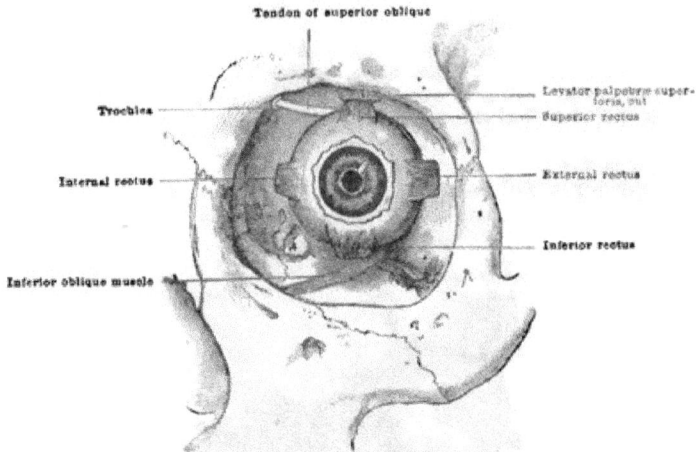

FIG. 99.—LEFT EYEBALL SEEN IN ITS NORMAL POSITION IN THE ORBIT, WITH VIEW OF THE OCULAR MUSCLES. (After Merkel, modified.)

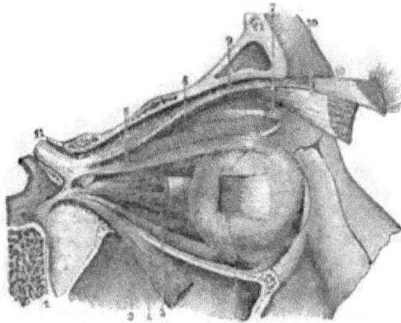

FIG. 100.—MUSCLES OF THE EYE. TENDON OR LIGAMENT OF ZINN.

1. Tendon of Zinn. 2. External rectus divided. 3. Internal rectus. 4. Inferior rectus. 5. Superior rectus. 6. Superior oblique. 7. Pulley for superior oblique. 8. Inferior oblique. 9. Levator palpebræ superioris. 10, 10. Its anterior expansion. 11. Optic nerve.

outward, and on its inner surface you will see the fine filaments of the sixth cranial nerve—the abducens.

The Superior Oblique Muscle and the Fourth Nerve (Fig. 97).—These form a group by themselves. You will trace the tendon of this muscle around the trochlea, where it turns at a right angle and passes outward under the superior rectus to be inserted into the sclerotica. You will find its nerve, the fourth, breaking up into a number of filaments on the ocular surface of the muscle.

The **nasal** nerve crosses the optic nerve, passes under the superior rectus muscle, then passes between the superior oblique muscle and the internal oblique muscle. Here it gives off its infratrochlear branch to meet the supratrochlear branch of the frontal nerve. These two are sensory nerves. The nerve leaves the orbit by the anterior ethmoid foramen with the anterior ethmoidal artery. In the cranial cavity it lies under the dura, on the cribriform plate of the ethmoid bone. It passes through the nasal slit to the nasal fossa. In the nasal fossa it

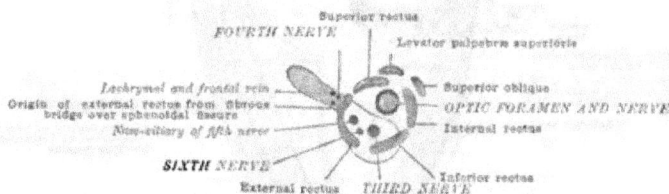

FIG. 101.—DIAGRAMMATIC REPRESENTATION OF ORIGINS OF OCULAR MUSCLES AT THE APEX OF THE RIGHT ORBIT.

(After Schwalbe, slightly altered.)

divides into branches: (1) Septal, to the mucous membrane of the septum; (2) turbinals, to the turbinated bones. A small twig from the latter passes between the lower end of the nasal bone and its cartilage, and appears on the face as the naso-lobular—a nerve of sensation, of course, to the wing and tip of the nose.

TABLE OF ORIGIN OF OCULAR MUSCLES.

NAME.	ORIGIN.
Levator palpebræ superioris,	Above optic foramen from lesser ala of sphenoid.
Superior rectus,	Upper margin of optic foramen.
Inferior rectus,	From ligament of Zinn.
Internal rectus,	From ligament of Zinn.
External rectus,	Upper head. Outer margin of optic foramen.
External rectus,	Lower head. From ligament of Zinn.
Superior oblique,	Above the optic foramen.
Inferior oblique,	Orbital plate of superior maxilla.

The **internal rectus** and the **inferior rectus** may be brought into view by cutting the optic nerve and pushing the eyeball and the fatty mass forward. You will see these muscles taking their nerve-supply from the **third** nerve. Trace each muscle to its insertion into the sclerotica.

The **inferior oblique muscle** arises from the orbital plate of the superior maxilla just external to the lower end of the lachrymal groove. It is inserted into the sclerotica, on the outer surface, in such a way as to antagonize the superior oblique muscle. Its nerve, the long branch of the third cranial, is of large size, and gives to the ciliary ganglion its motor root.

The **orbital fascia** invests the muscles. It passes forward on the tendons. In the vicinity of the globe it binds the tendons together by extending from one to another, thus forming, posterior to the ball, a loose double capsule—the capsule of Tenon. This double layer forms a socket for the ball to move in. It separates posteriorly the fatty contents of the orbit from the globe of the eye.

Describe the Ophthalmic Artery (Fig. 103).—It is a branch of the internal carotid artery. It enters the orbit by the optic foramen with the optic nerve. It is attended by the **ophthalmic vein,** which leaves the orbit by the sphenoidal

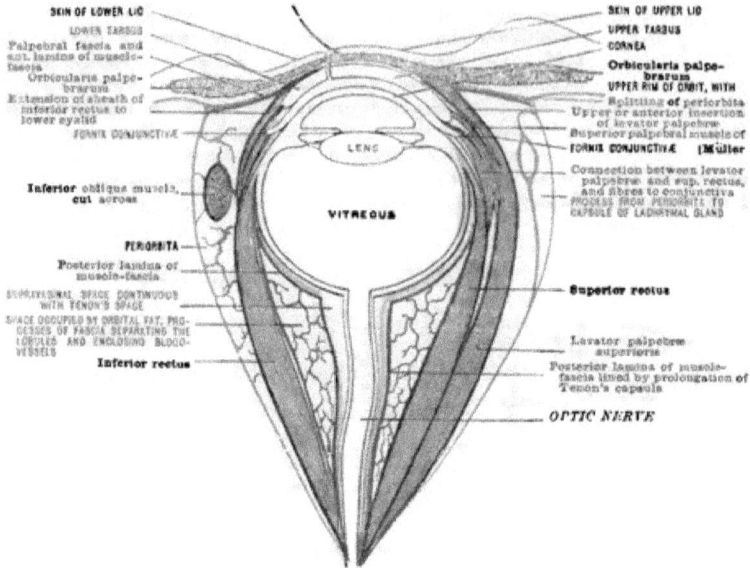

SKIN OF LOWER LID
LOWER TARSUS
Palpebral fascia and ant. lamina of muscle-fascia
Orbicularis palpebrarum
Extension of sheath of inferior rectus to lower eyelid
FORNIX CONJUNCTIVÆ
LENS
Inferior oblique muscle, cut across
VITREOUS
PERIORBITA
Posterior lamina of muscle-fascia
SUPRAVAGINAL SPACE CONTINUOUS WITH TENON'S SPACE
SPACE OCCUPIED BY ORBITAL FAT, PROCESSES OF FASCIA DEPARTING THE LOBULES AND ENCLOSING BLOODVESSELS
Inferior rectus

SKIN OF UPPER LID
UPPER TARSUS
CORNEA
Orbicularis palpebrarum
UPPER RIM OF ORBIT, WITH
Splitting of periorbita
Upper or anterior insertion of levator palpebræ
Superior palpebral muscle of
FORNIX CONJUNCTIVÆ [Müller
Connection between levator palpebræ and sup. rectus, and fibres to conjunctiva PROCESS FROM PERIORBITA TO CAPSULE OF LACHRYMAL GLAND
Superior rectus
Levator palpebræ superioris
Posterior lamina of muscle-fascia lined by prolongation of Tenon's capsule
OPTIC NERVE

FIG. 102.—VERTICAL SECTION THROUGH THE EYEBALL AND ORBIT IN THE DIRECTION OF THE ORBITAL AXIS, WITH CLOSED EYELIDS.
(Semi-diagrammatic. After Schwalbe, modified to show fasciæ.)
Periorbita *green*; muscle-fascia *red*; Tenon's capsule *yellow*.

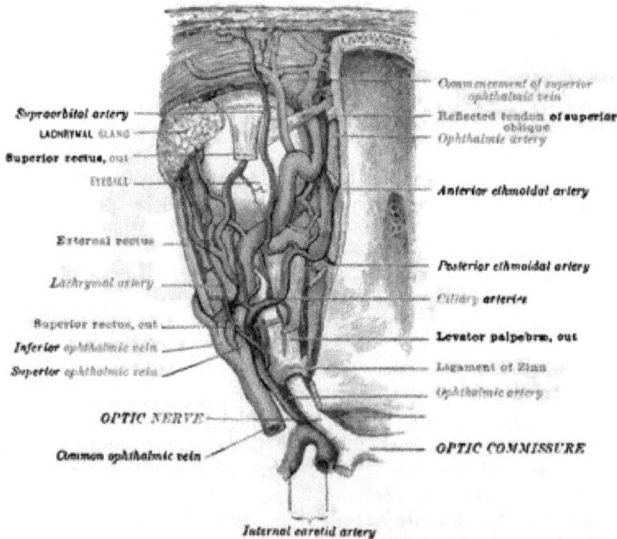

Supraorbital artery
LACHRYMAL GLAND
Superior rectus, cut
EYEBALL
External rectus
Lachrymal artery
Superior rectus, cut
Inferior ophthalmic vein
Superior ophthalmic vein
OPTIC NERVE
Common ophthalmic vein

Commencement of superior ophthalmic vein
Reflected tendon **of superior** oblique
Ophthalmic artery
Anterior ethmoidal artery
Posterior ethmoidal artery
Ciliary arteries
Levator palpebræ, cut
Ligament of Zinn
Ophthalmic artery
OPTIC COMMISSURE

Internal carotid artery
FIG. 103.—THE OPHTHALMIC ARTERY AND VEIN.

fissure and expands behind the sphenoidal fissure to form the cavernous sinus. Its branches are :

1. Lachrymal, with the external rectus, sixth nerve to lachrymal gland.
2. Supraorbital, with the frontal nerve, levator palpebræ, superior rectus.
3. Arteria centralis retinæ, in centre of optic nerve to the retina.
4. Anterior ethmoidal, with the nasal nerve to ethmoid cells.
5. Posterior ethmoidal, to the posterior ethmoid cells.
6. Palpebral arteries, to the upper and lower lids.
7. Frontal artery, to the inner angle of the eye.
8. Nasal artery, to the lachrymal sac and caruncula.
9. Muscular branches, to the muscles of the eye.
10. Ciliary arteries, to the iris and choroid coat.

How would you classify the nerves ?

1. Special sense : The optic nerve—special of vision.
2. { Motor : Third, to all except two muscles. / Motor : Fourth, to the superior oblique muscle. / Motor : Sixth, to the external rectus muscle.
3. { Sensory : Nasal branch of ophthalmic, to the nose and orbit. / Sensory : Lachrymal branch of ophthalmic, to lachrymal gland. / Sensory : Frontal branch of ophthalmic, to forehead.
4. Sympathetic, from cavernous plexus, to the ciliary ganglion.

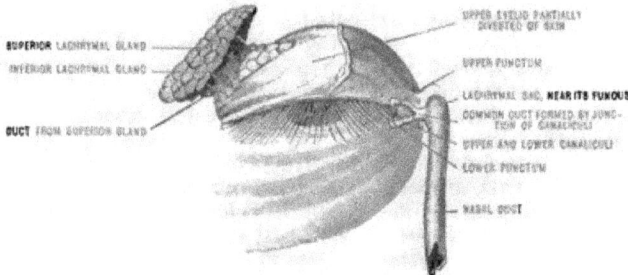

FIG. 104.—LACHRYMAL APPARATUS. (After Schwalbe.)

The lachrymal gland you will find under the external angular process of the frontal bone in the lachrymal fossa. (Fig. 103.) It has an orbital and a palpebral part, a capsule, an artery, and a nerve.

Define the periorbita and give its functions.

This is the orbital periosteum. It forms the tarsal ligaments and the limbs of the inner palpebral ligament, lines the floor of the lachrymal groove, forms the trochlea for the superior oblique muscle of the eyeball, assists the orbital fascia in forming a capsule for the lachrymal gland, and sustains the orbital fat.

What is the first structure you saw on removing the orbital plate of the frontal bone ?.

The periorbita or orbital periosteum, a continuation of the dura mater through the sphenoidal fissure and the optic foramen.

On removing the periorbita, what did you see ?

1. In the middle we saw the frontal artery, a branch of the ophthalmic, and the frontal nerve, a branch of the ophthalmic division of the fifth nerve, lying on the levator palpebræ muscle.
2. To the outer side we saw the lachrymal artery, a branch of the ophthalmic

artery, and the lachrymal nerve, a branch of the ophthalmic division of the fifth nerve, lying on the external rectus muscle.

3. To the inner side we saw the fourth nerve, called **also the** patheticus, lying on and supplying the superior oblique muscle with motion, in its course having passed above the levator palpebræ muscle and the frontal nerve.

Describe the frontal nerve, which you found on the levator palpebræ, and tell what structure intervened between the periorbita and this nerve.

The intervening structure was the peripheral part of the orbital fascia. It was so thin and translucent that we could see the nerve through the fascia. The frontal nerve is a branch of the ophthalmic part of the fifth nerve. About midway between the optic foramen and the supraorbital foramen it divides into a supraorbital and a supratrochlear branch. The supraorbital branch leaves the orbit through the supraorbital foramen and supplies the skin of the upper lid, the skin of the forehead, and the pericranium in this same region ; the supratrochlear branch communicates with the infratrochlear branch of the nasal nerve around the pulley **for** the superior oblique muscle, branches being given off from this loop to **supply** the upper lid, forehead, and nose.

Describe the lachrymal nerve, which you saw on the external rectus muscle.

It is a branch of the ophthalmic part of the fifth nerve. It gives branches **to the** lachrymal gland, and to the skin and mucous membrane of the upper lid, and **sends** a communicating branch to the orbital branch of the superior maxillary nerve.

Describe the nasal nerve.

It is a branch of the **ophthalmic.** It has a most complicated course, which to be learned thoroughly must be studied according to these stages :

1. The nerve enters the orbital cavity by the sphenoidal fissure.

2. In its course through the orbit it passes : (1) Between the optic nerve and superior rectus muscle ; (2) between the external rectus muscle and the superior oblique muscle, and arrives at the anterior ethmoidal foramen. In this, its orbital stage, the nerve gives off the long root to the ciliary ganglion, at the sphenoidal fissure, and the long ciliary nerves to the eyeball, as it crosses the optic nerve. At the anterior ethmoidal foramen the nerve gives off the infratrochlear branch, to communicate with the supratrochlear branch of the frontal nerve.

3. In company with the anterior ethmoidal artery it enters the cranial cavity, by the anterior ethmoidal foramen, crosses the cribriform plate of the ethmoid bone, and leaves the cranial cavity and gains the nasal cavity by the nasal slit, by the side of the crista galli.

4. The nerve divides in the nasal cavity into septal branches, which are distributed to the upper and front part of the septum nasi, and branches to the outer wall and anterior ends of the two lower turbinals. The end branch of this nerve passes between the cartilage and the end of the nasal bone, supplying the tip and lobe of the nose.

Describe the motor oculi, or third cranial nerve.

This nerve is the most important motor nerve to the muscles of the orbit. As the nerve emerges from the cavernous sinus it divides into a superior and an inferior division, which enter the orbit by the sphenoidal fissure. The superior division supplies the levator palpebræ and superior rectus muscles. The inferior division supplies the internal rectus, the inferior rectus, and the inferior oblique. The third nerve supplies with motion the circular fibres of the iris and the ciliary muscle, through the ciliary ganglion.

Describe the fourth cranial nerve.

It is called patheticus and trochlear It enters the orbit by the sphenoidal fissure, **passes above** the frontal nerve and levator palpebræ muscle, and is distributed to the superior oblique muscle. This is the smallest of the cranial nerves.

Describe the sixth cranial nerve.

It is called the abducent nerve. It enters the orbit by the sphenoidal fissure, between the two heads of the external rectus, to which muscle it is distributed.

Describe the optic nerve.

This is the nerve of the special sense of sight. It is the second cranial nerve. It enters the orbit by the optic foramen, internal to and above the ophthalmic artery. It pierces the ball of the eye and forms the retina. It is surrounded by the ciliary vessels and nerves. To its outer side is the ciliary ganglion. At the optic foramen it is surrounded by the four recti muscles. The arteria centralis retinæ pierces its under surface and goes to the interior of the eyeball. The process of dura that surrounds the nerve divides in such a manner as to form a sheath for the nerve, and also the orbital periosteum, inside the orbit. The optic sheath is continuous with the sclerotica.

What is the orbital fascia?

It is a variety of connective tissue, being to the contents of the orbit what the deep fascia is to the thigh. It forms a capsule, in conjunction with the periorbita, for the lachrymal gland, and sheaths for the muscles, vessels, and nerves. It is connected to the ocular conjunctiva close to the cornea. The muscular sheaths are firmly adherent to the muscles. The insertions of the ocular muscles into the sclerotica are connected together around the whole circumference of the globe. From this attachment this fascial connecting medium is reflected backward in a double layer, forming the fibrous basis of a serous membrane— the visceral layer covering the back part of the globe, the parietal lining the postocular fat in the orbit. A layer of endothelial cells completes the serous membrane. Here we have a visceral layer, a cavity, and a parietal layer, the three cardinal points in any serous apparatus. This reflected part of the orbital fascia permits free movement of the eyeball, as the head of a bone revolves in its socket, and is known in anatomy as the capsule of Tenon.

By what is the orbital fascia on the posterior part of the eyeball pierced?

It is pierced by the vessels and nerves that supply the globe.

What muscle did you find immediately under the levator palpebræ?

The superior rectus, and under this we found the optic nerve, surrounded by fat, vessels, and nerves.

Name the muscles of the eyeball and give their function.

The external rectus abducts the cornea.

The internal rectus adducts the cornea.

The superior rectus elevates, adducts, and rotates cornea inward.

The inferior rectus depresses, adducts, and rotates cornea outward.

The superior oblique depresses, abducts, and rotates cornea inward.

The inferior oblique elevates, abducts, and rotates cornea outward.

The levator palpebræ elevates the upper eyelid.

Name all the nerves that supply the orbital contents.

The optic, the nerve of special sense of sight.

The motor oculi, a motor nerve to all muscles except two.

The pathetic, a motor nerve to the superior oblique muscle.

The abducens, a motor nerve to the external rectus muscle.

The frontal, a branch of the ophthalmic of the fifth nerve.

The lachrymal, a branch of the ophthalmic of the fifth nerve.

The supratrochlear branch of the frontal.

The palpebral branches of the frontal.

The infratrochlear branch of the nasal.

The long ciliary branches of the nasal.

The short ciliary branches of the lenticular ganglion.

The cavernous sympathetic branches to lenticular ganglion.

The lenticular or ciliary ganglion.

Where is the ciliary ganglion located, and how is it formed?

It lies between the external **rectus** muscle and the optic nerve on the outer side of the ophthalmic artery. It is the size of a pin's head. It has three roots:

1. A sensory root from the nasal nerve.
2. A motor root from the motor oculi nerve.
3. A sympathetic root from the cavernous sympathetic.

How do the nerves gain access to the orbit?

The optic **nerve** passes through the optic foramen; all the others pass through the sphenoidal fissure with the ophthalmic vein.

Name all the structures seen on inspecting the eye of a patient.

1. The palpebral fissure, a slit between the lids.
2. The outer and inner canthi, the extremities of the fissure.
3. The tarsal cartilages, the free margins of the lids.
4. The cornea, behind which is the iris.
5. The pupil, surrounded by the iris.
6. The tendo oculi, or inner palpebral ligament.
7. The lachrymal caruncle, just within the inner canthus.
8. The superior palpebral fold **when** the eye is opened.
9. The inferior palpebral fold when the eye is opened.
10. The palpebral and ocular conjunctivæ.
11. The conjunctival cul-de-sac or fornix.
12. The sclerotica, or white of the eye.
13. The conjunctival sac. This is the space between the posterior surface of the lid and the eyeball. It is completely covered by conjunctiva.

A patient gets a cinder in his eye: What nerve or nerves report the pain to the brain?

The cornea and sclerotica are supplied by the ciliary nerves. The infratrochlear branch of the nasal nerve supplies the conjunctiva, the lachrymal sac, and the caruncula lachrymalis. The lachrymal nerve sends conjunctival branches to the upper lid. The palpebral branches of the superior maxillary send conjunctival branches to the lower lid. While, then, a number of branches are concerned locally, one nerve, the trigeminus or fifth, does terminal work for all, since the lachrymal, nasal, superior maxillary, and the sensory root of the ciliary ganglion are all derived from the fifth nerve. A *reflex circuit* is completed by the seventh nerve, acting on the orbicularis palpebrarum to close the eye.

Emissary Foramina and Their Veins.—

Define emissary foramina and emissary veins.

Emissary foramina derive their importance from the veins they transmit; emissary veins are both physiologically and pathologically important. The emissary veins are connected centrally to the sinuses in the dura mater; peripherally they communicate with veins both deep and superficial. From a physiological standpoint these veins regulate indirectly the arterial pressure in the brain, since, in cases of cerebral engorgement, when the internal jugular veins are unable to convey the required amount of blood in the dural sinuses from the brain, the emissary veins deliver large quantities of this venous blood to the superficial **or** to the deep veins with which they communicate peripherally. This action of the emissary veins has been compared to a safety-valve in mechanics. From a pathological standpoint these veins are conveyers of septic material from areas accessible to the surgeon and physician, to regions beyond the reach of either. Infection in the orbit, posterior nares, and scalp over the forehead reaches the cavernous sinus via the angular and ophthalmic veins. Infection in the face, nose, and teeth reaches the cavernous sinus via the Vesalian vein and pterygoid plexus. The conservatism of emissary veins is well illustrated in the profuse epistaxis of children, incident to cerebral congestion. In this case an emissary vein connects centrally the blood in the superior longitudinal sinus with the blood in the mucous membrane of the nose ; the latter, being the weaker structure, gives way, and the arterial pressure in the brain is relieved by a profuse hemorrhage. The emissary foramina seem to belong more especially to the physiological needs of childhood. The greater number of emissary veins and their foramina disappear as adult years come on. By osteological classification these foramina are inconstant.

The following table gives a list of the emissary foramina, the central sinuses, and the peripheral veins, between which the emissary veins stand, holding the two together like the handle connects the two globes of a dumb-bell :

TABULATED LIST OF EMISSARY STRUCTURES.

FORAMEN.	CENTRAL SINUS.	PERIPHERAL VEIN.
Cæcum,	Superior longitudinal,	Nasal mucous membrane.
Carotid canal,	Cavernous,	Internal jugular vein.
Parietal,	Lateral,	Occipital.
Vesalii,	Cavernous,	Internal jugular vein.
Posterior condylar,	Lateral,	Deep cervical.
Occipital,	Torcular Herophili,	Occipital veins.
Anterior condylar,	Occipital,	Deep cervical.
Mastoid,	Lateral,	Occipital veins.
Sphenoidal fissure,	Cavernous,	Angular and ophthalmic.
Ovale,	Cavernous,	Pharyngeal and pterygoid.

How Septic Thrombi Reach the Sinuses.

How may a carbuncle of the face, or facial erysipelas, or infection of the scalp over the forehead result in fatal blood-poisoning ?

Septic thrombi from these sources may reach the cavernous sinus and get beyond the reach of the surgeon. The frontal and supraorbital veins unite to form the angular, and this is one of the large tributaries of the facial vein. (Fig. 18.) Now, the angular vein communicates with the ophthalmic. The ophthalmic vein passes through the sphenoidal fissure, and is tributary to the cavernous sinus. The angular, facial, and ophthalmic veins contain no valves, hence the blood can pass either forward or backward from the superficial seat of infection to a region beyond the reach of surgical drainage and antisepsis.

Tell the course septic material from a carious tooth, alveolar abscess, or suppuration of the antrum follows in cases of fatal blood-poisoning to reach the sinuses in the dura mater beyond the reach of operative procedure.

To appreciate an answer to this question you must recall the distribution of the internal maxillary artery to the muscles of mastication, the teeth, the nasal fossæ, the meninges, and the palate. Now, the veins returning the blood from these areas take the course of their companion arteries, and the same name as the arteries. The veins from these regions, the small and great meningeal, the supraorbital and posterior dental, the palatine and spheno-palatine, the deep temporal, pterygoid, and buccal, the lower ophthalmic and Vesalian, all come together on the inner surface of the internal pterygoid muscle to form the *pterygoid plexus*. This plexus communicates with the cavernous sinus by the Vesalian vein. As these veins contain no valves, septic thrombi originating anywhere in the dental, alveolar, antral, nasal, or palatine regions may spread backward to the cavernous sinus via the Vesalian vein and pterygoid plexus.

SHOULDER, ARM, FOREARM, AND HAND.

I prefer to have my students begin a dissection of the upper extremity on the fingers and hand. The liability of these parts to become dry, to say nothing of the great practical importance of this region, is sufficient apology. Any dissection of the hand without a review of the salient osteological landmarks will be, in a measure, a poor dissection, unsatisfactory alike to teacher and student. The student should become master of the following points in osteology before he begins work on the cadaver:

OSTEOLOGICAL POINTS ON THE HAND.

The Carpus and Hand (Figs. 108 and 109).

Name the points on the bones of the carpus and hand of importance in practical anatomy.

The carpo-metacarpal articulations.
The base of the first or thumb metacarpal.
The base of the second metacarpal, radial side.
The base of the third metacarpal, radial side.
The base of the fifth metacarpal, ulnar side.
The pisiform bone of the carpus.
The groove in the os trapezium.
The unciform process of the unciform bone.
The base of the first thumb phalanx.
The base of the second thumb phalanx.
The dorsal bases of index, middle, and ring fingers.
The dorsal bases of second and third phalanges.
The palmar bases of second and third phalanges.

What is the importance of the groove in the os trapezium at the base of the thumb?

It transmits the tendon of the flexor carpi radialis to its insertion into the base of the index metacarpal bone.

What is the importance of the base of the fifth metacarpal?

On its palmar side is inserted the flexor carpi ulnaris; on its dorsal the extensor carpi ulnaris. These muscles are antagonized by radial and ulnar carpal extensors.

Name the radial and ulnar extensors of the carpus.

There are two radial extensors of the carpus: the extensor carpi radialis longior and brevior, inserted into the bases of the second and third metacarpals respectively, on the radial side. The extensor carpi ulnaris is inserted into the base of the fifth metacarpal on the ulnar side.

Give the importance of the pisiform bone and the unciform process of the unciform bone on the ulnar side of the carpus, in conjunction with the scaphoid and trapezium on the radial side.

It is these bones that give attachment to the anterior annular ligament of the

152

wrist, a band of deep fascia that binds down the strong flexor tendons of the fingers and thumb, and also preserves the transverse carpal arch.

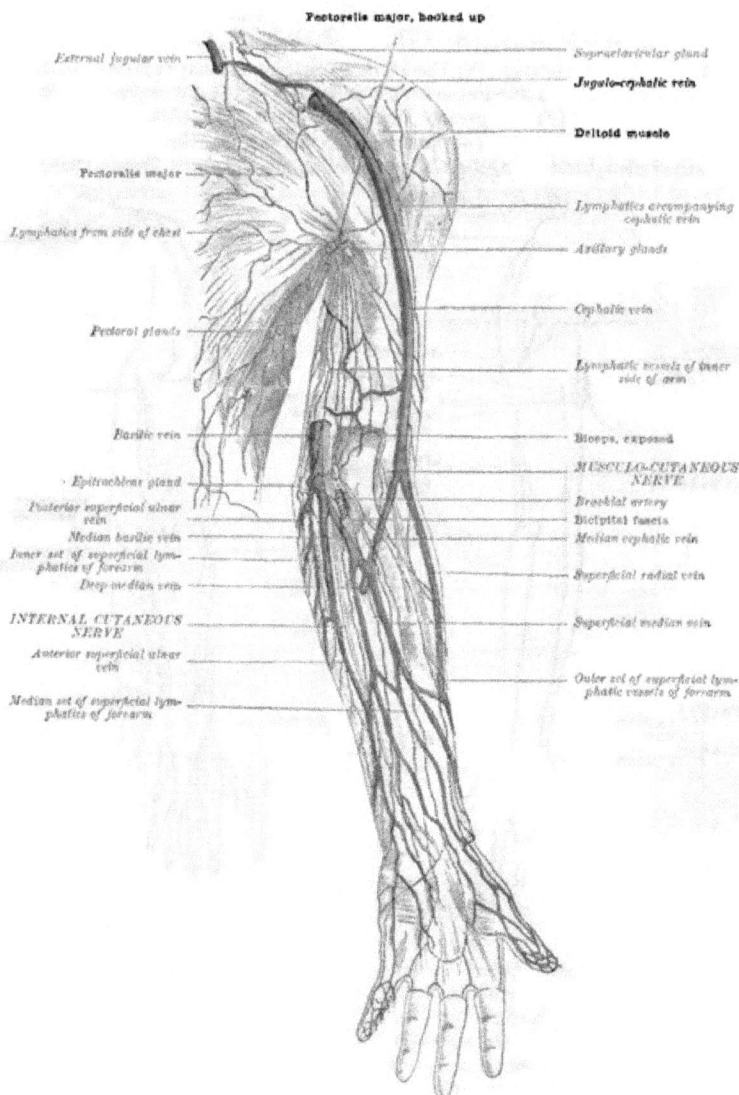

FIG. 105.—SUPERFICIAL VEINS AND LYMPHATICS OF THE FOREARM AND ARM.

What is on the palmar surface of the phalanges?

A groove limited laterally by a ridge of bone on each side. The groove is converted into an osseo-aponeurotic canal by a strong covering of deep digital

fascia, called the ligamentum vaginale. In this canal are found the flexor tendons, invested by a synovial membrane called theca.

ANTERIOR PART.

Locate on the cadaver : (1) The clavicle ; (2) the sternum ; (3) sterno-clavicular articulation ; (4) acromio-clavicular articulation ; (5) the acromion ; (6) **the**
coracoid process ; (7) the greater tuberosity of the humerus ; (8) the lesser
tuberosity of the humerus ; (9) the internal condyle ; (10) the external condyle ;
(11) the radial head ; (12) the clefts of the fingers ; (13) the thenar eminence ;

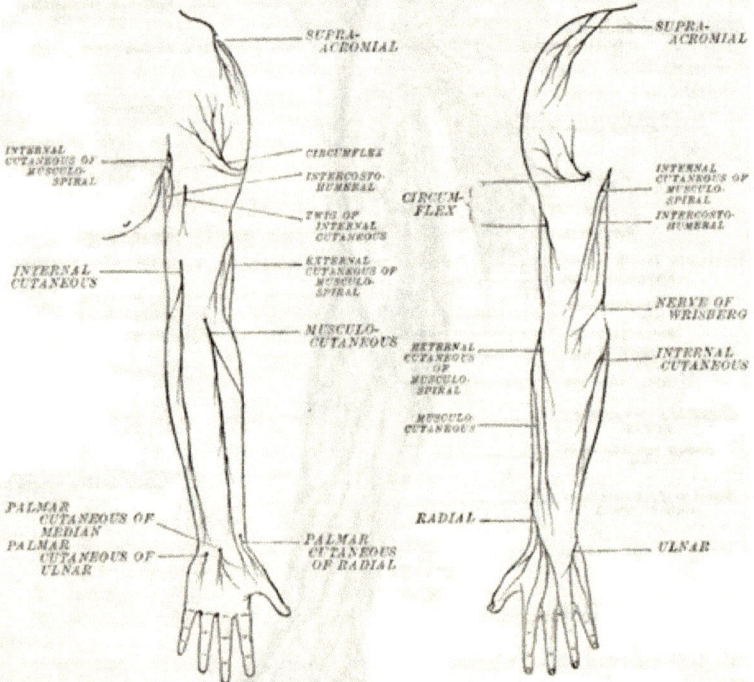

FIG. 106.—DISTRIBUTION OF CUTANEOUS NERVES ON THE ANTERIOR AND POSTERIOR ASPECTS
OF THE SUPERIOR EXTREMITY.

(14) the hypothenar eminence ; (15) the metacarpo-phalangeal articulation ;
(16) the phalangeal articulations ; (17) **the** pisiform ; (18) the unciform process of the unciform bone.

Incisions.—(1) From the acromion to the end of **the** middle finger ; (2) from
the sterno-clavicular articulation to the **acromion ;** (3) from condyle to condyle ;
(4 through the metacarpo-phalangeal crease.

NOTE.—In making these incisions care must be taken not to cut too deeply,
in order to avoid doing violence to the cutaneous vessels and nerves. Now
remove the skin very carefully and expose the following cutaneous nerves and
veins :

1. The *cephalic vein*—a continuation of the radial above the elbow. (Fig. 105.)

2. The *basilic vein*—a continuation of the ulnar above the elbow.
3. The *superficial radial vein*, on radial side of forearm.
4. The *superficial median vein*, on midfront of forearm.
5. The *anterior superficial ulnar vein*, on ulnar side of forearm.
6. The *posterior superficial ulnar vein*, on posterior part of forearm.
7. The *deep median vein*, a communicating vessel.
8. The *external jugular vein*, seen above the clavicle.

Cutaneous Nerves.

1. The *palmar cutaneous branch of the ulnar nerve*, supplying 1.5 fingers.
2. The *palmar cutaneous branch of the median nerve*, supplying 3.5 fingers.
3. The **palmar** *cutaneous branch of the radial* **nerve**, to ball of **thumb**.

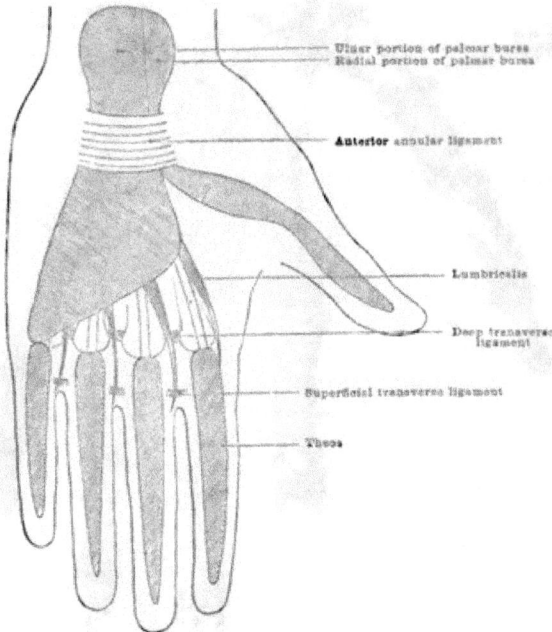

FIG. 107.—DIAGRAM OF THE GREAT PALMAR BURSA.

4. The cutaneous branch of the *musculo-cutaneous nerve.*
5. The *internal cutaneous nerve* on internal part of the forearm.
6. The *intercosto-humeral nerve* on internal part of the arm.
7. The *cutaneous branch of the circumflex nerve* over insertion of deltoid.
8. The *internal cutaneous branch of the musculo-spiral nerve.*

Directions for Dissecting the Cutaneous Vessels and Nerves.—Consult figures 105 and 106. Plow through the upper fat-bearing strata of the *superficial fascia* with the forceps—never with the scalpel. You will find all the vessels and nerves in the deep layer of the superficial fascia. Divide the fascia in the direction of the vessels and nerves.

The radial veins converge to form the cephalic vein above the elbow.

(Fig. 105.) The **ulnars** converge to form the basilic vein above the elbow. The **medians** converge to form one large vein. This vein breaks up into the *median cephalic* and *median basilic*, which join respectively the cephalic and basilic veins. (Fig. 105.)

Observe that the internal cutaneous nerve passes behind the median basilic vein; that the *cutaneous branch of the musculo-cutaneous nerve* passes behind the median cephalic vein. (Fig. 105.)

Find the deep median vein, piercing the deep fascia near the bifurcation of

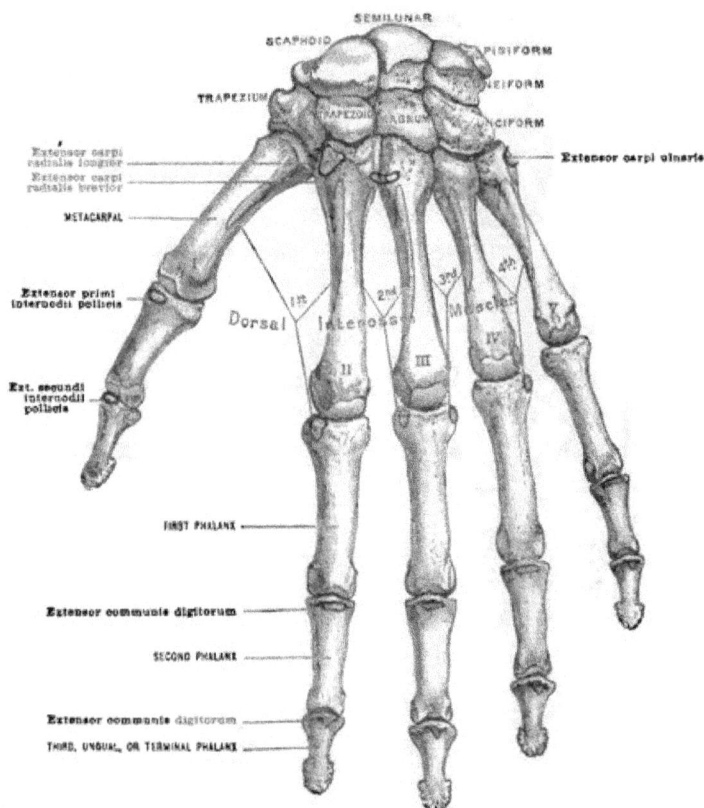

FIG. 108.—THE LEFT HAND. (Dorsal surface.)

the superficial median vein. Trace it to the ulnar vein below the deep fascia. (Fig. 105.)

Note that the median basilic vein crosses the lower part of the brachial artery. Notice also the bicipital fascia between these two structures. (Fig. 105.) This vein was formerly a favorite in bleeding. The basilic vein terminates in the axillary; the cephalic in the axillary vein also. Find the cephalic vein in a groove between the pectoralis major and deltoid muscles with the descending branch of the acromio-thoracic artery below the clavicle.

The cutaneous branch of the musculo-cutaneous nerve supplies the skin over the insertion of the biceps ; inasmuch as the supinator longus is also a flexor of the forearm, and receives a part of its nerve-supply from the musculo-cutaneous nerve, the skin over the insertion of this muscle is also supplied by the same nerve. See Hilton's law in the introductory chapter.

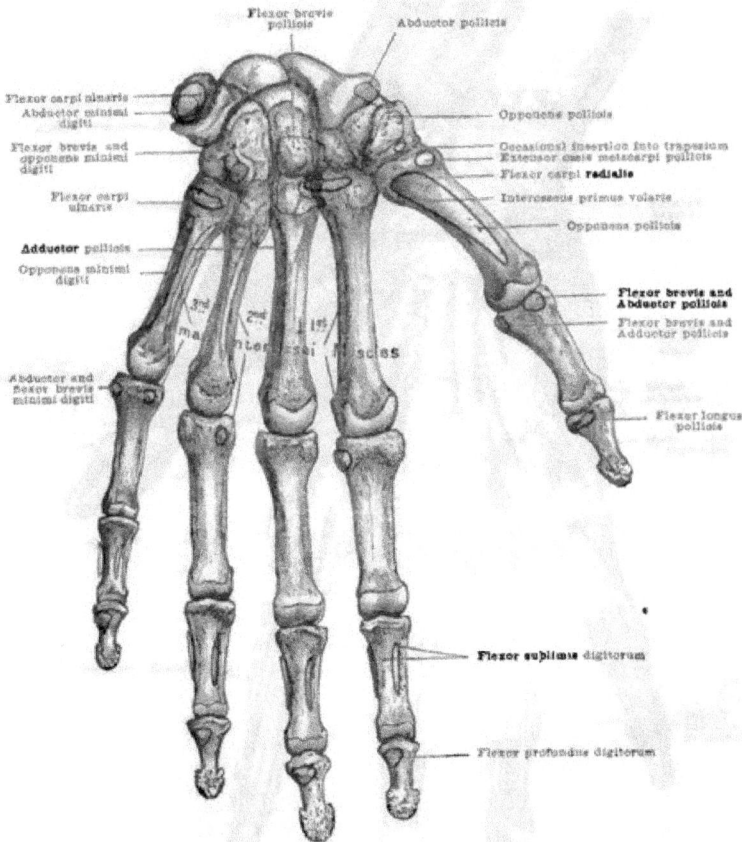

FIG. 109.—THE LEFT HAND. (Palmar surface.)

DEEP FASCIA OF THE HAND.

The deep fascia—fascia profunda—lies immediately under the superficial fascia, being in intimate trabecular relation with the deep layer of the superficial fascia. In the palm of the hand it is very dense and heavy ; over the thenar and hypothenar eminences it is very thin.

The deep fascia takes the following special names :

1. The anterior annular ligament—in front of wrist.
2. The palmar fascia—in palm of the hand.
3. The ligamenta vaginales—on flexor side of the digits.

4. Internal and external intermuscular fascia.
5. The posterior annular ligament.
6. The dorsal fascia of the hand.

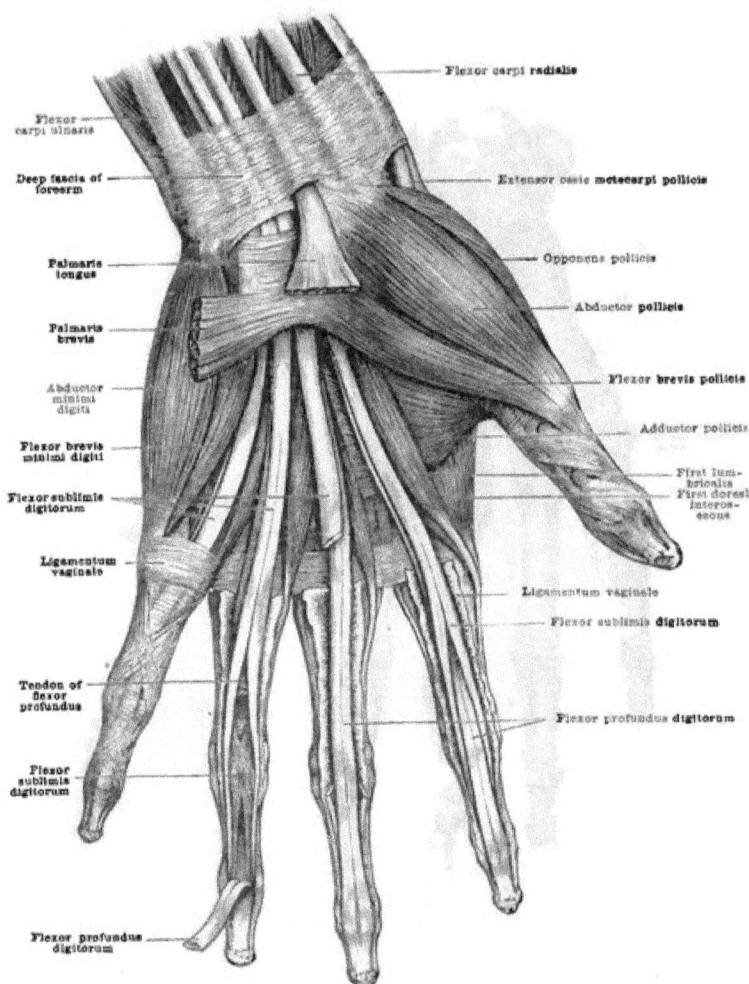

FIG. 110.—THE SUPERFICIAL MUSCLES OF THE PALM OF THE HAND.

Dissection of the hand—in which find the following:

1. The palmar fascia and palmaris brevis muscle.
2. The flexor tendons and their sheaths and vincula.
3. The digital nerves and arteries and digital veins.
4. The superficial palmar arch and its branches on flexor tendons.
5. The thecal culs-de-sac, synovial in the ligamenta vaginalis.

6. The anterior annular ligament on front of carpus.
7. The musculi lumbricales on flexor tendons.
8. The thenar and hypothenar muscles about thumb and little finger.
9. The deep palmar arch and branches below the flexor tendons.
10. The deep palmar branch of the ulnar nerve with deep arch.
11. The dorsal interossei muscles.
12. The ulnar nerve in the hand, supplying 1.5 fingers.
13. Deep palmar arch and branches of ulnar nerve in the hand.
14. The median nerve in the hand, supplying 3.5 fingers.
15. The radial and ulnar arteries in the hand, forming the arches.
16. The palmar interosseous muscles on the palmar surface.

Dissection.—The palmar fascia is the deep fascia of the flexor surface of the

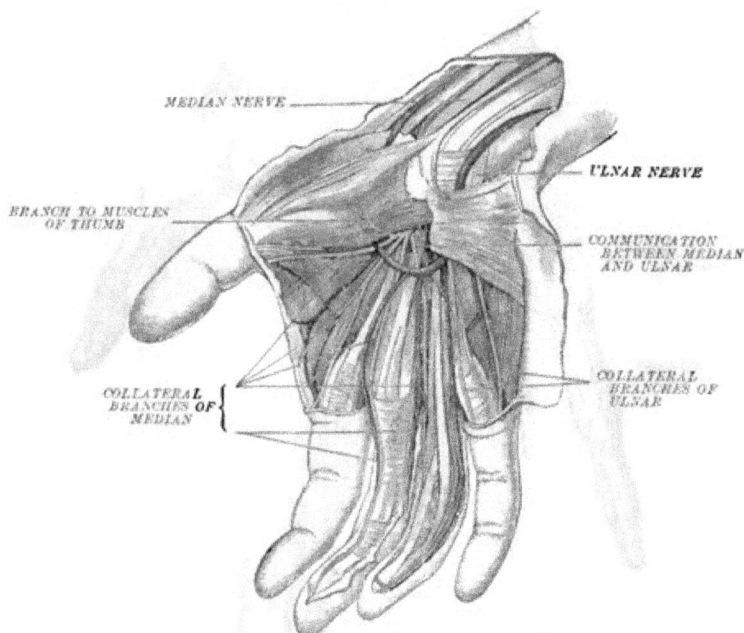

FIG. 111.—SUPERFICIAL NERVES OF THE PALM.   (Ellis.)

hand from the annular ligament above to the clefts of the fingers below. Having removed the skin of the palm, and noticed the granular fat in the superficial fascia thereof, you are now ready to study the palmar fascia. This fascia has three divisions : (1) An inner thin part that covers the muscles forming the ball of the little finger ; (2) an outer thin part that covers the muscles forming the ball of the thumb ; (3) a central strong portion, investing the central distributory region of the hand, in such a way as to protect the vessels and nerves for the supply of the fingers and adjacent structures. The palmar fascia is continuous above with the anterior annular ligament, and below with the ligamenta vaginales in the form of heavy protective sheaths for the flexor tendons of all the fingers. Figure 110 shows this ligamentum vaginale still intact in the little finger.

The Flexor Tendons and Ligamenta Vaginales.—Cut through the very centre of the dense *ligamenta vaginales* and you will come down on the two flexor tendons. Notice that the tendon of the *flexor profundus digitorum* passes through a slit in the tendon of the *flexor sublimis digitorum.* (Fig. 110.) The former is inserted into the base of the third, the latter into the base and sides of the second phalanx. Now gently separate these tendons and you will see (1) the visceral layer of synovial membrane on the tendons, and (2) the vincula— delicate tendinous cords.

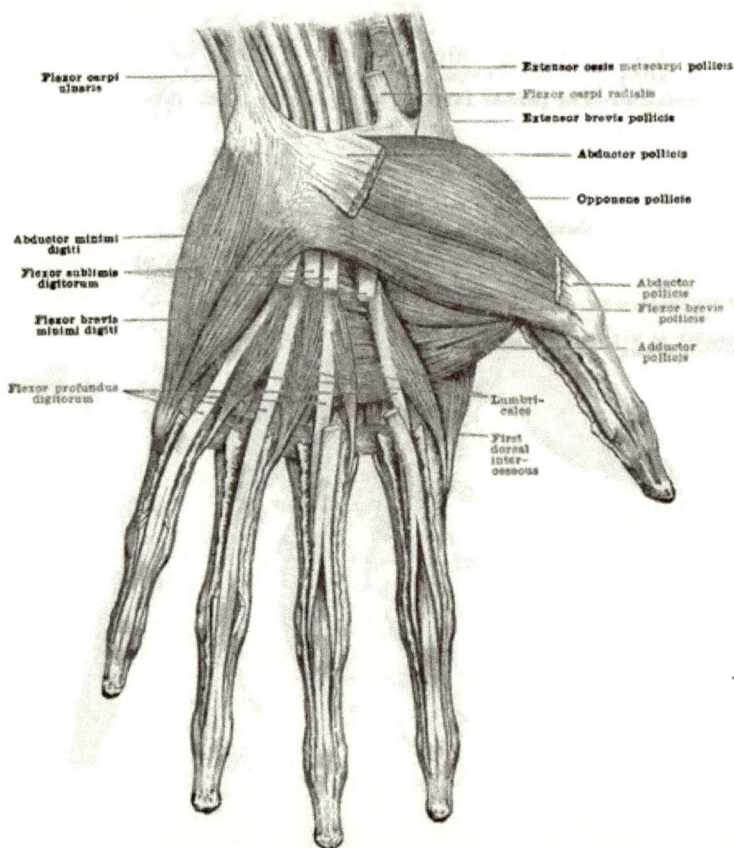

FIG. 112.—THE DEEPER MUSCLES OF THE PALM OF THE HAND.

These heavy, deep fascial sheaths you have just cut through are lined by a synovial vaginal membrane called the **theca.** In the thumb and little finger these synovial cavities communicate with the ulnar and radial **palmar bursæ** above the annular ligament. (Fig. 107.)

The thecæ for the remaining fingers terminate in thecal culs-de-sac opposite the metacarpo-phalangeal articulations. (Fig. 107.)

Infection in the thumb and little finger thus is liable to travel beyond the annular ligament, while in the other fingers it would be arrested at the culs-de-sac.

A case of unusual interest, bearing on this point, reported by Dr. David Loring, of Valparaiso, Indiana, is as follows: "Infection ensued where a patient thrust a nail through the anterior annular ligament. One by one all the carpal bones, the bones of the thumb and the little finger, and the greater part of the metacarpals disappeared by suppuration. At the present time the patient is well, and has the bones of all the fingers between the thumb and little finger." This case is of

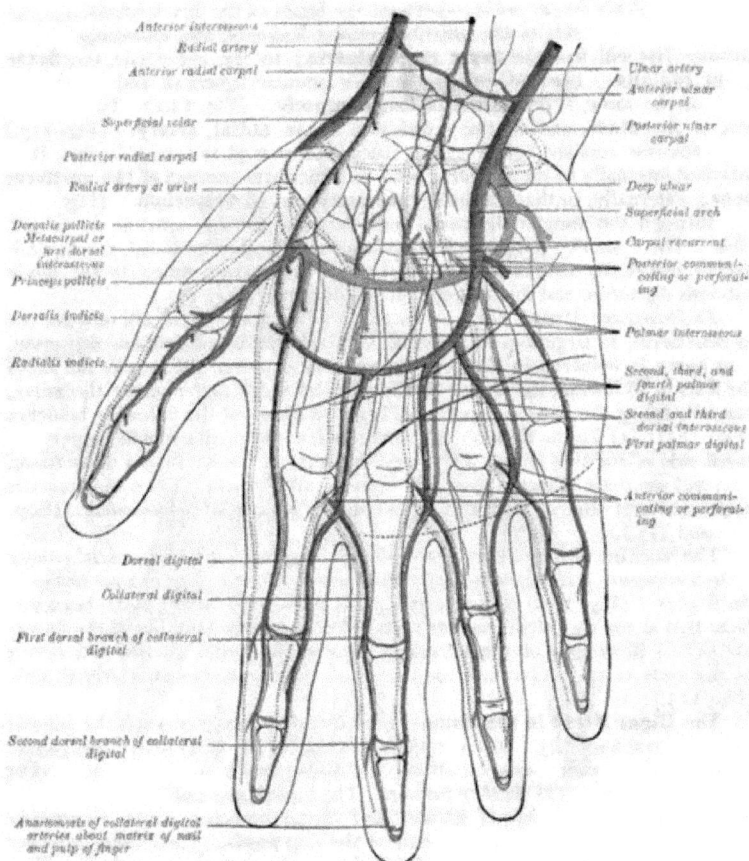

FIG. 113.—ANASTOMOSES AND DISTRIBUTION OF THE ARTERIES OF THE HAND.

interest because the pathological process followed to the letter the anatomical rule.

The vincula will be seen between the flexor tendons. They are delicate, silvery, tendinous threads. They aid the capillaries in passing from one tendon to another. They are, then, mechanical devices for the support of the capillaries that nourish the synovial membrane. In some cases the vessels which they support are injected, and can be seen by the unaided eye.

The palmaris brevis muscle (Fig. 110) arises from the annular ligament and palmar fascia, and is inserted into the skin of the hypothenar eminence.   It is a corrugator of the integument.

The anterior annular ligament (Fig. 111) has passing under it the following structures :

1. The *median nerve* and its nutrient artery—arteria comes nervi mediani.
2. The *flexor sublimis digitorum*—musculus perforatus.
3. The *flexor profundus digitorum*—musculus perforans.
4. The *flexor longus pollicis*—between the heads of the flexor brevis.

To the inner side of the anterior annular ligament, and intimately related thereto, you will find the **flexor carpi ulnaris ;** to the outer side, the **flexor carpi radialis.** Inserted into the anterior annular ligament, and passing in front of the same, is the **palmaris longus muscle.** (Fig. 110.) To the ulnar side is the **ulnar,** and to the radial side is the **radial, artery.** (Fig. 117.) This anterior annular ligament is a specialized part of the deep fascia. It is attached internally to the **pisiform bone** and **unciform process of the unciform bone ;** externally, to the **scaphoid tuberosity** and **os trapezium.** (Fig. 109.) Cut through the annular ligament ; pull the sides forcibly apart as far as possible. There are two portions of the palmar bursa : (1) The radial, corresponding to the flexor longus hallucis ; (2) the ulnar, corresponding to the flexor sublimis digitorum and flexor profundus digitorum. (Fig. 107.)

To Dissect the Median Nerve in the Hand (Fig. 111).—You will now see the median nerve, as large as a lead-pencil, lying on the flexor sublimis digitorum. This nerve is involved in the synovial membrane, and it will require the aid of the forceps to liberate the same.   Place your left index finger under the nerve, carefully lifting the same upward ; now begin to trace out the following branches (Fig. 111) : (1) To the thumb ; (2) to the index finger, the middle finger, and radial side of the ring finger ; (3) a communicating branch to the ulnar nerve. You will see these branches given off from an enlargement   Trace the branches downward and observe them passing behind the superficial palmar arch. (Figs. 111 and 113.)

The branches of the median nerve, after passing behind the superficial palmar arch, accompany arterial branches of the superficial and deep palmar arches to the fingers. (Fig. 111.) These arteries and nerves are called digital branches. Note that at the superficial palmar arch the nerves lie behind, but in the fingers the nerves lie in front of, the arteries. Trace these digital arteries and nerves to the ends of the finger, and see the distal digital anastomotic arterial arch. (Fig. 113.)

The Ulnar Nerve in the Hand.—Find the ulnar nerve crossing the annular ligament (not under it). It lies to the ulnar side of the ulnar artery, and somewhat deeper.   Trace it down and find it dividing into two branches : (1) The *superficial palmar ;* (2) *the deep palmar.*   The former gives off branches to the skin, to the palmaris brevis muscle, and digital branches which accompany arteries to the little finger and one-half of the ring finger. Trace these branches to the finger end and observe the distal digital anastomotic nerve loop. (Fig. 111.) The deep palmar branch of the ulnar nerve accompanies the deep arch, and will be described in its proper place on page 164.

The musculi lumbricales (Fig. 112) are four in number. Cut the tendons of the flexor sublimis as shown in figure 112. These muscles arise from the tendons of the flexor profundus digitorum muscle. Trace them on the thumb side, to the extensor communis digitorum, where they are inserted. The ulnar two interossei muscles are supplied by the ulnar, the radial two by the median, nerve.

The thenar muscles, or muscles of the thumb (Figs. 109 and 114) :

1. The abductor pollicis, abducts first phalanx.

2. The flexor brevis pollicis, flexes first phalanx.
3. The opponens pollicis, acts on metacarpal.
4. The adductor pollicis, adducts the first phalanx.

The abductor pollicis (Fig. 110) is easily isolated. Cut the same near its insertion, and turn it aside so as to preserve the nerve-supply. (Figs. 109 and 114.) This muscle arises from the bone at the base of the thumb—the os trapezium—and the annular ligament. It is inserted into the base of the thumb, proximal phalanx. (Fig. 109.)

The opponens pollicis is inserted into the whole length of the thumb metacarpal bone. (Fig. 109.) It arises from the os trapezium and annular ligament. (Fig. 114.)

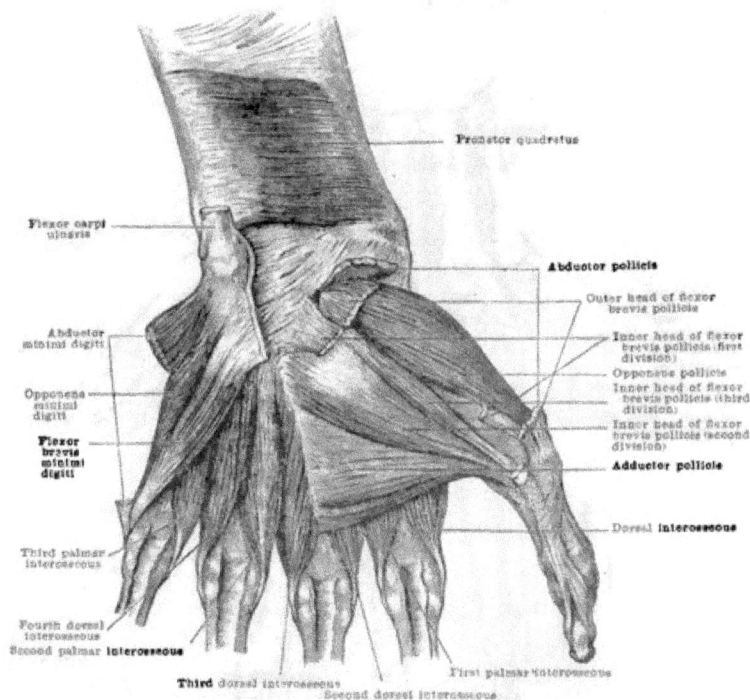

FIG. 114.—THE PRONATOR QUADRATUS AND DEEP VIEW OF THE PALM.

The flexor brevis pollicis has two heads, and between these two heads lies the tendon of the flexor longus pollicis muscle. The outer, or more superficial head (Fig. 114), arises from the os trapezium and annular ligament, and is inserted into the base of the first thumb phalanx on the outer side. (Fig. 109.) The deep or inner head has two parts: (1) A small part that arises from the first metacarpal bone; (2) a large part that arises from the annular ligament, the os magnum, and the second and third metacarpals. (Fig. 114.)

This latter is described by some authors as the oblique part of the adductor pollicis. The deep part of the flexor brevis is *inserted* into the base of the first thumb phalanx on the *inner side*.

The adductor pollicis is the remaining muscle. It arises from the *metacarpal bone* of the middle finger, the lower two-thirds. It is inserted into the base of the first thumb phalanx on the inner side. (Fig. 114.)

Nerve-supply.—The *median* nerve supplies the abductor pollicis, the opponens, and the outer head of the flexor brevis pollicis. The ulnar nerve, by its deep palmar branch, will supply all the remaining muscles of the hand, all the interossei muscles, the two ulnar lumbricals, the little finger group of small muscles, the adductor pollicis, and the inner head of the flexor brevis pollicis. Now trace out and review the branches of the median nerve to the thumb.

The hypothenar muscles, or muscles of the little finger (Fig. 114):

1. The abductor minimi digiti abducts the little finger.
2. The flexor brevis minimi digiti bends the first phalanx.
3. The opponens minimi digiti bends the metacarpal.

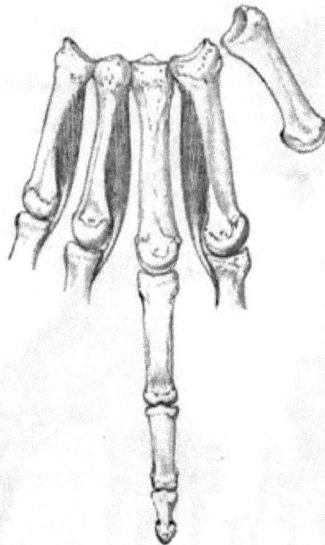

FIG. 115.—THE PALMAR INTEROSSEI.

The abductor minimi digiti arises from the pisiform bone. (Fig. 109.) It is inserted into the base of the first phalanx of the little finger. (Fig. 109.) Cut **and turn** aside, and expose the two remaining muscles—the flexor brevis minimi digiti and the opponens minimi digiti.

Note, as you turn this muscle aside, the abductor minimi digiti—the deep palmar nerve—going to join the deep palmar arterial arch from the radial artery (Fig. 116), whose course it will now follow.

The flexor brevis minimi digiti arises from the unciform process of the unciform **bone and** annular ligament. It is inserted into the base of the first phalanx of the little finger. (Fig. 109.)

The opponens minimi digiti arises from the unciform process of the unciform **bone** and annular ligament. It is inserted into the whole length of the metacarpal bone of the little finger. Trace branches from the deep palmar branch of the ulnar nerve to this **group of muscles.** (Fig. 111.)

The **Palmar Interossei Muscles** (Figs. 114 and 115).—These are three in number. They are physiological adductors. They arise from one side of the metacarpal bone corresponding to the finger, into the base of whose proximal phalanx they are inserted. They are inserted into the bases of the proximal phalanges of the index, little, and ring fingers in such a way as to adduct these fingers to the middle finger. Trace branches to them from the deep **palmar** branch of the ulnar nerve.

The **Deep Palmar Branch of the Ulnar Nerve.**—You have seen the ulnar

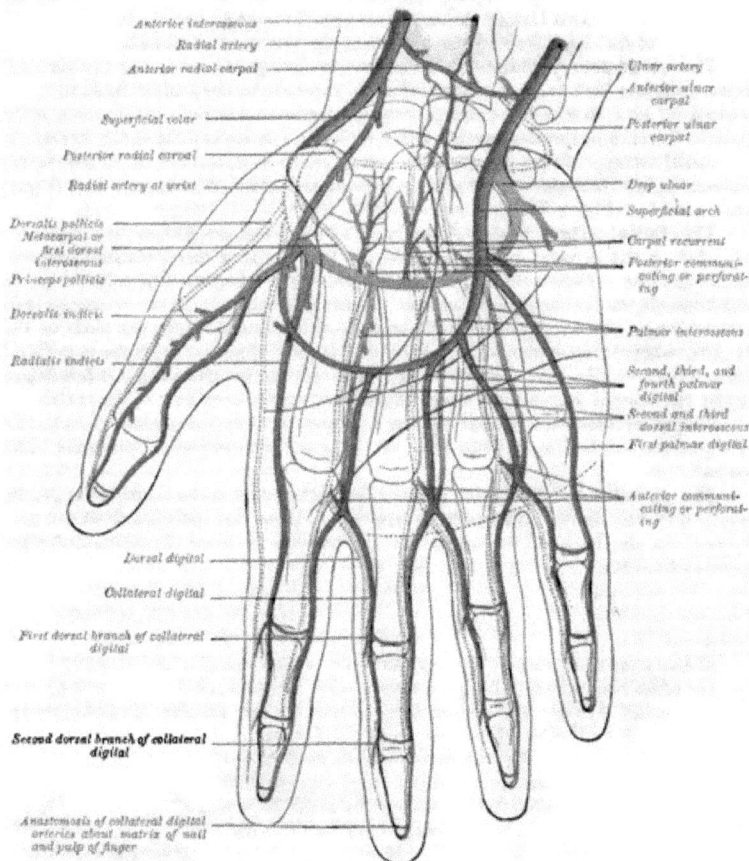

FIG. 116.—ANASTOMOSES AND DISTRIBUTION OF THE ARTERIES OF THE HAND.

nerve crossing the anterior annular ligament, and dividing, below the pisiform bone, into superficial and deep palmar branches. (Fig. 111.) You traced the superficial branches to the palmaris brevis, the skin over the hypothenar eminence, and to the little finger and half the ring finger. (Fig. 111.) You will now trace in review the deep palmar branch, between the abductor minimi digiti and flexor brevis minimi digiti, and see it join company with the deep palmar arterial arch, and send off branches to these muscles:

1. The abductor minimi digiti muscle.
2. The flexor brevis minimi digiti muscle.
3. The opponens minimi digiti muscle.
4. The two ulnar lumbricales muscles.
5. The four dorsal interossei muscles.
6. The three palmar interossei muscles.
7. The adductor pollicis muscle.
8. The inner head of the flexor brevis pollicis.

THE RADIAL AND ULNAR ARTERIES IN THE HAND (Fig. 116).—It now remains to reduce to its *simplest terms* the blood-supply of the whole hand.

The ulnar artery crosses the anterior annular ligament. Below the pisiform bone it divides into (1) the superficial arch and (2) the deep ulnar branch. The *superficial arch* crosses the palm, resting on the flexor tendons under cover of the palmar fascia, and terminating by anastomosis in the superficial volar branch of the radial artery. From the arch are given off the first, second, third, and fourth palmar digital branches that supply the second, third, fourth, and fifth fingers with blood. (Fig. 116.)

The radial artery passes to the back of the hand, under the three extensor muscles of the thumb; passes between the two heads of the first dorsal interosseous muscle; crosses the palm beneath the flexor tendons, and terminates, by anastomosis, in the deep ulnar artery. From this arch are given off five palmar interosseous arteries. Three of these—the three inner—go to the clefts of the fingers, where they unite with the *palmar digital branches* from the superficial palmar arch. The two outermost branches go to the thumb and index finger under the special names of *princeps pollicis* and *radialis indicis.* (Fig. 116.)

Carpal arteries are four in number: two from the radial artery—the anterior and posterior radial carpal; two from the ulnar—the anterior and posterior ulnar carpal.

The anterior and posterior perforating arteries communicate, through the hand, with the dorsal interosseous arteries. These are arteries from an arch formed on the back of the carpus by anastomosis between the radial and ulnar posterior carpals.

Where would you look for the cephalic vein in the vicinity of the shoulder?
Between the deltoid and pectoralis major muscles; the vein is in a groove with a small artery—the descending branch of the acromio-thoracic artery. (Fig. 105.)
How and where do the superficial, radial, median, and ulnar veins terminate?
They terminate opposite the elbow by changing their names.
The radial above the elbow is called the cephalic vein.
The ulnar above the elbow is called the basilic vein.
The median breaks up near the elbow into the median basilic and cephalic.
What are the venæ comites?
The two companion veins attending deep arteries in the extremities of the body, below the knee and below the elbow.
How do the superficial veins communicate with the venæ comites?
By perforations in the deep fascia. By these communicating veins the circulation internal to or below the deep fascia is equalized to the circulation external to the deep fascia.
If you were going to draw blood from a patient's arm, why would you select the median cephalic vein in preference to the median basilic vein for venesection?

The median basilic vein lies on the brachial artery, which latter would be endangered were this vein selected.

Suppose you employ the median basilic vein in venesection, and accidentally cut the brachial artery, and your patient dies from hemorrhage: In this case what ground for action for damages would relatives of deceased have that you could not possibly circumvent?

The ground of *want of ordinary care.* The cephalic vein would have answered the venesection purpose, and its employment would not have endangered your patient's life. Under the law, a physician must possess *ordinary* skill and exercise *ordinary* care.

Can you say anything regarding any cutaneous nerves liable to be injured in phlebotomy of the median cephalic and median basilic veins?

The internal cutaneous nerve crosses the median basilic vein ; the musculo-cutaneous nerve terminal crosses the median cephalic—either may be injured.

Name the great nerve-trunks that supply the skin of the hand with sensation.

The median gives palmar and digital branches to 3.5 fingers, radial side.

The ulnar gives palmar and digital branches to 1.5 fingers, front and back. The musculo-spiral, radial nerve, supplies 3.5 fingers dorsally.

Have the median, ulnar, and musculo-spiral nerves any articular distribution in the joints of the hand?

Yes ; they supply joints between the phalanges—the metacarpo-phalangeal, the carpo-metacarpal, the intermetacarpal, the medio-carpal, and the radio-carpal.

Is there any systematic distribution of articular nerves?

Yes ; a nerve-trunk that supplies a muscle that moves a joint supplies the joint as well. The inferior radio-ulnar joint is pronated by the pronator quadratus, and supinated by the supinator longus. To determine the nerve-supply to this joint, you ascertain the nerve-supply to the muscles that move the joint. In this particular case the median and musculo-spiral nerve-trunks supply the pronator quadratus and supinator longus muscles respectively ; these same nerves supply the joint.

Give attachments of the anterior annular ligament and tell what structures pass under the same.

This ligament is a specialized part of the deep fascia. It is attached externally to the scaphoid bone and os trapezium ; internally to the pisiform bone and the unciform process of the unciform bone. Under this ligament pass the median nerve, with its median artery, the flexor longus pollicis, the flexor profundus digitorum, and the flexor sublimis digitorum.

Name, locate, and give importance of three surgical areas of the fingers.

(1) The central, in which are found the tendons of the flexors in their synovial sheath, called theca. (2) On each side of this tendinous area is a neuro-vasal area, in which are found the digital arteries and nerves that supply the fingers with blood and sensation respectively.

Why is infection of the little finger or thumb more dangerous than infection of the fingers between these two extremes.

The vaginal thecal sheaths of synovial membrane of these extremes are in communication with the great palmar bursæ under the anterior annular ligament.

Give the symptoms in paralysis of the ulnar nerve.

There would be loss of sensation and motion in those parts to which this nerve is distributed, as follows : (1) Loss of sensation in the skin of the little finger and the ulnar half of the ring finger and in the corresponding articulations ; (2) loss of motion in the flexor carpi ulnaris and in half of the flexor profundus digitorum muscles ; (3) loss of motion in all the interossei muscles and in the two lumbricals on the ulnar side of the hand ; (4) loss of motion in the flexor brevis minimi digiti, in the abductor minimi digiti, and in the opponens minimi digiti ;

(5) loss of motion in the adductor pollicis and in the ulnar head of the flexor brevis pollicis ; (6) loss of sensation in all the intermetacarpal joints.

Give symptoms in paralysis of the median nerve.

(1) Loss of sensation in the thumb, index, middle, and half the ring finger, on the palmar surface, and about the roots of the nails of the dorsal surface ; (2) loss of motion in all the muscles of the anterior part of the forearm, except the muscle and a half supplied by the ulnar nerve, as previously explained ; (3) loss of motion in the two lumbrical muscles on the radial side of the hand, and loss of motion in the outer head of the flexor brevis pollicis, in the opponens pollicis, in the abductor pollicis.

Describe the blood-supply of the hand.

The ulnar and radial arteries terminate in the superficial and deep palmar arches respectively. These arches give off branches to the hand and fingers. (Fig. 116.)

Describe the superficial palmar arch.

This arch is a continuation of the ulnar artery across the hand. It anastomoses with the superficialis volæ—a branch of the radial artery—to complete the arch. The arch may be completed by anastomosis with the radialis indicis or princeps pollicis, both branches of the radial artery also. The superficial arch gives off four branches—called first, second, third, and fourth palmar digital branches—when enumerated toward the thumb. The first palmar digital artery supplies the ulnar side of the little finger ; each of the three others divides into two collateral digital branches. The radial side of the index finger is supplied by the radialis indicis ; the thumb by the princeps pollicis. The digital branches from the superficial arch are joined, opposite the clefts of the fingers, by (1) the palmar interosseous, branches of the deep palmar arch ; (2) by the inferior perforating arteries from the dorsum of the hand.

Describe the deep palmar arch.

This arch is a continuation of the radial artery across the hand. It anastomoses with the deep branch of the ulnar artery. (Fig. 116.) It begins at the first interosseous space, and rests on the metacarpal bones close to their carpal ends. The arch is attended by the deep branch of the ulnar nerve as described on page 164. The deep arch gives off three palmar interosseous arteries, which inosculate with the digital branches of the superficial arch at the clefts of the fingers. All the digital branches are attended by collateral nerve branches of the median and ulnar nerves.

THE FOREARM.

In this region you must review the osteology of the radius and ulna, and become familiar with eminences, depressions, that are associated with the origin, insertion, and location of muscles. You must be able to name all the articular surfaces, and give the rule of occupancy for their names. You must name technically all articulations, and give the rule for writing compound words.

The Radius (Fig. 124).

Name the important points on this bone concerned in practical anatomy.

The radial head and its concave and convex articular surfaces.

The bicipital tuberosity for the tendon of the biceps and for a bursa.

The ulnar articular surface for the lesser sigmoid cavity of the ulna.

The humeral articular surface for articulation with the capitellum.

The neck of the radius located above the bicipital tuberosity.

The oblique line of the radius—a very important structure.

The styloid process, with base and apex for ligament and muscle.

The sigmoid cavity, for articulation with the head of the ulna.

The semilunar and scaphoid articular surfaces.

The nutrient foramen, directed upward according to rule.

The interosseous border, for the interosseous membrane.

The anterior surface, occupied by deep flexor muscles.

The posterior surface, occupied by deep extensor muscles.

The outer surface, occupied by muscles.

Describe the anterior surface of the radius.

The oblique line of the radius extends from the bicipital tuberosity to the insertion of the pronator radii teres on the outer border of the bone. (Fig. 124.) This line may be viewed as consisting of three lips: (1) An upper one, into which the supinator brevis is inserted; (2) a middle one, from which the radial head of the flexor sublimis digitorum muscle takes its origin; (3) an inferior one, from which the flexor longus pollicis takes its origin. This surface gives origin below the oblique line to the flexor longus pollicis and insertion to the pronator quadratus.

What can you say of the external surface of the radius?

Its upper one-half is occupied by the supinator brevis and pronator radii teres; its lower one-half is overlapped by the radial extensors of the carpus, and crossed by the extensor ossis metacarpi pollicis and extensor brevis pollicis; this latter muscle is also called extensor primi internodii pollicis.

What can you say of the posterior surface of the radius?

Its upper one-third gives origin to the radial part of the extensor ossis metacarpi pollicis and the extensor primi internodii pollicis; the part of the bone below this is covered by the tendons of the two preceding muscles.

The Ulna. (Fig. 124.)

Name the important bony parts of the ulna and give their importance in practical anatomy.

The olecranon process, for the insertion of the triceps muscle.

The coronoid process, for the insertion of brachialis anticus muscle.

The greater sigmoid fossa, for articulation with the humerus.

The lesser sigmoid fossa, for articulation with the radius.

The oblique line, for attachment of the supinator brevis muscle.

The interosseous ridge, for attachment of the interosseous membrane.

The styloid process (apex), for internal lateral ligament of wrist.

The head articulates with the ulnar sigmoid cavity of the radius.

The base of the styloid process, for the insertion of the supinator longus.

The anterior surface of the ulna is for muscular origin.

The inner surface of the ulna is for muscular origin.

The posterior surface of the ulna is for muscular origin.

Explain the anterior surface of the ulna.

It has the nutrient foramen, which, according to the rule, is directed toward the elbow. It gives origin to the flexor profundus digitorum and pronator quadratus muscles. This surface is limited by the anterior border, the interosseous border, and an oblique line.

What can you say of the internal surface of the ulna?

It is occupied by the flexor profundus digitorum in its upper three-fourths; the remainder of this surface of the ulna is subcutaneous.

What can be said of the posterior surface of the ulna?

This is the most difficult part of the bone, but must be mastered before you can appreciate the attachment of certain muscles. This surface is subdivided by two lines, an oblique and a vertical one, into three smaller surfaces. **Above the** oblique line, which runs from the lesser sigmoid cavity to the posterior border of the bone, is the anconeus muscle. To the oblique line the supinator brevis muscle **is** attached. The surface below the oblique line is divided by a vertical line into an internal and an external portion.

What can you say of the portion of the posterior surface of the ulna internal to the vertical line?

It gives **origin to** the extensor carpi ulnaris muscle.

What can you say of the portion of the posterior surface of the ulna external to the vertical line?

It gives attachment to the supinator brevis, the extensor ossis metacarpi pollicis, the extensor longus pollicis, and the extensor indicis muscles.

Does the head of the ulna enter into the formation of the wrist-joint?

No; the triangular fibro-cartilage intervenes. The head of the ulna articulates with the sigmoid fossa in the lower end of the radius, forming the inferior radio-ulnar articulation.

The Humerus. (Fig. 129.)

Locate, classify, and describe geometrically the humerus.

It is the arm bone, located between the shoulder and elbow, and long by classification. It has an outer, an inner, and a posterior surface. It has an anterior, an external, and an internal border. It has a superior extremity or upper one-third; an inferior extremity or lower one-third; a middle one-third.

Name the bony parts of the humerus concerned in practical anatomy, and give their practical importance.

The head of the humerus articulates with the head of the scapula.

The anatomical neck gives attachment to the capsule.

The surgical neck is very often the seat of fracture.

The bicipital groove lodges the long head of biceps muscle.

The bicipital lips surmount and deepen the bicipital groove.

The greater tuberosity gives tendinous insertion to three muscles.

The lesser tuberosity gives tendinous insertion to one muscle.

The internal condyle is developed by traction of flexors and pronators.

The internal condylar ridge has the internal intermuscular septum.

The external condyle is developed by extensors and supinators.

The external condylar ridge has the external intermuscular septum.

The capitellum articulates with the concave surface of the radius.

The trochlea articulates with the greater sigmoid of the ulna.

The coronoid fossa is produced mechanically by the coronoid process.

The olecranon fossa is produced mechanically by the olecranon process.

The radial depression is produced by the head of the radius.

Name the muscles inserted into the greater tuberosity of the humerus.

The supraspinatus into the upper facet; the infraspinatus into the middle facet; **the teres** minor into the lower facet.

How is the lesser tuberosity occupied?

By the tendinous insertion of the subscapularis muscle. It has one facet only.

Where is the transverse humeral ligament and what is its function?

It extends from the greater to the lesser tuberosity of the humerus, converting the bicipital groove into a canal for the lodgment of the long or scapular head of the biceps muscle.

How would you locate the surgical neck of the humerus?

It is the constricted portion of the bone below the tuberosities. It is the portion of the humerus most frequently-fractured in the upper one-third.

How do the three fossæ of the humerus, in the vicinity of the elbow enhance the gravity of a fracture passing through them?

They may become filled with provisional callus and bone, when a stiff joint will be the result.

How is the posterior surface of the humerus occupied?

By the long humeral head of the triceps above the musculo-spiral groove; by the short humeral head below the groove; by the musculo-spiral groove, in which are the musculo-spiral nerve and superior profunda artery.

How is the outer lip of the bicipital groove occupied?

By the tendinous insertion of the pectoralis major muscle.

How are the outer and inner surfaces of the humerus occupied?

They are occupied by the pectoralis major, deltoid, and brachialis anticu on the outer; the teres major, latissimus dorsi, coraco-brachialis, and brachialis anticus on the inner surface.

How is the external condylar ridge occupied?

By the supinator longus and the extensor carpi radialis longior; the former occupies the upper two-thirds, the latter the lower one-third of the ridge.

Name all the muscles attached to the middle one-third of the humerus whose action would tend to displace the fragments in case of fracture in this locality.

The long and short humeral heads of the triceps, the deltoid, the coraco-brachialis, the brachialis anticus muscles.

What might be the most serious complication of a fracture in the middle one-third of the humerus?

Compression and paralysis of the musculo-spiral nerve, in which there would be inability to extend the carpus and digits and to supinate the radius.

What is the direction taken by the nutrient foramen of the humerus?

It conforms to the rule governing the direction of these foramina in the long bones of the extremities: to the elbow and from the knee being the rule.

How many muscles and ligaments are attached to the humerus?

There are twenty-five muscles and eight ligaments attached.

In this region you will find the following arrangement of—

Muscles *Physiologically* Grouped.

Carpal Flexors.	Flexor carpi radialis.
	Flexor carpi ulnaris.
	Palmaris longus.
Digital Flexors.	Flexor sublimis digitorum.
	Flexor profundus digitorum.
	Flexor longus pollicis.
Radial Pronators.	Pronator radii teres.
	Pronator radii quadratus
Forearm Flexors.	Flexor biceps cubiti.
	Flexor cubiti brachialis anticus.
	Flexor brachio-radialis.
Nerves.	Median nerve and its branches.
	Ulnar nerve and its branches.
	Musculo-cutaneous nerve and its branches.
Arteries and Veins.	Radial artery and the deep palmar arch.
	Ulnar artery and the superficial palmar arch.
	Common interosseous artery and vein.
	Anterior interosseous artery and vein.
	Posterior interosseous artery and vein.
Surgical Areas.	The cubital fossa and its contents.
	The radial groove and its contents.
	The ulnar groove and its contents.

The cubital fossa has geometrically a—

1. *Roof*, formed by the skin and fasciæ—superficial and deep.
2. *Floor*, formed by the brachialis anticus and supinator brevis.
3. *Superior border*, formed by an imaginary line from condyle to condyle.
4. *Outer border*, formed by the brachio-radialis muscle.
5. *Inner border* formed the pronator radii teres muscle.

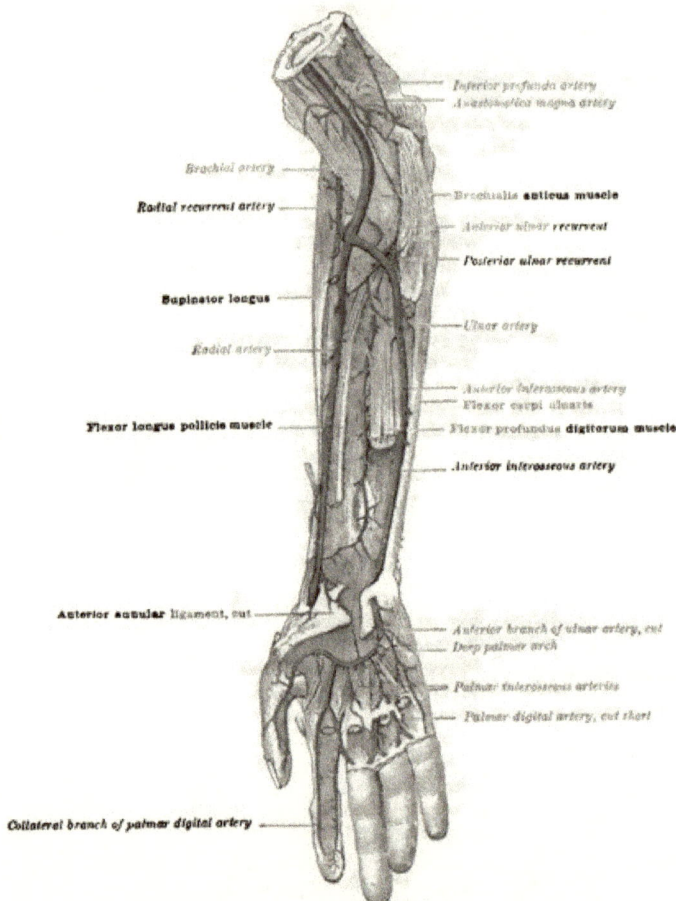

Inferior profunda artery
Anastomotica magna artery
Brachial artery
Radial recurrent artery
Brachialis anticus muscle
Anterior ulnar recurrent
Posterior ulnar recurrent
Supinator longus
Ulnar artery
Radial artery
Anterior interosseous artery
Flexor carpi ulnaris
Flexor longus pollicis muscle
Flexor profundus digitorum muscle
Anterior interosseous artery
Anterior annular ligament, cut
Anterior branch of ulnar artery, cut
Deep palmar arch
Palmar interosseous arteries
Palmar digital artery, cut short
Collateral branch of palmar digital artery

FIG. 117.—THE ARTERIES OF THE FOREARM WITH THE DEEP PALMAR ARCH.

The fossa contains the following structures—
(1) The brachial artery and veins ; (2) the radial artery ; (3) the ulnar artery ; (4) the tendon of the biceps muscle ; (5) the median nerve ; (6) the musculo-spiral nerve ; (7) the radial recurrent artery ; (8) the ulnar recurrent artery ; (9) the common interosseous artery.

The radial groove has these geometrical parts :

1. *Roof*, formed by skin and fasciæ—superficial and deep.
2. *Floor*, formed by flexor longus pollicis and pronator quadratus.
3. *Outer border*, formed by supinator longus or brachio-radialis muscle.
4. *Inner border*, formed by the flexor carpi radialis and pronator radii teres.
5. *Contents*, consisting of the radial artery and its venæ comites.

The ulnar groove has these geometrical parts:

1. *Roof*, formed by the skin and fasciæ, superficial and deep.
2. *Floor*, formed by the flexor profundus digitorum muscle.

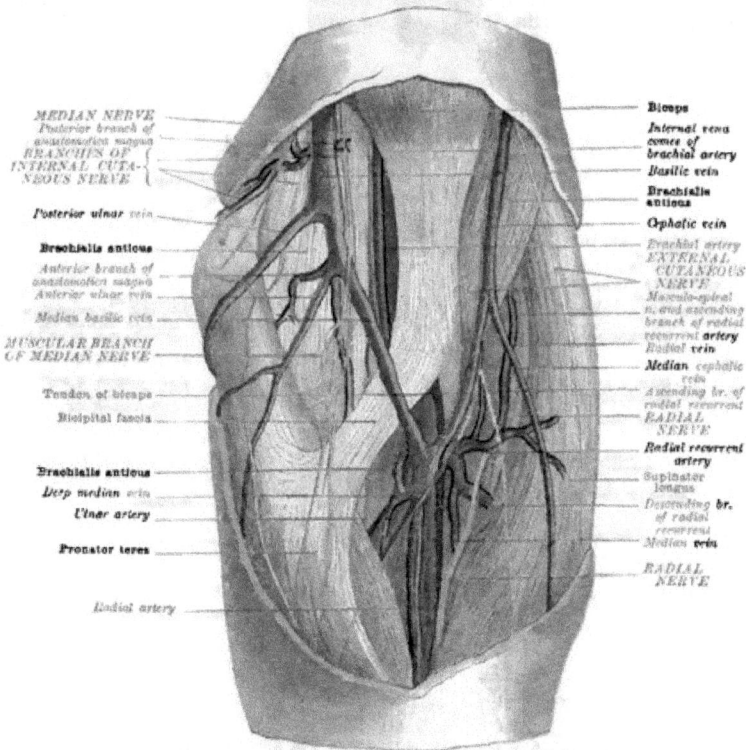

FIG. 118.—THE BEND OF THE ELBOW WITH THE SUPERFICIAL VEINS.
(From a dissection by Dr. Alder Smith in the Museum of St. Bartholomew's Hospital.)

3. *Outer border*, formed by the flexor sublimis digitorum muscle.
4. *Inner border*, formed by the flexor carpi ulnaris muscle.
5. *Contents*, which are: ulnar artery, venæ comites, and ulnar nerve.

Dissection (Fig. 117). **The Radial Artery.**—Cut through the roof of the groove. Then, with your scissors and forceps or director, follow the artery up to the *bifurcation of the brachial artery*. Observe the numerous small muscular branches that are given off by this artery both to muscles and skin.

The Ulnar Artery.—Cut through the roof of its groove and follow it up as far as you can without cutting into the flexor sublimis digitorum muscle. Note the presence of the ulnar nerve on its inner or ulnar side. Find also where this nerve gives off its dorsal ulnar cutaneous branch, to be distributed to the dorsum of the little finger and half of the ring finger.

The Cubital Fossa.—Figure 118 shows the superficial dissection of this area you have already made. Now cut the bicipital fascia with scissors, and develop the brachial artery and its branches, and make your dissection look like figure 119. Note (1) the brachial artery ; (2) to the inner side of this artery the median nerve ; (3) to the outer side of the artery the tendon of the biceps muscle ; (4) to the outer side of the tendon of the biceps, see the musculo-spiral nerve, deeply located between the brachialis anticus muscle and the supinator longus muscle, dividing into the radial and posterior interosseous nerves. Find also

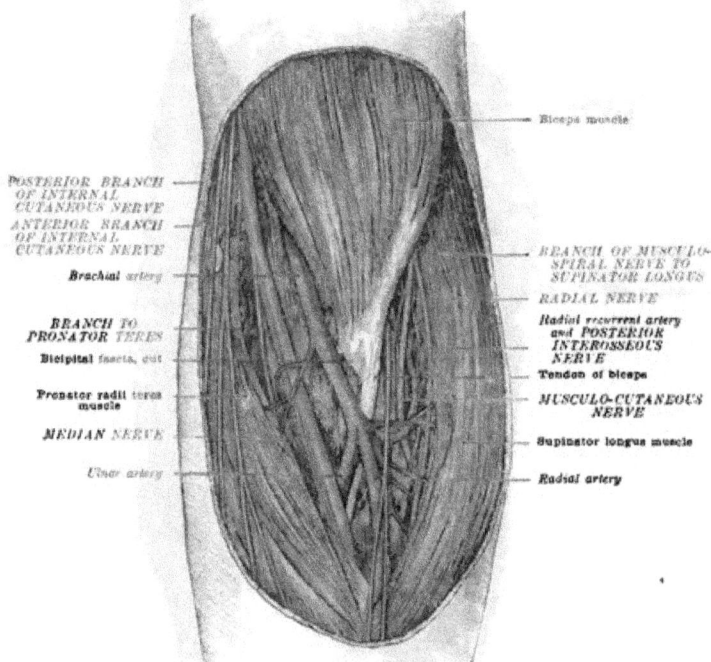

FIG. 119.—THE BRACHIAL ARTERY AT THE BEND OF THE ELBOW.

the muscular branches to the pronator radii teres and supinator longus muscles, from the median and musculo-spiral nerves respectively.

Carefully remove the deep fascia from the muscles and your dissection of the cubital fossa will look like figure 120. You will now find each tendon individually, as in figure 120, near the annular ligament ; trace it up to the muscular mass from which it comes ; then refer to the figure for the name of the structure. Caution : As you thus in cleaning the muscles one by one remove the deep fascia from the muscles, do not destroy the superficial vessels and nerves—divide only the connective tissue.

For convenience of dissection develop the muscles of first group (Fig. 120) :

1. *Pronator radii teres ;* origin, inner condyle and coronoid process.

2. *Palmaris longus ;* origin, inner condyle by the common tendon.

3. *Flexor carpi radialis ;* origin, inner condyle by the common tendon.

4. *Flexor carpi ulnaris;* origin, inner condyle and olecranon. The flexor carpi ulnaris has two heads, with the ulnar nerve between them. This muscle also

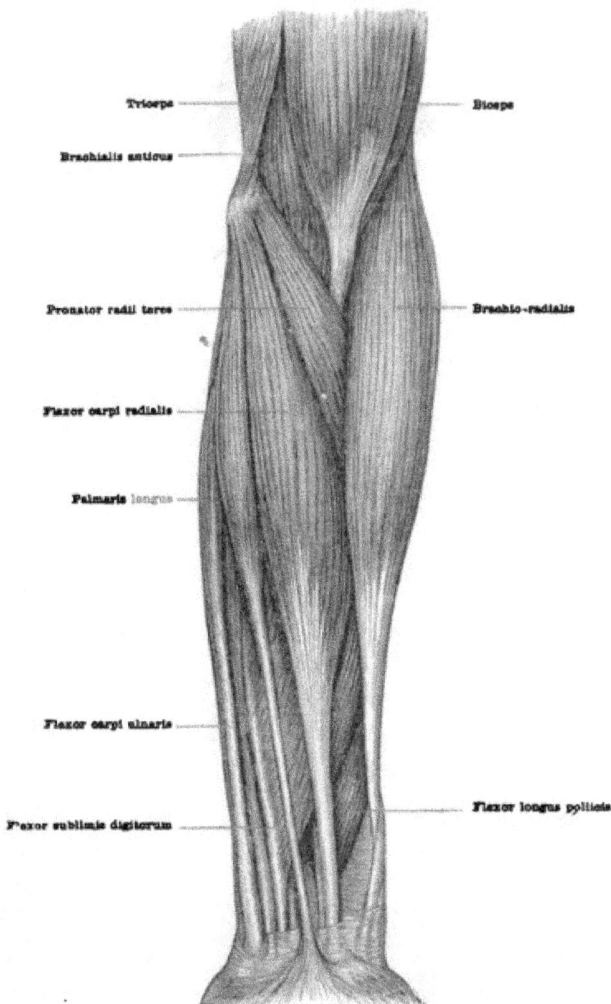

FIG. 120.—FRONT OF THE FOREARM: FIRST LAYER OF MUSCLES.

forms one of the boundaries for the ulnar groove. Notice carefully the extensive insertion of the flexor carpi ulnaris muscle into (1) the pisiform bone, (2) the unciform bone, (3) the anterior annular ligament, (4) the base of the fifth metacarpal bone.

The flexor carpi radialis muscle (Fig. 120) is seldom properly dissected by the student to its specific insertion. Trace its tendon through two canals, and to two bones for insertion; one canal is on the outer side of the **annular ligament,** the other is in the bone at the base of the thumb—the **os trapezium.** The muscle is inserted into the bases of the second and third metacarpals.

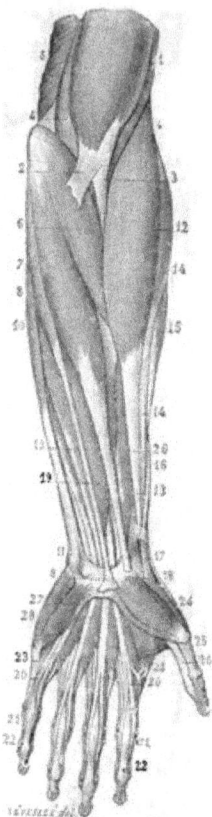

FIG. 121.—SUPERFICIAL MUSCLES OF PALMAR ASPECT OF FOREARM.

1. Lower portion of biceps. 2. Bicipital fascia. 3. Tendon of insertion into radius. 4, 4. Brachialis anticus. 5. Internal head of triceps 6. Pronator radii teres. 7. Flexor carpi radialis. 8. Palmaris longus. 9. Its termination in palmar ligament. 10. Flexor carpi ulnaris. 11. Its attachment to pisiform bone. 12. Supinator longus. 13. Its attachment to styloid process of radius. 14, 14. Extensor carpi radialis longior. 15. Extensor carpi radialis brevior. 16. Extensor ossis metacarpi pollicis. 17. Its tendon of insertion into base of first metacarpal bone. 18. Tendon of extensor secundi internodii pollicis. 19, 19. Flexor sublimis digitorum. 20, 20. Tendons of this muscle. 21, 21. Their attachment to second phalanges of fingers. 22, 22. Attachment of tendons of flexor profundus digitorum to last phalanges of fingers. 23, 23. Lumbricales. 24. Abductor pollicis. 25. Its insertion into first phalanx of thumb. 26, 26. Flexor longus pollicis. 27. Flexor brevis minimi digiti. 28. Abductor minimi digiti.

The **palmaris longus muscle** is properly a tensor of the palmar fascia. The muscle is inserted into the annular ligament and also into the palmar fascia. It is often absent. It is a decided flexor of the carpus.

The **pronator radii teres muscle** must first be separated from the flexor carpi radialis, by dividing the connective tissue with scissors and forceps. Next pull the brachio-radialis outward, and expose the insertion of the pronator radii teres into the middle third of the outer surface of the radial shaft. Now trace the

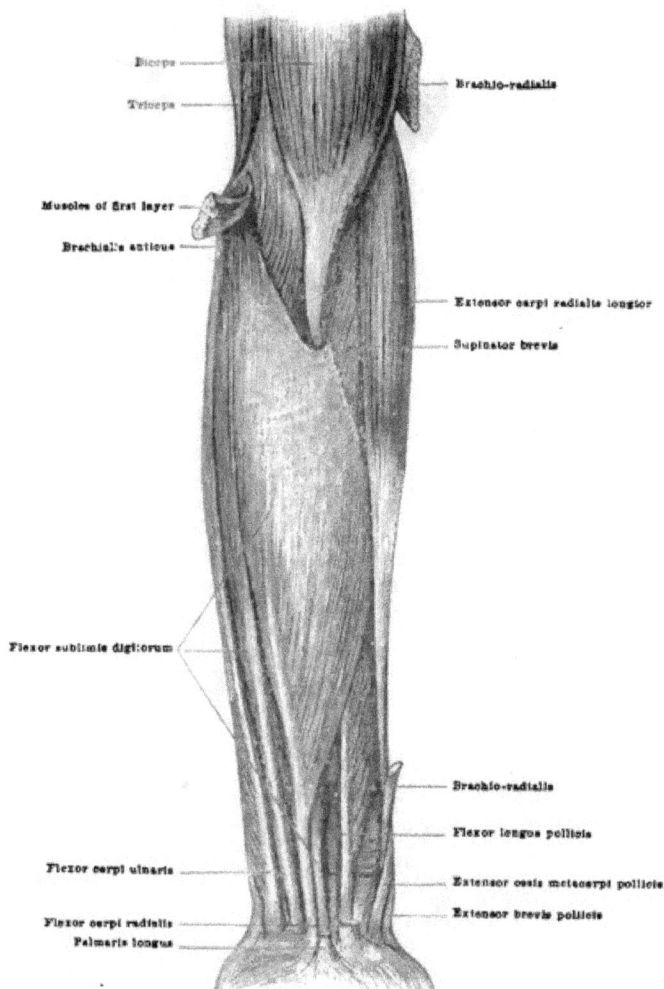

FIG. 122.—FRONT OF THE FOREARM: SECOND LAYER OF MUSCLES.

median nerve down, and see it pass between the two heads of the pronator radii teres, to reach the under surface of the flexor sublimis digitorum. (Fig. 119.)

The **ulnar artery's** course (Fig. 117) to reach its groove as described on page 172 : Cut the origins at the internal condyle, of the pronator radii teres,

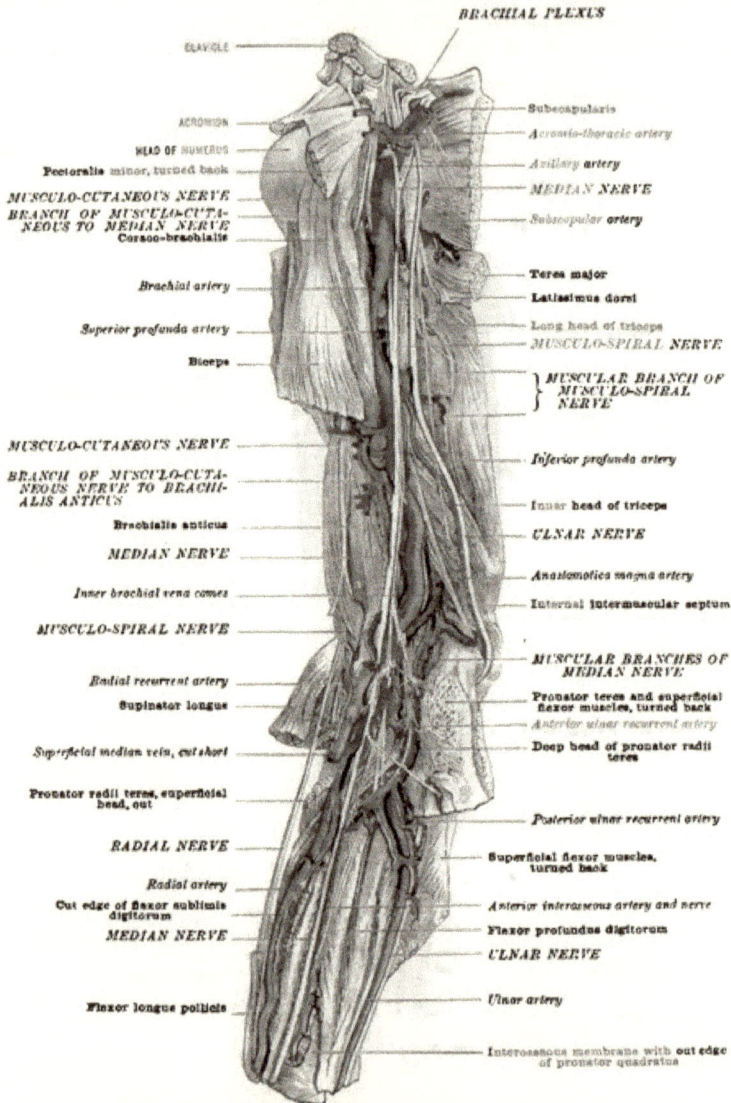

FIG. 123.—THE LOWER PART OF THE AXILLARY, THE BRACHIAL, AND THE RADIAL
AND ULNAR ARTERIES.
(From a dissection in the Hunterian Museum.)

the flexor carpi radialis, the palmaris longus, flexor sublimis digitorum, and turn them aside, but do not injure the nerves and vessels to the same. Find the ulnar artery, and you will see it passes behind all these three muscles. Now trace it a little further, and see it pass behind the flexor sublimis digitorum and median 'nerve. Here you will find it has reached the ulnar groove. Remem-

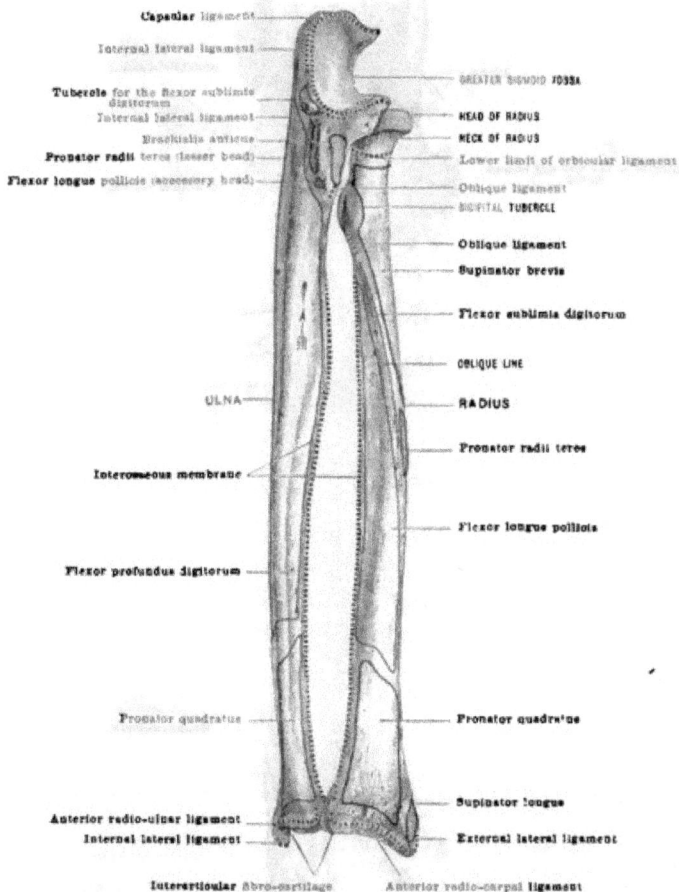

Capsular ligament
Internal lateral ligament
Tubercle for the flexor sublimis digitorum
Internal lateral ligament
Brachialis anticus
Pronator radii teres (lesser head)
Flexor longus pollicis (accessory head)
GREATER SIGMOID FOSSA
HEAD OF RADIUS
NECK OF RADIUS
Lower limit of orbicular ligament
Oblique ligament
BICIPITAL TUBERCLE
Oblique ligament
Supinator brevis
Flexor sublimis digitorum
OBLIQUE LINE
ULNA
RADIUS
Pronator radii teres
Interosseous membrane
Flexor longus pollicis
Flexor profundus digitorum
Pronator quadratus
Pronator quadratus
Supinator longus
Anterior radio-ulnar ligament
Internal lateral ligament
External lateral ligament
Interarticular fibro-cartilage
Anterior radio-carpal ligament

FIG. 124.—THE LEFT ULNA AND RADIUS. (Antero-internal view.)

ber the course: behind the median nerve and all the muscles originating from the inner condyle, except the flexor carpi ulnaris.

The radial's course (Fig. 117): You will find it very superficial. It begins at the bifurcation of the brachial artery in the cubital fossa. Now demonstrate the fact that this artery lies on: (1) Biceps tendon; (2) supinator brevis; (3) flexor sublimis digitorum; (4) pronator radii teres; (5) flexor longus hallucis; (6) pronator quadratus. These structures form the floor of the radial groove.

The flexor sublimis digitorum (Fig. 122) is brought into view by turning the above muscles well aside. You will see the radial artery lying on its outer part; the ulnar artery and median nerve above, passing behind the arch connecting the two heads of the flexor sublimis digitorum muscle. This muscle has three heads : (1) A condylar, arising from the inner condyle of the humerus ; (2) a

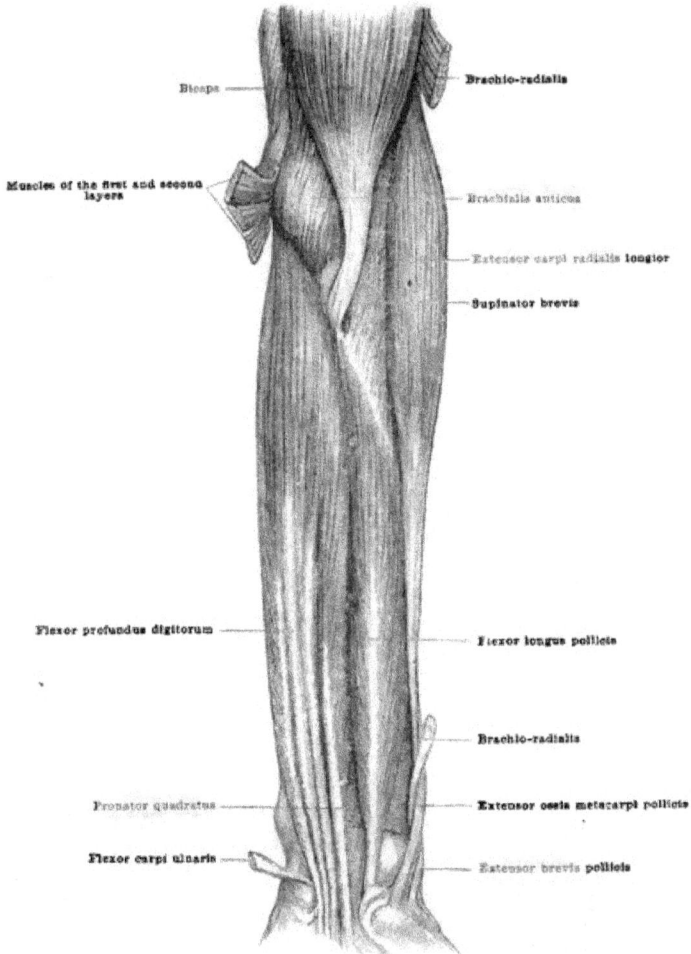

Bicaps

Brachio-radialis

Muscles of the first and second layers

Brachialis anticus

Extensor carpi radialis longior

Supinator brevis

Flexor profundus digitorum

Flexor longus pollicis

Brachio-radialis

Pronator quadratus

Extensor ossis metacarpi pollicis

Flexor carpi ulnaris

Extensor brevis pollicis

FIG. 125.—FRONT OF THE FOREARM: THIRD LAYER OF MUSCLES.

coronoid, from the coronoid process of the ulna : (3) a radial head, from the oblique line of the radius. (Fig. 124.) Cut the radial origin, and turn the muscle toward the ulna, without injury to the radial artery, and see : (1) The median nerve, on the under surface of the muscle; (2) the insertion of the brachialis anticus muscle into the base of the coronoid process of the ulna ; (3)

the origin of the flexor longus pollicis muscle; (4) the origin of the flexor profundus digitorum; (5) the anterior interosseous artery and nerve on the interosseous membrane, but lying deeply between two muscles—the flexor profundus digitorum and the flexor longus pollicis.

The Median Nerve and its Branches.—Take the nerve up, and you can easily trace muscular branches with the forceps to all the muscles on the front

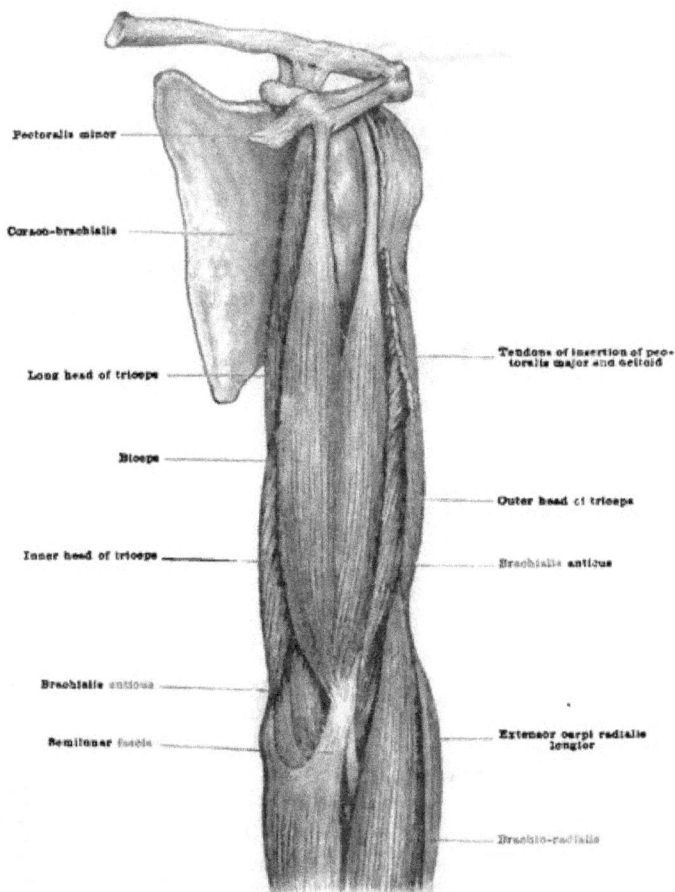

Pectoralis minor

Coraco-brachialis

Long head of triceps

Biceps

Inner head of triceps

Brachialis anticus

Semilunar fascia

Tendons of insertion of pectoralis major and deltoid

Outer head of triceps

Brachialis anticus

Extensor carpi radialis longior

Brachio-radialis

FIG. 126.—SUPERFICIAL VIEW OF THE FRONT OF THE UPPER ARM.

part of the forearm except the flexor carpi ulnaris and half of the flexor profundus digitorum muscles; the flexor carpi ulnaris and the ulnar half of the flexor profundus digitorum muscles are supplied by the ulnar nerve. You will observe that some of the branches of this nerve are given off above the articulation. (Fig. 123.) Remember, the shape of a muscle is an index to nerve distribution. See introductory chapter.

The ulnar nerve (Fig. 123) you have already seen in the ulnar groove with its accompanying artery and venæ comites. Now see how it gains this groove. Trace it behind the inner condyle, between the condylar and olecranon heads of the flexor carpi ulnaris muscle. Trace its branches to this muscle and to the ulnar half of the flexor profundus digitorum.

The following muscles remain to be dissected (Fig. 125):

1. The *flexor profundus digitorum*, on the ulnar side.

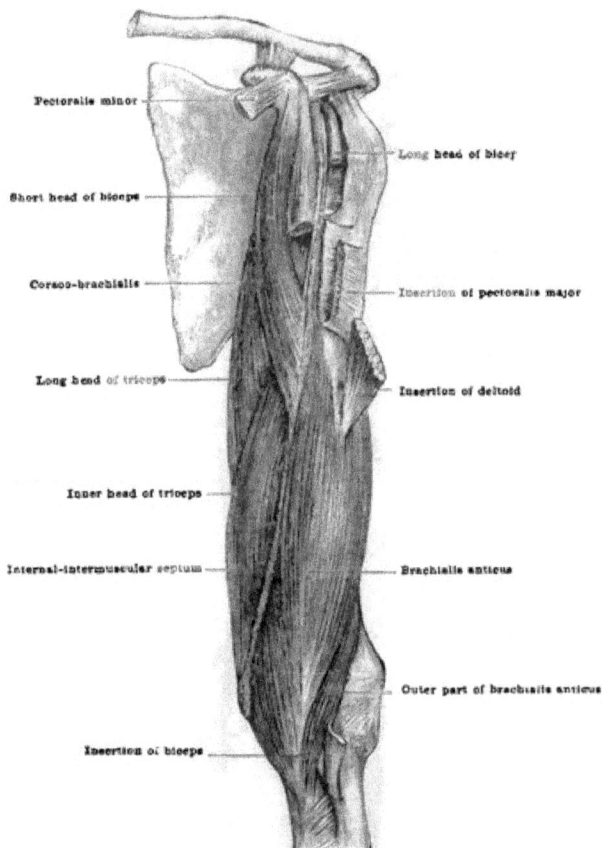

FIG. 127.—DEEP VIEW OF THE FRONT OF THE UPPER ARM.

2. The *flexor longus pollicis*, on the radial side.
3. The *pronator radii quadratus.*
4. The *supinator radii brevis.*

You are expected to dissect these muscles and study critically their specific attachments to bone, and make your dissection tally with the origins as indicated on the bones in figure 124.

The flexor profundus digitorum is seldom well learned, because students

5444445

forget its aponeurotic origin. The muscle originates (1) from the anterior surface, upper two-thirds of the ulna; (2) from the upper three-fourths of the posterior border of the ulna, with the flexor carpi ulnaris and extensor carpi ulnaris muscles. This latter is called the aponeurotic origin. Trace nerves to this muscle from the median and ulnar.

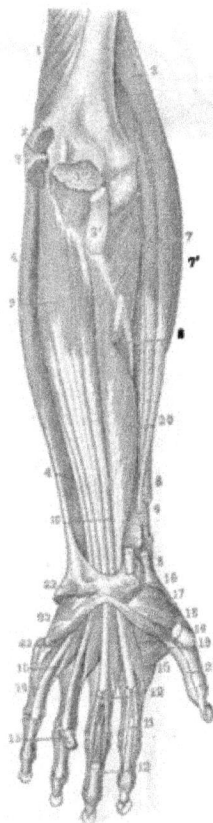

FIG. 128.—DEEP MUSCLES OF PALMAR ASPECT OF FOREARM.

1. Lower portion of triceps. 2, 2. Attachments of pronator radii teres. 3. Attachment of flexor carpi radialis, palmaris longus, and flexor sublimis digitorum. 3'. Tendon of biceps. 3''. Tendon of brachialis anticus. 4, 4. Flexor carpi ulnaris. 5. Supinator longus. 6. Its distal attachment. 7. Supinator brevis. 7'. Extensor carpi radialis longior. 8, 8. Extensor ossis metacarpi pollicis. 9. Flexor profundus digitorum. 10. Its four tendons. 11. Tendon for index finger. 12, 12. Tendon for middle finger. 13. Tendon of flexor sublimis. 14. Tendon of flexor profundus for little finger. 15, 15. Lumbricales. 16, 16. Attachments of abductor brevis. 17. Opponens pollicis. 18. Flexor brevis pollicis. 19. Adductor pollicis. 20. Flexor longus pollicis. 21. Its tendon. 22, 22. Attachments of flexor brevis and adductor minimi digiti. 23. Opponens minimi digiti.

Detach the extreme upper origin of this muscle and study (1) the insertion of the brachialis anticus; (2) the lesser or coronoid head of the pronator radii teres; (3) the accessory head of the flexor longus pollicis; (4) the coronoid origin of the flexor sublimis digitorum.

The Pronator Quadratus.—Observe the anterior interosseous nerve terminating in this muscle. Detach the muscle from its ulnar origin, turn the same aside, and see the anterior interosseous artery piercing the membrane and gaining the posterior surface ; also see the anastomosis between a branch of this artery and the anterior carpals.

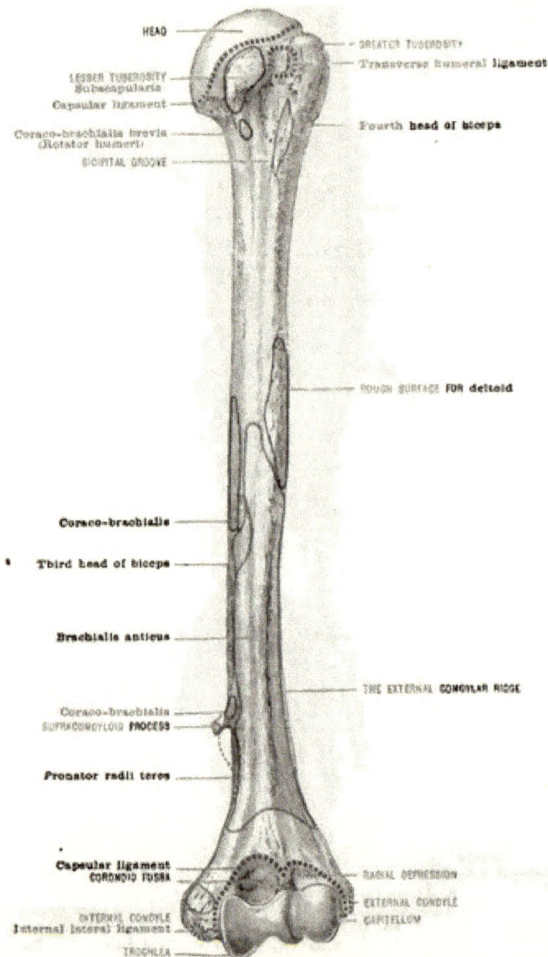

HEAD

GREATER TUBEROSITY
Transverse humeral ligament

LESSER TUBEROSITY
Subscapularis
Capsular ligament

Fourth head of biceps

Coraco-brachialis brevis
(Rotator humeri)
BICIPITAL GROOVE

ROUGH SURFACE FOR deltoid

Coraco-brachialis

Third head of biceps

Brachialis anticus

THE EXTERNAL CONDYLAR RIDGE

Coraco-brachialis
SUPRACONDYLOID PROCESS

Pronator radii teres

Capsular ligament
CORONOID FOSSA

RADIAL DEPRESSION

INTERNAL CONDYLE
Internal lateral ligament

EXTERNAL CONDYLE
CAPITELLUM

TROCHLEA

FIG. 129.—THE LEFT HUMERUS WITH A SUPRACONDYLOID PROCESS AND SOME IRREGULAR
MUSCLE ATTACHMENTS. (Anterior view.)

The Flexor Longus Pollicis (Fig. 124).—Notice and demonstrate on your dissection that the origin of this muscle is limited above by the tuberosity and oblique line of the radius ; that the lower part of the same is limited by the pronator quadratus, the inner by the interosseous membrane. Trace this muscle under the anterior annular ligament, and between the two heads of the flexor brevis pollicis, to the base of the distal phalanx of the thumb, where it is tendinously inserted.

You are to study the **radial oblique line,** and let your dissections show the following points:

1. It terminates above in **the bicipital tuberosity, and into this tuberosity is** inserted the tendon of the biceps muscle.

2. It terminates below in a round depression, and into this is inserted the pronator radii teres muscle. (Fig. 128.)

3. It has three lips: an outer, into which is inserted the supinator brevis muscle; a middle lip, that gives origin to the radial head of the flexor sublimis digitorum, and location to the nutrient foramen of the radius; an inner lip, that gives origin to the flexor longus pollicis. (Fig. 124.)

The Forearm Flexors.

In the order of their strength they are as follows:

1. The *musculus biceps cubiti* or *biceps muscle.*
2. The *musculus brachio-radialis* or *supinator longus muscle.*
3. The *musculus brachialis anticus*—very broad and fleshy.

Dissect and study the biceps with reference to:

1. Its origin by a coracoid or short, and a scapular or long head.
2. Its insertion by a tendon and by an aponeurosis.
3. Its fusiform belly and the rule for the nerve-supply of **muscles.**
4. Its inferior, external, and internal relations.
5. Its synergists and antagonists, and nerve-supply.
6. Its limiting intermuscular fasciæ.
7. Its fibrous arch and the rule for fibrous arches.

The biceps (Fig. 126) you have found inserted aponeurotically, by the bicipital or semilunar fascia, into the deep fascia, over the pronator radii teres muscle. It is tendinously inserted into the bicipital tuberosity of the radius. It arises by two heads, called long and short. The long head arises from the bicipital tubercle, above the glenoid cavity of the scapula; it passes under the transverse humeral ligament of the shoulder, between the greater and lesser tuberosities of the humerus, in the bicipital groove. The short head you will find arising from the apex of the coracoid process of the scapula, with the coraco-brachial muscle. These two heads of the biceps are connected by a fibrous arch. See rule for fibrous arches in the introductory chapter. The biceps muscle rests on the brachialis anticus muscle. Under the biceps muscle you will find the musculo-cutaneous nerve which supplies it and also the brachialis anticus, coraco-brachial, and supinator longus muscles. Externally are the deltoid and triceps and pectoralis major. Internally are the triceps, the latissimus dorsi, teres major, and coraco-brachial muscles. Lift the biceps from its bed, pull the same outward, and you will expose the brachialis anticus muscle below. In the groove, between the triceps and the inner part of the biceps, are to be found the large branches of the brachial plexus and the brachial artery and its venæ comites.

The brachio-radialis or supinator longus muscle (Fig. 128) arises from the upper two-thirds of the *outer condylar ridge* of the humerus. (Fig. 129.) It is inserted into the base of the *radial styloid process.* It has the double function of flexion and supination. Trace to it nerves **from the flexor trunk,** the musculo-cutaneous, and from the extensor trunk, the musculo-spiral.

The brachialis anticus (Fig. 127) arises from the outer and inner surfaces of **the humerus, limited above by the insertion** of the deltoid and coraco-brachial **muscles. (Fig. 129.)** Its insertion you have already seen into the coracoid pro-

13

cess of the ulna. Its nerve-supply comes from the musculo-cutaneous. This muscle and the preceding one act synergistically with the biceps as flexors of the forearm ; they are all three antagonized by the triceps.

The middle one-third of the **humerus** (Fig. 129) shows the following muscles exerting traction : (1) The brachialis anticus ; (2) the deltoid ; (3) the coraco-brachial ; (4) the upper and lower humeral heads of the triceps, on the posterior surface.

The capsule of the elbow may now be studied. Remove the insertions of the biceps and brachialis anticus, and expose and cut through the anterior ligament of the elbow-joint.

THE ANTERIOR REGION OF THE SHOULDER.

Dissection of this region must be preceded by a review of its osteology. Remember, the surgical importance of the shoulder-joint is second to none.

The Scapula.

Locate, classify, and describe geometrically this bone.

It is located on the upper, posterior, and outer aspect of the thorax, limited above by the second and below by the seventh rib. It is **a flat bone.** It has a dorsal, or posterior, and a ventral, or anterior, surface. It has **superior, axillary,** and **vertebral** borders ; an anterior, a posterior superior, and **a posterior inferior angle.** It has a **spine,** a coracoid, and an acromion process.

Give the importance of the following parts of the scapula in practical anatomy :

The anterior angle, or head, articulates with the humerus.

The anatomical neck gives attachment to the capsular ligament.

The surgical neck is often **the seat** of fracture.

The spine divides the dorsum into supra- and infraspinous fossæ.

The posterior inferior angle is crossed by the latissimus dorsi muscle.

The supraspinous fossa is occupied by the supraspinatus muscle.

The infraspinous fossa is occupied by the infraspinous muscle.

The subscapular fossa is occupied by the subscapularis muscle.

The superior border has the suprascapular foramen and omo-hyoid muscle.

The suprascapular foramen transmits the suprascapular nerve.

The coracoid process gives attachment to three muscles and three ligaments.

The acromion process gives attachment to the trapezius and deltoid.

The scapular notch transmits the infraspinous vessels and nerve.

The scapular angle is the deepest part of the subscapular fossa.

Name the muscles and ligaments you are expected to find attached to the coracoid process of the scapula?

The short head of the biceps, the coraco-brachialis, and the pectoralis minor muscles ; the coraco-acromial, conoid, and trapezoid ligaments.

The supra- and infraglenoid tubercles are where situated and how occupied?

They are immediately above and below the glenoid cavity of the scapular head respectively : the supraglenoid tubercle gives origin to the long head of the biceps ; the infraglenoid to the long head of the triceps muscle.

What can you say specifically about the origins of the three muscles from the three scapular fossæ?

The whole of the fossa in each case does not give origin to the muscle, only the outer two-thirds thereof. The one-third next the shoulder is occupied by fatty connective tissue, vessels, and nerves.

Name the muscles attached to the middle lip of the axillary border of the scapula.

The triceps, teres minor, and teres major. The rule governing the relation of anatomical minors to majors is : minors occupy a high, majors a low, level.

Name the bony points of the scapula traversed by a fracture of the so-called surgical neck of the scapula, and the important structures endangered by such a fracture.

The suprascapular notch, in which is the suprascapular nerve, and above which are the suprascapular vessels ; the scapular notch, in which are found the vessels and nerves for the infraspinatus muscle ; the scapular angle, in which are the vessels and nerves for the subscapularis muscle.

How is the anterior lip of the vertebral border of the scapula occupied?

By the subscapularis and serratus magnus muscles. The latter muscle depresses the shoulder-girdle, being antagonistic to the trapezius and levator anguli scapulæ muscles.

Name the muscles inserted into the middle lip of the vertebral border of the scapula.

The levator anguli scapulæ, rhomboideus minor, rhomboideus major. Notice here the high origin of the minor, and note the rule previously mentioned governing the relation of anatomical majors and minors.

How are the three lips of the spine of the scapula occupied?

The superior by the insertion of a part of the trapezius ; the inferior by origin of part of the deltoid ; the middle is subcutaneous, and is also called the crest.

The Clavicle.

Name parts of this bone of importance in practical anatomy.

The sternal end, articulating with the manubrium.

The acromial end, articulating with the acromion.

The anterior surface of the inner two-thirds of the bone.

The superior surface of the outer one-third of the bone.

The inferior surface of the bone.

What is the importance of the superior surface of the outer one-third of the clavicle?

The attachment of the trapezius muscle posteriorly, and the deltoid anteriorly.

What is the importance of the anterior surface of the inner two-thirds of the clavicle?

The origin of the pectoralis major and the sterno-cleido-mastoid muscles.

What can you say of the posterior surface of the clavicle?

It forms an arch over the brachial plexus and subclavian artery and vein.

Tell what you expect to find on the inferior surface of the clavicle.

The subclavius muscle, in a groove in the middle one-third ; the origins of the sterno-hyoid and sterno-thyroid muscles, and the impress for the rhomboid or costo-clavicular ligament ; the oblique line for the trapezoid ligament and a small tuberosity for the conoid ligament.

1. Intermuscular groove between the pectoralis major and deltoid.
2. The pectoralis major muscle and pectoral fascia.
3. Suprasternal and supraclavicular nerves.
4. The cephalic vein and descending branch of acromio-thoracic artery.
5. The pectoral and deltoid origins on the clavicle.
6. The perforating branches of the internal mammary artery.
7. The axillary space geometrically.

The Axillary Space.

1. *A Base.*—Skin, superficial, and deep fascia, called axillary.
2. *An Apex.*—Formed by clavicle, first rib, and scapula.
3. *Anterior Border.*—Pectorales major and minor, and clavi-pectoral fascia.

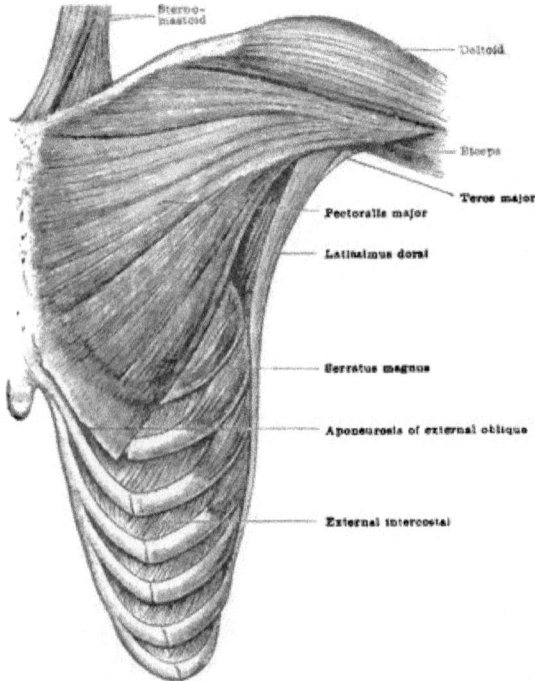

FIG. 130.—THE PECTORALIS MAJOR AND DELTOID.

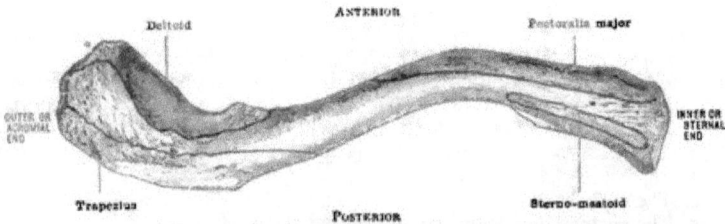

FIG. 131.—THE LEFT CLAVICLE. (Superior surface.)

4. *Inferior Border.*—Ribs, intercostal spaces (six), and serratus magnus.
5. *Anterior Thoracic Angle.*—Long thoracic artery.
6. *Posterior Thoracic Angle.*—Subscapular artery and Bell's nerve.
7. *External or Humeral Angle.*—Axillary artery and nerves.

8. *Contents.*—Axillary connective tissue and glands ; the axillary artery and vein and their branches ; the large branches of the brachial plexus.

9. *The posterior boundary* is formed by the subscapular, teres major, and latissimus dorsi muscles.

The cephalic vein and descending branch of the acromio-thoracic artery. The vein (Fig. 105) is the guide to the groove. The groove is the boundary-line between the clavicular origins of the pectoralis major and deltoid muscles, The vein opens into the subclavian.

The Pectoralis Major Muscle.—Remove the skin and find in the superficial fascia two cutaneous nerves—the supraclavicular and suprasternal ; the one supplying the skin over the deltoid is the supraacromial These are the descending branches from the superficial part of the cervical plexus.

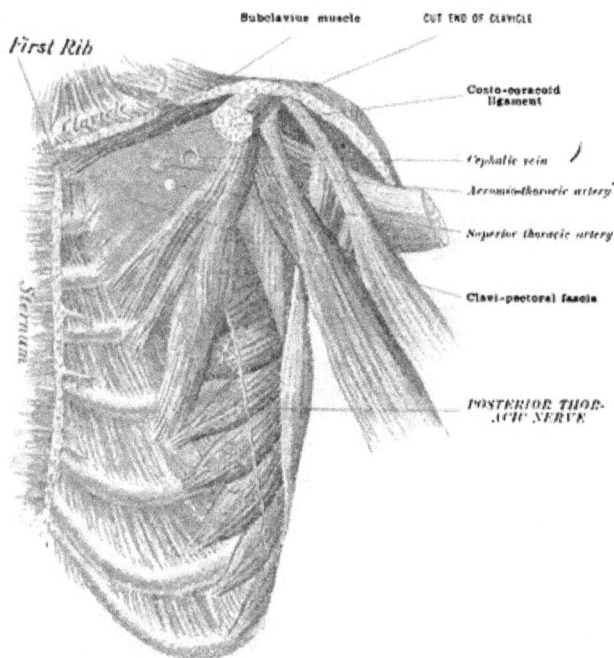

FIG. 132.—THE CLAVI-PECTORAL FASCIA.

The deltoid aponeurosis is the deep fascia covering the muscle of the same name. It is continuous with the pectoral in front, the infraspinous behind, and the axillary below.

The Deltoid Muscle.—Detach this muscle from the clavicle (Fig. 131) and acromion process of the scapula. Turn the muscle outward. Follow the muscle to its insertion into the deltoid ridge, on the outer surface of the humerus. (Fig. 127.) Cut this insertion and turn the muscle farther back, so as to expose the insertion of the pectoralis major.

The Pectoralis Major.—Remove the fascia with a sharp scalpel. Place the arm at a right angle to the body ; notice that the deep fascia you are now

removing is continuous with the deep fascia covering the base of the axilla.
Observe the sternal and clavicular origins of the muscle. See some cutaneous
nerves and arteries piercing the thoracic wall, near the sternum. (Fig. 132.) The
arteries are the perforating branches from the internal mammary ; the nerves are
the anterior cutaneous branches of the intercostals. Cut the insertion of the pectoralis major. Note that the fibres that form the lower margin of the tendon
of the muscle are inserted high ; those that form the upper border of the muscle
are inserted low, into the posterior lip of the bicipital groove of the humerus.

Now cut the clavicular origin of the pectoralis major, and carefully pull the
muscle forward. On the under surface you will see the external anterior thoracic
nerve, a branch of the outer cord of the brachial plexus, coming through the
pectoralis minor muscle to supply the major pectoral muscle. Now cut the
sternal origin of the pectoralis major and turn the same out of the way without
injuring the nerve-supply

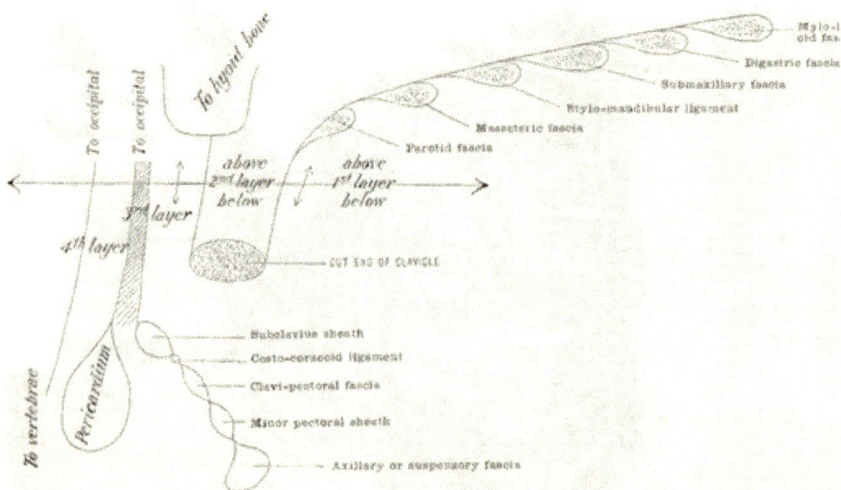

FIG. 133.—TO SHOW SCHEMATICALLY THE DISTRIBUTION OF THE DEEP CERVICAL FASCIA ABOVE
AND BELOW THE CLAVICLE.

Four very important structures, easily understood, quickly developed, and
seldom appreciated by the student, are now before you :

1. The *subclavius muscle* and its sheath of cervical fascia.
2. The *costo-coracoid ligament*, a part of subclavius sheath.
3. The *clavi-pectoral fascia*—triangular and thin.
4. The *pectoralis minor muscle* and its sheath.

The third layer of deep cervical fascia passes behind the clavicle, forms a
sheath for the subclavius muscle, and unites along the lower margin of the subclavius muscle to form a thick, strong band—the costo-coracoid ligament. The
fascia then bridges the triangular space between the subclavius muscle and the
pectoralis muscle, under the name of clavipectoral fascia. (Fig. 132.) At this
point the fascia again delaminates, to form a sheath for the pectoralis minor
muscle. At the lower border of this muscle the fascia is continuous with the
deep fascia of the base of the axilla, and is known technically as the axillary

fascia. Let figure 133 illustrate in this connection schematically the distribution of the four layers of deep cervical, both below and above.

It will be seen from figure 133 that the first layer of deep cervical fascia is specialized above the clavicle as parotid fascia, masseteric fascia, stylo-mandibular ligament, submaxillary fascia, digastric fascia, and mylo-hyoid fascia ; below, the first layer is attached to the clavicle. The second layer is attached above to the hyoid bone, having ensheathed the depressor muscles of this bone (Fig. 24) ; below, this fascia is attached to the clavicle. The fourth layer is connected to the occipital bone and the muscles in front of the vertebral column.

The third layer is of greatest importance surgically, and the most complex. Passing behind the clavicle and sternum, this layer divides into a thoracic portion, which expands to form the fibrous part of the pericardium, and also into an axillary portion, which has been previously described.

Pus formed between the first and second layers of deep cervical fascia would be arrested at the clavicle. Pus formed in the carotid sheath might easily find its way to either the axilla or pericardium. The student must take schematic representations with a grain of allowance, remembering that fascial development, like muscular development, is a variable thing, and in some cases would not direct pus whither the scheme might indicate. After all, the scheme just

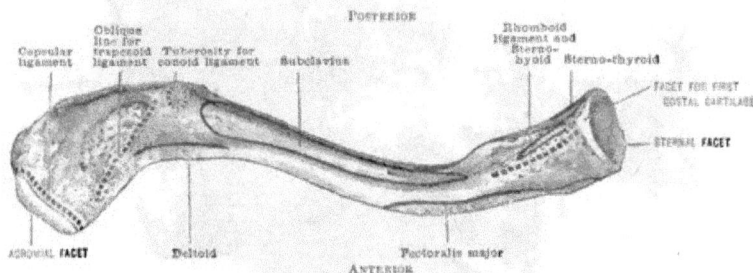

Fig. 134.—THE LEFT CLAVICLE. (Inferior surface.)

given is to teach that many fascial entities are definitionally only a certain segment of deep cervical fascia. For example : What is the subclavius sheath ? It is that part of the third layer of the deep cervical fascia that invests the subclavius muscle. What is the parotid fascia ? It is that part of the first layer of deep cervical fascia which is continued above the angle of the jaw to form a capsule for the parotid gland. See in the introductory chapter the rule for naming specialized modifications of deep fascia in any region of the body.

The subclavius muscle arises from the first rib, near the sternum, and is inserted into the groove on the under surface of the clavicle (Fig. 134), occupying the middle two-fourths of the surface. It is surrounded by a sheath—**the subclavian sheath.**

The pectoralis minor muscle arises, as a rule, from the third, fourth, and fifth true ribs, and is inserted into the anterior border of the coracoid process of the scapula. Its nerve is the internal anterior thoracic, a branch of the inner cord of the brachial plexus.

The clavi-pectoral space is the triangular interval you see between the upper border of the pectoralis minor muscle and the subclavian muscle. As you will see on your cadaver, this space is filled in by a delicate fascia, attached above to the subclavius muscle and below to the pectoralis minor muscle. This is the **clavi-pectoral fascia.** (Fig. 132.)

THE AXILLA, OR AXILLARY SPACE.

The axilla, or **axillary space,** is of interest anatomically because important surgical operations are performed here. Armpit and axilla are not synonymous terms. The physician places a clinical thermometer in the armpit—not in the axilla. The axillary lymphatic glands become secondarily enlarged in cases of infection, and may require extirpation. The head of the humerus may become dislocated into the axilla and require reduction. Pus may form in the neck and find its way to the axilla and make its liberation incumbent. A wound of the axillary artery makes its ligation a necessity. Amputation at the shoulder-joint is sometimes done.

The apex of the axilla or inlet has bony boundaries: (1) *The outer surface of the first rib;* (2) *the clavicle;* (3) *the superior costa or superior border of the scapula.* It is by the apex that this space communicates with the thorax and neck.

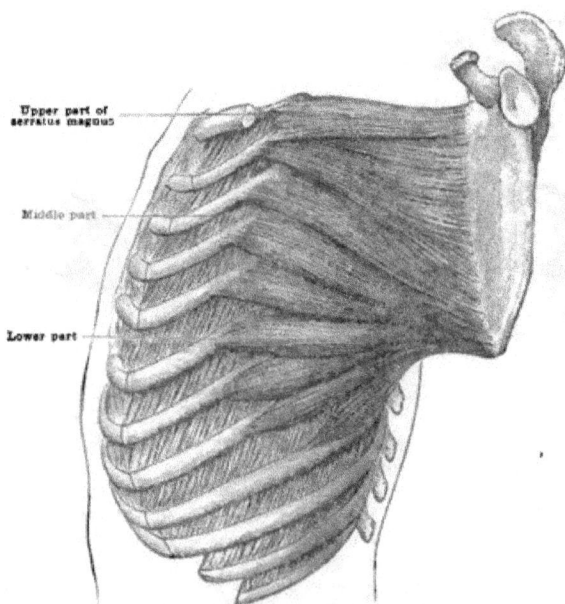

FIG. 135.—SERRATUS MAGNUS. THE INNER WALL OF THE AXILLA.

The base of the axilla is formed by the skin, the superficial and the deep fascia. The *deep fascia* in this locality is known specifically as *axillary fascia;* it is also called *suspensory fascia.* The deep fascia is a derivative of the third layer of the deep cervical fascia, passing under the clavicle. In front it is continuous with the pectoral fascia; behind, with the fascia covering the latissimus dorsi muscle.

The anterior boundary is formed by the skin, fasciæ, pectorales major and minor muscles, and clavi-pectoral fascia. This boundary is very well defined.

The posterior boundary is formed by the subscapularis, the teres major, and latissimus dorsi muscles.

The inner boundary is formed by six or eight ribs, their intercostal muscles and fasciæ, and six serrations of the serratus magnus muscle. (Fig. 132.)

The angles are three in number, and the surgical importance of each is that in these angles are found some structure or structures to be avoided, or reached in surgical operations. (1) The *anterior thoracic* angle contains the long thoracic artery; (2) the *posterior thoracic* angle contains the subscapular artery and the external phrenic nerve; (3) the *outer or humeral* angle contains the axillary artery and nearly all the important branches of the brachial plexus.

The Axillary Contents.—On removing the anterior wall of the axilla, as you may now complete doing, by cutting the pectoralis minor at its insertion, and turning the same back, with its attached clavi-pectoral fascia, you will see (1) a mass of axillary connective tissue containing some fat and blood-vessels, the alar thoracic vessels, and axillary lymphatic glands. In health these glands are not larger than a small pea, and are bluish in color. When diseased, as is frequently the case in dissecting-room material, they are often very large; they may be as

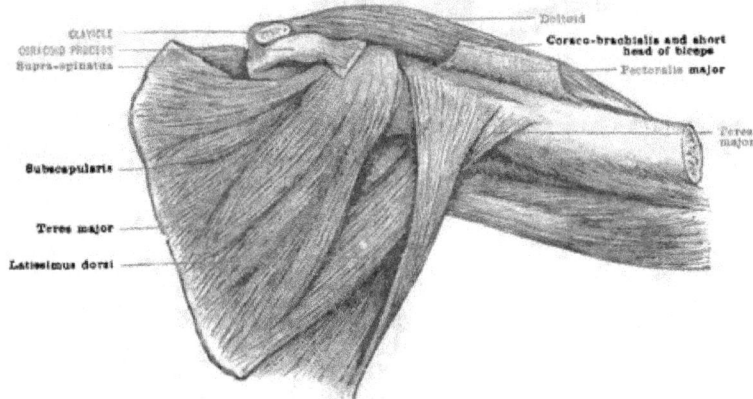

FIG. 136.—FRONT VIEW OF THE SCAPULAR MUSCLES. THE POSTERIOR WALL OF THE AXILLA.

large as a hazelnut, or even larger. Each student should visit every table in the room, when this area is being dissected, to inspect the condition of these glands.

To clean out the axilla means, in surgery, to remove the diseased axillary glands. This operation is attended by slight hemorrhage. The vessels that should bleed are the alar thoracics. These supply the glands and fat in this space. The term unavoidable hemorrhage would be an excellent expression by which to designate the bleeding incident to the removal of these glands, if this expression were not already the stereotyped property of obstetric nomenclature.

Systematic Examination of the Walls and Contents of the Axillary Space. —1. *Examine the inner wall* and find: (1) The outer surface of the six upper ribs. (2) The serrations of the serratus magnus muscle. (Fig. 135.) (3) Coming through two or three intercostal spaces,—the second, third, fourth,—some cutaneous nerves will be seen. These nerves cross the space and seem to become lost in the mass of fat you have just examined. These nerves are the posterior lateral cutaneous branches of the intercostal nerves. Now trace them out as follows: One passes out under the pectoralis major muscle to the skin of this muscle, in front; a second passes out behind, around the latissimus dorsi, to the skin of this muscle; a third passes down the center of the space, joins a branch of the

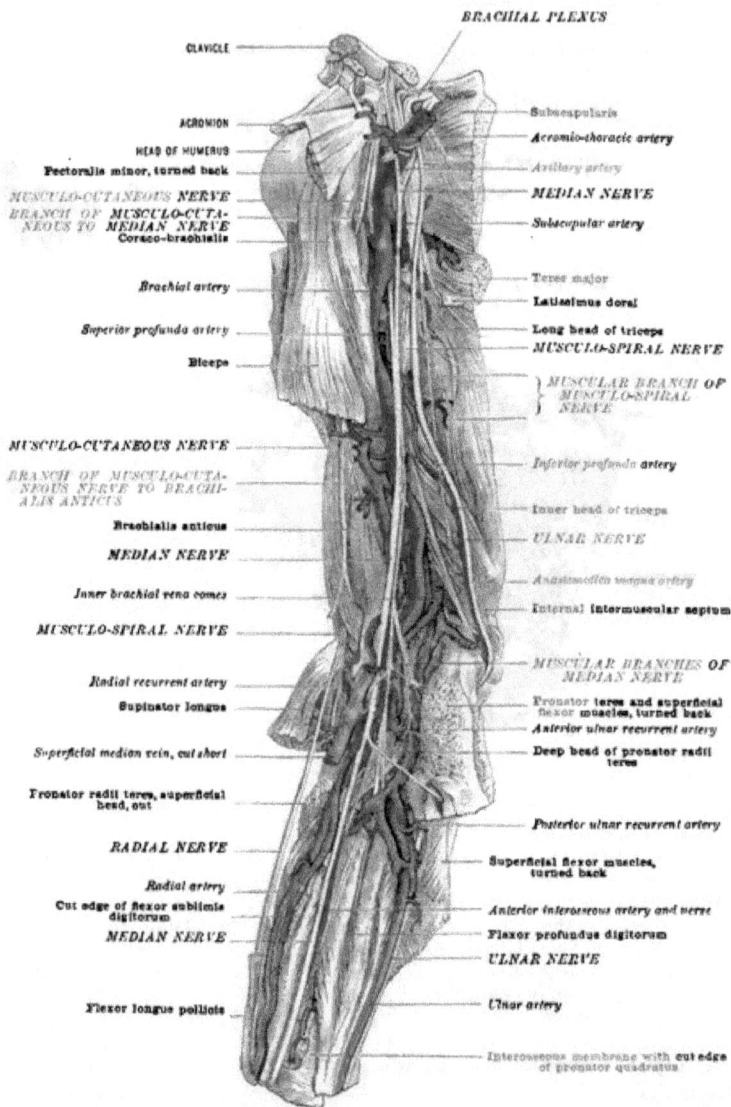

FIG. 137.—THE LOWER PART OF THE AXILLARY, THE BRACHIAL, AND THE RADIAL
AND ULNAR ARTERIES.
(From a dissection in the Hunterian Museum.)

lesser internal cutaneous nerve (Wrisberg's), and is distributed to the skin of the inner surface of the arm. This nerve is called the intercosto-humeral nerve. It comes through the second intercostal space.

2. *Examine the posterior wall* (Fig. 136) and find the (1) subscapularis muscle; (2) *teres major* muscle; (3) latissimus dorsi muscle; (4) the three subscapular nerves supplying these muscles. They are branches of the posterior cord of the brachial plexus.

Notice while dissecting the posterior wall of the axilla the following structures shown in figure 136: (1) The relation between the insertions of the teres major and latissimus dorsi. Clean off the fascia and fat carefully, and see the lower muscle crossing the higher, to be inserted above this. (2) Trace out and study carefully the specific tendinous insertion of the pectoralis major. (3) Clean the subscapularis muscle, notice its nerve- and blood-supply, and the relation of this muscle to the insertion of the serratus magnus.

3. *Examine the Anterior Thoracic Angle.*—In this angle you will find the long thoracic or external mammary artery. Trace its branches to the pectoral muscles, the serratus magnus, the subscapularis, and the lymphatic glands.

4. *Examine the Posterior Thoracic Angle.*—In this you will find the long thoracic nerve and the subscapular artery. Trace branches from this nerve to the serratus magnus. (Fig. 132). Trace the subscapular artery to the muscles of the posterior wall. Find its largest branch, leaving the axillary space by the triangular space, between the teres major and minor and internal to the scapular head of the triceps.

5. *Examine the External or Humeral Angle* (Fig. 137).—Find first the median nerve, lying on the front of the axillary artery. Trace this nerve upward and find its inner and outer heads, from the inner and outer cords of the brachial plexus.

The ulnar nerve lies to the inner side of the axillary artery. Trace this up to the inner cord of the brachial plexus; also trace it down behind the inner condyle of the humerus. See where it joins company with the inferior profunda artery.

The musculo-cutaneous nerve lies to the outer side of the axillary artery. You will find this nerve close to the insertion of the pectoralis minor. Trace it through a hole in the coraco-brachial muscle. It is a branch of the outer cord of the brachial plexus.

The musculo-spiral nerve lies behind the axillary artery. (Fig. 137.) Trace it downward and it will soon disappear between the two humeral heads of the triceps muscle. It is a branch of the posterior cord of the brachial plexus.

The circumflex nerve lies behind the axillary artery. It is a branch of the posterior cord of the brachial plexus. It soon escapes to the posterior part of the shoulder by the quadrangular space, with the posterior circumflex artery.

The subscapular nerves, three in number, lie the most deeply of all the nerves in this region. They are given off to the muscles that form the posterior boundary of the axillary space. The nerve to the latissimus dorsi is called the long subscapular.

The anterior thoracic nerves are two in number; they are called external and internal. The internal anterior thoracic nerve is from the inner cord of the brachial plexus; it supplies the pectoralis minor. The external anterior thoracic nerve is from the outer cord of the brachial plexus; it pierces the clavi-pectoral fascia and goes to the pectoralis major. The internal anterior thoracic nerve also sends a branch to the pectoralis major muscle. This pierces the clavi-pectoral fascia.

The internal cutaneous and lesser internal cutaneous nerves are internal to the axillary artery. They must not be mistaken for the ulnar.

· **The Axillary Artery and Vein.**—This artery has three stages, as follows
1. *First stage,* from first rib to the pectoralis minor muscle.
2. *Second stage,* behind the pectoralis **minor** muscle.
3. *Third stage,* below the pectoralis **minor** muscle.

Note that the axillary vein lies internal to the artery in all three stages; that it receives two large tributaries : (1) the cephalic, (2) the basilic veins.

Branches of the axillary artery according to stages :

First Stage.—(1) The *superior thoracic* passes between the pectoral muscles to the chest walls. Trace branches from it to these muscles and to the thoracic walls. (2) The *acromio-thoracic* perforates the *clavi-pectoral fascia* (Fig. 132), and sends branches to the deltoid and pectoral muscles. Its descending branch you found deeply buried in a groove with the *cephalic vein,* between the deltoid and pectoralis major muscles.

Second Stage.—(1) *Alar thoracic* branches have been seen supplying the axillary glands and fat. (2) The *long thoracic* you will find in the anterior thoracic angle. Trace its numerous branches to all the walls of the axillary space.

Third Stage.—(1) The *anterior circumflex artery* you will trace behind the coraco-brachial and biceps muscles. It sends a branch to the shoulder-joint through the bicipital groove. (2) The *posterior circumflex* passes through the *quadrangular space,* with the circumflex nerve, to the deltoid muscle and shoulder-joint. The space referred to is bounded externally by the humerus, internally by the scapular head of the triceps, superiorly by the teres minor, inferiorly by the teres major and latissimus dorsi. (3) The *subscapular* is the largest branch of the axillary artery. You will find it in the posterior thoracic angle. It passes through the *triangular space,* and takes the name of *dorsalis scapulæ.* This space is bounded by the triceps and the teres minor and major.

The muscles in this locality are :

1. **The Serratus Magnus.**—This arises from the outer surface of the upper eight ribs by nine fleshy processes. (Fig. 135.) It is inserted into the anterior lip of the vertebral border of the scapula. It is supplied by the long thoracic nerve, a branch of the brachial plexus. (Fig. 42.)

2. **The Subscapularis Muscle.**—This arises from the subscapular fossa and is inserted into the lesser tuberosity of the humerus.

3. **The teres major** arises from the middle lip of the axillary border of the scapula. It is inserted into the posterior lip of the bicipital groove of the humerus with its synergist, the latissimus dorsi.

4. **The latissimus dorsi** is a muscle of the back, and will be described in the proper place. The student will now trace the subscapular nerves, from the posterior cord of the brachial plexus to the subscapular, the teres major, and the latissimus dorsi muscles. Also study the relation and manner of insertion of the two latter muscles into the humerus.

The Brachial Artery.—This vessel begins as a continuation of the axillary artery, at the lower border of the teres major muscle. (Fig. 137.) It ends at the elbow, by dividing into the radial and ulnar arteries. In its course you can show the triceps muscle on one side and the biceps and coraco-brachial on the other side of this artery. The median basilic vein lies in front of it, near the elbow (Fig. 105), the bicipital fascia intervening. The median nerve crosses it in the middle of its course. As you must demonstrate on your work, this artery lies superficially in its entire course, and may be easily reached for compression or operation. Find the ulnar and internal cutaneous nerves, internal to the artery. Examine every **cadaver in** the room to find high bifurcations of the brachial artery.

The branches of the brachial artery are :
1. *The muscular branches* to the flexor muscles of the forearm.

2. *The nutrient artery* to the humerus, about the middle.   Find the nutrient foramen directed toward the elbow on the inner surface of the humerus, near the insertion of the coraco-brachial muscle.

3. *The Superior Profunda Artery.*—Find this artery on a level with the major tereal muscle.   It accompanies the musculo-spiral nerve in the groove bearing the same name.   This artery supplies the triceps and gives off an artery to the upper one-third of the humerus.   It anastomoses with the radial recurrent, the posterior ulnar recurrent, and the posterior interosseous recurrent, and with the inferior profunda and anastomotica magna arteries.

4. *The inferior profunda* you will find a little below the middle of the arm.   It joins the ulnar nerve, accompanies the same, and terminates by anastomosing with the anastomotica magna and posterior ulnar recurrent arteries.

THE THORAX.

ANTERIOR AND LATERAL WALLS.

Locate (1) the sternum, presternum, mesosternum, metasternum. (2) The suprasternal notch and the interclavicular ligament. (3) Find the pronounced elevation between the first and second pieces of the sternum. This is the manu-brio-gladiolar joint, and is the point concerned in the rule for counting the ribs.

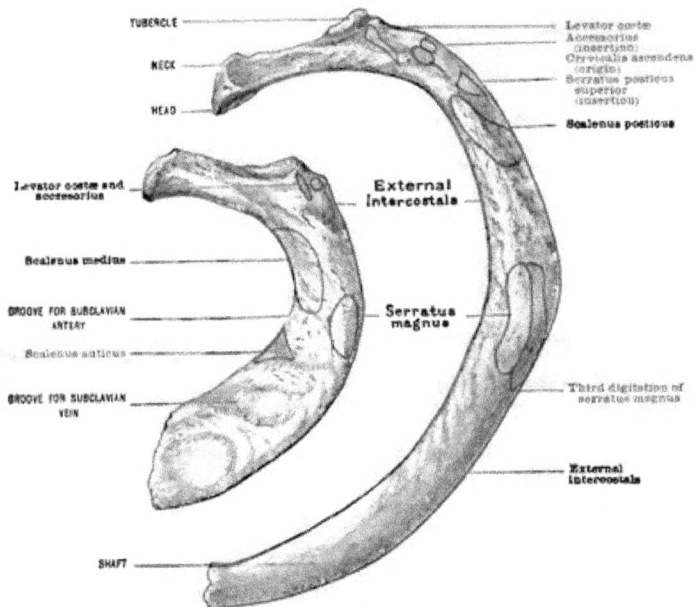

FIG. 138.—FIRST AND SECOND RIBS.

Rule : this ridge corresponds to the cartilage of the second rib. (4) Find the sterno-clavicular articulation. See if it is freely movable, as it should be. Note when you dissect this region the presence of an interarticular fibro-cartilage, dividing the synovial cavity into two compartments—an inner and an outer. (5) Note that seven true ribs articulate by their cartilages with the sternum ; that all except that of the first rib are movable. (6) Examine the metasternum or ensiform process to see its deflection, whether lateral, forward, or posterior. (7) Count the ribs and study the nature of the chondro-costal articulation.

Dissect the following structures and study them carefully :

1. The subclavius muscle and the costo-clavicular ligament.
2. The external intercostal muscles and their fasciæ.
3. The internal intercostal muscles and their fasciæ.
4. The intercostal fasciæ, internal, middle, and external.

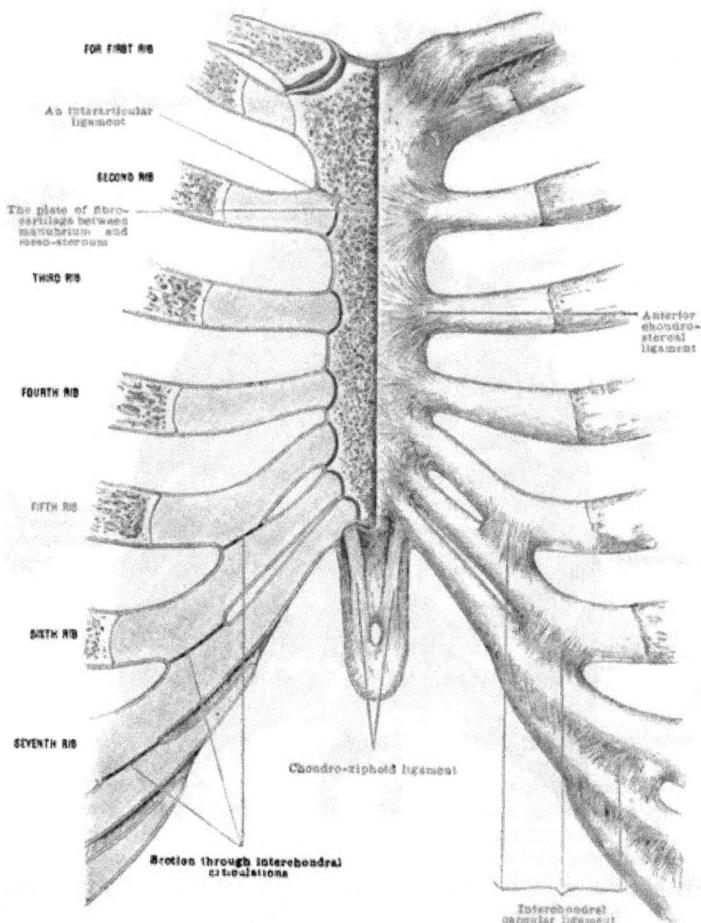

FIG. 139.—THE STERNUM.
(Left side, showing ligaments; right side, the synovial cavities.)

5. The anterior chondro-sternal ligaments.
6. The anterior common sternal ligament of periosteal derivation.
7. The scalenus anticus, inserted into the scalene tubercle of the first rib.
8. The scalenus medius, inserted into the scalene depression of the first rib.
9. The subclavian groove, between the scaleni muscles on the first rib.
10. The groove for subclavian vein on the outer surface of the first rib.

11. The relation of subclavian artery and vein on first rib.

12. A digitation of serratus magnus muscle on the first rib.

13. Study the insertion of the scalenus posticus muscle on the outer surface of the second rib. Also the digitations for the serratus magnus muscle on the second.

The Subclavius Muscle.—Dissection of this muscle requires much care. Cut the costo-clavicular ligament and the sterno-clavicular ligaments and then lift the clavicle. This will bring the muscle into view. The muscle has a strong sheath. Notice that the space between this muscle and the pectoralis minor is bridged by the clavi-pectoral fascia. (Fig. 132.) Note the origin of the subclavius from the first rib at the junction of its cartilage; its insertion into a groove

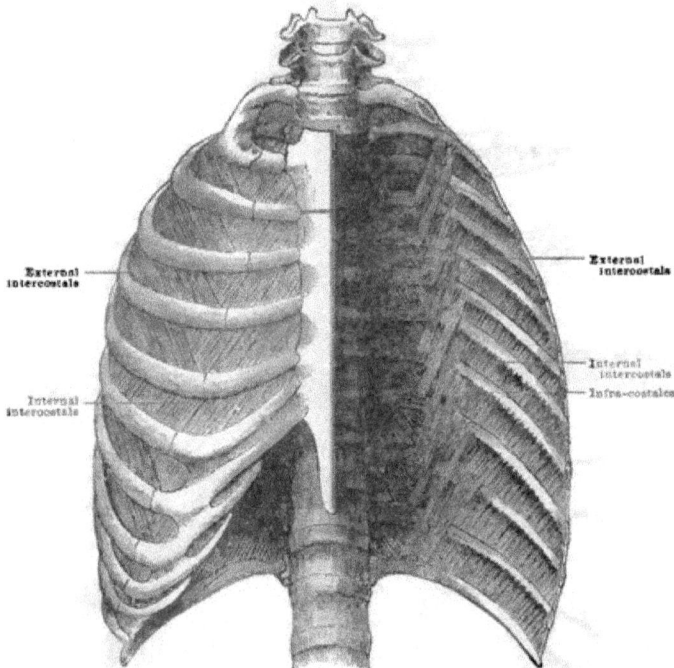

FIG. 140.—THE INTERCOSTAL MUSCLES.

on the under surface of the clavicle, middle two-fourths. The structures you will now see under the clavicle are: (1) The subclavius muscle and its sheath; (2) the subclavian artery and vein; (3) the brachial plexus of nerves.

The scaleni muscles are three in number—anticus, medius, and posticus. A few movements of the forceps will develop their insertions—two into the first and one into the second rib.

The intercostal muscles and fasciæ occupy the intercostal spaces. The intercostal spaces are eleven on each side. Each space is occupied by: (1) External and internal intercostal muscles; (2) external, middle, and internal intercostal fasciæ; (3) intercostal arteries, veins, and nerves.

The external intercostal muscles are descending—*i. e.*, their fibres pass downward and inward, toward the median line of the body. (Fig. 140.) They extend from the tubercle of the rib behind to the costo-chondral joint in front.

The internal intercostal muscles ascend ; their fibres pass inward, upward, and toward the mid-line of the body. They extend from the sternum in front to the angle of the rib behind.

The osteology of the ribs must be reviewed. A typical rib has :

1. A *head*, or articular vertebral extremity
2. A *neck*, or constiction, anterior to the head, extending to the tubercle.
3. An *articular tubercle*, articulating with the transverse process.
4. A *non-articular tubercle*, for ligamentous attachment.
5. A *sternal extremity*, for articulation with its cartilage.
6. An *outer surface*, covered by numerous muscles.
7. An *inner surface*, covered by pleura mostly
8. A *superior border*, for muscular attachment.
9. An *inferior border*, possessing two lips and a subcostal groove. You must become familiar with all these points on the dry skeleton.

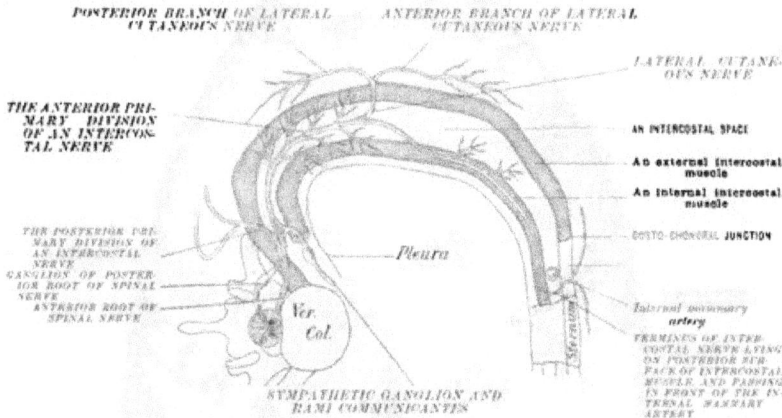

FIG. 141.—THE INTERCOSTAL NERVES.

Figure 140 shows the *musculi infracostales*. These muscles arise near the angle of the rib, from the inner surface and lower border ; they are inserted into the second rib above. They are also called *subcostals* or *musculi subcostales*.

Intercostal Fascia.—Remove every vestige of the pectoralis major, and expose the intercostal muscles. Find a delicate layer of fascia covering the outer surface of the external intercostal muscle. This is the *external intercostal fascia.* Trace this fascia to the sternum. You can now see, showing through this fascia, the internal intercostal muscle. The internal intercostal fascia covers the inner surface of the internal intercostal muscle. The middle intercostal fascia is between the two intercostal muscles.

The Intercostal Nerves.—Figure 141 represents the course and divisions and relations of an intercostal nerve, such as may be seen in the second, third, fourth, fifth, and sixth intercostal spaces. Students frequently graduate unable to locate properly an intercostal nerve. You are to note the following stages of an intercostal nerve :

1. The nerve is formed by an anterior and a posterior root.
2. The posterior or sensory root has a ganglion.

14

3. The intercostal nerve lies between the pleura and external intercostal muscle.

4. The nerve throws off muscular branches.

5. The nerve throws off lateral cutaneous branches.

6. The nerve is in the substance of the internal intercostal.

7. The nerve is on the posterior surface of the internal intercostal.

8. The nerve passes in front of the internal mammary artery.

9. The nerve throws off anterior cutaneous branches.

The branches of the intercostal nerve are : (1) *Muscular,* to the internal and external intercostals, the infracostales, the serratus posticus superior, the triangularis sterni, and levatores costarum ; (2) *cutaneous* branches, lateral and anterior. Now you may find the nerves according to the diagram. You will find the

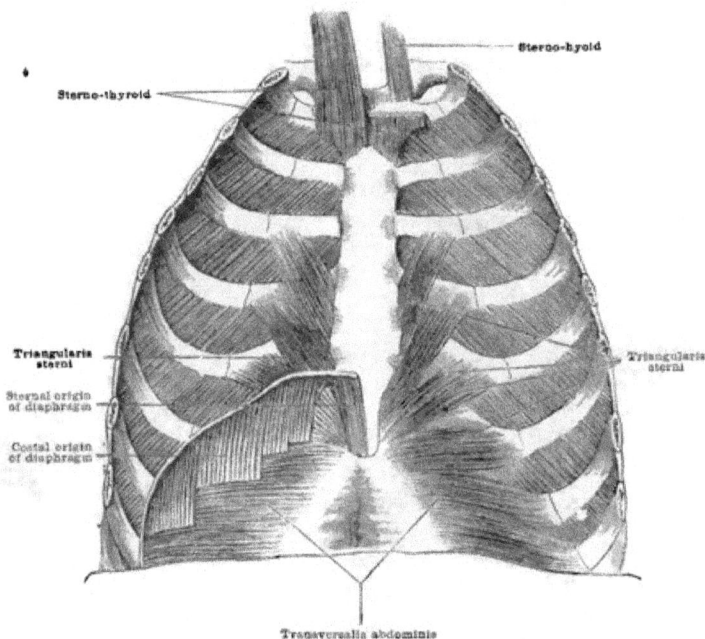

FIG. 142.—THE MUSCLES ATTACHED TO THE BACK OF THE STERNUM.

intercostal arteries with the nerves to some extent. In this dissection care must be taken not to injure the pleura.

Where to Find the Internal Mammary Artery and Veins.—With your forceps go through the soft structures one-half of an inch to the outer margin of the sternum on either side. Try to find this artery, and ligate the same, without injuring the pleura. In reaching this artery, do you cut through the external intercostal muscle? No. Do you cut through the three intercostal fasciæ and internal intercostal muscle? Yes. Give the location of the internal mammary artery. It lies between the pleura and internal intercostal fascia. It has two companion veins.

The Inner Surface of the Sternum and Ribs (Fig. 142). See :

1. The *internal intercostal muscles,* eleven pairs on each side.

2. The *origin of the sterno-hyoid muscle*.

3. The *origin of the sterno-thyroid muscle*.

4. The *origin and insertion of the triangularis sterni*.

5. The *costal and sternal origins of the diaphragm*.

6. The *parietal layer of the pleura*, through which you can see **the muscles**.

7. The *internal mammary artery and veins*, on each side of the sternum, giving off their anterior intercostal and other arteries. (Fig. 143.)

Dissection.—Cut through the ribs in the mid-axillary line. (Fig. 145.) Cut through the abdominal walls along the **margin of the** false ribs. **Cut through**

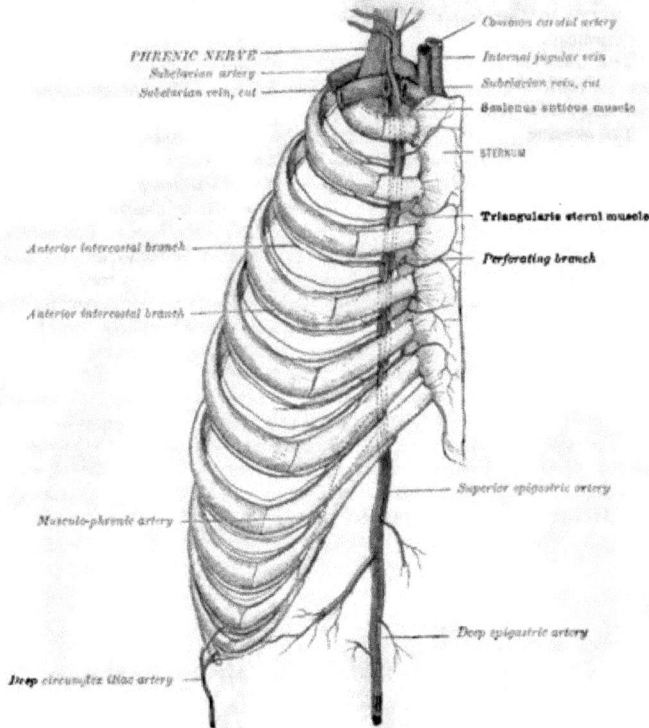

FIG. 143.—Scheme of the Internal Mammary Artery.

the diaphragm close to the ribs and sternum. Then elevate from below the section of thoracic wall you have made loose. As you turn this section up you will see a large amount of anterior mediastinal connective tissue behind the sternum. Elevate the section slowly; let an assistant divide the connective tissue, and do not injure the numerous **structures** that enter or leave the apex of the thorax.

Figure 141 shows the *infracostales*. These muscles arise near the angle of the ribs, from the inner surface and lower border; they are inserted into the second rib above. They are also called *subcostals*.

The internal mammary artery has the following branches, all of which you can easily find:

1. The *perforating branches* that come through the five or six upper intercostal spaces and supply the major pectoral muscle and the mammary gland.

2. The *mediastinal branches* to the mediastinal connective tissue and glands under the sternum. They also supply the thymus gland in the foetus or its remains in the adult, hence they are called also *thymic* arteries.

3. The *sternal branches*, quite small and numerous, to the sternum and triangularis sterni muscle.

4. *Pericardiac branches* to the anterior surface of the pericardium.

5. The *comes nervi phrenici* is a long, slender artery that accompanies the phrenic nerve to the diaphragm. You will find this nerve between the pleura and pericardium.

6. The *anterior intercostals* supply the six upper intercostal spaces. There are two in each space, one above and one below. They anastomose with the *posterior intercostal arteries* from the aorta.

7. The *musculo-phrenic*, that furnishes the anterior intercostal arteries to the remaining spaces. It also supplies branches to the diaphragm.

8. The *superior epigastric* passes through the diaphragm between its costal and sternal parts, enters the sheath of the rectus muscle, finally the muscle itself, and anastomoses with the *deep epigastric branch* of the external iliac artery.

The internal mammary artery is a branch of the subclavian, as you saw when you dissected the neck. Its parietal branches are to the anterior walls of the thorax what the parietal branches of the aorta are to the posterior walls of the thorax. Now go down between the sixth and seventh cartilages and find where the artery divides into the musculo-phrenic and superior epigastric.

The thoracic walls are supplied with blood, then, by intercostal arteries. These arteries come: (1) From the *aorta*; (2) from the *internal mammary*; (3) from the *subclavian*; this latter furnishing the artery to the superior intercostal space. The blood from this area is collected and delivered as follows: the first or superior intercostal vein is tributary to the vertebral vein; the others to the azygos veins, in the posterior mediastinal space, to be presently dissected.

The triangularis sterni muscle arises from the lateral aspect of the sternum, and is inserted into the cartilages of the third, fourth, fifth, and sixth ribs.

INTERIOR OF THE THORAX.

Geometrically the thorax possesses:

1. An *apex*, through which pass: (1) The œsophagus; (2) the trachea; (3) the common carotid artery on the left, and the innominate artery on the right side; (4) the phrenic, pneumogastric, sympathetic, and recurrent laryngeal nerves; (5) the internal jugular and subclavian veins; (6) the thoracic duct; (7) the sterno-hyoid and sterno-thyroid muscles; (8) the third and fourth layers of deep cervical fascia form the pericardium and invest the longus colli muscle respectively; (9) the subclavian artery on the left side.

2. A *base*, formed by the diaphragm. This is a musculo-aponeurotic structure, in relation with the pleuræ and pericardium above and below with the peritoneum. It transmits: (1) The aorta; (2) the ascending vena cava; (3) the œsophagus and pneumogastric nerves; (4) the superior epigastric branch of the internal mammary artery; (5) the thoracic duct; (6) the communicating branch from the lumbar veins to the azygos veins.

3. An *anterior wall*, formed by the sternum and cartilages of the ribs, with their investing soft parts. On this we find the internal mammary arteries and their veins, and the sterno-hyoid, sterno-thyroid, and triangularis sterni muscles.

4. A *posterior wall*, formed by the thoracic portion of the vertebral column and the ribs outward from their heads to the angles, with their investing soft parts.

5. *Two lateral walls*, formed by the ribs and intercostal spaces, between the angles and the costo-chondral articulations.

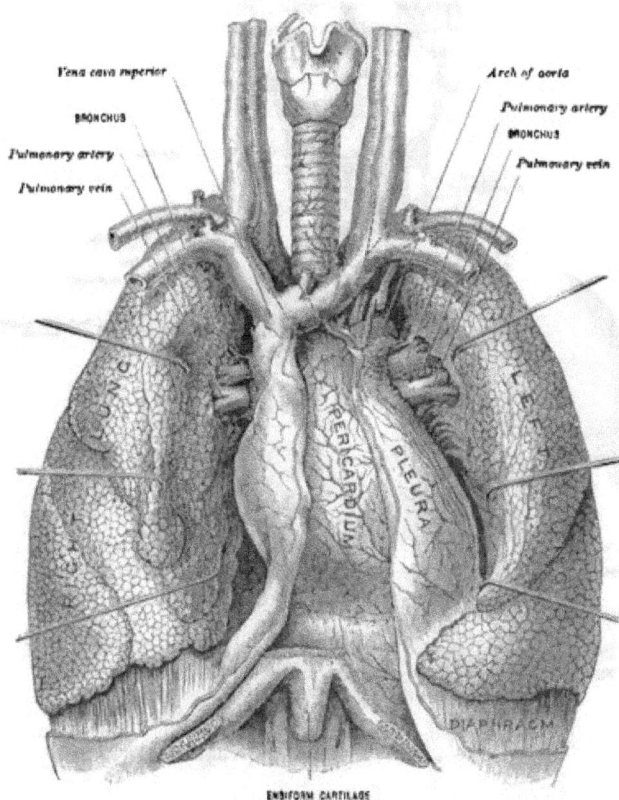

FIG. 144.—ANTERIOR VIEW OF THE LUNGS: PERICARDIUM. (Modified from Bourgery.)

The **inlet or apex of the thorax** is bounded by the sternum, first rib, and first thoracic vertebra. The diaphragm is attached to, or rather has its origin from, the sternum, ribs, ligamenta arcuata interna and externa. There are twenty-four ribs, twenty-two intercostal spaces, and forty-four intercostal muscles.

The interior of the thorax consists of:

1. A *pulmonary portion*, a compartment containing the lungs.

2. A *mediastinal portion*, or non-lung-containing compartment.

The pulmonary portion of the thoracic cavity contains the lungs. The inner surfaces of ribs are covered by the pleura costalis. The lung is invested by pleura, just as the abdominal organs are invested by peritoneum. The layer of pleura covering the lungs is called visceral layer, or pleura pulmonalis. The pleural cavity is the space between the two layers of pleura. This cavity contains only a small amount of serum for lubrication. You will often find adhesions between the parietal and visceral layers of the pleura, the result of inflammation. Frequently you will find many ounces of fluid, the result of hypersecretion : this condition is *hydro-thorax*. Occasionally you will find pus in the pleural cavity : this is *pyo-thorax*. You may find fœtid gas : this is *pneumo-thorax* ; or pus and gas, *pyo-pneumo-thorax*.

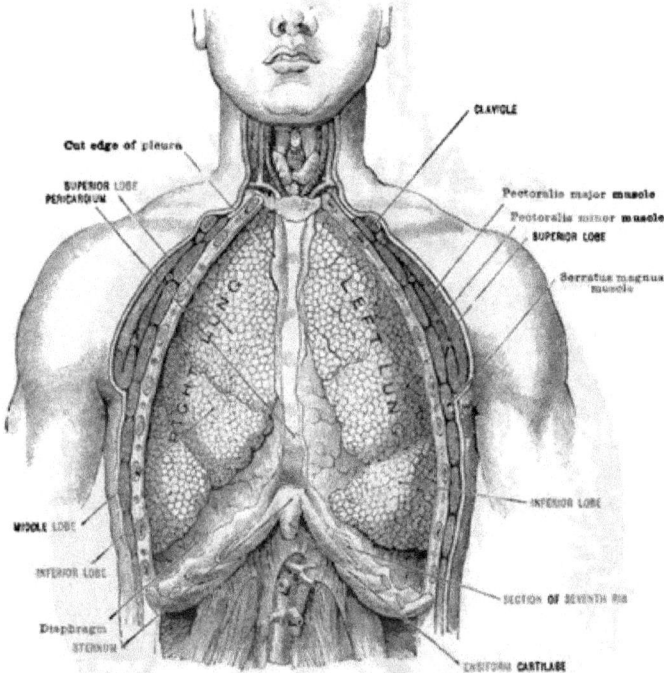

FIG. 145.—ANTERIOR VIEW OF THE THORAX WITH CHEST WALL REMOVED, SHOWING THE LUNGS. (Modified from Bourgery.).

Cavities of tubercular origin are often found in the lungs. In these cases notice the cavity, while an assistant inflates the lung. A large percentage of the cases you see in the dissecting-room have *pulmonary tuberculosis ;* some have *croupous pneumonia ;* others, *pulmonary gangrene.*

Anthracosis is carbuncular disease of the lung. This you will find occasionally. Every adult lung you will see on the tables will be literally covered by carbonaceous spots. This pigmentation is undoubtedly produced by impure air ; still, you fail to find it in domestic animals at the stock-yards. Possibly if these animals lived forty years, they would show the same spots on their lung surfaces that are found in man's lungs.

The *apex* of the lung extends one and one-half inches above the first rib. The *base* of the lung rests on the pulmonary surface of the diaphragm. The lung has a pleuritic attachment to the vertebral column, called the *ligamentum latum pulmonis.*

Anatomical Root of the Lung.—Like the root of the liver, the root **of the lung** transmits structures whereby the lung carries on—(1) the functional activity ; (2) **the** nutritive activity of the lung. By the latter the lung as an organ **lives ; by** the former the lung as an organ aerates the blood. Pull the lung carefully over the cut margin of the ribs (Fig. 145) and you will see the root structures.

Inflate the lungs and study the lobes and fissures. Introduce a one-half inch rubber tube into the trachea and inflate the lungs. You will then see a space, corresponding to one-third the anterior surface of the heart, that is not covered by the lungs when they are inflated. If you could mark on the chest-wall the size of this space, that would give you the area of precordial dullness on percussion. As you inflate the lungs you notice the right lung has three, the left two, lobes.

The root structures of the lungs are :

1. The bronchial tubes, air-conveying conduits.
2. The bronchial arteries, nutritive conduits to the lungs.
3. The pulmonary artery, bearing blood laden with CO_2.

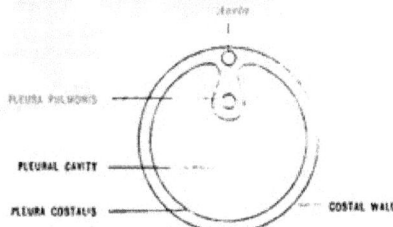

FIG. 146.—INVAGINATION OF PLEURA.

The three structures produced by invagination of pleura may be schematically represented as the above figure shows. 1. The visceral layer, or pars pleuræ invaginata. 2. The parietal layer, or pars pleuræ costalis. 3. The pleural cavity. The same scheme gives like results in accounting for the cavity and layers in connection with peritoneum and pericardium.

4. The pulmonary veins, bearing blood $+O$, and $-CO_2$.
5. The pulmonary sympathetic nerves, from the pulmonary part of the cardiac plexus.

You are to learn that the thoracic, abdominal, and pelvic viscera are innervated by the *sympathetic nerve.* The *cardiac gangliated plexus* supplies the thoracic organs ; the *solar,* the abdominal organs ; the *hypogastric,* the pelvic organs. Each plexus, however, has a primary element, *sympathetic,* and a secondary *pneumogastric* element ; hence we say, in general, the nerve-supply is from the sympathetic. Analytically, we speak of the nerve-supply as sympathetic and pneumogastric.

The Pulmonary Trio.—The sympathetic nerve always accompanies the **artery** to an organ. The artery that nourishes the lungs is the bronchial ; but **this** artery accompanies the bronchial tube. The pulmonary trio, then, consists of the bronchus, the bronchial artery, and the sympathetic nerve-supply.

Relation of bronchus, artery, and veins on the two sides are as follows (1) On the right side, from above down—bronchus, **artery,** and vein. From before back—vein, artery, bronchus.

On the left side, from above down—artery, bronchus, vein. From before back—vein, artery, bronchus.

What is the relation of the visceral and parietal pleuræ at the root of the lung?
They are continuous, since the pleura is an invaginated sac. (Fig. 146.)
Name the three structures produced by a simple invagination of the pleura.
1. The visceral layer, or the pars pleuræ invaginata—pleura pulmonis.
2. The parietal layer, or the pars pleuræ costalis—parietal pleura.
3. The space between the two layers, the pleural cavity.
What are the grand divisions of the parietal pleura?
The diaphragmatic, the mediastinal, the external or costal. The diaphragmatic rest on the diaphragm; the costal on the ribs and internal intercostal muscles; the mediastinal is in relation with the contents of the mediastinum.
Has the upper extremity of the pleura pulmonis any important relations?
Yes; it extends an inch above the first rib, and is in relation to the subclavian vessels, which lie in front and internal to the apex of the pleura.
Is there any special provision made for strengthening or protecting that part of the dome of the pleura that projects above the first rib?
Yes; there is a heavy layer of subpleural connective tissue, called Sibson's fascia, that descends from the scaleni muscles to the first rib and gives strength to the dome.
How is the root of each lung secured to the diaphragm?
By a fold of pleura called the ligamentum latum pulmonis.
Name the geometrical parts of the lung.
The apex, base, outer surface, inner surface, posterior border, anterior border, and pulmonary root.
What is the inner surface of the lung?
That part in contact with the mediastinum and pericardium; it is concave.
Where is the base of the lung and what is its shape?
It is in contact with the diaphragm; it is concave.
What can you say about the anterior border of the lung?
It is sharp and thin, and separates the external from the inner border. The posterior border is thick and round, and occupies the deep groove on each side of the vertebral column.
Is there any provision made for possible physiological increase of breathing space in the pleural spaces?
Yes; the same principle is here seen as is observed in the tortuous arteries of some localities—a condition apparently foreseen to meet an emergency. The lung, under ordinary conditions of man, is less extensive than the pleural sacs. The parietal pleura is really tortuous in the region of the anterior border of the lung. The emergency that would take the kinks out of this pleura would be physical training, where wind is the prime desideratum, as in pugilistic mills. The same principle is seen in a pregnant uterus, reducing the tortuosity of uterine arteries.

THE MEDIASTINAL SPACES.

1. *Anterior*—Is the space between the heart and sternum.
2. *Middle*—Is the space occupied by the heart and its root structures.
3. *Posterior*—Is the space behind the heart.

The Contents of the Anterior Mediastinal Space:

1. The *remains of the thymus gland*—a fœtal structure.
2. The *left brachio-cephalic or left innominate vein.*
3. The *internal mammary artery and its companion veins.*
4. The nearly obsolete *triangularis sterni muscle.*
5. The origin of the *sterno-hyoid* and *sterno-thyroid muscles.*
6. The large amount of *connective tissue* you noticed when you opened the thorax.

Contents of Middle Mediastinal Space :

1. *Pericardium* and *heart*, and the *cardiac root structures.*
2. *Phrenic nerve* and its *artery—arteria comes nervi phrenici.*
3. The *aorta, pulmonary artery,* and *pulmonary vein.*
4. The *venæ cavæ,* superior and inferior, ascending and descending.
5. The *tracheal bifurcation* and the two bronchial tubes.
6. *Bronchial lymphatic glands* and their vessels and nerves.
7. The *cardiac plexus of the sympathetic nerve.*

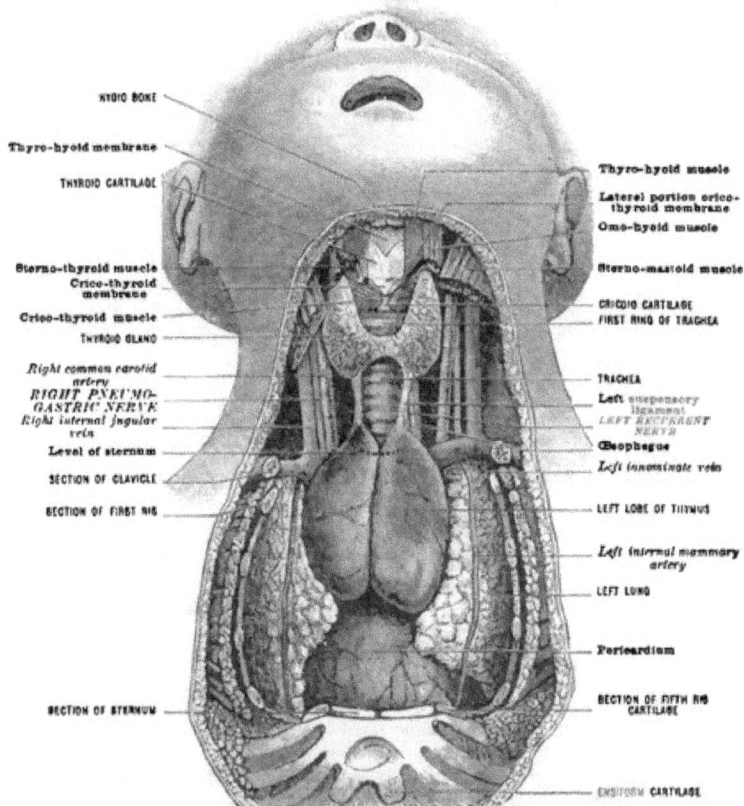

Fig. 147.—Thymus Gland in a Child at Birth.

The Posterior Mediastinal Contents :

1. The *aorta* and *posterior intercostal arteries.*
2. The *œsophagus* and *pneumogastric nerves.*
3. The *vena azygos* and *intercostal veins.*
4. The *thoracic duct* and its fatty bed.
5. The *thoracic gangliated sympathetic cord.*
6. The greater, lesser, and smallest *splanchnic nerves.*

7. The *intercostal nerves*, each giving off two communicating branches—*rami communicantes*—to the ganglia of the sympathetic nerve in the thorax.

Dissection of these regions must be done with the forceps. No cutting instrument should be used. Proceed in this order : (1) Find the phrenic nerve, a branch of the third and fourth cervical, and the arteria comes nervi phrenici, a branch of the internal mammary artery, on either side of the pericardium, between this structure and the pleura. This nerve comes from the third and fourth cervical plexus. It lies on the scalenus anticus muscle in the neck. It enters the mediastinal space between the subclavian artery and vein. It passes anterior to the root of the lung. It communicates with the sympathetic, with the nerve

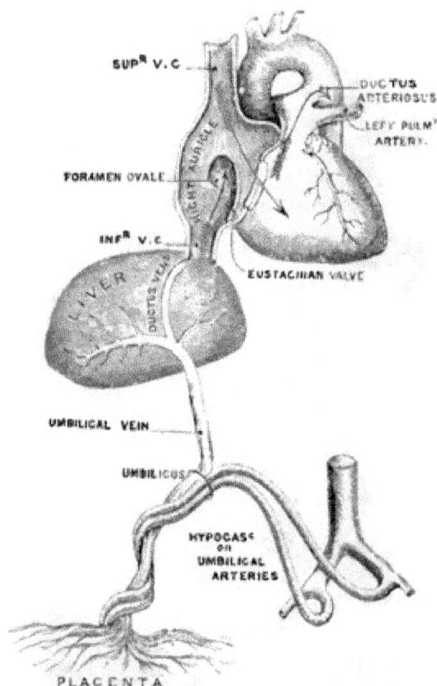

SUP^R V.C

DUCTUS ARTERIOSUS

LEFT PULM^Y ARTERY.

FORAMEN OVALE

INF^R V.C

EUSTACHIAN VALVE

LIVER

DUCTUS VENOSUS

UMBILICAL VEIN

UMBILICUS

HYPOGASS OR UMBILICAL ARTERIES

PLACENTA

FIG. 148.—SCHEME OF THE FŒTAL CIRCULATION.

to the subclavius, and with the ansa hypoglossal loop. It is distributed to the diaphragm, pleura, pericardium, and diaphragmatic peritoneum.

The pericardium has : (1) A base, attached to the diaphragm ; (2) an apex, continuous above with the third layer of deep cervical fascia. It may have much fat. It has a visceral layer, a parietal layer, and a cavity. The latter is the space between the two layers, and contains a small quantity of fluid for lubrication. It may contain many ounces of fluid—this is hydro-pericardium. Note the relation of the inflated lung to the pericardium and poststernal connective tissue. (Fig. 110.) Make traction on the diaphragm where the pericardium is attached, and observe that this cannot be depressed because of its continuity with the third layer of the cervical fascia. Inflate the pericardiac sac ; then cut the parietal layer of the pericardium.

The Thymus Gland or Body.—You will find the fibrous remains of this gland on the anterior part of the pericardium in the adult, in the anterior mediastinal space. This gland is a structure of fœtal life; in children two years old it is quite large. (Fig. 147.) In the four months' fœtus the thymus is in appearance like the lung, on casual examination, and is frequently mistaken for this organ by the novice.

The cardiac root structures are :

1. The *coronary*, or *nutrient*, *artery* to the heart, from the aorta.
2. The *venæ cavæ*, coming from the whole body with venous blood.
3. The *pulmonary arteries*, conveying blood to the lungs.
4. The *pulmonary veins*, returning blood from the lungs.
5. The *aorta*, distributing blood to the whole body.
6. The *cardiac branches*, from the sympathetic and vagus nerve.

How to Study Relations of These Structures Which You have now Exposed.

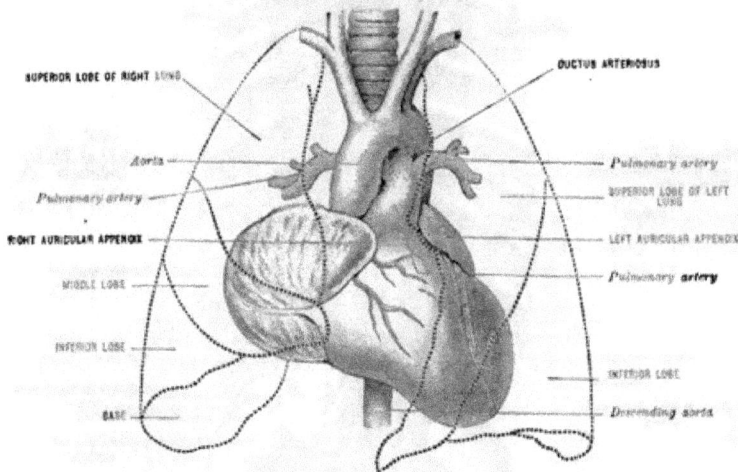

FIG. 149.—ANTERIOR VIEW OF FŒTAL HEART, VESSELS, AND LUNGS.

Regardless of the Particular Arbitrary Space (Fig. 145).—(1) Locate in the middle the ascending part of the arch of the aorta (Fig. 153); (2) to the right of this is the descending vena cava above and the ascending vena cava below; (3) to the left of the aorta is the common pulmonary artery. The remaining structures are the pulmonary veins, which may be seen by turning the apex of the heart toward the right shoulder.

Notice now particularly the pulmonary **artery** (Fig. 148), for it is the one structure in this root that complicates relations. At its origin it overlaps the first part of the aorta; separate these vessels—the aorta from the pulmonary artery. Next it divides into a right pulmonary and a left pulmonary. The *left pulmonary* passes to the left lung, in front of the descending aorta and left bronchus. The *right pulmonary artery* passes to the right lung, behind (1) the aorta, (2) the vena cava, (3) the phrenic nerve and its artery, and (4) below the right bronchus.

The ductus arteriosus (Fig. 149) in the fœtus is a vessel connecting the pulmonary artery and the under surface of the arch of the aorta ; in the adult you must find here in this place the fibrous remains of this fœtal structure.

The ascending vena cava passes through the caval opening in the diaphragm. You see it when you open the pericardium. Its length is about one-half of an inch, above the diaphragm in the pericardium.

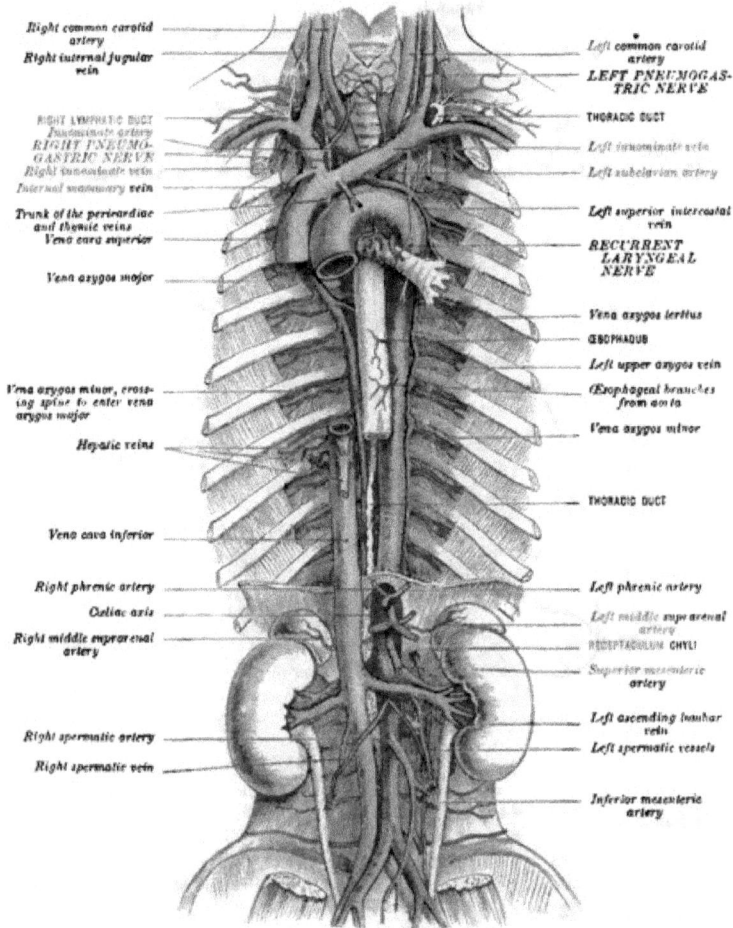

Right common carotid artery
Right internal jugular vein

RIGHT LYMPHATIC DUCT
Innominate artery
RIGHT PNEUMO-GASTRIC NERVE
Right innominate vein
Internal mammary vein

Trunk of the pericardiac and thymic veins
Vena cava superior

Vena azygos major

Vena azygos minor, crossing spine to enter vena azygos major

Hepatic veins

Vena cava inferior

Right phrenic artery
Cæliac axis
Right middle suprarenal artery

Right spermatic artery
Right spermatic vein

Left common carotid artery
LEFT PNEUMOGAS-TRIC NERVE

THORACIC DUCT
Left innominate vein
Left subclavian artery

Left superior intercostal vein
RECURRENT LARYNGEAL NERVE

Vena azygos tertius
ŒSOPHAGUS
Left upper azygos vein
Œsophageal branches from aorta
Vena azygos minor

THORACIC DUCT

Left phrenic artery
Left middle suprarenal artery
RECEPTACULUM CHYLI
Superior mesenteric artery
Left ascending lumbar vein
Left spermatic vessels

Inferior mesenteric artery

FIG. 150.—THE ARCH OF THE AORTA, THE THORACIC AORTA, AND THE ABDOMINAL AORTA, WITH THE SUPERIOR AND INFERIOR VENA CAVA AND THE INNOMINATE AND AZYGOS VEINS.

The descending vena cava is made up by the confluence of the right and left brachio-cephalic or innominate veins; and the vena azygos major, the pericardiac, thymic, and œsophageal veins are its subsequent tributaries. It returns the blood to the right auricle of the heart from above the diaphragm.

The brachio-cephalics—right and left, also called the innominate veins—

are formed by the confluence of the internal jugular veins and the subclavian and the lymphatic ducts. (Fig. 150.)

The left brachio-cephalic vein (Fig. 150) is the one structure that complicates relations at the root of the neck. This vein lies in the anterior mediastinal space, behind the first piece of the sternum. Great care must be taken not to rupture this vessel or its tributaries when you remove the sternum and subsequently dissect this region. It is formed by the confluence of the left internal jugular vein, the thoracic duct, the left subclavian vein, the left superior intercostal, and the inferior thyroid veins. It crosses (1) the left subclavian artery ; (2) the left common carotid artery ; (3) the two pneumogastric nerves and the left phrenic nerve ; (4) the innominate artery ; (5) the left recurrent laryngeal nerve ; (6) the trachea ; (7) the œsophagus.

The thoracic duct (Fig. 150) in many cases is filled with blood, the valves at its terminus not being able to resist the pressure, used in embalming the dissecting material. The many cases reported by students of absence of this vessel have no foundation in fact, but belong to a host of finds and absences which may collectively be known as *cases of first dissection.*

Develop now the origin from the aorta of the innominate, the left common carotid, and left subclavian arteries. You have dissected the branches of these structures in your work on the neck ; now review their relations. See the innominate artery dividing behind the right sterno-clavicular articulation into the right common carotid and right subclavian arteries.

The thoracic **part of the** aorta has two divisions. The student will have noticed that this structure is in both the middle and posterior mediastinal spaces. You must remember that classification of structures in reference to specific regions is at best a very arbitrary institution ; that a knowledge of anatomy consists in a thorough acquaintance with organs wherever found, rather than in an ability to recite instanter an artificial, arbitrary classification of structures found *in parte* in a given region.

1. The aortic **arch**—three propositions : (1) It begins at second sterno-chondral articulation on the right side ; (2) it ends at the fifth thoracic vertebra on left side ; (3) its branches are : coronary, innominate, left common carotid, left subclavian arteries.

2. The thoracic aorta—three propositions (1) It begins at the fifth thoracic vertebra on left side. (2) It ends at the twelfth thoracic in the mid-line. (3) Its branches are : (1) Pericardiac to the pericardium ; (2) œsophageal to the œsophagus ; (3) bronchial to the bronchi, and lung-substance ; (4) the posterior mediastinal branches, to the glands and connective tissue ; (5) the intercostals, ten in number, to the muscles and fasciæ in the intercostal spaces and to the skin covering these structures.

Practical Dissection Observations.—(1) Note that the branches to the bronchi, œsophagus, glands, and pericardium are given off from the anterior part of the aorta, or some derivative of the same. (2) Note that the mediastinal glands, in nearly every case, contain a fluid of inky-black color. This dark color is produced by absorption of anthracite particles When enough has accumulated to produce *carbuncular disease* of the lung, the patient has *anthracosis.* You must make this practical **distinction** : Every lung you find will have carbonaceous pigmentations ; not every lung will have enlarged, inky-fluid-containing lymphatics —only some will show this condition. Some lungs will possess carbuncular nodules and nodes ; these always have enlarged lymphatics, filled with inky fluid, if the disease is anthracosis.

Conclusion.—In every adult cadaver you will see pigmented lungs. In some cadavers you may see anthracotic lungs—that is, chronic interstitial pneumonia, produced by inhalation of atmospheric impurities, the basis of which is carbon.

A Posterior View of the Thoracic Viscera (Fig. 151).—Turn the lung still further to the left and you can see the following structures even without cutting structures as shown in the figure : (1) The aorta giving off its intercostals behind, and its visceral branches in front; (2) the trachea, left bronchus, œsophagus, and thoracic duct behind the transverse arch of the aorta,; (3) the pulmonary veins, right and left, leading to the left auricle of the heart; (4) the recurrent laryngeal branch of the left pneumogastric nerve ; (5) the trachea, bifurcating into the right and left bronchi, opposite the fifth thoracic vertebra ; (6) the right and left pneu-

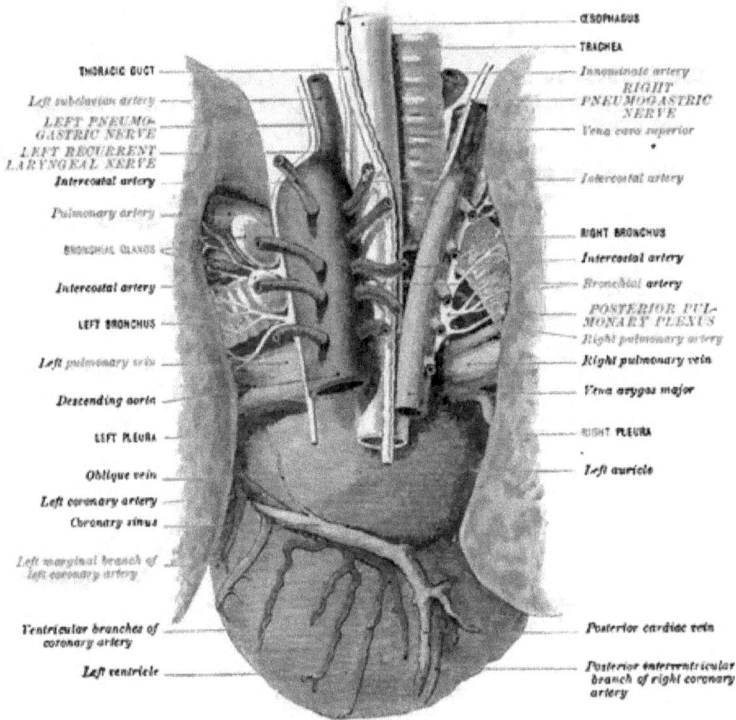

FIG. 151.—THE HEART AND GREAT VESSELS, WITH THE ROOT OF THE LUNGS, SEEN FROM BEHIND
(St. Bartholomew's Hospital Museum.)

mogastric nerves; (7) the pulmonary sympathetic plexus ; (8) some bronchial lymphatic glands.

The Cardiac and Pulmonary Plexuses.—On the posterior part of the bronchi you will find some nerves; these form the *posterior pulmonary plexus.* On the front of the bronchi you will also find some nerves ; these form the *anterior pulmonary plexus.* Now cut through the trachea two inches above its bifurcation, and you will see, immediately in front of the bifurcation, the *deep cardiac plexus.* Look under the arch of the aorta and see the *superficial cardiac plexus.*

The above plexuses furnish nerves that supply the heart and lungs. The pulmonary plexus is a subdivision of the cardiac plexus. The terms anterior and posterior pulmonary are used for convenience in describing the particular regions occupied by the plexuses. The terms superficial and deep cardiac are used for the same reason. The cardiac plexus is formed as follows:

1. By cervical and thoracic branches of the pneumogastric nerve.
2. By branches from the recurrent laryngeal nerve.
3. By branches from the cervical sympathetic nerve.

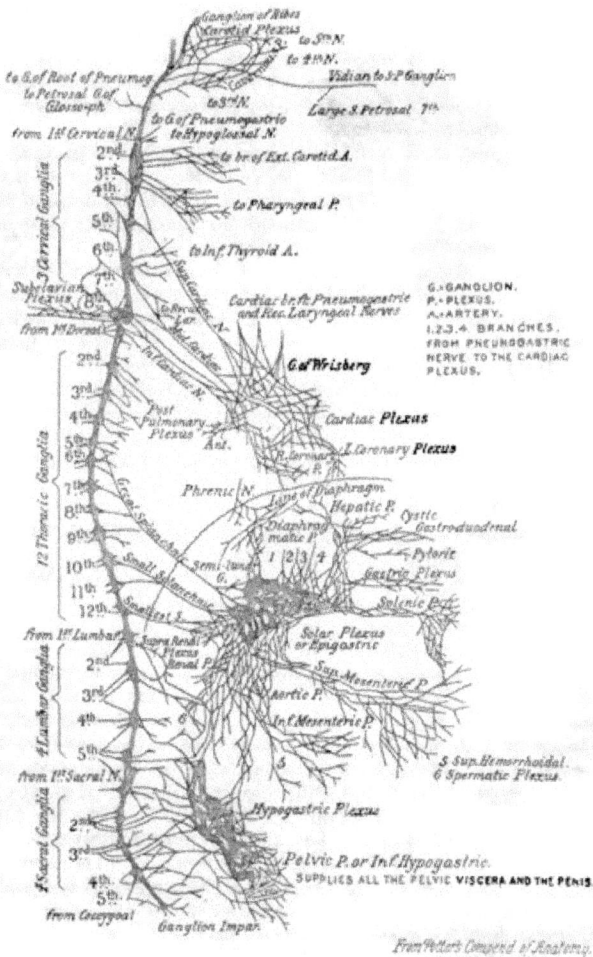

Fig. 152.—The Sympathetic System of Nerves.

Dissection of the Posterior Mediastinal Space (Fig. 150).—Remember, you are **not** to remove the heart and lungs in order to dissect the space. Modus operandi : (1) Turn the right lung to the left side. This is easily done, provided there are no pleural adhesions. Where adhesions exist, these must be broken up. (2) Separate the parietal pleura from the ribs, at the cut margin of **the ribs,** and you can then remove it from its parietal attachment, toward the mid-line of the body, to the place where it is reflected on to the lungs, just as easily as a paper-hanger removes his paper from the wall, for purposes of proper adjustment, before his paste becomes set. (3) You now see the subpleural connective tissue—the medium intervening between the chest walls and the pleura.

The structures now seen are (Figs. 150 and 152): The *fibres of the internal intercostal* muscles, terminating near the angle of the ribs posteriorly. The space between this pleura and the vertebral column is occupied by fatty connective tissue, which must be dissolved in gasolene, ether, or benzine. Now you will see, as in figure 150, the intercostal arteries, veins, and nerves lying on the external intercostal muscles. You soon lose sight of these structures, for they pass between the intercostal muscles.

The thoracic gangliated cord of the sympathetic nerve (Fig. 152) is easily dissected, but requires a delicate touch to preserve its integrity, such is the delicacy of its structure. You can find a ganglion opposite the head of each rib. Trace two little branches from this ganglion to the spinal nerve ; these branches are the rami communicantes.

Figure 151 shows the spinal nerves near their exit from the intervertebral foramina, giving communicating branches—rami communicantes—to the ganglia of the thoracic sympathetic cord. It also shows the manner in which the *six upper* distribute their branches to the thorax, and the *six lower* their branches to the abdomen by the *splanchnic nerves.*

Branches of the Thoracic Gangliated Cord (Fig. 152).—(1) To the spinal nerves—rami communicantes ; (2) branches connecting the ganglia ; (3) branches of distribution. Notice the six upper ganglia give off branches to the aorta and its branches in the thorax, to the bodies of the vertebræ and their ligaments, and to the pulmonary and cardiac plexuses. Notice the six lower give off branches, (1) to the aorta and its thoracic branches, (2) and unite to form the splanchnic.

The splanchnics are the great abdominal visceral branches from the thoracic part of the gangliated cord ; they are named—

1. The *greater splanchnic* nerve (sixth, seventh, eighth, and ninth).
2. The *lesser splanchnic* nerve (tenth and eleventh).
3. The *smallest or renal splanchnic* nerve (twelfth).

Can these nerves be seen?

Yes, easily. The scheme, as set forth in figure 152, can be verified by any student on any cadaver, with the proper care. Note that the three splanchnic nerves pass through the diaphragm. When we study the thoracic aorta we will see what becomes of the branches of distribution from the upper six thoracic ganglia ; we must ask, Why are they so small, when the six lower are so large?

The vena azygos major lies on the vertebral column. (Fig. 150.) It is composed of the intercostal veins from the thoracic walls. The blood from the intercostal spaces of the left side is collected by the vena azygos minor. Sometimes two small azygos veins will be found on the left side. The minor is tributary to the major azygos vein. Remember, the left superior intercostal vein is tributary to the left brachio-cephalic ; the right to the vertebral vein.

The thoracic duct lies in a bed of fatty connective tissue, between the aorta and the vena azygos major, on the vertebral column. It is confluent to the left brachio-cephalic vein. It originates in the receptaculum chyli, on the second lumbar vertebra. It passes through the aortic opening in the diaphragm. This is easily inflated, in the recently-dead, before the material has been embalmed. To do this inflation, cut into the brachio-cephalic and find the orifice of the duct.

The œsophagus extends from the pharynx to the stomach. It leaves the thorax by the œsophageal opening in the diaphragm. Find it passing behind (1) the arch of the aorta, (2) the trachea, (3) the left bronchus, (4) the left common carotid and left subclavian veins, and (5) pericardium. Find (6) the vena azygos major to the right, and (7) the aorta to the left. On each side find (8) the pneumogastric nerves. Near the diaphragm find (9) the left pneumogastric in front, (10) the right behind the œsophagus—a result of rotation of the stomach.

The pneumogastric nerves (Fig. 153) have been thoroughly considered in the dissection of the neck. Still, there are some practical relations of this nerve

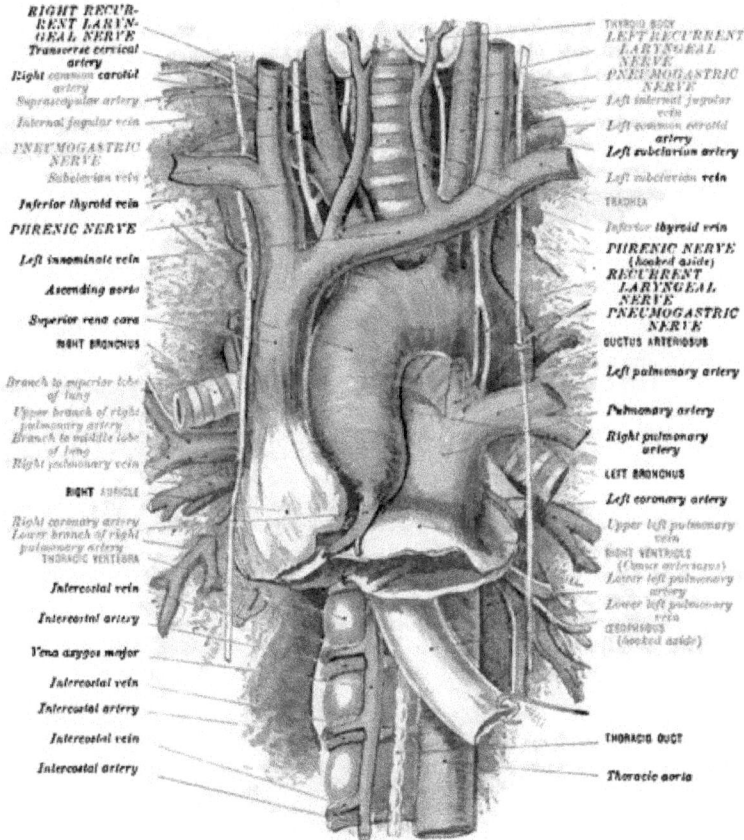

FIG. 153.—THE ARCH OF THE AORTA, WITH THE PULMONARY ARTERY AND CHIEF BRANCHES OF THE AORTA.
(From a dissection in St. Bartholomew's Hospital Museum.)

in the thorax I wish the student to understand. The pneumogastric nerve leaves the cranium by the jugular foramen. You found it in the carotid sheath, between the internal jugular vein and common carotid artery in the neck. The two pneumogastric nerves differ in their relations somewhat, hence dissect them separately. The *right nerve* (Fig. 153) crosses the right subclavian artery, and at once gives off its recurrent or inferior laryngeal nerve, which passes upward and inward, behind the subclavian and common carotid arteries, to the larynx,

15

lying, in its course, between the œsophagus and trachea, in a bed of fatty connective tissue. In the lower part of the neck, cervical and thoracic cardiac branches are given off to the cardiac plexus. A little lower are given off pulmonary branches, called, from their distribution, anterior and posterior ; these supply the lungs through the pulmonary gangliated plexus. Œsophageal branches from the pneumogastric form the plexus gulæ. The pneumogastric nerve leaves the thorax by the œsophageal opening in the diaphragm, and sends branches to the solar plexus, from which, possibly, all the abdominal organs receive pneumogastric influence. This, however, is purely physiological speculation.

On the left side, the pneumogastric nerve passes behind the left brachiocephalic vein and in front of the arch of the aorta. As it is crossing the arch it gives off its recurrent laryngeal branch. This nerve passes to the outer side of the ductus arteriosus and gains the side of the left bronchus and trachea. The inferior or recurrent laryngeal nerves supply branches to the trachea, œsophagus, and to all the intrinsic muscles of the larynx except the crico-thyroid muscle.

The coronary arteries supply the heart. They are given off from the ascending aorta. They embrace at their origin the common pulmonary artery by their divergence. (Fig. 153.) Blood is returned **from** the heart **by** two sources : (1) By minute veins which end in the foramina of Thebesius in the right auricle of the heart. (2) Numerous veins come together to form the great coronary sinus. This sinus opens into the right auricle of the heart, between the auriculoventricular opening and the ascending vena cava. You will **demonstrate** this opening, and also the valve, called the coronary, on your dissection.

The Right Auricle of the Heart.—Cut through the wall of the auricle from the ascending to the descending vena cava in a direction from above downward. Now, having thoroughly washed the parts, you will see : (1) The *appendix auriculæ*, and (2) the *atrium, sinus,* or principal cavity of the auricle. In the atrium see and locate : (1) The *fossa ovalis*, surrounded by a rim called the *annulus ovalis* ; (3) the caval and auriculo-ventricular openings (Fig. 154) ; (4) the *coronary valve* or valve of Thebesius and the *Eustachian valves*, at the respective openings for the coronary sinus and ascending vena cava ; (5) the musculi pectinati, which you will see on cutting through the appendix auriculæ. These latter are mechanical devices for giving strength to the walls of the auricle, analogous to the chordæ tendineæ in the ventricles.

The Semilunar Valves (Fig. 154).—Cut into the aorta **and** pulmonary artery an inch above where they leave the heart, and demonstrate these valves. Fill the **vessels** with water and show the action of **these** valves in preventing regurgitation. In the central point of each free margin of the valve find the corpus Arantius. This is a mechanical device, whereby **the** valves occlude more perfectly. Also find in the *aorta* the sinuses of Valsalva, **or** aortic sinuses, opposite to which are **found** the attached margins of the valves. Above the free margin of the valves find the coronary arteries, **left** and right, given off from the aorta.

The auriculo-ventricular valves, right and left, guard the respective openings between **the** auricles and ventricles. Cut into the right ventricle (Fig. 154) and study **the** mechanical devices known as columnæ carneæ and **chorda** tendineæ ; also **the** left ventricle **study in** the same manner.

The septa are : (1) The interauricular. In this you see the fossa ovalis. In the fœtus this was a communication between the auricles, called the foramen ovale. It is in the posterior wall of the right auricle. (2) The interventricular, separating the ventricles.

Steps in Adult Circulation.—(1) Blood is taken from below the diaphragm by the inferior vena cava, and from above the diaphragm by the descending vena cava, to the right auricle of the heart. (2) Blood flows from the right auricle, through the auriculo-ventricular opening, to the right ventricle. This opening is guarded by the right auriculo-ventricular valves. (3) Blood flows from the right ventricle, by the pulmonary artery, to the lungs. This artery is guarded by the pulmonary semilunar valves. (4) Blood flows from the lungs to the left auricle of the heart by the pulmonary veins. These veins have no valves. (5) Blood passes from the left auricle, by the left auriculo-ventricular opening,

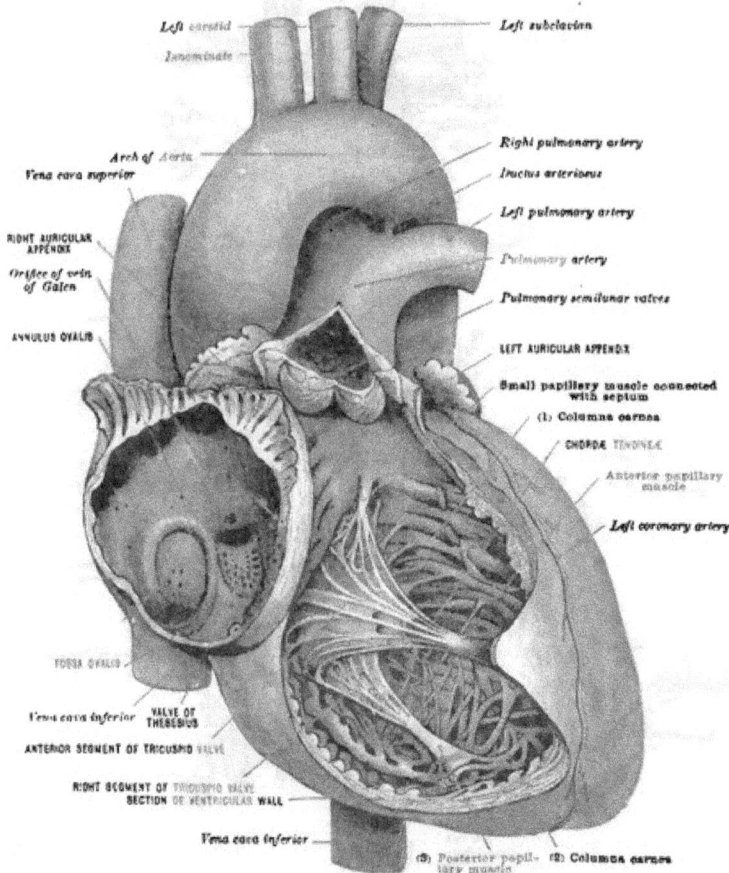

Fig. 154.—Anterior View of the Right Chambers of the Heart, with the Great Vessels.

to the left ventricle. This opening is guarded by the left auriculo-ventricular valves. (6) Blood passes from the left ventricle through the aorta to all vascular parts of the body. The aorta is guarded by the aortic semilunar valves. Systole is a term by which contraction of the auricles and ventricles is designated; diastole is its opposite. The one extrudes, the other draws in, blood.

The Fœtal Circulation.—The lungs are essentially organs whose use in distributing the blood for purification begins after birth. Before the product of

conception reaches the ninth month of utero-gestation, its blood is distributed
for purification in an organ called the placenta. This placenta, while structurally very different from the lungs, accomplishes for the blood of the fœtus just

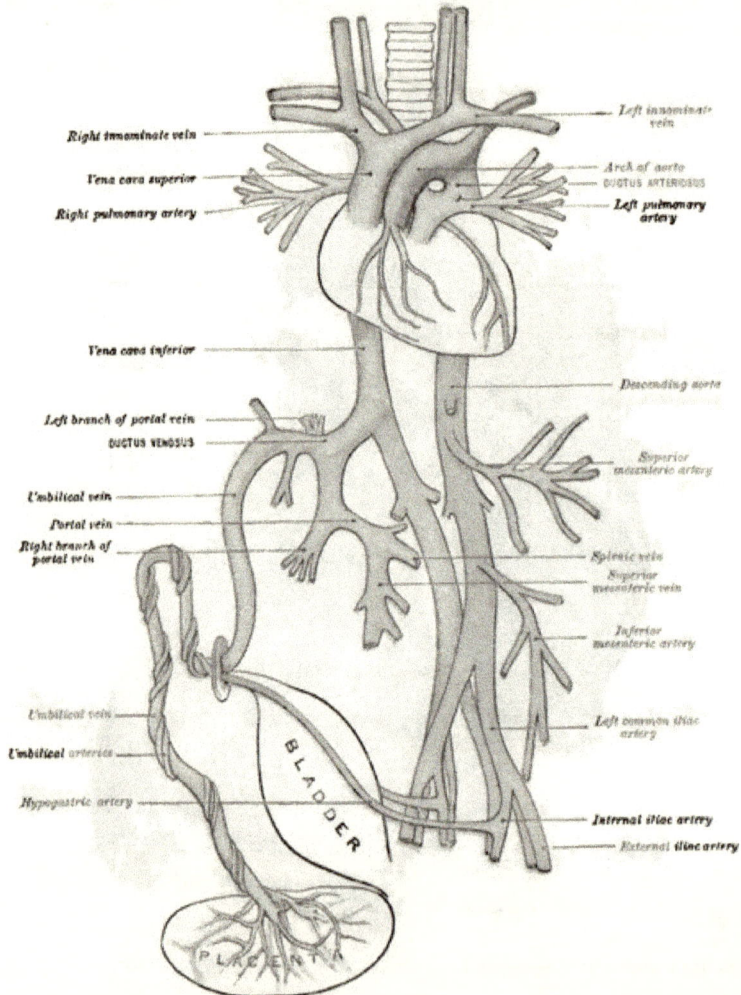

FIG. 155.—THE HEART, WITH THE ARCH OF THE AORTA, THE PULMONARY ARTERY, THE DUCTUS
ARTERIOSUS, AND THE VESSELS CONCERNED IN THE FŒTAL CIRCULATION.
(From a preparation of a fœtus in the Museum of St. Bartholomew's Hospital.)

the same results as are accomplished by the lungs *post partem.* The purification
of the blood in either case is in accordance with the law of physics regulating
the diffusion of gases. Blood, on entering lung or placenta, is laden with CO_2;

on leaving the same it is laden with O. The difference, then, between the ante partem and post partem purification of blood is not a physiological one; but a difference is to be found in certain anatomical routes taken by this blood, in the one case to and from the placenta, in the other case to and from the lungs. A study of the metamorphosis of parts after functional activity ceases is replete with interest, since it shows the converse of the law of physiology that growth is the correlation of function. Study the following steps in the fœtal circulation, in order that in your dissection you may be able to appreciate the several fœtal vestiges you shall find there.

1. **In the fœtus** the blood is purified in the placenta of the mother. (Fig. 155.) It is taken to that organ by **the** hypogastric arteries, from the internal iliac arteries.

2. It is returned **from the placenta by the umbilical** vein, as follows: (1) **To** the ascending **vena cava by the ductus** venosus; (2) **to** the portal vein to traverse the liver, and **finally reach the** ascending vena cava by the hepatic veins.

3. The blood from above **the** diaphragm reaches the heart by the descending **vena cava, as** in the adult.

4. The blood from the descending vena cava passes through **the auricle to the ventricle, and** thence **out** through the pulmonary **artery; the** lungs **not being** yet **fitted for** purifying, **the blood** is taken by the **ductus arteriosus to the** aorta.

5. The blood from the ascending vena cava passes **through the foramen ovale** (Fig. 154) to the left **auricle,** thence to the left ventricle, and **out by the aorta.**

The *circulatory apparatus* of the fœtus possesses:

1. The *umbilical* vein brings blood from the placenta.
2. The *hypogastric arteries* conveys blood **to** the placenta.
3. The *ductus venosus* takes blood to the vena cava.
4. The *ductus arteriosus* connects the pulmonary artery and aorta.
5. The *foramen ovale* is an opening between the auricles. All of these structures forming the circulatory apparatus of the fœtus are set aside **at birth, when** the lungs become the aerating organ for the blood. By non-use these **structures** lose their specific characters—they become obsolete. They obtain in **the adult as:**

1. **Remains of** the umbilical vein, the round ligament **of** the liver.
2. **Remains of** the hypogastric arteries near the urachus.
3. **Remains of** the ductus venosus on under surface of liver.
4. **Remains of** the ductus arteriosus between **aorta** and pulmonary artery.
5. Remains **of** the foramen ovale in posterior **wall of** auricle.

What is a fœtal vestige? The fibrous **remains in the adult** of an organ that under the conditions of the fœtus in utero **was a necessary** organ or part. The umbilicus and urachus **are** fœtal **vestiges.**

What are the grand divisions of the thorax?

The thorax may be subdivided into a pulmonary part, occupied by the lungs, and a mediastinal part, *not occupied by the lungs.*

Is the word thorax synonymous with trunk?

No; thorax is the cavity above the diaphragm; trunk means the whole body except the head, neck, and extremities.

Explain the superior mediastinal space.

This is a space bounded above by the superior aperture of the thorax; below by a plane limited in front by the manubrio-gladiolar articulation and behind by the body of the fourth thoracic vertebra; in front by the manubrium; behind by the four upper thoracic vertebræ; laterally by the pleural sacs. This space contains the following structures, which you may find on your work : (1) The thoracic portion of the trachea ; (2) the thoracic portion of the œsophagus ; (3) the thoracic portion of the thoracic duct ; (4) the transverse part of the aortic arch ; (5) the imal and innominate arteries ; (6) the thoracic portion of the left subclavian artery ; (7) the innominate veins, and the superior vena cava ; (8) the terminations of the internal mammary veins ; (9) the terminations of the inferior thyroid veins ; (10) the superior intercostal vein of the left side ; (11) the two pneumogastric or vagi nerves ; (12) the two phrenic and cardiac nerves ; (13) the left recurrent or inferior laryngeal nerve ; (14) thymus gland in the fœtus and remains of the thymus in the adult.

Describe the pericardium.

It is a fibro-serous sac surrounding the heart. The pericardium is cone-

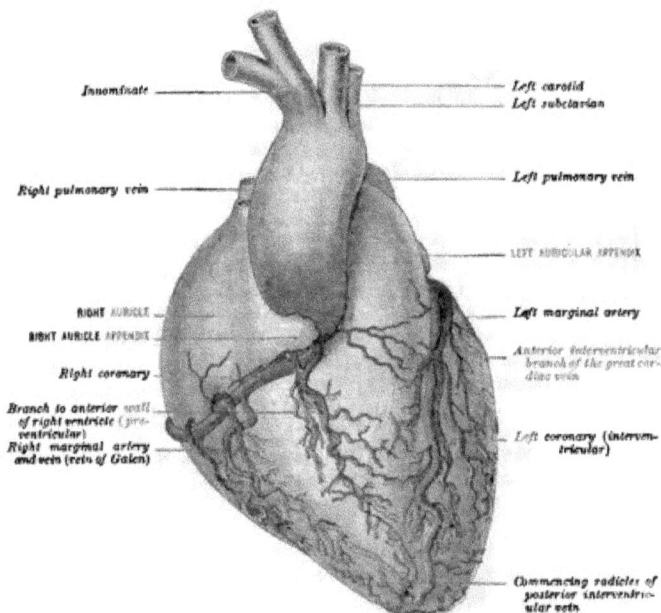

FIG. 156.—ANTERIOR VIEW OF THE HEART, SHOWING ITS ARTERIES AND VEINS.

shaped, the base being attached to the central tendon of the diaphragm, the apex being lost in the fibrous sheath of the vessels that arise from the base of the heart. Structurally the pericardium is inelastic. Above it is continuous with the third layer of deep cervical fascia. The pericardium is attached in front to the sternum by sterno-pericardial bands.

From what source does the pericardium receive its blood-supply ?

From the phrenic below ; from the internal mammary in front ; from the œsophageal, pericardiac, and bronchial branches of the thoracic aorta behind.

From what source does the pericardium receive its nerve-supply ?

From the phrenic, vagus, and sympathetic.

Name the structures at the base of the heart that derive strengthening bands from the pericardium.

The aorta, the pulmonary arteries and veins, the ductus arteriosus.

What can you say of the investment of the aorta and pulmonary artery?

They are invested by a common sheath; behind these vessels is a passage called the great sinus of the pericardium.

Name the valves of the heart and the valves of its vessels, and tell how they are formed.

The coronary, the Eustachian, the aortic and pulmonary semilunar, the right and left auriculo-ventricular. They are formed by folds of endocardium and subserous connective tissue.

What are the grand divisions of the right auricle?

The auricular appendix and sinus venosus, or atrium. This auricle forms the front part of the base of the heart. Into the atrium the following structures

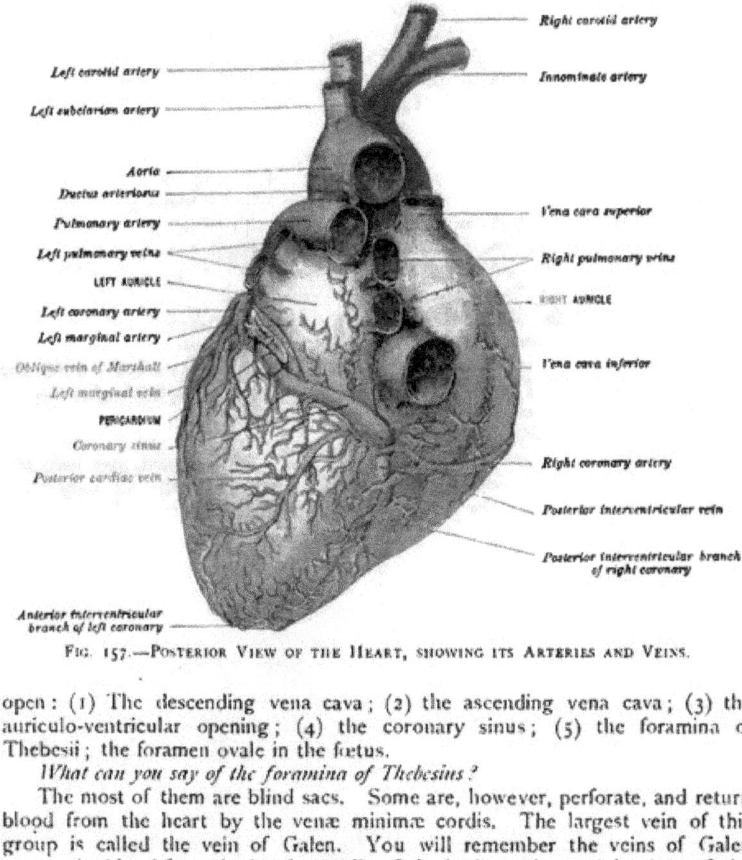

Left carotid artery
Left subclavian artery
Aorta
Ductus arteriosus
Pulmonary artery
Left pulmonary veins
LEFT AURICLE
Left coronary artery
Left marginal artery
Oblique vein of Marshall
Left marginal vein
PERICARDIUM
Coronary sinus
Posterior cardiac vein
Anterior interventricular branch of left coronary

Right carotid artery
Innominate artery
Vena cava superior
Right pulmonary veins
RIGHT AURICLE
Vena cava inferior
Right coronary artery
Posterior interventricular vein
Posterior interventricular branch of right coronary

FIG. 157.—POSTERIOR VIEW OF THE HEART, SHOWING ITS ARTERIES AND VEINS.

open: (1) The descending vena cava; (2) the ascending vena cava; (3) the auriculo-ventricular opening; (4) the coronary sinus; (5) the foramina of Thebesii; the foramen ovale in the fœtus.

What can you say of the foramina of Thebesius?

The most of them are blind sacs. Some are, however, perforate, and return blood from the heart by the venæ minimæ cordis. The largest vein of this group is called the vein of Galen. You will remember the veins of Galen return the blood from the basal ganglia of the brain to the straight sinus of the dura mater. To be specific in speech, then, speak of the veins of Galen of the brain and of the heart.

Name the openings in the right auricle having valves.

The auriculo-ventricular, the coronary sinus, and the ascending vena cava.

Name and give the importance of the openings of the right ventricle.

The auriculo-ventricular opening, guarded by valves of like name; the pulmonary opening, at the beginning of the pulmonary artery, guarded by the pulmonary semilunar valves, three in number.

What are the sinuses of Valsalva and where do you find them in practical anatomy?

They are pouches or dilatations behind the semilunar valves of both the aorta and pulmonary artery.

What are the columnæ carneæ?

They are muscular columns found in the ventricles of the heart. They are mechanical devices for strengthening the cardiac walls. They also give a yielding attachment to the auriculo-ventricular valves.

What are the chordæ tendinæ?

Strong, glistening, tendinous threads attached by one extremity to the auriculo-ventricular valves, by the other to the walls of the ventricle by the medium of the columneæ carneæ. They inhibit the movement of the valve during systole and aid recoil of the valve in diastole of the ventricles.

Describe the blood-supply of the heart.

The heart is supplied with blood by two arteries, called the left and right coronary arteries. These arteries arise from the sinuses of Valsalva of the aorta. The left arises from the posterior sinus, the right from the anterior sinus. The auriculo-ventricular grooves and the interventricular grooves are where you will find the main trunks of the arteries that supply the heart.

Two anastomotic circles, a horizontal and a vertical, corresponding to the grooves separating the auricles from the ventricles, and the ventricles from each other respectively, may be traced out. The right coronary artery divides into two main branches: the horizontal occupies the right auriculo-ventricular groove; the vertical, the right interventricular groove. The left coronary artery divides likewise into two branches: the horizontal one occupies the left auriculo-ventricular groove, and the vertical the left interventricular groove. In these grooves the opposite arteries anastomose.

How is blood returned from the heart?

Veins called coronary accompany the arteries just described, and open into the right auricle by the coronary sinus, which is guarded by the coronary valve. This opening is between the inferior vena cava and the auriculo-ventricular opening.

ABDOMEN.

THE ABDOMINAL WALLS.

Dissection.—Locate on the cadaver (1) the *ensiform*, or *metasternum* ; (2) the *symphysis pubis ;* (3) the *spine of the pubes ;* (4) the *crest of the pubes ;* (5) the *crest of the ilium ;* (6) the *anterior superior iliac spine ;* (7) the *lower eight ribs ;* (8) the *umbilicus ;* (9) *Poupart's ligament ;* (10) the *mid-line of the abdomen.*

The abdominal walls include (1) skin ; (2) superficial fascia ; (3) panniculus adiposus ; (4) deep fascia ; (5) the external or descending oblique muscle ; (6) the internal or ascending oblique muscle ; (7) the transversalis muscle ; (8) the rectus abdominis ; (9) the pyramidalis ; (10) the sheath of the rectus ; (11) the transversalis fascia ; (12) the linea alba ; (13) the lineæ transversæ ; (14) the lineæ semilunares ; (15) peritoneum.

1. *Define panniculus adiposus.*

The great mass of fat in the upper strata of superficial fascia is thus designated. It is of variable thickness. In it are found the cutaneous vessels and nerves to the skin covering the abdomen.

2. *What can you say of the deep fascia ?*

It is beyond doubt the same as the aponeurosis of the external oblique muscle. The thin fascia described by some authors is not continuous with the fascia lata, nor has it attachments to bone ; hence it must be the deep layer of the superficial fascia. The aponeurosis of the external oblique muscle is continuous with the fascia lata ; hence it must represent the deep fascia.

3. *Define linea alba.*

It is in the mid-line, extending from the ensiform cartilage to the symphysis pubis. It is the place where the muscles of the two sides of the abdomen meet.

4. *How is the sheath of the rectus formed?*

By a delamination or splitting of the aponeurosis ofthe internal oblique muscle. In front of the muscle, then, is the entire thickness of the aponeurosis of the *external oblique*, and half of the thickness of the *internal oblique ;* behind the muscle is the entire thickness of the transversalis, and half of the thickness of the aponeurosis of the internal oblique.

5. *Does the muscle extend through the sheath from end to end in this manner ?*

No ; it perforates the posterior wall of its sheath midway between the umbilicus and the pubic crest, thus leaving, from this point down to the pubic crest, the three aponeuroses all in front of the muscle.

6. *By what name is the lower margin of the posterior part of the sheath of the rectus known ?*

It is called the semilunar fold of Douglas.

Preparation and Incisions.—Make a circular incision through the skin, three inches in diameter, around the umbilicus as the central point. Begin at the circumference and dissect the skin toward the umbilicus for about one inch all around. Now, with a small pair of scissors puncture into the peritoneal cavity through the center of the umbilicus ; insert a tube and inflate the cavity to its

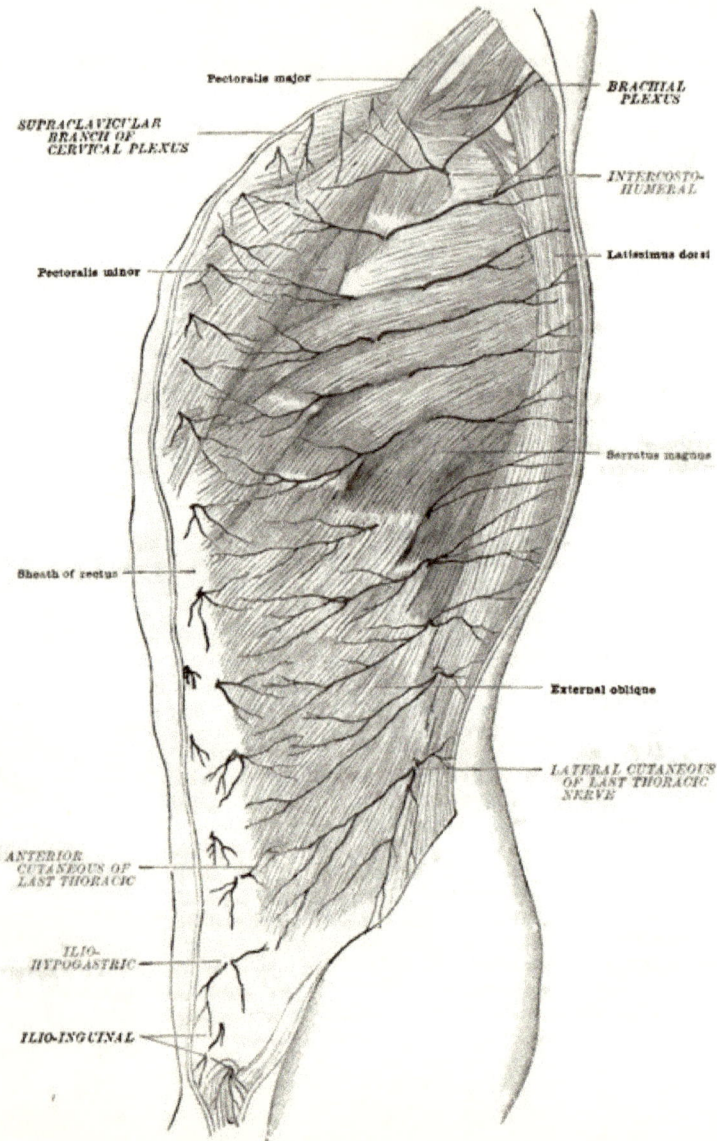

FIG. 158.—CUTANEOUS NERVES OF THE THORAX AND ABDOMEN, VIEWED FROM THE SIDE. (After Henle.)

fullest extent, and let an assistant throw a string about the umbrella of skin just turned upward, and while you withdraw the tube he will tie the cord. The

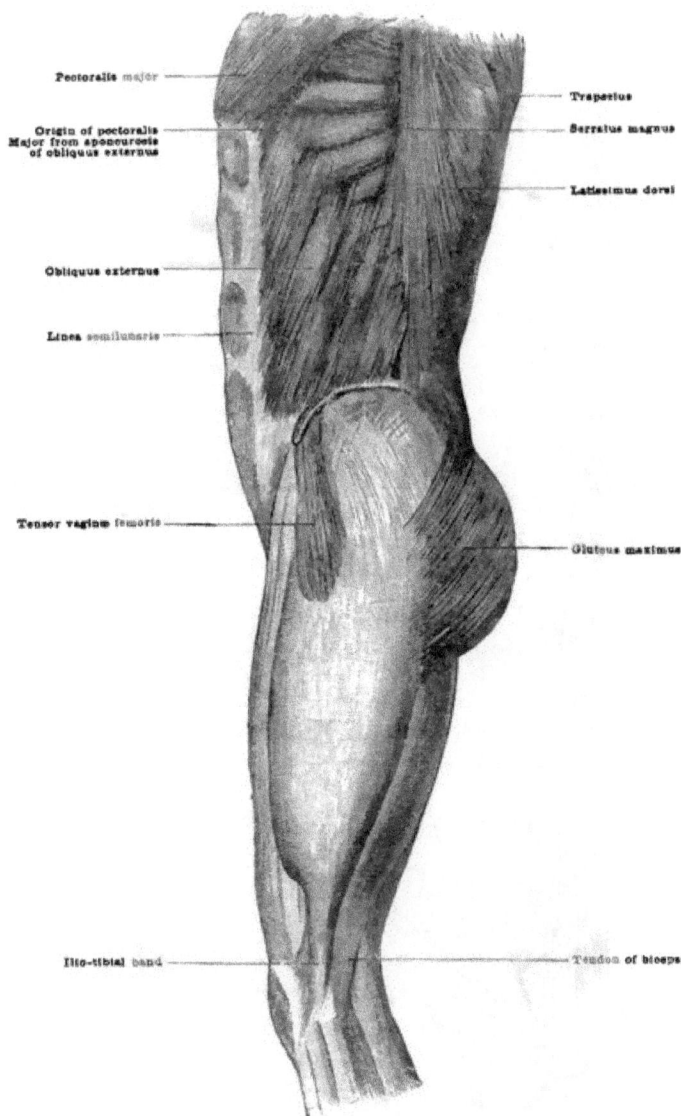

FIG. 159.—EXTERNAL OBLIQUE AND ILIO-TIBIAL BAND.

cavity distended in this way, you can proceed properly with your dissection as follows :

Incisions.—(1) Cut from ensiform to symphysis pubis ; (2) cut from umbilicus horizontally outward, at right angles to the first incision. Remove the skin carefully, beginning at the umbilicus.

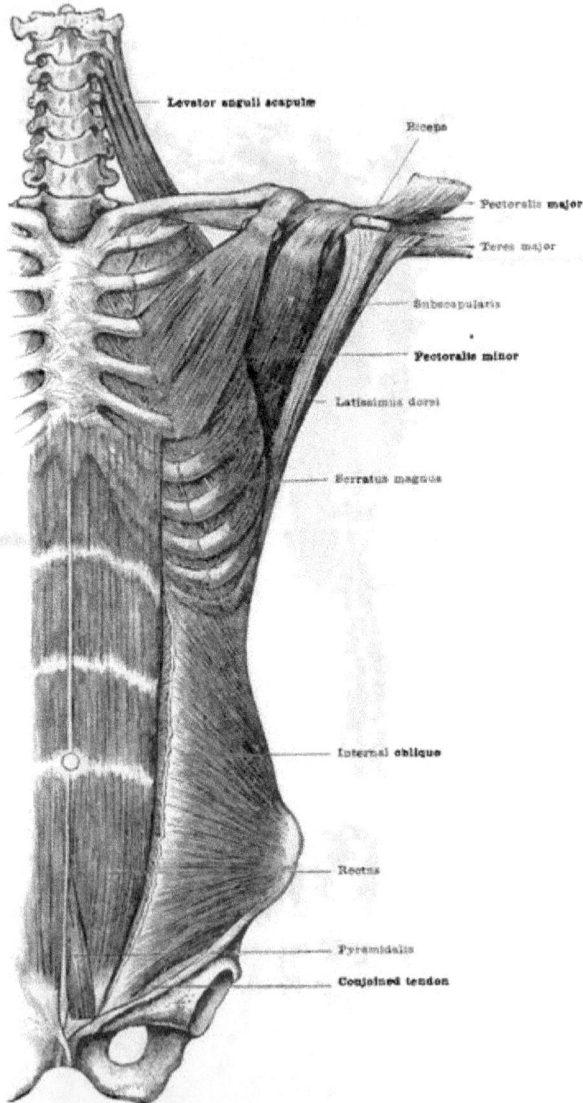

FIG. 160.—THE PECTORALIS MINOR, OBLIQUUS INTERNUS, PYRAMIDALIS, AND RECTUS ABDOMINIS.

In this region you will find: (1) The cutaneous branches of the six lower intercostal nerves; (2) the ilio-hypogastric nerve; (3) the ilio-inguinal nerve;

(4) the spermatic cord with the genital branch of the genito-crural nerve on its posterior surface and the ilio-inguinal nerve in front ; (4) you will see also with the above abdominal intercostal nerves some small arteries. They are the lower intercostals and lumbars.

The nerves you see near the mid-line (Fig. 158) are the terminal branches, while those represented more externally are the lateral cutaneous branches of the intercostals. Search in the fat until you find the point of emergence of the nerve ; then trace out its branches by dividing the connective tissue in the direction of the branches.

External Oblique (Fig. 159).—(1) Dissect the fat and fascia off and expose the muscle as in the figure. At the upper part of the muscle develop the digitations of this muscle with the serratus magnus. Trace each digitation to the bone. In developing these origins use the forceps and scissors. (2) Now trace

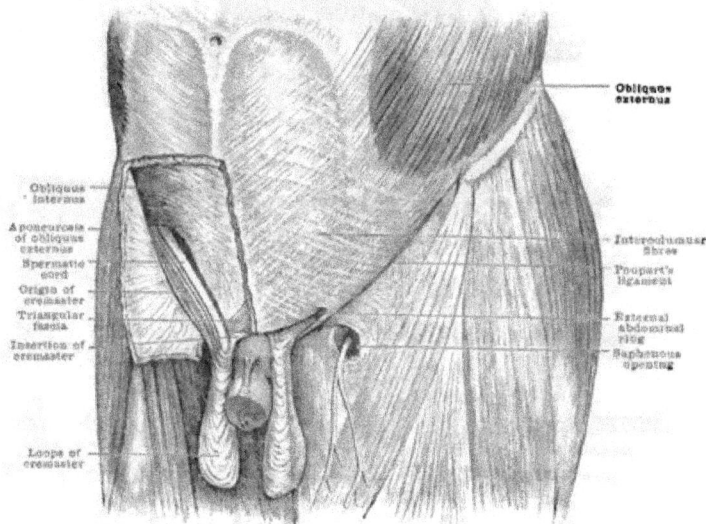

FIG. 161.—OBLIQUUS EXTERNUS AND FASCIA LATA.

the muscle downward and forward in the direction of the fibres to the outer lip of the iliac crest. Notice that the muscular fibres disappear at the anterior superior iliac spine, and from this point onward to the mid-line of the abdomen, and downward to the pubes, the muscle is continued as an aponeurosis—*i. e.*, the muscle minus the lean meat! (3) Remove the fat and fascia, and demonstrate the linea semilunaris, the lineæ transversæ, and the linea alba. The last is in the mid-line. (4) Examine the lower margin of the aponeurosis of this external oblique muscle. It extends in a sagging manner from the anterior superior iliac spine to the pubic spine under the name of Poupart's ligament, or crural arch. This arch is continuous below with the fascia lata of the thigh. (5) Locate the spine of the pubes, and on the under part of the same find the spermatic cord. (Fig. 161.) Trace this cord upward and outward to a point where it enters the external abdominal ring. (Fig. 161.) Above and below the ring, see the pillars or columns of the ring. Notice and develop some transverse

fibres, called the intercolumnar fibres.   (6) Insert the forceps between the pillars
just mentioned, parallel to Poupart's ligament, for about two inches.   They are
now in the inguinal canal. (Fig. 163.)   Below them is the floor of the canal,—the
upper surface of Poupart's ligament,—and on this floor is the spermatic cord ;
external to them is the outer wall of the canal—the aponeurosis of the external
oblique muscle ; above them is the roof of the canal—the arched fibers of the
internal oblique and transversalis muscles : internal to them is the inner wall of
the canal—the conjoined tendon of the internal oblique and transversalis muscles
and the transversalis fascia.   If you gently pull the spermatic cord, you will see
where it comes into view.   This spot where the cord appears is the internal

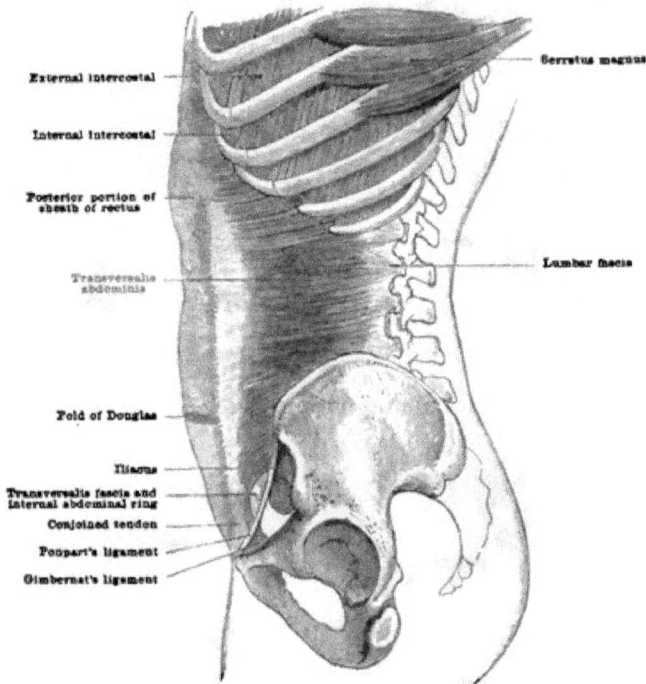

FIG. 162.—TRANSVERSALIS ABDOMINIS AND SHEATH OF RECTUS.

abdominal ring in the transversalis fascia.   Do not remove the forceps until you
have re-read (6),—*i. e.*, the above geometrical description of the inguinal canal,
—and carefully studied and learned the following :

  The inguinal canal has the following points :
  1. An *external ring* in the aponeurosis of the external oblique.   (Fig. 161.)
  2. An *internal ring* in the transversalis fascia.
  3. A *floor*, the upper inner surface of Poupart's ligament.
  4. A *roof*, the arched fibres of the internal oblique and transversalis.
  5. An *outer wall*, the aponeurosis of the external oblique.
  6. An *inner wall*, conjoined tendon, triangular ligament, transversalis fascia.
  7. *Contents :* spermatic cord, male ; round ligament, female.

8. *Extent*, from external to internal abdominal ring.

9. *Location of internal ring*, one-half of an inch above Poupart's ligament and midway between symphysis pubis and anterior superior iliac spine.

10. *Length of canal*, one and one-half inches.

Having learned the above thoroughly, cut through the **external wall and** expose the canal, as in figure 161.

Coverings of the Spermatic Cord and Testicle.—Pass your finger **down** into the scrotum anterior to the cord ; cut through the skin and fascia and expose the testicle and find the following coats :

1. The *skin* and superficial fascia.
2. The *dartos*, consisting of muscular and contractile tissue.
3. The *intercolumnar fascia*, part **of** external oblique.
4. The *cremasteric fascia*, part of the internal oblique.
5. The *infundibulum*, part of transversalis fascia.
6. The *tunica vaginalis*, part of the peritoneum.

You will then look upon these coats as small parts of the individual constituents of the abdominal walls pushed ahead of the testicle in its descent. The tunica vaginalis was pushed ahead, a part of the peritoneum ; the infundibulum, a part of the transversalis fascia ; the cremaster, a part of the internal oblique ; the intercolumnar fascia, a part of the external oblique ; the dartos, a part of the superficial fascia, slightly modified by the presence of elastic fibres and muscle fibres ; the skin to form the scrotum.

The Spermatic Cord.—In front of the cord find the ilio-inguinal **nerve ;** behind the cord the genital branch of the genito-crural nerve. Find the excretory duct of the testicle—the vas deferens ; the spermatic artery, a branch of the aorta ; the spermatic veins ; a little artery—the deferential—in the sheath of the vas deferens. **These** structures are all bound loosely together by connective tissue. You will trace all these structures to the testicle.

Next expose the internal oblique muscle by turning aside the **external ob-lique in this manner :** Divide the digitations at their attachments **into the eight lower ribs. Turn the** whole muscle forward and see the internal **oblique.** Its **fibres run upward** and inward. This is also called ascending oblique.

The Internal Oblique (Fig. 160).—This muscle has attachments · (1) To the outer surface of the four lower ribs ; (2) to the middle lip of the crest of the ilium ; (3) to the outer one-half of Poupart's ligament ; (4) to the pubic crest and ilio-pectineal line ; (5) to the linea alba. Notice, first, the arch formed by this muscle and the transversalis. In figure 160 you see these muscles are attached to Poupart's ligament for about the outer one-half the length of this ligament ; then they leave this ligament, arch over the cord, form the roof of the inguinal canal, and are inserted into the pubic crest by the conjoined tendon. Trace **this** muscle to the middle lip of the iliac crest, anterior two-thirds.

Make an incision, parallel to the iliac crest, through this muscle, and you will come down upon the neuro-vascular area of the abdominal walls. These vessels and nerves are between the internal oblique and transversalis, in a small quantity of fatty connective tissue, and are as follows :

1. Deep circumflex iliac artery, from the external **iliac.**
2. Ilio-hypogastric nerve, from the lumbar plexus.
3. Ilio-inguinal nerve, from the lumbar plexus.
4. Lower intercostal nerves, the twelfth thoracic.

It will require care to separate these muscles. **The** guide is **this :** Keep the nerves in sight and follow them. Now trace the internal oblique muscle to the outer surface of the four lower ribs. Notice the direction taken by the muscular fibres. See **also** that the fibres of this muscle are attached to the outer one-half **of** Poupart's ligament, with the transversalis, and that both are inserted by a conjoined tendon into the pubic crest and ilio-pectineal line for a variable distance.

The transversalis muscle is now fully exposed to view. (Fig. 162.) On its outer surface you see the plexus of nerves and the vessels previously described. On its inner surface you will find the transversalis fascia. The fibres of this muscle extend transversely across the wall. The muscle arises (1) from the inner surfaces of the six lower ribs, interdigitating with the diaphragm ; (2) from the lumbar fascia ; (3) from the inner lip of the iliac crest ; (4) from the outer two-thirds of Poupart's ligament. The muscle is inserted into (1) the linea alba ; (2) the crest of the pubes ; (3) the ilio-pectineal line. This muscle assists the internal oblique in forming the arch and the conjoined tendon.

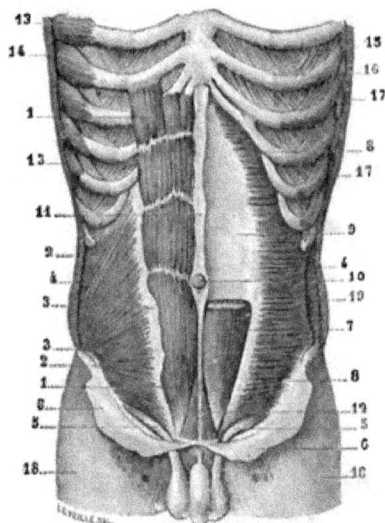

FIG. 163.—INTERNAL OBLIQUE AND TRANSVERSALIS ABDOMINIS MUSCLES.
1, 1. Rectus abdominis. 2, 2. Internal oblique. 3, 3. Anterior leaflet of aponeurosis of internal oblique 4, 4. Divided external oblique. 5, 5. Spermatic cords. 6, 6. Inferior portion of aponeurosis of external oblique. 7. Lower portion of left rectus abdominis ; upper portion removed. 8, 8. Muscular portion of transversalis abdominis. 9. Aponeurotic portion. 10. Umbilicus. 11. Supra-umbilical portion of linea alba. 12. Infra-umbilical portion. 13. Serratus magnus. 14. Divided right latissimus dorsi. 15. Divided left latissimus dorsi. 16. Divided serratus magnus. 17, 17. External intercostals. 18, 18. Femoral aponeurosis. 19. Divided internal oblique.

The Rectus and Pyramidalis.—Make an incision through the sheath of the rectus, extending from the pubic spine to the fifth rib, parallel with the linea alba. Carefully dissect the sheath both ways—i. e., turn the sheath to the right and left—until you have fully exposed the muscular contents. (Fig. 160.) The white lines crossing the rectus are the lineæ transversæ. You will find the pyramidalis below, arising from the pubic crest in front of the rectus. Trace it to its insertion into the linea alba. (Fig. 160.) Now you can easily lift the rectus from its bed, and look down on the posterior part of its sheath. (Fig. 163.) A description of the formation of the sheath is given in the beginning of this section. Cut through the rectus one inch below the umbilicus and see : (1) The *deep epigastric artery* and its veins ; (2) the *semilunar fold of Douglas*. Turn the divided ends of the rectus aside and trace out the branches of the deep epigastric artery.

Hesselbach's triangle is of surgical importance because direct inguinal hernia passes through it. (Fig. 165.) It is bounded internally by the rectus, externally by the deep epigastric artery, below by Poupart's ligament. See, now, that the conjoined tendon of the internal oblique and transversalis stretches across the inner two-thirds of this triangle. Hence this tendon may become one of the coverings of a direct inguinal hernia.

1. *Points of surgical interest.*

The inguinal canal, since, physiologically, it transmits the male spermatic cord and its homologue, the round ligament of the uterus; since, pathologically, inguinal hernia and diseases of the cord are interrogated here.

2. *Importance of the umbilicus.*

This is a physiological cicatrix or scar, marking the aperture through which passed the vessels that made up the umbilical cord, or funis, in the child before birth. These vessels were the right and left hypogastric arteries (branches of the internal iliacs) and the umbilical vein. The arteries took blood to the placenta for aeration; the vein returned this blood aerated. You will find internally in adult dissections remains of these three vessels centering at the umbilicus.

3. *Further importance of umbilicus.*

In **both adult** and child it may be the location of umbilical hernia.

4. *Give nerve-supply of the abdominal muscles.*

The six lower thoracic nerves, through **their anterior** primary divisions, assisted by the ilio-hypogastric and ilio-inguinal **nerves from** the lumbar plexus. They also supply the skin covering these muscles.

5. *Give the blood-supply of the abdominal walls.*

This is both abundant and important. (1) The deep epigastric; (2) the internal mammary; (3) the lumbar arteries; (4) the intercostals; (5) the deep circumflex iliac; (6) the superficial circumflex iliac; (7) the superficial epigastric. The deep epigastric, a branch of the external iliac, anastomoses in the substance of the rectus muscle with the superior epigastric branch of the internal mammary artery, a branch of the subclavian artery.

6. *Function of abdominal muscles.*

(1) They **protect from violence** and temperature changes the organs in **the abdominal cavity; (2) they** assist internal organs to discharge their contents, in **that they excite peristalsis—a** physiological Credé; (3) they, by their various **contractions, alter the relations** between the thorax and abdomen; (4) they are strongly analogous to structures above the diaphragm. The external oblique is analogous to the external intercostals; the internal oblique to the internal intercostals; the transversalis to the triangularis sterni; the rectus to the sternum; the lineæ semilunares to the chondra, vertically; the lineæ transversæ to the costal cartilages; their nerve- and blood-supply **are** strongly analogous, while functions conform to local demands and structure is modified accordingly.

Important Attachments and Relations.—Notice carefully the following, and demonstrate on your dissection the following points:

1. Each of the three broad muscles has (1) an outer muscular part, and (2) an inner aponeurotic part; the latter are inseparably united, the former may be separated from each other.

2. The conjoined aponeuroses of all three planiform muscles form the vertical mid-line of the abdominal walls, known as linea alba abdominis.

3. The two inner muscles are attached to the upper surface of Poupart's ligament for about one-half the length of this (the outer one-half), and then leaving the same arch over the spermatic cord, forming the roof of the inguinal canal, and are inserted into the pubic crest and inner part of the ilio-pectineal line.

4. Between the muscular parts of the internal oblique and transversalis muscles is located a meagre plexus of nerves, from which the abdominal muscles

16

are supplied. The deep internal circumflex artery and its veins are also found here.

5. *Give the origin and insertion of the external oblique.*

Origin.—The outer surface of the eight lower ribs about their middle by a series of nearly horizontal lines which, after crossing each rib obliquely downward and backward, extend for a short distance along its lower border.

Insertion.—(1) By a strong aponeurosis along the whole of the linea alba ; (2) the front of the os pubis close to the symphysis ; (3) the spine of the pubes and the adjacent part of the ilio-pectineal line ; (4) the deep fascia of the thigh in a thickened band which stretches from the spine of the pubes to the anterior superior spine of the ilium ; (5) the anterior half of the outer lip of the crest of the ilium ; (6) at the lower part of the linea alba some of the fibres (the *triangular fascia*) stretch across the middle line, and are inserted into the front of the crest of the pubes and the inner part of the ilio-pectineal line of the other side.

6. *Give origin and insertion of internal oblique.*

Origin.—(1) The outer half of Poupart's ligament ; (2) the anterior two-thirds of the space intervening between the inner and outer lips of the crest of the ilium ; (3) the outer and posterior aspect of the aponeurosis of the transversalis abdominis (which aponeurosis is also called the lumbar fascia).

Insertion.—(1) For about one inch into the inner extremity of the ilio-pectineal line ; (2) the anterior border of the crest of the pubes ; (3) the whole length of the linea alba ; (4) the lower borders of the cartilages of the last three ribs.

7. *Give origin and insertion of transversalis.*

Origin.—(1) The inner surface of the cartilages of the last six ribs, close to their junction with the ribs, by processes which interdigitate with the attachments of the diaphragm ; (2) the strong aponeurosis called the lumbar fascia, which arises (*a*) by its anterior layer from the front of the transverse processes of the five lumbar vertebræ, (*b*) by its middle layer from the tips of the transverse processes of the five lumbar vertebræ, (*c*) by its posterior layer from the general vertebral aponeurosis which is attached to the spines of the thoracic, lumbar, and sacral vertebræ ; (3) the anterior two-thirds of the inner lip of the crest of the ilium ; (4) the outer third of Poupart's ligament.

Insertion.—(1) The whole length of the linea alba ; (2) the anterior border of the crest of the pubes ; (3) the inner end of the ilio-pectineal line for about one inch and a half.

8. *Give origin and insertion of the rectus.*

Origin.—By two tendons : (1) The larger or outer head arises from the whole of the crest of the pubes ; (2) the inner head crosses the middle line of the body, and arises from the fibrous structures lying in front of the symphysis pubis.

Insertion.—(1) The anterior surface of the tip of the fifth rib ; (2) the front of the costal cartilages of the fifth, sixth, and seventh ribs ; sometimes also (3) the deep posterior surface of the ensiform cartilage near its outer border.

Anatomy of Inguinal Hernia.—There are two varieties of inguinal hernia, founded on the idea of internal to or external to the deep epigastric artery, a structure already seen in your dissection. The name direct inguinal is given when a hernia comes through Hesselbach's triangle. In this case the coverings are : **(1)** Peritoneum ; (2) transversalis fascia ; (3) conjoined tendon ; (4) intercolumnar fascia ; (5) superficial fascia ; (6) skin. The name oblique is given when the hernia enters the canal by the internal abdominal ring, in the transversalis fascia, external to the deep epigastric artery. It has these coats : (1) Peritoneum ; (2) subserous fatty connective tissue ; (3) infundibulum ; (4) cremasteric muscle ; (5) intercolumnar fascia ; (6) superficial fascia ; (7) skin. This variety of hernia follows the course taken by the testicle in its descent, and receives the same kind of investments. As before stated, these coats are simply certain parts of the walls which have receded before the hernia in its descent.

Describe the ilio-hypogastric nerve.

It is a branch of the lumbar plexus. It is known also as the superior musculo-cutaneous nerve. It crosses the quadratus lumborum muscle, gains the space between the transversalis and internal oblique muscles, and divides into (1) an iliac branch, which supplies the skin over the gluteal region ; (2) a terminal hypogastric branch, that passes between these two muscles—*i. e.*, the internal oblique and transversalis—to near the mid-line, where it comes through the abdominal walls an inch above the external abdominal ring.

Describe the ilio-inguinal nerve.

You find this nerve anterior to the spermatic cord. It is distributed to the

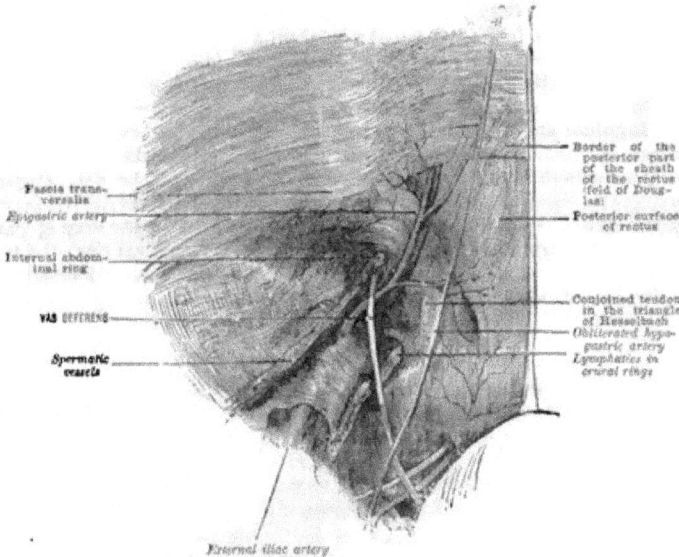

Fasoia transversalis
Epigastric artery
Internal abdominal ring
VAS DEFERENS
Spermatic vessels

Border of the posterior part of the sheath of the rectus (fold of Douglas)
Posterior surface of rectus

Conjoined tendon in the triangle of Hesselbach
Obliterated hypogastric artery
Lymphatics in crural rings

External iliac artery

FIG. 164.—DISSECTION OF THE LOWER PART OF THE ABDOMINAL WALL FROM WITHIN, THE PERITONEUM HAVING BEEN REMOVED. (Wood.)

scrotum and labia majora, and to the inner and upper part of the thigh. It is a branch of the lumbar plexus. It runs near to and plexifies with the preceding nerve, between the internal oblique and transversalis.

Describe the genito-crural nerve.

This is seen behind the spermatic cord, in the canal ; after the cord emerges from the external abdominal ring, the genital branch of the genito-crural nerve is seen behind the cord. This supplies the cremaster muscle with motion. The crural branch of the genito-crural nerve goes to the skin over the mid-front of the thigh half way to the knee.

INTERIOR VIEW OF ABDOMINAL WALLS.

Make an incision on each side, from the lower part of the umbilicus to the anterior superior spine of the ilium, through the entire remaining abdominal wall. Turn the V-shaped flap thus formed forward, and at the same time lift it upward and put it on the stretch as much as possible. You are to see and **study the** following structures through the peritoneum (do not remove this yet):

1. **The plica urachi**—a peritoneal fold covering the remains of the **urachus.** This is in the mid-line from the summit of the bladder to the umbilicus.

2. **The plica hypogastrica**, covering the remains of the hypogastric arteries.

3. **The deep epigastric arteries and veins**, passing upward and inward to enter the sheath of the rectus muscle by passing under the fold of Douglas.

4. **Poupart's ligament** and the external iliac vessels leaving the pelvis, to be called femoral in the thigh. Here, too, see the external iliac artery giving off its two branches, the deep epigastric and the internal circumflex iliac branch.

5. **The internal abdominal ring, to** which you may see plainly the constituents of the spermatic cord coming—viz., the vas deferens, the spermatic vessels. Locate this ring just **to the outer side of the** deep epigastric artery.

6. **The femoral canal, called** also *crural canal*. Find this to the inner side **of the** femoral vein, between Poupart's ligament above and **the** bone below. To **its inner** side you will feel the sharp falciform margin of Gimbernat's ligament. **This canal** is in the femoral sheath, and occupied by fat and connective tissue **called** the septum crurale or septum femorale.

7. **Inguinal and Femoral Fossæ.**—There are three of these. The *internal* is between the remains of the urachus **and** hypogastric artery ; the *middle* inguinal fossa is between the remains of the hypogastric artery and the deep epigastric artery. The importance of these two fossæ is they permit *direct inguinal hernia* to form. The *external* inguinal fossa is external to the deep epigastric artery and corresponds to the situation of the internal abdominal ring, and is the location of oblique inguinal hernia. In this connection notice the femoral fossa. This corresponds to the femoral ring, and marks the location of femoral hernia.

DISSECTION OF FEMORAL HERNIA.

Find the femoral fossa, covered by peritoneum, just below Poupart's ligament, and internal to the femoral vein. Now remove the peritoneum, by pulling the same gently backward and downward. Take **the** forceps and break up the connective **tissue in the** femoral canal, the depression corresponding to the femoral fossa **just mentioned** and internal **to** the femoral vein. We will consider femoral hernia **with reference to** (1) the **femoral** sheath ; (2) the femoral canal ; (3) the coverings of **femoral hernia ;** (4) contents of the canal ; (5) the relations to other structures ; (6) anatomical factors concerned in reduction ; (7) the deep crural arch.

The femoral sheath is formed under Poupart's ligament by the meeting and union of the fascia transversalis in front of the femoral vessels and of the iliac fascia behind the femoral vessels. As you will demonstrate on your cadaver, these fasciæ unite on the outer side of the femoral artery very close to the vessel ; on the inner side of the vein, however, they leave an interval, one-half of an inch in width, between the femoral vein and Gimbernat's ligament. (Fig. 165.) This interval is the femoral canal.

The deep crural arch extends in an archiform manner across the femoral sheath, strengthening thereby the transversalis fascia. It is inserted into the spine of the pubes, and may be looked upon as a slip of contribution from the transversalis fascia.

The femoral sheath contains the following structures : (1) The femoral artery on the outer side ; (2) the femoral vein in the middle ; (3) the femoral canal internal to the vein. The beginning of the femoral canal—not the beginning of the sheath—is the femoral ring. The femoral canal contains some connective tissue that you just broke up with your forceps. The special name for this is the septum femorale or septum crurale. Make note of this, for it forms one of the coverings of femoral hernia. I wish to make emphatic the difference now between the contents of the femoral sheath and the contents of the femoral canal. The canal itself is in the sheath.

The femoral ring and its relations are the beginning of the femoral canal, just as the brim is the beginning of a cup. The ring has certain definite boundaries and relations, which you must both demonstrate on your work and commit thoroughly to memory. In figure 165 you can study these. Internal to the ring is the sharp falciform edge of Gimbernat's ligament. This, as you can easily demonstrate on your work, is a reflection of Poupart's ligament on to the ilio-pectineal line. To the outer side of the ring is the femoral vein. Above the

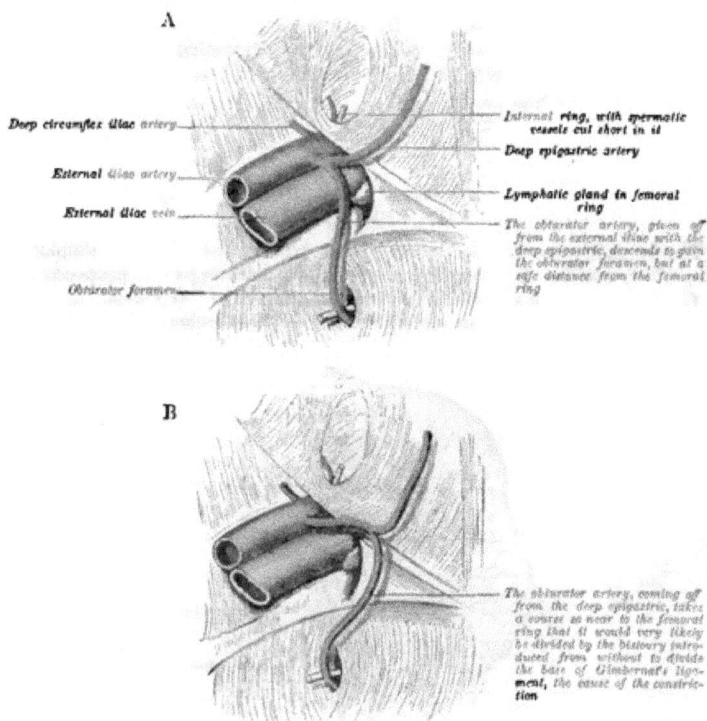

Deep circumflex iliac artery

Internal ring, with spermatic vessels cut short in it

Deep epigastric artery

External iliac artery

External iliac vein

Lymphatic gland in femoral ring

The obturator artery, given off from the external iliac with the deep epigastric, descends to gain the obturator foramen, but at a safe distance from the femoral ring

Obturator foramen

The obturator artery, coming off from the deep epigastric, takes a course so near to the femoral ring that it would very likely be divided by the bistoury introduced from without to divide the base of Gimbernat's ligament, the cause of the constriction

FIG. 165.—IRREGULARITIES OF THE OBTURATOR ARTERY. (After Gray.)

ring is Poupart's ligament. Below the ring is the horizontal part of the os pubis, covered by the pectineus muscle and its aponeurosis. You will also see the deep epigastric artery arising from the femoral artery and passing upward and inward across the upper part of the femoral ring. In a certain number of cases the deep epigastric artery gives off the obturator artery, which passes either to the outer side of the ring, as in A, figure 165, or to the inner side, as in B. In either case it would complicate an operation for reduction of femoral hernia. You will note on the cadaver that the obturator artery and nerve lie on the

outer wall of the pelvis.   The artery is, as a rule, a branch of the internal iliac.
It forms an exception to the rule in the figures given here.

The coverings of femoral hernia from within outward are: (1) The peritoneum; (2) the subperitoneal connective tissue; (3) the septum femorale; (4) the
femoral sheath; (5) the cribriform fascia—the deep layer of superficial fascia that
covers the saphenous opening in the fascia lata; (6) the superficial fascia; (7) the
skin.

Factors concerned in reduction of hernia are those that tend to tighten the
canal.   These are : (1) The saphenous opening; (2) the iliac and pubic portions
of the fascia lata; (3) the external oblique muscle and its lower portion—Poupart's ligament.   These structures that tend to strengthen a succession of anatomical weak points, determining the course of a femoral hernia, may all be dis-
empowered by flexion of the thigh on abdomen; flexion of leg on thigh; adduction of thigh.   Demonstrate this on your work by placing your finger in the
femoral canal, and let an assistant produce first abduction and extension, and
adduction and flexion, when the scissor-like action of Poupart's and Gimbernat's
will be appreciated.

·THE PERITONEUM.

The inquisitive student instinctively asks himself the following simple questions, which I will here answer fully enough for dissecting-room purposes :

1. *What kind of a structure is peritoneum and where is it found?*

Peritoneum is a serous membrane found in the abdominal cavity only.

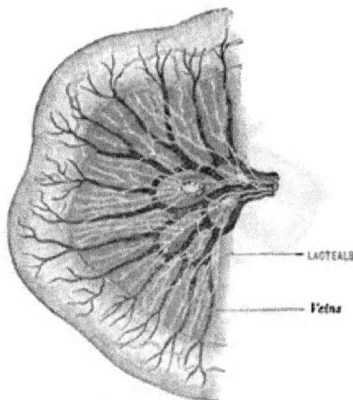

LACTEALS

Veins

FIG. 166.—VESSELS OF THE SMALL INTESTINE.

2. *What is the appearance of peritoneum and how may I recognize it?*

Peritoneum is smooth, moist, glistening.   You can recognize it by its location and its appearance.   Its most distinctive feature is its location in the abdominal cavity; removed from this cavity you could not distinguish it from other
erous membranes—pleura and pericardium.

3. *Where will I see peritoneum on opening the abdominal cavity?*

You will see it forming the innermost layer of the wall you cut and turned back—smooth, moist, glistening, and attached; you will see it covering every organ in the abdominal cavity partially or completely. If by force of circumstance organs lose all or some of their peritoneal covering, this is no fault of the peritoneum. Remember, the primitive relation of every organ to peritoneum is behind—partially or completely covered thereby. (Figs. 167 and 168.)

Figure 167 shows aorta giving three branches to three organs. The organ in the centre is for practical purposes completely invested by peritoneum. Between the two parts of the peritoneal fold is—(1) The organ; (2) the vessels that take blood to and from the organ; (3) the nerves that innervate the organ; (4) the

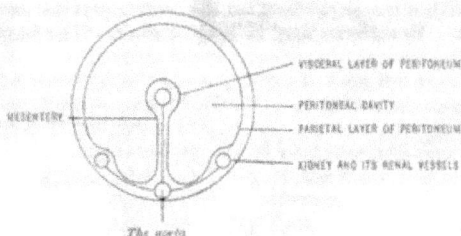

FIG. 167.—SHOWING PARTIAL AND COMPLETE INVESTMENT OF ORGANS BY PERITONEUM.

lymphatics that scavenger the organ. The organ has only one place where it is accessible—only one communication with the wall.

Examine the cadaver, take up a portion of the small intestine, find the superior mesenteric artery, and trace it down between its two parts of the peritoneal folds. Now, with your forceps, make a rent in one side of the mesentery, examine, and find an artery, a vein, nerves, glands, and fat, as represented in figure 166.

4. *Explain implantation of meso-structures.*

That part of the alimentary canal which remained in its original mid-line of the body—as the greater part of the small intestine and rectum—may be said to be an example of primitive implantation; those parts which contracted adhesions

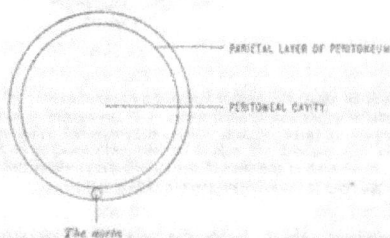

FIG. 168.—SHOWING RETRO-PERITONEAL LOCALITY.

to localities other than the mid-line, may be said to be examples of acquired implantation. The colon and duodenum are examples of this latter variety.

5. *From what source does the peritoneum derive its blood-supply?*

It will be presently seen that the abdominal aorta has parietal branches and visceral branches. The parietal branches, as the lumbar and phrenics, supply the peritoneum covering these walls; the visceral branches supply the peritoneum covering the organs. Peritoneal arteries are, then, of two classes—visceral and parietal.

6. *Explain peritoneal investment of organs.*

This is as follows : (1) Organs may simply be *behind* the peritoneum, as arteries and veins, in which case we say they are covered by this membrane. (Fig. 168.) They have not grown sufficiently in thickness to produce any appreciable displacement of the peritoneum. Figure 168 shows the aorta behind the peritoneum, but producing no bulging forward of the peritoneum. (2) Organs may produce appreciable bulging forward, as in figure 167, when we say they are partially invested. Now examine the kidney and you will find its relation to the peritoneum represented in figure 167 in the two organs on the two sides.

7. *What determines the degree of peritoneal investment ?*

The height to which any organ grows determines this. The peritoneum pushed ahead of the organ is called the *mesentery* of the organ and the peritoneal investment. Mesenteries may be long or short. The height to which an organ

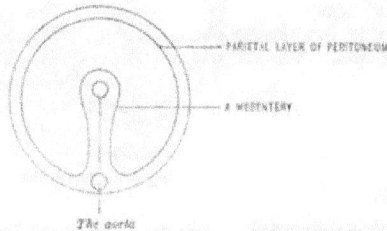

The aorta
FIG. 169.

In this figure an organ has grown to a considerable height and a long mesentery is the result.

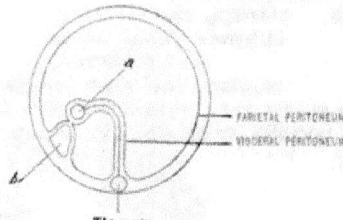

The aorta
FIG. 170.

In this figure, *a* is an organ completely invested by peritoneum. Its mesentery grows high and falls on *b*, an organ partially invested by peritoneum. At the contact point of *a* and *b* loss of epithelium occurs. Parts of organs, or entire organs, which once possessed peritoneal investment, but subsequently lost the same in this manner, are said to be retro-peritoneal, or to have no peritoneal investment. An example is to be seen in the case of the right kidney, where the duodenum and colon bear the above relation (Fig. 170) to the anterior surface of this organ.

grows determines, then, both the *degree of investment* and the *length of the mesentery.*

8. *How are the abdominal muscles classified ?*

As vertical and horizontal. The pyramidalis and rectus abdominis are vertical ; the external and internal oblique and the transversalis are horizontal.

9. *Name the abdominal weak points in the walls of the abdominal cavity and give their practical importance in surgery.*

(1) The inguinal canal, limited internally by the internal abdominal ring, and externally by the external abdominal ring. This canal lodges the spermatic cord of the male and the round uterine ligament of the female. It is the succession of weak points that formed the line of least resistance to the descending testicle.

This canal acquires surgical importance in inguinal hernia, of which there are two arbitrary divisions—direct and oblique : the former being internal to, the latter being external to, the deep epigastric artery.

(2) The femoral canal, being that part of the initial three-quarters of an inch of the femoral sheath unoccupied by the femoral vessels. (Fig. 165.) Normally, the canal is occupied by a fatty connective tissue, with or without a lymphatic gland. The entire amount of this connective tissue—called technically the septum crurale, or septum femorale—weighs three grains. The surgical importance of this septum is due to the fact that it is inferred to form one of the coverings of a femoral hernia. Just why rational medical literature should continue to dignify this frail structure, and place it in the same category with such structures as Gimbernat's and Poupart's ligaments, the femoral sheath, and the skin itself, is difficult to determine.

(3) The umbilicus is the point where the two hypogastric branches of the internal iliac arteries and the umbilical vein came together to form the umbilical cord, a fœtal arrangement that took impure blood to the placenta and returned in its stead the same blood plus O and minus CO_2.

10. *Where is the spermatic cord made up and of what structures does it consist?*

It is made up at the internal abdominal ring, to the outer side of the deep epigastric artery, by the coming together of the spermatic artery, the spermatic veins, the deferential artery, and the excretory duct of the testicle—the vas

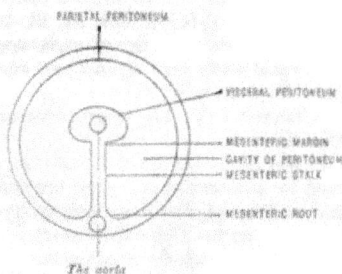

FIG. 171.—SHOWING SIMPLE VISCERAL AND PARIETAL LAYERS OF PERITONEUM WITH CAVITY BETWEEN THEM.

deferens. These structures are loosely bound together by connective tissue. In practical dissection the cord is found below and external to the pubic spine, not as a cord, but as a flattened band. The cord lies on the shelved inner and upper part of Poupart's ligament, in the inguinal canal. The function of the spermatic artery is to supply the testicle with blood, from which the testicle secretes semen. The spermatic veins return the blood to the general circulation, the right being tributary to the ascending vena cava, the left to the left renal vein. The vas is the conduit that leads the semen to the vesiculæ seminales at the base of the bladder. The artery of the vas nourishes this conduit and inosculates with the spermatic artery.

11. *Tell all you can about the femoral sheath.*

It is composed of two fasciæ that come down out of the abdominal cavity, one in front of and the other behind the femoral vessels. These fasciæ surround the vessels and take the name "femoral sheath."

12. *What are the fasciæ that form the femoral sheath and why are they so called?*

The one in front of the femoral vessels is the transversalis ; the one behind the vessels, the iliac fascia. They are named in accordance with the rule governing the naming of deep fasciæ—*i. e.*, according, in this case, to the name of the muscles.

13. *Do the lumbar arteries assist in furnishing the peritoneum with blood?*

Yes; the same arteries that supply the walls, supply the peritoneum covering the walls.

14. *What can you say of the lumbar arteries?*

They are four in number on each side. They are analogous to the intercostal arteries. They arise from the abdominal aorta. The right are longer than the left lumbar arteries. These arteries are small when compared to their homologues, the intercostals. The lumbar arteries lie behind the abdominal sympathetic chain. According to their distribution, the lumbar arteries are classified as dorsal branches and abdominal branches. Each has specific relations and course. The dorsal branches pass out between the transverse processes of the lumbar vertebræ, with the posterior primary divisions of the lumbar nerves, and attended by their corresponding lumbar veins, to be distributed to the spinal cord, passing through the intervertebral foramina; to the muscles and skin of the back; and to the lumbar vertebræ in this region. The abdominal branches of the lumbar arteries pass behind the sympathetic chain and behind the quadratus lumborum muscle. They are distributed to the quadratus lumborum and psoas magnus muscles and to the abdominal walls.

15. *What can you say of the anastomosis of the lumbar arteries?*

The dorsal branches anastomose with each other and with the intercostals. The abdominal branches of the lumbar arteries anastomose: (1) Above with the lower intercostal arteries; (2) below with the ilio-lumbar and circumflex iliac; (3) in front with the epigastric and internal mammary.

16. *If the abdominal aorta were ligated just above* the bifurcation, *how could the blood reach the lower extremities?*

Through two channels (1) The epigastro-mammary arch; (2) the lumbo-circumflex anastomotic arch.

17. *Describe the epigastro-mammary anastomotic arch.*

An arch formed by anastomosis between the deep epigastric artery, a branch of the external iliac, and the internal mammary artery—a branch of the subclavian. This union occurs in the rectus abdominis muscle.

18. *Describe the lumbo-circumflex anastomotic arch.*

It is an arch formed by anastomosis between the lumbar arteries, branches of the abdominal aorta, and the deep internal circumflex iliac artery—a branch of the external iliac.

The arteria sacra media is the sacral representative of the lumbar arteries. It is to the aorta what the coccyx is to the vertebral column, what the ensiform cartilage is to the sternum—the exhaustion. It is the representative of the caudal artery of some animals. Its small branches are distributed in the same manner as those of the lumbar region.

19. *What can you say of the lumbar veins?*

They accompany the lumbar arteries and are tributary to the ascending vena cava.

20. *How would you explain the nerve-supply of the peritoneum?*

The same nerves that supply muscles that enclose a serous cavity supply the serous membrane enclosed. This principle was pointed out by Hilton and Van der Kolk. The full law, as given previously, applies not only to serous but to mucous membranes as well.

21. *What are the functions of peritoneum?*

It secretes serum for lubrication of opposed surfaces. It is a highly absorptive structure. It forms ligaments for holding organs in place. It protects organs against violence. It gives strength to organs which it invests. It conserves heat and prevents sudden changes of temperature.

22. *Is serum ever secreted in abnormally large quantities?*

Yes, and the terms by which such conditions are designated are hydro-pericardium, hydro-thorax, hydro-cephalus, and hydro-peritoneum.

23. *What is a peritoneal cavity?*

The space between the visceral and parietal layers of the peritoneum. The same is true of any serous cavity—viz., it is the space between a visceral and a parietal layer of serous membrane.

24. *What do we understand by mesenteric contents?*

All the structures between the two surfaces or folds of the mesentery—i. e., in the mesenteric space. The intestine is, of course, the largest structure. (Fig. 166.) The vessels, nerves, and glands to and from the intestine are all called mesenteric. They are all embedded in a variable amount of mesenteric fat. Sometimes this fat is so abundant as to completely hide the vessels from view; at other times the vessels may be seen.

25. *Explain the primitive alimentary canal.*

In figure 172 you have a view of the primitive canal and its mesentery in longitudinal section. Will you remember that the mesentery has *two layers*

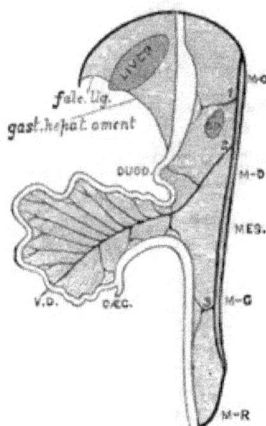

Fig. 172.—Diagram of the Primitive Alimentary Canal.

with the arteries, that make the canal grow, between them? You see in the figure only *one surface* of the mesentery. At an early period there is only one gut and one mesentery. This gut is straight, and located in the mid-line and on the posterior abdominal wall. The gut grows in proportion to the blood-supply it receives. In figure 172 this primitive mesentery had three important organs between its two layers: (1) The stomach, in the middle; (2) the liver, between the two layers, in front of the stomach; (3) the spleen, between the two layers, behind the stomach.

We see parts of this primitive partition unoccupied (1) in front of the liver; (2) between the liver and stomach; (3) between the stomach and spleen; (4) between the spleen and posterior wall.

26. *Can we find evidence in our dissection of the unoccupied parts just referred to?*

Yes; the one in front of the liver is the *broad* or *falciform ligament* of the liver. The one between the liver and stomach is the *gastro-hepatic* or *lesser omentum*. The one between the stomach and spleen is the *gastro-splenic omentum*.

27. *Give the difference between omentum and mesentery.*

In reality there is no difference except an arbitrary one. Each is of peritoneal origin ; each has two layers ; each escorts vessels and nerves to organs. Omenta connect organ to organ ; mesenteries connect organ to wall.

28. *Usage arbitrary.*

The broad ligament holds the liver to the diaphragm ; the broad ligament of the uterus holds the uterus to the lateral pelvic walls and floor. They are the *mesenteries* of these viscera. Usage, however, restricts the use of the word *mesentery* exclusively to the intestines. As you will see, the ascending colon lost its mesentery. The descending colon lost its mesentery, except the rectum, in part, and the sigmoid. The transverse colon retained its mesentery.

29. *Define mesenteric attachment—primitive and acquired.*

By primitive attachment of the mesentery we mean attachment in the midline, or nearly in the original line of the straight gut. The mesentery of the small intestine is, for practical purposes, an example of primitive attachment ; the rectum is primitive in attachment. The duodenum, stomach, and colon have acquired attachments.

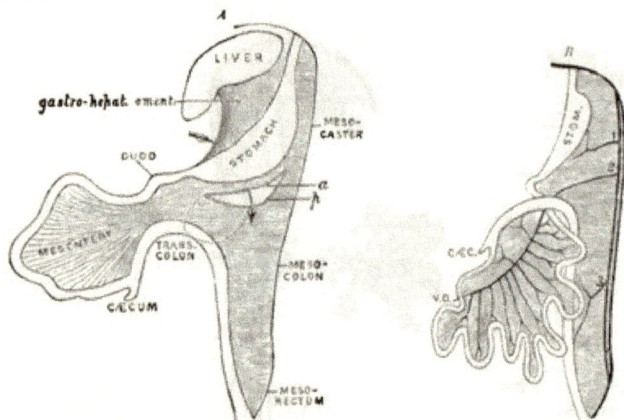

FIG. 173.—DIAGRAMS SHOWING (*A*) THE FORMATION OF THE GREAT OMENTUM, AND (*B*) THE ROTATION OF THE INTESTINAL CANAL.

30. *What is differentiation of the alimentary canal and its physiological significance ?*

In the evolution of the alimentary canal, differentiation is the process whereby the primitive straight gut undergoes a change which determines large and small intestines. The ileo-cæcal junction is the place where this change took place. (Fig. 172.) The physiological significance seems to be simply a physiological division of labor, whereby the small intestine is specialized for digestive purposes, the large intestine for disinfecting, deodorizing, storage, and extrusive purposes.

31. *What is rotation ?*

It is the process by which the upper border of the small intestine (Fig. 173) is turned from you, from left to right, and brought out toward you under the arch of the transverse colon (Fig. 173), so as to be, when completely turned, as figure 173. In figure 172 both large and small intestines have a common continuous mesentery, as you see. In figure 173 they *still* have the same mesentery, with this difference : that for the small intestine is twisted on itself one-half around. The pivot about which this rotation took place is seen in figure 172— the superior mesenteric artery. Examine figure 172 and see the jejunum. Trace this down to the cæcum.

This is an explanation of the fact you can demonstrate on the cadaver : that the small intestine is almost completely surrounded by the ascending transverse and descending parts of the colon. Show this on your work.

32. Formation of greater and lesser peritoneal cavities.

In figure 172 you have the primitive peritoneal partition dividing the abdominal cavity into a right and a left half. In front of the stomach you see the gastro-hepatic omentum ; behind the stomach the mesentery, which now takes the specific name meso-gaster. You see the initial end of the small intestine (duodenum), and below this the remainder of the small intestine and all the large intestine, attached by a common mesentery ; that peritonium, however, which corresponds to the colon is called meso-colon, and that which corresponds to the small intestine is called mesentery. Can you now distinguish the three meso-structures ? Can you remember that each has two layers, between which are located the mesenteric arteries and other contents ? Can you remember that the attached margin of this meso-gut now extends from the floor of the pelvis to the diaphragm, in the mid-line ? Can you remember that the intestinal margin of the meso-gut must become longer, in proportion to the growth of the intestine, large and small ? Now you are on the left of this partition. Thus far the space on the right is as great as that on the left. The arrow in figure 173 is represented as simply coming through an artificial opening, to aid you in remembering the space on the other side of the partition. Now imagine you took hold of the partition, the meso-gaster, and pulled toward you in the direction of the dotted line. Would you not, by continued pulling, produce a bagging-out on your side of the partition, and a consequent depression or bagging-in on the other side? Can you imagine, now, if you pulled long enough, the beginning of the depression or sac on the other side must become smaller and smaller? The hole on the other side would admit you to this little sac thus formed by your pulling. In anatomy the name of this hole is the foramen of Winslow. It leads to the sac you have just formed. This sac is the lesser cavity of the peritoneum ; all the space outside of this is the greater cavity of the peritoneum.

33. What parts of the abdominal viscera, in the adult, have lost their peritoneal covering by pressure atrophy, as referred to in a preceding paragraph?

No two cases are exactly alike in this regard. Still, the following represents a general average, as you will meet these cases in the dissecting-room : (1) The posterior part of the descending colon rests on the posterior wall of the abdomen, consequently there is adhesion, with loss of the specific epithelial element. (2) The ascending colon, the second and third parts of the duodenum, the colic and duodenal elbows to be presently explained, rest on the peritoneum, partially investing the right kidney, and here is peritoneal loss. The second part of the rectum is adherent to the anterior surface of the sacrum, and here is peritoneal loss. Each case is a law unto itself, in a measure ; but the student with an inkling of the developmental principles, as given in the foregoing, and as elaborated in Morris, can soon acquire a facility in locating on his dissections, those peritoneal ruins, whose presence forms one of the mainstays for the receptacular part of the alimentary canal, whose absence belongs to the rarest of rare developmental freaks.

34. Does rotation occur in every case?

Faulty rotation, and even failure to rotate, may occur. One interesting case, found recently by the author, showed the cæcum and the entire colon on the left side, the small intestine in or near the right iliac fossa.

35. Are any special organs developed from the alimentary canal?

The liver and pancreas are developed from the duodenum ; the former grows into the mesentery in front of the stomach, the latter into the posterior meso-gaster. You will see their ducts open in common into the second part of the duodenum.

Short Summary of Peritoneal Considerations.

1. The complex adult alimentary canal **and** its complex peritoneal invest-ing membrane **are** evolved from a simple straight gut and a simple straight mes-entery.

2. The abdominal part of this canal in the fœtus extends from the diaphragm to the anus on the posterior wall of the abdominal cavity in front of the vertebral column. It is attached to the anterior wall, as far downward as the umbilicus, under the name of ventral mesentery.

3. The stomach appears as an enlargement in the canal, and its dorsal mesen-tery is called meso-gaster.

4. The upper part grows much more rapidly than the lower, receiving a more liberal blood-supply. In round numbers, the small intestine is the product of the superior mesenteric artery ; the large intestine is the product of the inferior.

5. At an early period the liver is in front of the stomach and the pancreas behind. The liver grows into the ventral mesentery, the pancreas into the pos-terior meso-gastrium. The primary mesentery of the pancreas fuses with the posterior body wall.

6. As the liver and spleen become larger, and as the bowel begins to rotate about the superior mesenteric artery, the stomach assumes a transverse position, the anterior border becomes the lesser curve, posterior border becomes the greater curve, the left surface becomes the anterior, and the right surface becomes the posterior.

7. The small intestine rotates from left to right in such a way that the large intestine extends from right to left and crosses the small intestine near the stomach.

8. Rotation accounts for the duodenum passing under the second or trans-verse stage of the colon and apparently through its transverse meso-colon.

9. The meso-gaster becomes the greater omentum, consists of four layers, grows out from the greater curvature of the stomach.

10. The duodenum and ascending and descending colon lose their mesen-teries.

ABDOMINAL CONTENTS.

Cut slightly to the left of the umbilicus, then through the linea alba to the ensiform cartilage. It may be necessary to divide the cartilages of the two or three lower true ribs to make ample room. Make the following observations before any further dissecting is done :

1. Find the broad ligament of the liver in the longitudinal fissure of this organ. See the round ligament of the liver in the free margin of the broad. This is the remains of the umbilical vein that brought aerated blood from the placenta during intrauterine life.

2. Locate the *stomach, liver, spleen,* and *transverse colon.* Lift the lower margin of the liver upward, and at the same time pull the stomach downward ; now see the peritoneal connection between the liver and stomach—the *gastro-hepatic omentum* or *lesser omentum.* Notice the peritoneal connection between the stomach and the colon. This is the *gastro-colic omentum* or *great omentum.* See the peritoneal connection between the left end of the stomach and the spleen. This is the *gastro-splenic omentum.* Folds of peritoneum that connect organ to organ are called omenta, and the three above given are the omenta.

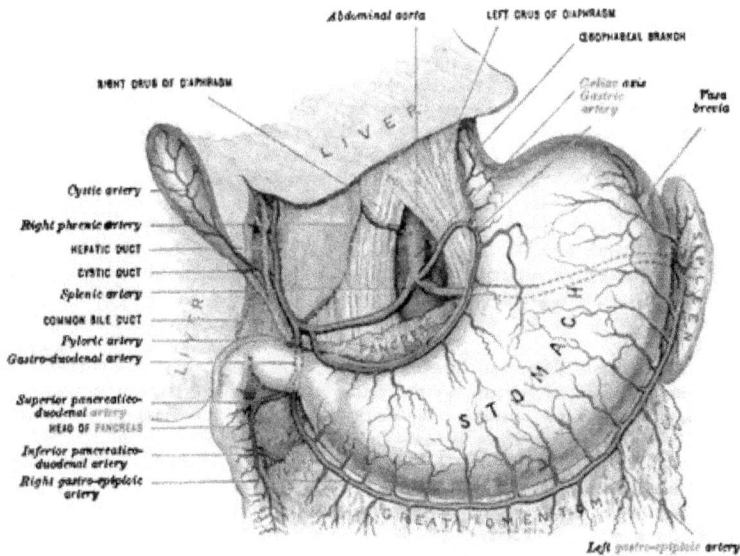

Fig. 174.—The Cœliac Artery and its Branches.

Abdominal aorta
LEFT CRUS OF DIAPHRAGM
ŒSOPHAGEAL BRANCH
Cœliac axis
Gastric artery
RIGHT CRUS OF DIAPHRAGM
LIVER
Vasa brevia
Cystic artery
Right phrenic artery
HEPATIC DUCT
CYSTIC DUCT
Splenic artery
COMMON BILE DUCT
Pyloric artery
Gastro-duodenal artery
Superior pancreatico-duodenal artery
HEAD OF PANCREAS
Inferior pancreatico-duodenal artery
Right gastro-epiploic artery
STOMACH
GREAT OMENTUM
Left gastro-epiploic artery

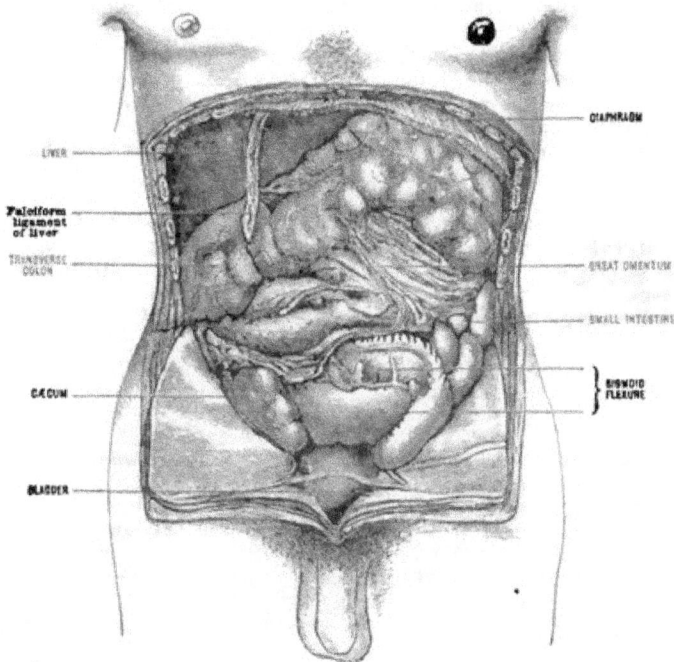

Fig. 175.—The Viscera as seen on fully opening the Abdomen without Disarrangement of the Internal Parts. (After Sarazin.)

DIAPHRAGM
LIVER
Falciform ligament of liver
TRANSVERSE COLON
GREAT OMENTUM
SMALL INTESTINE
SIGMOID FLEXURE
CÆCUM
BLADDER

247

3. *Three Divisions of the Colon.*—The divisions of the colon are ascending, transverse, descending, sigmoid, and rectum. The ascending colon extends from the right iliac fossa to the hepatic flexure. The transverse colon extends from the hepatic flexure to the lower margin of the spleen—the splenic flexure. The descending colon extends from the splenic flexure to the brim of the pelvis.

Notice the manner in which the three parts of the colon are held in place. The meso-colon of the transverse is quite long; that of the ascending and descending is very short; in fact, the nature of the union is one of adhesion rather than mesenteric, in the *two latter.* Folds of peritoneum that bind organs to walls are called mesenteries. The continuation of the descending colon below the pelvic brim, to a point opposite the second piece of the sacrum, is called the sigmoid. Lift it up and see the great length of its mesentery—called meso-sigmoid.

4. *Differential Diagnosis Between Large and Small Intestine.*—Inflate the large intestine (Fig. 175) and see the following differential points possessed by large intestine but not possessed by small intestine : (1) *Fatty masses.* (2) *Longitudinal muscular bands*, three in number, and at equal distances from each other. Each will lead to the appendix. (3) *Sacculations*, produced by contraction exerted by the longitudinal bands.

5. *Jejunum, Ileum, and their Mesentery.*—Compare your dissection with figure 175 and you will observe that the three parts of the colon almost surround the large central mass of small intestine. Collect this mass in your hands and find the upper end of the jejunum going through the transverse meso-colon opposite the third and fourth lumbar vertebra to become duodenum ; the lower end you will find near the right iliac fossa, ending in the cæcum, forming therewith a junction called the ileo-cæcal junction. Lift the whole mass of small intestine upward, to estimate the length of the mesentery.

6. *The Duodenum.*—Turn the transverse colon and great omentum upward. Find the beginning of the jejunum and shut the same off the intestine below this point by a ligature. Now inflate the alimentary canal above this point—*i. e.*, duodenum and stomach. You will now see the transverse part of the duodenum passing *behind the superior mesenteric* artery and vein.

7. *Cæcum and Appendix.*—You will find the cæcum, as a rule, in the right iliac fossa. Its meso-cæcum may be long in one case and short in another. The longitudinal bands traced downward will lead to the appendix. This organ has a peritoneal ligament—the meso-appendix. It may occupy a variety of locations. It may hang down in the pelvis across the brim. Its usual position is in an angle between the ileum and the pelvic side of the cæcum.

A general review will now be made of what you have examined *in situ* and studied in the normal position without cutting.

1. You saw, on opening the abdomen, the walls and also the viscera (organs), covered by a very thin, smooth, and glistening membrane—*peritoneum.* The layer covering the walls above, below, in front, behind, and on the sides is called *parietal peritoneum ;* that covering the organs themselves is called *visceral peritoneum.* The space between these two layers is the *peritoneal cavity*, which contains only a small amount of serum for lubrication. Henceforth you will define a peritoneal cavity as a space between a visceral and a parietal layer of peritoneum. (Fig. 171.)

2. On the interior of the abdominal walls you saw, through the thin parietal

peritoneum, the urachus in the mid-line extending from the summit of bladder to the umbilicus. This is the fœtal remains of the stalk of the allantois; it is covered by peritoneum, and the particular name given the ridge or fold in this peritoneum is plica urachi. This opening sometimes fails to close.

3. You saw on each side of the plica urachi the fœtal remains of the hypogastric arteries; these are covered by peritoneum, called plica hypogastrica. The triangular depression between these two plicæ just noticed is the internal inguinal fossa, and is associated with direct inguinal hernia.

4. You located the *deep epigastric artery and veins.* To the inner side of this

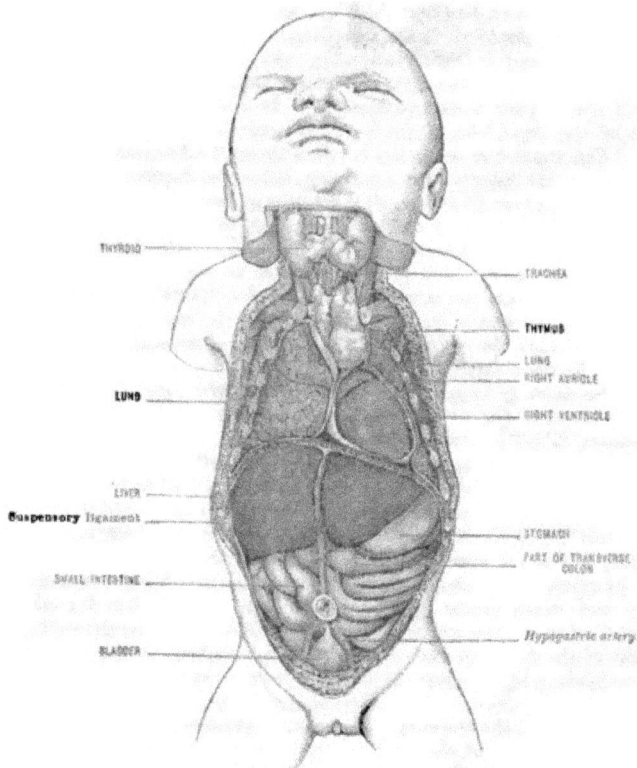

Fig. 176.—The Viscera of the Fœtus. (Rüdinger.)

artery you saw the *middle inguinal fossa,* associated with direct inguinal hernia; to the outer side the external inguinal fossa, associated (1) with the beginning of the spermatic cord; (2) with *indirect* or *oblique inguinal hernia.* You saw also that this deep, epigastric artery forms the outer boundary of Hesselbach's triangle.

5. You saw the *femoral sheath,* containing the femoral artery, vein, and femoral canal. These structures you saw leaving the pelvis below Poupart's ligament. The peritoneal depression corresponding to the beginning of the femoral canal—the femoral ring—is the *femoral fossa,* and is associated with femoral hernia.

17

6. *The round ligament of the liver*, you will remember, extends from the umbilicus to the liver. It is in the free border of the broad ligament of the liver. It is the fœtal remains of the umbilical vein. (Fig. 176.) You will remember that in the child before birth, the placenta purifies the blood, as do the lungs, after birth. The blood is taken in an impure state to the placenta by the hypogastric arteries, and returned pure by the umbilical vein. (Fig. 155.)

7. *Three omenta* are the gastro-hepatic, or small; the gastro-colic, or great; the gastro-splenic. You learned to define omentum as peritoneal folds connecting organ to organ. In your dissection you saw the great omentum as a heavy, fatty veil, covering in all the small intestine, and hanging down even into the true pelvis. This may even become a hernia,—an *omentocele*,—and appear at any one of the four fossæ described in the foregoing review.

8. *The Colon and Its Subdivisions* (Fig. 175).—The point where differentiation takes place is usually in the right iliac fossa; still, it may occur above this point. You will see on your work two flexures in the colon—the hepatic, at the lower margin of the right lobe of the liver; the splenic, at the lower margin of the spleen. The transverse colon lies between these two flexures. The descending colon is below the splenic; the ascending, below the hepatic flexure.

9. You will examine and recall the nature of the colic attachments. That of the ascending, and also of the descending, is in the nature of one structure adherent to another; their meso-colons, in other words, are short; hence you are not able to move them freely from place to place, as you can move the transverse colon. On the other hand, the meso-colon of the transverse colon is very long, due to which fact the transverse is the most movable part of the colon. Also recall the definition for mesentery: peritoneal folds binding organ to wall. The meso-sigmoid is also long.

10. *Diagnostic of large intestine*, you will remember, are the three longitudinal muscular bands of fibres; the consequent sacculations produced by these; the fatty masses called *appendices epiploicæ*. The latter are variable in size; in one case they may be very large, in another quite small; they are always present on large intestine. Any one of the three bands will lead to the appendix.

11. As you will remember, you inflated the duodenum and studied the third part of this organ as it passed behind the superior mesenteric vessels. The remaining part of the small intestine is arbitrarily divided into an upper two-fifths, called jejunum, and a lower three-fifths, called ileum. Jejunum means empty or hungry, and ileum means coiled. There is no practical histological difference between the two. They are held by their mesentery to a line extending from the left side of the body of the second lumbar vertebra, to the right sacro-iliac synchondrosis. In its course it lies upon, as you will presently see, the *aorta, vena cava inferior, transverse part of the duodenum*, and *vertebral column*. You will see, then, that the mesentery of the small intestine forms a partition between the right iliac fossa and the true pelvis; that fluid to the left of this partition would be directed into the pelvis, to the right of the same, into the right iliac fossa. Make this experiment.

You will now, by dissection, analyze the following localities and structures:

1. The foramen of Winslow.
2. The root-structures of the liver.
3. The lesser cavity of the peritoneum.
4. The greater cavity of the peritoneum.
5. The ascending, descending, and transverse duodenum.

6. The pancreas and splenic **artery.**
7. The stomach and gastric artery.
8. The liver, hepatic artery, and portal **vein.**
9. The superior and inferior mesenteric arteries and veins.
10. The anterior relations of the right kidney.
11. **The** anterior relations of the left kidney.
12. The root structures of the kidney.
13. Descriptive renal terms.

The foramen of Winslow is the communicating passage between the greater and lesser peritoneal cavities. The foramen has definite boundaries, which must be located by each student, and then learned. Surgical operations on the gall ducts of late years have made it imperative for the student of anatomy to have such a thorough knowledge of structures in this locality, that by the sense of touch alone, he may know perfectly both his longitude and latitude. The foramen is variable in shape; it may be round, semilunar, or triangular. As a rule, it will admit two fingers. Introduce your left index finger and find, in front of your finger, the front boundary—the free border of the lesser omentum containing the hepatic root-structures; behind, the posterior boundary—the ascending vena cava and a ligamentous band of peritoneum extending **from the**

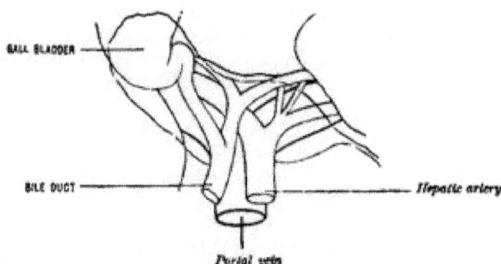

FIG. 177.—RELATION OF STRUCTURES AT AND BELOW THE TRANSVERSE FISSURE. (Thane.)

liver to the right kidney; above your finger, the caudate lobe of the liver; and below, you will recognize the hepatic artery and the duodenum.

The Root-structures of the Liver (Fig. 174).—Insert your finger into the foramen of Winslow, and with your forceps plow through the anterior layer of the gastro-hepatic omentum. You will now find three large structures, which a few moves of the forceps will liberate from their bed of connective tissue: (1) To the right side and below, and corresponding to the gall-bladder, the *common bile duct;* (2) to the left side and above, the *hepatic artery;* (3) between these two, but on a deeper plane, the *portal vein.* (Fig. 177.) Carefully examine the artery **and** you will find numerous nerves surrounding the same. They come from the pneumogastric and sympathetic. The connective tissue embedding these structures and escorting them to the interior of the liver is the *capsule of* **Glisson.**

The Bile Ducts (Fig. 174).—These are three in number. The hepatic brings bile from the liver; the cystic is the duct between the gall-bladder and the point where the hepatic meets it. The common duct begins at this point, passes behind the first part **of the** duodenum, and under the head of the pancreas, to reach the retiring duodenal elbow, where, with the pancreatic duct, you will find it opening into the duodenum. This duct is about four inches long.

The Hepatic Artery (Fig. 174).—Trace this vessel to the cœliac axis. It lies, as you will see, between the two layers of the lesser omentum. **Follow its**

cystic artery to the gall-bladder; its pyloric branch to meet the gastric; its superior pancreatico-duodenal branch downward behind the duodenum; here one branch follows the second part of the duodenum and anastomoses with the inferior pancreatico-duodenal branch of the superior mesenteric artery; the other follows the greater curvature of the stomach and, under the name of right gastro-epiploica, anastomoses with the left gastro-epiploica, a branch of the splenic artery Finally, trace the main branch of the hepatic artery to the transverse fissure of the liver.

The portal vein lies between the common duct and the hepatic artery. (Fig. 177.) It is formed by veins from the abdominal organs of digestion. Its blood is laden with bile, glycogen, and urea. The bile is stored up in the gall-bladder; the glycogen is stored up in the liver; the urea is thrown off. Find the hepatic veins discharging into the ascending vena cava, as this vessel passes through the diaphragm.

The hepatic veins take all blood from the liver brought to that organ by the hepatic artery and portal vein. The vein discharges into the ascending vena cava, just as that vessel is passing through the diaphragm.

The Greater and Lesser Peritoneal Cavities.—A knowledge of these cavities must be gained by study of the evolution of the peritoneum and alimentary canal, from a straight gut and single simple mesentery (See Morris.) The foramen of Winslow connects all the cavity you can see thus far, with the lesser cavity, behind the stomach. By mechanical devices your teacher will demonstrate the modus operandi of rotation, and this demonstration will interpret the description given in the larger text. Permit me to add in this place that every organ, with all its adnexa, grows up behind peritoneum, pushes this ahead of itself, and thereby becomes invested by the same, partially or completely.

To Dissect the Duodenum.—Thoroughly inflate the stomach and duodenum, as previously directed. Then divide the gastro-colic omentum and turn the transverse colon down toward the pelvis. Lift the stomach up, and observe the lesser cavity of the peritoneum behind the same. In cutting the great omentum, care must be taken not to injure the arteries along the greater curve of the stomach. (Fig. 174.) Now remove the colon from the second stage of the duodenum. The stages of the duodenum are: (1) From the pylorus of the stomach to the gall-bladder, two inches; (2) from the gall-bladder to the hilum of the kidney, three inches; (3) from the hilum of the kidney to the duodeno-jejunal angle, five inches. These stages are called ascending, descending, and transverse, respectively.

Relations of Transverse Duodenum.—You will find, above, the superior mesenteric vessels and the head of the pancreas; in front, the superior mesenteric vessels, the mesentery, and small intestine. Behind you will now dissect down with the forceps, and find the aorta, vena cava inferior, the crura of the diaphragm, and the fourth lumbar vertebra. On the left, the transverse duodenum terminates in the jejunum; on the right, in the descending duodenum, on the middle one-third of the kidney.

The Pancreas and Splenic Artery.—A very careful dissection of the pancreas must be made. The organ is delicate. It lies behind the stomach (Fig. 180); hence turn the stomach up to expose the same. Steps: (1) Locate the splenic artery by gently lifting the upper border of the pancreas. Follow the artery to the spleen. You will now find it necessary to divide the gastro-splenic omentum, but in doing this do not injure the arteries! Notice and save the pancreatic arteries, large and small, given to this organ—the pancreas—by the splenic artery. Trace out the branches given to the stomach. Find the gastro-epiploica sinistra artery. (2) Now dissect out the head of the pancreas. Trace the superior pancreatico-duodenal artery between the pancreas and the duodenum, taking notice of branches to each. Find the excretory duct of this organ—the pancreatic duct —opening into the duodenum with the common bile duct.

The Stomach (Fig. 174).—This organ has an *œsophageal* opening ; a *pyloric* opening ; an *anterior surface*, on which you will find anastomosing the transverse arteries from the gastric artery and the right and left gastro-epiploic arteries ; a *posterior surface*, likewise occupied by arteries ; a *greater curvature*, occupied by the greater omentum ; a *lesser curvature*, occupied by the lesser omentum ; a *fundus*, in relation with the spleen by the gastro-splenic omentum. Now trace the gastric artery from the cœliac axis to the gastro-œsophageal junction ; here you will see the artery divide into two branches—one to the walls of the œsophagus, the other to the cardiac end of the stomach. The main artery follows the lesser curve of the stomach and anastomoses with the pyloric branch of the hepatic artery.

The liver must be studied and dissected with reference to :

1. Lobes and their visceral impressions.
2. Fissures and their occupants.
3. Ligaments and their derivation.
4. Capsule of Glisson and its function.
5. Blood- and nerve-supply.
6. Fœtal remnants and their location.
7. The portal vein and its formation.
8. The hepatic veins and their escape.
9. The gall-bladder and bile ducts.
10. The descriptive terms used.
11. The relations to other structures.

Lobes.—(1) Right lateral ; (2) left lateral ; (3) quadrate ; (4) caudate ; (5) Spigelian.

Impressions are produced by continuous contact with certain adjacent organs. On the under or visceral surface find : (1) the renal impression ; (2) the colic impression ; (3) the duodenal impression ; (4) the gastric impression.

Fissures.—(1) *The transverse fissure* is the most important. It transmits the hepatic artery, the portal vein, the hepatic ducts, and the hepatic branches of the vagus and sympathetic nerves. In front of it is the quadrate lobe ; behind it the caudate and Spigelian lobes. (2) *The left longitudinal.* This separates the left from the right lobes, and has an *anterior* and a *posterior* part. The anterior part contains the round ligament ; the posterior part contains the remains of the ductus venosus. (3) *The fissure for the gall-bladder* and (4) the fissure for the ascending vena cava. Note, too, and show on your dissection, that on the posterior surface of the liver are found two fissures already mentioned—(*a*) fissure for the ductus venosus, (*b*) fissure for the vena cava ; while on the inferior or visceral surface are found three fissures—(*a*) transverse, (*b*) umbilical for round ligament, (*c*) vesicle for the gall-bladder. (5) *The right longitudinal.* This also consists of an anterior branch that contains the gall-bladder and a posterior branch that contains the vena cava. The two are interrupted by the caudate lobe. Compare your dissection with figure 178.

The letter H includes the five fissures of the liver in this manner : the transverse fissure forms the cross-bar ; the fissure for gall-bladder and fissure for vena cava form one vertical bar ; the fissure for the ductus venosus and the fissure for the round ligament form the other vertical bar.

The hepatic ligaments are five in number. One of these is a false ligament, a fœtal remnant called the round ligament, the ligamentum teres hepatis. It occupies the anterior part of the left longitudinal fissure. The other four ligaments are of peritoneal origin, and are called : (1) The suspensory, broad, or falciform ; (2) the right lateral ; (3) the left lateral ; (4) the coronary. The latter binds, as you will see, the posterior surface of the liver to the diaphragm. The lateral ligaments, which you will see on lifting the extremities of the liver, are simply outward prolongations of the coronary ligament. The liver is firmly fused with the diaphragm, a perfect appreciation of which you can only gain from study of the embryology of the alimentary canal.

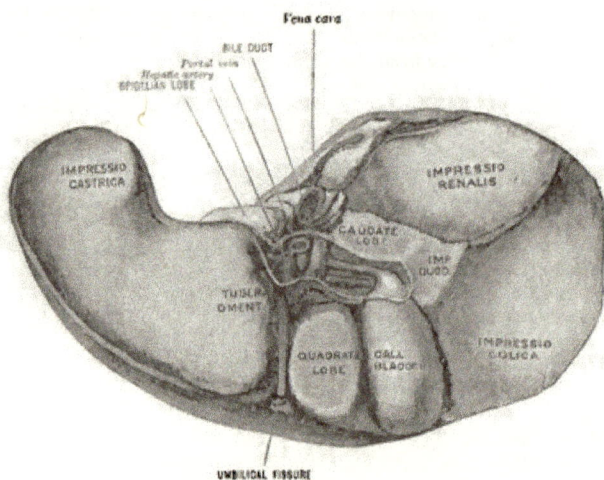

FIG. 178.—THE INFERIOR SURFACE OF THE LIVER.

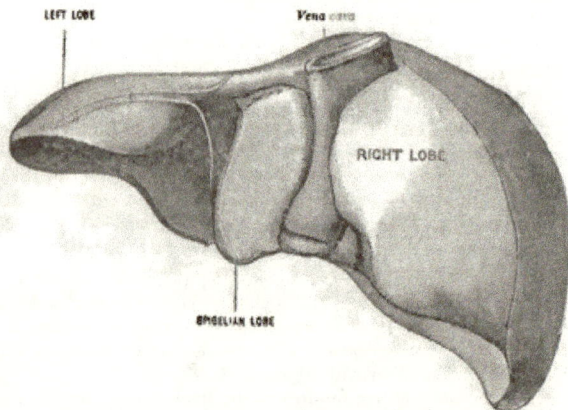

FIG. 179.—POSTERIOR SURFACE OF THE LIVER.

Glisson's Capsule.—You will recall your dissection of the root-structures of the liver. You found the *hepatic artery* dividing into two branches, the *portal vein* the same, and two *hepatic ducts* coming down to join the cystic; you saw nerves which are called hepatic branches from the vagus and sympathetic, surrounding the hepatic artery. Now recall this: the very intimate manner in which these structures were bound together by connective tissue. This connective tissue is the capsule of Glisson—the framework for the hepatic root-structures.

Hepatic Nerves and Arteries.—The nerves reach the liver through the transverse fissure, are protected in transit by the capsule of Glisson, are distributed to the substance of the liver on the branches of the portal vein and hepatic artery. These nerves come from the hepatic plexus, one of the divisions of the cœliac plexus. This plexus is made up of branches from the sympathetic, the right phrenic, and the left pneumogastric or vagus nerve. The nutrient artery to the liver is the hepatic. The liver receives blood for purification from all the other abdominal organs of digestion, through the portal vein; all the blood in the liver escapes by the hepatic veins into the vena cava. To find these veins, turn the liver down into its normal position, locate the emergence of the vena cava from the abdominal cavity through the diaphragm, and into the pericardium. Now cut through the diaphragm at this point and plow through liver substance, and you will find the hepatic veins. Another method is this: having located the vena cava above the diaphragm, cut a hole in its walls and follow the vessel downward until you see its hepatic tributaries.

Fœtal Remains.—These are the *round ligament* and the *ductus venosus.* The former is the connective-tissue remnant of the *umbilical vein,* that in the fœtus brought pure blood from the placenta. (Fig. 155.) This vein, on reaching the liver, broke up into three channels: (1) The greater part of the blood joined the portal vein and traversed the liver; (2) a small amount entered the left lobe; (3) a small quantity passed through a vessel—the ductus venosus—to meet the left hepatic vein, as that vessel was entering the vena cava. After the lungs became the organ of respiration, the placental circulation lost its specific character as a conduit. The main vessel, the umbilical vein, dwindled to a mere cord—the round ligament. The three termini of the vein did the same, and are known as remnants. The ductus venosus dwindled, and is consequently in the adult a fœtal remnant.

The portal vein in its relation to the root-structures of the liver will be seen in figure 177. In your dissection you have found it embedded in Glisson's capsule, and occupying a place between the hepatic artery and common bile duct, and posterior to both. Its tributaries come from all the organs in the abdominal cavity associated with the digestion of food, except the liver. The veins which you must demonstrate on your dissection as forming the portal vein are: (1) The superior mesenteric; (2) the inferior mesenteric; (3) the splenic; (4) the gastric. Notice particularly on your dissection that three veins—the gastric, splenic, and superior mesenteric—come together to form the portal vein *directly;* the inferior mesenteric is tributary to the splenic, hence it is not a primary, but a secondary tributary to the portal vein.

The hepatic veins must be located according to the two methods given under caption of Hepatic Nerves and Arteries. (Page 253.) Having found the hepatic vein, look on its under surface for the remains of the ductus venosus, for, you will remember, one branch of the umbilical vein in the fœtus terminated here. (Fig. 155.)

The gall-bladder and its system of ducts must now be carefully dissected. Any degree of rough manipulation will defeat a perfect dissection. Follow these steps: (1) Take the organ in the left hand, and with your forceps in your right carefully dissect the connective tissue, holding the gall-bladder in the cystic fissure. (2) Having now liberated the organ from its bed, and having exercised every care not to injure the vessels and ducts attached thereto, you may (3) cut a slit in the fundus of the organ, introduce a blowpipe, and inflate. (4) Have an assistant ligate as you inflate. (5) Now follow downward, and liberate from the capsule of Glisson the ducts : hepatic (two), cystic, and common. Follow this latter, the common bile duct, three and one-half inches to its confluence with the pancreatic duct, in the receding angle of the duodenum.

The descriptive terms used in speaking of the liver are :

1. *The root, transverse fissure,* or porta hepatis. This is where the vessels, nerves, and ducts enter, surrounded by the capsule of Glisson.

2. *The superior or phrenic surface* is in relation to the inferior surface of the diaphragm. It is covered by peritoneum, except a small space between the two layers of the suspensory ligament.

3. *The inferior or visceral surface* rests on the stomach, duodenum, colic elbow, right kidney. Two places are not covered by peritoneum : the fissure in which the gall-bladder rests and the transverse fissure. Surgically, this is the most important part of the liver, and should be most faithfully studied by the student.

4. *The posterior surface* is a territory you can see only when you remove the liver. It is mostly uncovered by peritoneum, and is limited by the anterior and posterior layers of the coronary ligament.

5. *The anterior border* is the one interrogated in physical examinations of the liver. It is quite thin and is marked by two notches, one for the suspensory ligament and one for the fundus of the gall-bladder. This latter is, in health, opposite the ninth costal cartilage. This border is about in line with the ribs.

6. *The extremities, right and left,* are associated with the lateral ligaments, but have no special importance in physical diagnosis. The hepatic elbow—a coinage—will be explained when the anterior relations of the right kidney are considered.

7. *Peritoneal investment* of the liver is for practical purposes complete ; nevertheless, on each surface we have found areas not completely covered. This has been explained under the head of peritoneum on page 238.

The relation of the liver to other organs is one of the most fascinating and instructive parts of the work thus far encountered. (1) Now show on your dissection that the superior surface of the liver is related to the under surface of the diaphragm, and separated by the diaphragm from the bases of the lungs, base of the pericardium, anterior abdominal wall, and six or seven lower ribs on the right side. (2) Show that the inferior surface of the liver lies on the right kidney and its adrenal ; on the hepatic flexure of the colon ; on the descending duodenum ; on the gall-bladder and its duct ; on the root-structures of the liver ; on the right end of the stomach ; on the upper curve of the duodenum ; on the anterior surface and lesser curve of the stomach ; on the lesser omentum. (3) Show that posteriorly the liver is in relation with the diaphragm and its crura, the tenth and eleventh thoracic vertebræ and their ribs, the œsophagus, aorta, vena cava, and thoracic duct. To show these posterior relations separate the liver from the diaphragm and pull it far forward, but do not injure the vena cava ; you may ligate and then cut the hepatic veins. This will give you an opportunity to study the coronary ligament also. All the above relations are to be studied without cutting anything except possibly the hepatic vein, as above indicated.

The Superior Mesenteric Artery.—To make a good dissection of this artery and its branches, turn the transverse colon up and pull the mass of small intestine forming the jejunum and ileum far down over the left iliac region, as shown in figure 180. Now, remember the mesentery contains, between its two layers, all the vessels, glands, nerves, and lymphatics that go to or come from the small intestine. Remember all these structures are embedded in connective tissue, just as are the root-structures of the liver enclosed by Glisson's capsule ; that the function of this mesenteric connective tissue is, like Glisson's capsule, protective and supportive, and contains, in some persons, large quantities of fat, technically designated mesenteric fat.

Dissection.—Locate the main stems of the artery and vein as they emerge from the under border of the pancreas, gently thrust your forceps—never the scalpel—through the upper layer of the mesentery, then use your fingers instead of the forceps and you can easily strip off the remainder of the peritoneum out to the intestinal margin of the mesentery. Trace out with your forceps, by removing the mesenteric connective tissue, always in the direction of the vessel, the following branches: (1) The ileocolic branch to the appendix, cæcum, ascending colon, and ileum; (2) the right colic artery to the ascending colon; (3) the middle colic artery to the transverse colon; (4) the main bulk of the artery to the jejunum and ileum under the name of intestinal branches or vasa

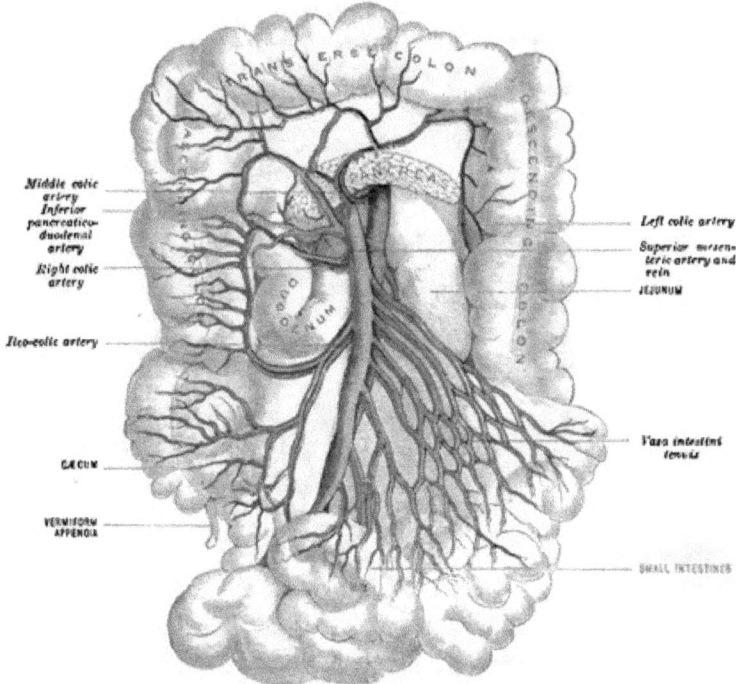

Fig. 180.—THE SUPERIOR MESENTERIC ARTERY AND VEIN.
(The colon is turned up, and the small intestines are drawn over to the left side.)

intestini tenuis; (5) a small branch already found—the inferior pancreatico-duodenalis. The veins must be dissected at the same time. If now you will thoroughly wash the mesenteric fat away with ether, and permit the specimen to dry, the sympathetic nerves accompanying the arteries may be seen.

The Inferior Mesenteric Artery.—To dissect this artery and its accompanying vein, take the mass of jejunum and ileum far over to the right side. (Fig. 181.) Locate the main stem of the artery at its origin from the abdominal aorta, make a hole through the peritoneum, and, as in the dissection above described, strip off the peritoneum. Now trace out the following branches: (1) The sigmoid artery to the sigmoid flexure of the descending colon; (2) the superior hæmorrhoidal artery to the rectum ; (3) the left colic artery to the descending colon.

Summary.—(1) The descending colon and rectum are supplied with blood by the inferior mesenteric artery. The blood from this area is returned by the inferior mesenteric vein, which passes behind the pancreas and is tributary to the splenic vein. (2) The remaining part of the intestinal canal, below the stomach, including the small intestine and the ascending and transverse colons, is supplied by the superior mesenteric artery. The blood from this area is collected by veins which come together to form, with the gastric and splenic veins, the portal vein. The name portal circulation is given to the veins coming from all the abdominal

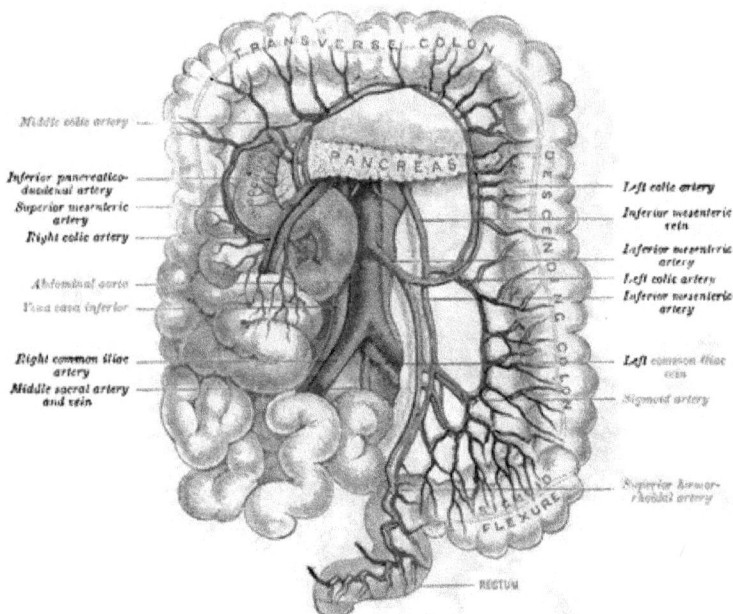

FIG. 181.—THE INFERIOR MESENTERIC ARTERY AND VEIN.
(The colon is turned up, and the small intestines are drawn to the right side.)

organs of digestion, except the liver. This blood is laden with urea, glycogen, and bile, all of which are removed from the blood in the liver.

Anastomosis of Arteries to the Abdominal Organs of Digestion.—(1) The gastric artery from the cœliac axis supplies the abdominal part of the œsophagus, the side of the stomach represented by the lesser curve, and anastomoses with the hepatic, at the pyloric end. (2) The hepatic artery supplies the liver and communicates, through the gastro-duodenal artery, with both the splenic and superior mesenteric arteries. (3) The several branches of the mesenterics successively anastomose, so you can trace from œsophagus to rectum the continuous blood-supply to all the viscera of this system.

Anterior Relations of the Right Kidney.—The anterior surface of the right kidney is occupied, from the kidney forward, (1) by the anterior part of the fatty capsule ; (2) by the following organs : liver, duodenum, colon. For the purpose of aiding the memory, call the visceral relation that of the three elbows: the hepatic elbow, the duodenal elbow, the colic elbow. Now demonstrate as follows: Carefully pull the hepatic flexure of the colon downward, and notice that this is made up of the upper end of the ascending and the proximal end of the transverse colon, and rests approximately on the lower third of the kidney. (Fig. 182.) Next trace down the second part of the duodenum (inflated), and see that it forms a flexure where it joins the transverse duodenum. Turn this off and call it the duodenal elbow, observing that it occupies the middle third of the kidney. Lastly, observe the anterior border of the liver come down and suddenly recede backward and upward. It covers the upper third of the anterior surface of the kidney, and forms the hepatic elbow.

Notice, too, that the areas occupied by the duodenal and colic elbows have lost their peritoneum, in accordance with a law explained under the head of peritoneum. The left kidney has two non-peritoneal areas.

Anterior Relations of the Left Kidney.—The first relation is the fatty capsule. Then the following viscera you will remove in this order: (1) The fundus

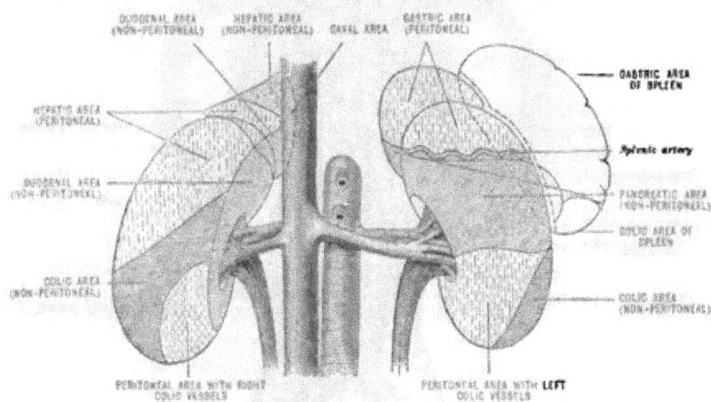

FIG. 182.—DIAGRAM SHOWING ANTERIOR RELATIONS OF KIDNEYS AND SUPRARENAL BODIES.

of the stomach; (2) the tail of the pancreas; (3) the descending colon; (4) the last part of the duodenum.

Posterior Relations of Each **Kidney.**—Carefully remove the kidney from its bed of fat and connective tissue, and turn the same to the mid-line with all its root-structures attached. Great care must be taken in handling the left one not to injure the left spermatic vein (ovarian, if a female), which discharges into the left renal vein. Then, having removed the posterior part of the fatty capsule, you will find three nerves: (1) The last thoracic; (2) the ilio-hypogastric; (3) the ilio-lumbar. Then you will be able to make out the (1) quadratus lumborum muscle, covered by the anterior lamella of the lumbar fascia; (2) the diaphragm; (3) the psoas magnus muscle; and (4) the lowest intercostal artery.

The Root-structures of the Kidney (Fig. 183).—These are, from before backward, as met in your dissection: (1) The vein; (2) the renal artery and its sympathetic nerves; (3) the expansion of the ureter, known as the pelvis of the kidney. You will find these structures loosely bound together by connective tissue, like the capsule of Glisson or the mesenteric connective tissue. This you must divide in the direction of the, structures. You will now puncture the ureter in ·

the region of the brim of the pelvis and inflate the same to see the beauty of the pelvis of the kidney.

Descriptive Terms.—The kidney has a *superior* and an *inferior* extremity; an *anterior* and a *posterior* surface; a strong fibrous capsule, which may be easily peeled off. The place where the nerves, vessels, and duct pass into the kidney is called the *hilum.* (Fig. 183.) Cut through the kidney longitudinally, from the outer to the inner border, and see the *cortical* and *pyramidal* structure of the

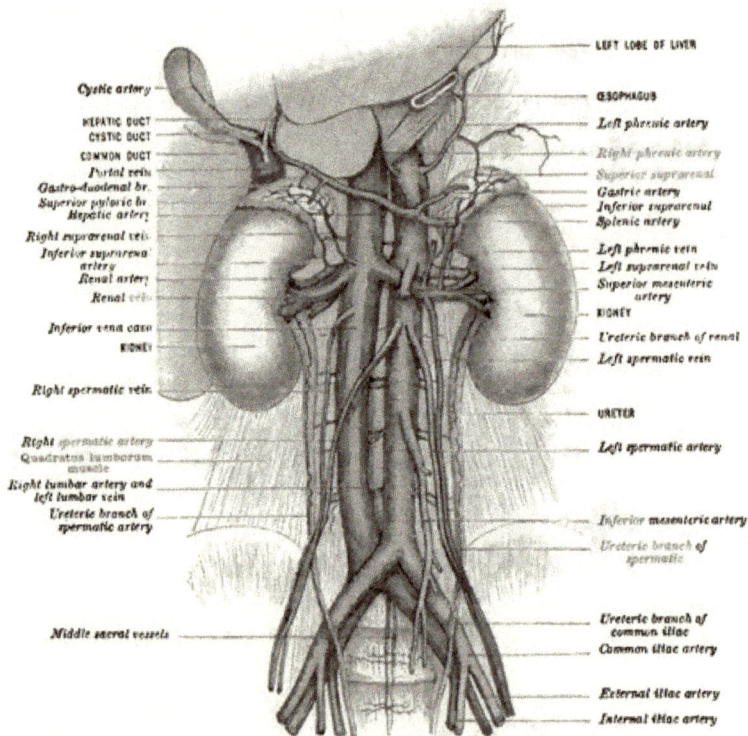

FIG. 183.—THE ABDOMINAL AORTA AND ITS BRANCHES, WITH THE INFERIOR VENA CAVA AND ITS TRIBUTARIES.

kidney. Figure 613 (Morris) shows the interior of the sinus and the formation of the pelvic portion of the ureter. The ureter having been inflated, examine for one or more small ureteric arteries, branches of the spermatic, superior vesical, and common iliac arteries on the ureter. Connected with the upper extremity of the kidney by connective tissue, find the *adrenals* or *suprarenal capsules.* These are ductless glands. Trace out their arteries to the renal and phrenic arteries and to the aorta.

THE SYMPATHETIC NERVE.

In the cranium, neck, thorax, abdomen, and pelvis you will have to do with arteries, nerves, viscera, and localities where the sympathetic nerve, while perhaps anatomically the smallest structure, in these localities is nevertheless the most important structure. It would be just as wanton to ignore and destroy, or, what is worse, fail to see, the hairspring of a watch, if you were taking a watch to pieces, as it would be to dissect a human body and not have a clear understanding of the sympathetic nerve.

The necessity of at least familiarity with the rudiments of this system of nerves will be apparent, when I remind you that you will hear more or less about the sympathetic nerve from every chair in your medical course. The physiological action of drugs depends on vaso-contraction and dilatation to a great extent. The rationale of the obstetrician's Credé manipulation, is stimulation of sympathetic uterine contraction. The physiologist invokes the sympathetic system when he would explain the function of viscera. The pallors and hyperæmias, the exanthemata, and the thousand and one strange phenomena in skin diseases are, to a greater or less degree, temporarily or permanently amenable to remedies that act in some way through the sympathetic. In general and special medicine, in surgery and all its special departments, you will hear of parts played by the sympathetic nerve. Are you, then, to remain in ignorance, or are you to gain a comprehensive knowledge of this subject by intelligent dissection?

An erroneous opinion prevails among students that the sympathetic cannot be seen, on account of its smallness. No; this is not the reason students often fail to find and become familiar with these nerves. You can see at a distance of six feet the cervical, thoracic, and abdominal ganglia, of the gangliated cord. You can see at a distance of ten feet the solar and cardiac plexuses. You can see at a distance of four feet the hypogastric plexus. You can see at a distance of two feet the communicating branches extending from one ganglion to another of the gangliated cord. You can see the greater and lesser splanchnics five feet away. At a distance of fourteen inches you can see the rami communicantes—the little nerves connecting the sympathetic cord and the anterior primary divisions of the spinal nerves. At a distance of twelve inches you can see the gastric, hepatic, splenic, and mesenteric sympathetic nerves. At a distance of ten inches you can see the sympathetic nerves on the internal carotid artery in its cavernous, petrosal, or cerebral stages. You can see the Vidian nerve, the petrosal nerves, the ophthalmic ganglion. At the closest normal visual range you can even see the sympathetic nerves that accompany the ovarian and uterine arteries. You can even see the long and short ciliary nerves to the eyeball.

I saw a man last summer who could not see the moon through a Yerkes telescope. In this case the moon was on exhibition, and the powerful lens was in working order, but the man in question did not know where to look. So, in anatomy, the sympathetic nerve is everywhere on exhibition. Its dimensions are not great, it is true; still, if you know where to look, you will have no difficulty in finding this nerve. The peculiar sympathetic lustre makes its identity certain, after you have dissected and studied this nerve a short time. The object of this chapter is to teach you where to look, to find the sympathetic nerve in all regions of the body; to teach you how to let the nerve alone, having once found the same—and this is its dissection; to furnish you with an outline, embracing the rudiments of what is known of the sympathetic to-day—and this by the following questions and their answers

1. *What is a sympathetic nerve?*

Cranial and spinal nerves are collectively designated somatic nerves. These preside over the special senses and supply muscles with motion and skin and membranes with sensation. These muscles and parts, constituting so much of the bulk of the body, depend on certain organs for air, blood, for the products of digestion ; in other words, there are certain organs in the body concerned in the preparation of nutriment for the tissues. This nutriment must be distributed

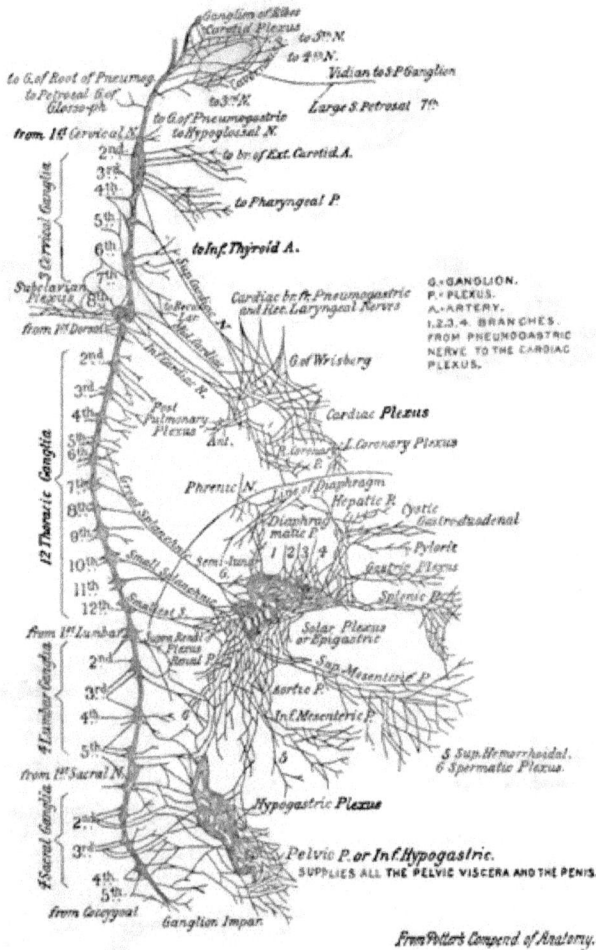

FIG. 184.—THE SYMPATHETIC SYSTEM OF NERVES.

to the tissues, at times in maximal quantities, at other times in minimal quantities. The action of the sympathetic is, among other things, to dilate or contract the vessels bearing this blood, for the nutrition of the tissues. In this sense then a feeder of organs, the sympathetic looks after the lives of organs—hence its synonym, the nerve of organic life.

2. *From where to where does the gangliated cord extend?*

From the ganglion of Ribes above, on the anterior communicating artery, to the ganglion impar below, on the coccyx.

3. *How many ganglia are in each region of the spine?*

Three in the cervical; twelve in the dorsal; four in the lumbar; four in the sacral region.

4. *Where will we find these ganglia in the cervical region in dissection?*

In the neck, behind the carotid sheath, the **superior** cervical ganglion; on the **inferior** thyroid artery, the middle cervical ganglion; on the inner side of the superior intercostal artery, the inferior cervical ganglion. This one is somewhat difficult to find, as it lies in fatty connective tissue. Look between the transverse process of the seventh cervical vertebra and the neck of the first rib, and you will find it. Dissolve the fat in ether and pack the region with a two per cent. solution of formaline, to develop well the branches of this ganglion, which branches are somewhat numerous.

5. *Where are the ganglia located in the thorax?*

You will find them in fatty connective tissue, behind the pleura costalis, on the heads of the rib. These are smaller than the cervical, but easily found.

6. *Where will we find the ganglia in the lumbar portion of the gangliated cord?*

Along the inner margin of the psoas magnus muscle, behind the peritoneum, in fatty connective tissue.

7. *Where will we find the ganglia in the sacral region of the gangliated cord?*

Along the inner side of the anterior sacral foramina, behind the peritoneum, in fatty connective tissue.

8. *How are these ganglia, in the regions above mentioned, connected?*

By interganglionic cords or nerves. These can all be seen, and must be demonstrated on the cadaver by the student.

9. *What are the rami communicantes?*

They are two little nerves that extend from the spinal nerve, shortly after its emergence, to the ganglia of the gangliated cord. These are the somatic communications referred to above.

10. *How do sympathetic nerves reach the organs or parts they are destined to supply?*

In two general ways: (1) The majority of sympathetic nerves travel with the artery to the organ, and take the same name as the artery. (2) Some few accompany somatic nerves to a part; the sympathetic nerves to the pharynx accompany the vagus and glosso-pharyngeal nerves, and then take the name of the part supplied.

11. *Are there any important nerves given off from the superior cervical ganglion? If so, what are they, and where are they distributed?*

Yes; from the ganglion are given off branches that accompany the branches of the internal carotid artery to the brain; the ophthalmic artery to the orbit; communicating branches to the third, fourth, fifth, and sixth cranial nerves.

12. *What branches are given off from the middle cervical ganglion?*

Nerves to all the branches of the external carotid artery; nerves to the pharyngeal plexus; a nerve—the superior cardiac nerve.

13. *What nerves are given off from the inferior cervical ganglion?*

From this are given off the principal nerves to the cardiac plexus.

14. *Explain the sympathetic distribution from the twelve thoracic ganglia.*

In round numbers, the six upper are given to the cardiac plexus; the six lower to the solar and renal plexuses, by the splanchnics.

15. *How is the hypogastric plexus formed?*

By branches from the lumbar and sacral parts of the gangliated cord.

16. *What is the function of the cardiac prevertebral plexus?*

It supplies lungs and heart. It is formed by sympathetic nerves, supplemented by filaments of the phrenic and vagus nerves.

17. *How is the solar plexus formed, and to what organs are its branches distributed?*

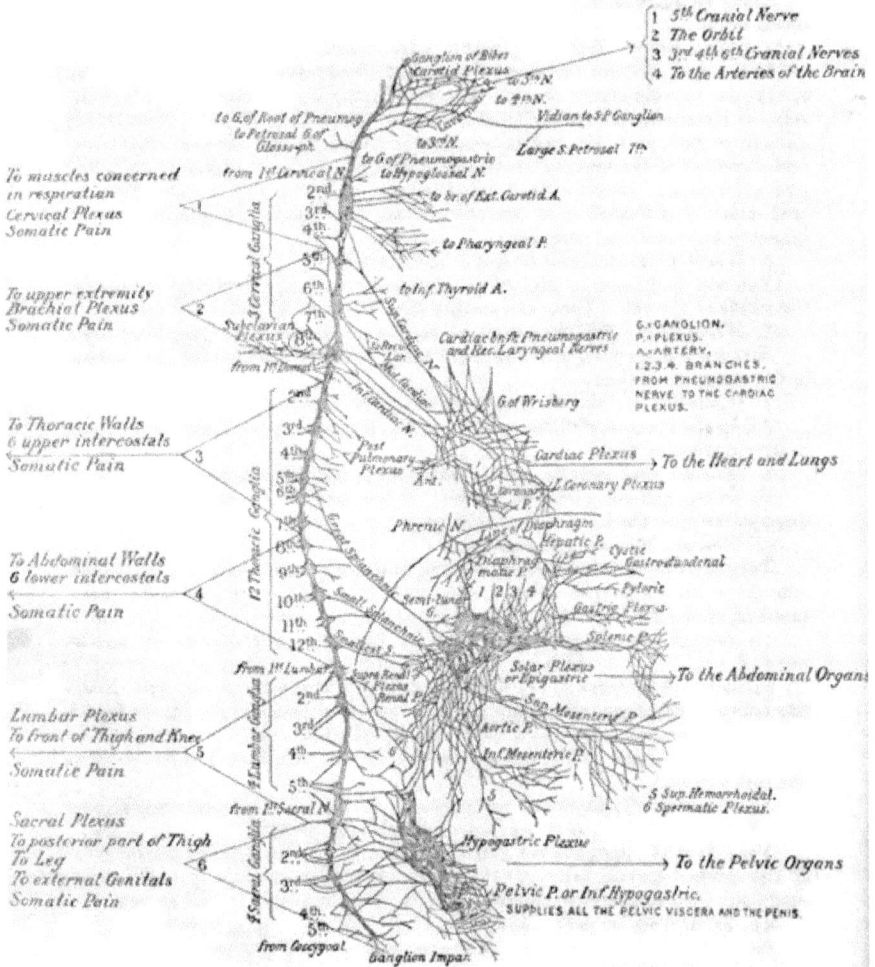

FIG. 185.—THE SYMPATHETIC SYSTEM OF NERVES.

By the sympathetic, as above indicated, in union with filaments from the vagus nerves. It is distributed to the abdominal organs, exclusive of the organs in the pelvis.

18. *To what is the hypogastric plexus distributed?*

To the organs in the pelvis. It will be remembered the nerves reach organs by accompanying the artery **to the part.** Hence it occurs that the ovary and testicle receive their nerve-supply from the renal plexus.

In the subjoined figure the tendency of plexuses to supply areas below their own level will be seen.

19. *What is meant by automatic action in sympathetic ganglia?*

The ganglia of the heart and intestine have the power of independent action, still, the sympathetic is no longer considered a separate system.

20. *Is there any definite means of judging of the location of disease founded on* **the nature or** *location of pain? Or, conversely, may pain incident to disease* **of the** *various organs in the three great cavities—thorax, abdomen, and pelvis—be rationally accounted for in the distribution of cerebro-spinal or somatic nerves?*

Yes; for, as the following pages will show, somatic nerves often report pain that has its origin **in an organ** far away from the origin. These somatic nerves also transmit both **pain and motion** more rapidly.

For an **example, in valvular** lesion the pain in the chest behind the sternum, **so long as** reported by **the** sympathetic nerve to the patient as a subjective symptom of disease, is constant and aching. When, however, the case assumes a grave aspect, and the pain is reported to the patient's sensibility by the somatic nerves, then an aching pain is replaced by one described as darting and stabbing. Then, too, the location of the pain will have changed. The aching sympathetic pain behind the sternum is now felt in the little and ring fingers. Consult the characters of sympathetic and somatic pain in the following paragraphs. Consult figure **185 and see:** (1) The heart is supplied by sympathetic nerves by the cardiac **plexus;** (2) **the** sympathetic **nerves forming** the cardiac **plexus communicate with the** somatic nerves in the **area where** are given **off the nerves** forming the brachial plexus; (3) as sensory nerves report pain peripherally, **we may** logically account **for** the digital pain in valvular lesions in the distribution of the brachial plexus **in** general, or in the specific distribution of the ulnar nerve in particular. The same is true of all diseases of viscera in which pain figures conspicuously as a subjective symptom. To find the possible somatic area where pain may occur, use figure 185 according to the following steps: (*a*) Where is the diseased organ located? (*b*) Are the organs in this area supplied by **the cardiac,** solar, or hypogastric plexus? (*c*) At what place do the sympathetic **nerves** forming the plexus, supplying the organ under consideration, communicate with the somatic or spinal nerves? (4) What becomes now of the spinal nerves given off where the sympathetic communicate **with them?** Do these spinals form either the cervical, brachial, lumbar, sacral, or coccygeal plexus, or do they form the thoracic nerves? (5) Trace out the distribution of the spinal nerve, and you will have the somatic route pursued by pain originating in the sympathetic area. The steps then would be: (*a*) sympathetic area; (*b*) communicating branches with the somatic nerves; (*c*) transfer centre in the cord. (*d*) Somatic distribution of nerves over which pain may travel. Articulations may be involved in reflex pain. In the case of pain in the little finger incident **to** valvular lesion, the student must be prepared to answer the question, **Where may** the pain have been in this same valvular lesion, aside **from in the skin of the** little finger and one-half **the** ring finger? (6) The same nerve, the ulnar, also sends sensory branches **to every** articulation, which any of the **muscles supplied** by its motor filaments moved **or** assisted in moving; hence in the **elbow or** any of the following joints a reflex pain, darting and stabbing in character, **may be a sequel** of valvular heart lesions. (7) Distribution of cranial **sympathetic nerves;** (8) Distribution of external carotid sympathetic **nerves;** (9) **Distribution of cardiac sympathetic** nerves; (10) Distribution **of solar sympathetic nerves;** (11) **Distribution** of hypogastric sympathetic nerves.

18

1. Local in the organ supplied by the sympathetic.—In this case the pain is a sensation of heat, fullness, distress, tenderness, oppression, burning, weight, and uneasiness.

2. Somatic.—In this case the pain is reflected out over the somatic nerves nearest to the sympathetic area supplying the irritated organ. This is indicated by the arrows on the left. (Fig. 185.)

3 The long arrow indicates transference of pain from any of the visceral areas through the gangliated sympathetic cord to the cranial nerves. In this manner an ovarian, uterine, intestinal, or cardiac irritation may be even many times more severe in the distribution of the fifth cranial nerve than at the seat of disease. Through the auriculo-temporal branches of the fifth, pain in the scalp may be reported.

Anatomical Classification of Pain.

To aid your memory and to teach you to be expert in tracing the transmission routes of pain from sympathetic system to somatic system, and to enable you to determine, from the nature of the pain, whether a somatic or sympathetic region is involved, learn the subjoined lists of adjective expressions, and study carefully the preceding figure, showing in a comparative schematic manner the respective distribution of somatic and sympathetic nerves.

Character of Pain in the Sympathetic.

As the sequel will show, pain in the sympathetic nerve, pure and simple, is designated by writers and lecturers on medicine by the following adjective expressions: A sensation of weight, constriction, fullness, uneasiness, heat, dullness, pricking, distress, tenderness, oppression, or burning. To be more explicit, the above terms are used descriptively, while our task now is to show that in those very diseases where these terms are used descriptively we find a sympathetic nerve-supply only involved.

Character of Pain in Somatic Nerves

Is designated by the following adjectives: Cutting, shooting, gnawing, darting, tearing, intense, sharp, severe, aching, griping, gnawing, boring, and fulgurating. It is our task here to show that where such pains occur a somatic area is either pathologically involved or a somatic nerve is reporting pain that originates in a sympathetic area. In other words, the character of the pain, as the sequel will show, is a reliable index to the nature of the region involved by the pain-producing agent.

TABLE SHOWING COMMON DISEASES,

The Nature of their Pain, when Local, and the Nature of their Pain and the Anatomical Route of its Transmission, when Reflex.

Pneumonia. . . . .
- Sympathetic pain, dull; when in center of lung, no pain.
- Somatic pain, sharp and lancinating.
- Reflex, lumbago on sound side.
- Nerve route, *via* cardiac plexus and intercostal nerves.

Pleurisy, ...
- Sympathetic pain, heavy and uneasy.
- Somatic pain, intense, sharp, cutting, lancinating.
- Reflex, epigastric region, subaxilla, chest wall.
- Nerve route, cardiac plexus and intercostals.

TABLE SHOWING COMMON DISEASES.—(*Continued.*)

Bronchitis, ...
- Sympathetic pain, not severe, dull, heavy behind sternum.
- Somatic pain, sharp soreness, acute.
- Reflex, epigastrium, short ribs.
- Nerve route, cardiac plexus *via* intercostals.

Pericarditis, ...
- Sympathetic **pain,** distress, in incipiency.
- Somatic pain, acute, darting, paroxysmal.
- Reflex, precordial tenderness and pain.
- Nerve route, cardiac plexus *via* intercostals.

Endocarditis, . . .
- Sympathetic pain, distress rather than pain.
- Somatic pain, acute when complicated by pericarditis.
- Reflex, precordial tenderness and pain.
- Nerve route, cardiac plexus *via* intercostals.

Valvular Lesions, .
- Sympathetic pain, constant and aching.
- Somatic pain, darting, stabbing.
- Reflex, chest, left shoulder, arm, forearm, fingers.
- Nerve route, cardiac plexus *via* circumflex, internall cutaneous, ulnar.

Angina Pectoris, . .
- Sympathetic pain, burning pain under sternum.
- Somatic pain, intense and lancinating.
- Reflex, chest, arm—right, neck—lower extremity.
- Nerve route, cardiac plexus *via* somatic system.

Thoracic Aneurysm,
- Sympathetic pain, aching locally.
- Somatic pain, intense and boring.
- Reflex, back and chest.
- Nerve route, cardiac plexus *via* intercostals.

Renal Colic, ...
- Sympathetic pain, pricking locally.
- Somatic pain, severe, excruciating.
- Reflex, chest, abdomen, penis, testicles.
- Nerve route, solar plexus *via* intercostals, internal pudic.

Aneurysm of Thoracic Aorta, .
- Sympathetic pain, local aching and burning.
- Somatic pain, severe and radiating.
- Reflex, chest walls and abdomen.
- Nerve route, cardiac plexus *via* intercostals.

Aneurysm of Abdominal Aorta, ...
- Sympathetic pain, heavy in back.
- Somatic pain, intense in abdominal walls.
- Reflex, abdominal walls.
- Nerve route, solar plexus *via* intercostals.

Spinal Meningitis, .
- Sympathetic pain, locally deep-seated and boring.
- Somatic pain, severe and radiating.
- Reflex, upper and lower extremities.
- Nerve route, sympathetic and somatic systems.

Locomotor Ataxia, .
- Sympathetic pain, sympathetic ataxia.
- Somatic pain, fulgurating pain in extremities.
- Reflex, girdle pain in abdominal walls.
- Nerve route, somatic and sympathetic.

Renal Inflammation,
- Sympathetic pain, dull pain locally.
- Somatic pain, severe, lancinating.
- Reflex, penis, perineum, and testicle.
- Nerve route, renal plexus *via* internal pudic nerves.

Cancer of Stomach, .
- Sympathetic pain, locally gnawing.
- Somatic pain, lancinating.
- Reflex, absent in eight per cent. of recorded cases.
- Nerve route, solar and vagus.

Gastric Ulcer, .
- Sympathetic pain, burning and gnawing.
- Somatic pain, pain often absent for days.
- Reflex, epigastric tenderness.
- Nerve route, solar plexus *via* intercostals.

Intestinal Colic, . .
- Sympathetic pain, griping.
- Somatic pain, severe character.
- Reflex, epigastrium.
- Nerve route, solar plexus *via* lower intercostal.

Enteralgia . . . . .
- Sympathetic pain, uneasiness.
- Somatic pain, tearing and cutting.
- Reflex, umbilicus and right iliac regions.
- Nerve route, solar plexus *via* lumbar plexus.

TABLE SHOWING COMMON DISEASES.—(Continued.)

Acute Enteritis,	Sympathetic pain, dull and aching. Somatic pain, umbilical tenderness. Reflex, abdominal walls. Nerve route, solar plexus *via* lumbar plexus.
Duodenal Ulcer,	Sympathetic pain, local burning. Somatic pain, gnawing Reflex, right hypochondrium. Nerve route, solar plexus *via* intercostals.
Intussusception,	Sympathetic pain, local pain set aside. Somatic pain, severe in umbilical region. Reflex, abdominal walls. Nerve route, solar plexus *via* lower intercostals.
Strangulated Hernia,	Sympathetic pain, set aside locally, or colicky. Somatic pain, severe in umbilicus. Reflex, abdominal walls. Nerve route, solar plexus *via* intercostals.
Parenchymatous Hepatitis,	Sympathetic pain, dull locally. Somatic pain, mild somatically. Reflex, under right scapula, in shoulder. Nerve route, solar plexus *via* circumflex and intercostals.
Suppurative Hepatitis,	Sympathetic pain; when deep-seated, no pain. Somatic pain, severe when superficial. Reflex, right shoulder. Nerve route, solar plexus *via* circumflex nerve.
Cirrhosis,	Sympathetic pain, weight in right hypochondrium. Somatic pain, very seldom any pain. Reflex, no reflex. Nerve route, solar plexus *via* vagus.
Cancer of Liver,	Sympathetic pain, darting in various directions. Somatic pain, severe. Reflex, the abdominal walls. Nerve route, solar plexus *via* intercostals.
Biliary Calculi,	Sympathetic pain, uneasiness and distress. Somatic pain, severe, cutting, tearing. Reflex, chest and abdomen. Nerve route, solar plexus *via* intercostals.
Ovaritis	Sympathetic pain, dull and distressing. Somatic pain, severe and annoying. Reflex, scalp, back, thigh. Nerve route, hypogastric plexus *via* fifth nerve, **lumbar plexus.**

Rationale of reflex pain and the anatomical factors involved in a simple physiological reflex circuit. The rationale of reflexes is to be found in the law of projectiles. Pain as a projectile having reached a transfer centre, pursues the line of least resistance or the point of greatest traction. The factors involved in a reflex circuit are: (1) A sensory nerve leading to a transfer centre. (2) A sensory nerve leading from a transfer centre to a sentient area. 3. A transfer centre communicating centrally with a sympathetic area, and peripherally with a somatic area. In figure 186 the pain is in the end of the penis; the disease is renal colic. The sensory nerve to the penis, the internal pudic, a branch of the sacral plexus; the nerve-supply to the kidney is from the renal plexus. The pain is more severe in the distribution of the somatic nerves than at the seat of the disease.

In like manner figure 186 shows the same. This we will then call a **sensory reflex pain**, in which the pain is greatly exaggerated by its transmission over the somatic part of the reflex circuit. A **motor reflex pain** may be represented in figure 187. Here also is a reflex circuit, the constituents of which are (1) a sensory nerve; (2) a motor nerve; (3) a transfer centre. The phenomena produced are called **motor reflexes**.

Remember, these schematic figures are introduced here, not for the purpose of attempting to teach the physiology of reflexes, but simply to impress on the student the imperative necessity of being able, in dissection, to show the place where somatic and sympathetic nerves communicate, by the *nervi communicantes.*

Is there any difference in the transmissional power of somatic and **sympathetic** *nerves?*

From the decided difference in the character of **pain reported by these two** nerves we would conclude there is a difference.

What is the difference in transmissional power?

This is conjectural only. Simply to aid the memory, we might view the subject in the light of logical reasoning. For physiological facts, however, you must consult physiologists. Structure is the correlative of function ; hence, marked differences in function entail consequent marked differences in structure.

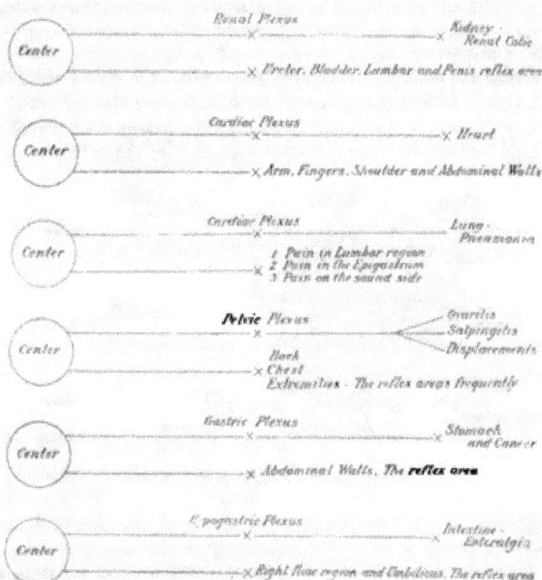

FIG. 186.—SHOWING SCHEMATICALLY A SENSORY REFLEX CIRCUIT.

The converse is true. The sympathetic individual fibre is smaller ; its fibres undergo numerous interruptions, not only in the ganglia of the gangliated cord, but also in numerous other plexuses, of which the plexuses of Auerbach and Meissner may be taken as the type.

Somatic and sympathetic nerves are routes by which nerve-impressions travel. Nerve-impulses, of whatever kind, are, then, projectiles, and amenable to the law of projectiles : "Projectiles follow the line of greatest traction, the point of least resistance, or the resultant of the two." If, then, the small calibre of sympathetic nerves, and their frequent interruptions, form a resistance to nerve-impulses, then **somatic nerves**, possessing neither of these assumed impediments, logically furnish lines of least resistance.

Our conclusion would be, then, that in the case of a given irritation in a sympathetic area, as in cardiac valvular lesion, the sympathetic reports the pain to

the cardiac plexus ; this plexus, however, transmits the impulse to the nearest
somatic nerves, which greatly intensify the pain, and refer the same, peripherally, to the chest, left shoulder, forearm, and fingers, by the intercostals,
descending branches of the cervical plexus, internal cutaneous, median, ulnar,
and radial nerves. Thus, a pain that was reported as simply "aching," in the
locality of the lesion, under poor transmissional conditions, becomes, **under the**
better transmissional conditions of somatic nerves, a pain *darting, stabbing*, and
excruciating.

Branches of the Abdominal Aorta and Sympathetic Nerves.—I now
desire you to dissect the branches of the abdominal aorta with special reference
to the sympathetic nerves. This is logical, since the sympathetic accompanies
every artery to its visceral end, takes the very name of the artery, and regulates
the amount of blood each organ shall receive, both in health and disease. The
sympathetic reports to the cortical sensory areas in the cerebrum sensations, both
of pain and well-being, through the vagus and all the spinal nerves. It carries
on the function of organs during suspension of volition.

The following explanation of arrangement will give you a **clear idea of the**
meaning of terms. Before you proceed further, master this scheme :

1. Every artery is surrounded by a plexus of small nerves, which accompany
the artery in all its branches to the organ supplied by the artery.

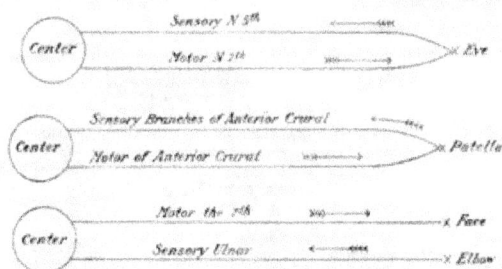

FIG. 187.—SHOWING SCHEMATICALLY A MOTOR REFLEX CIRCUIT.

2. This plexus, which always takes the name of the artery it accompanies,
has its origin in a ganglion near the beginning of the artery on the aorta ; the
plexus also takes the name of the artery which its branches accompany.

3. The ganglion—*i. e.*, all the ganglia situated at the beginning of all the
branches of the abdominal aorta—is fed by nerves from the solar plexus.

4. The solar plexus is situated in front of the beginning of the abdominal
aorta. It consists of two large semilunar ganglia, which embrace the cœliac
axis. The two ganglia are connected both above and below the artery which
they embrace. The plexus is formed by branches from the vagus, uniting here
with the sympathetic. In this way, then, may be explained the manner in which
the vagus reaches all the abdominal organs with motor fibres.

5. The solar plexus, then, is that part of the sympathetic nerve from which
all the abdominal viscera derive their nerve-supply. It is situated high, but is
prolonged downward under the name of the aortic plexus. From this the ganglia situated at the beginning of the branches of the aorta take their origin ;
these ganglia throw off plexuses which accompany the artery. The solar plexus
is a sort of clearing-house for the abdominal sympathetic nerves.

6. From above down, then, the succession of names is as follows : (1) The
solar plexus, formed of two semilunar ganglia ; (2) the downward prolonga-

tion of the solar plexus, called aortic plexus ; (3) ganglia situated at the origin of every artery from the aorta, which take the name of the aorta ; (4) branches or plexuses from the ganglia, which take also the name of the artery and accompany the same to its distribution.

In view of the foregoing, you are prepared now to find, locate, explain, and understand :

1. *Diaphragmatic artery*, nerves, plexus, ganglion, and vein.
2. *Splenic artery*, nerves, plexus, ganglion, and vein.
3. *Hepatic artery*, nerves, plexus, ganglion, and vein.
4. *Gastric artery*, nerves, plexus, ganglion, and vein.
5. *Suprarenal artery*, nerves, plexus, ganglion, and vein
6. *Renal artery*, nerves, plexus, ganglion, and vein.
7. *Spermatic artery*, nerves, plexus, ganglion, and vein.
8. *Ovarian artery*, nerves, plexus, ganglion, and vein.
9. *Superior mesenteric artery*, nerves, plexus, ganglion, and vein.
10. *Inferior mesenteric artery*, nerves, **plexus**, ganglion, and vein.

Dissection of the above nerves **requires care.** Work in the direction of the

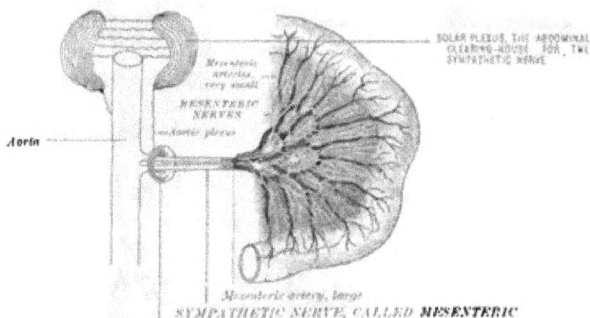

Fig. 188.—To Show Relation of Abdominal **Sympathetic Nerves** to (1) Solar Plexus and (2) to the Ganglion at the Root of the Artery.

vessel, with forceps. Consult the figures heretofore given to find name and location of vessels. For solar and aortic plexuses see figure 186.

Make special note : (1) *The renal veins* lie in front of their arteries, thereby forming an exception to the rule governing the relation of arteries to veins. (2) *The left spermatic vein* opens into the left renal vein, and this latter passes in front of the aorta. (3) *The right spermatic vein* opens into the ascending vena cava just below the renal. (4) *The ovarian veins* follow the course of their homologues. (5) *The spermatic arteries* arise from the aorta below the renals, pass in front of the ureter about opposite the bifurcation of the aorta. The ovarian take the same course. (6) *The four lumbar arteries* are analogous to the intercostals. (7) *The ureteric arteries* accompany the ureters ; they are branches of the common iliac and superior vesicle. (8) *The arteries to the vas deferens*, to the *urachus*, and to the *ureter* are branches of the *superior vesicle*. (9) *The sacra media*, the smallest.

Essential Points on the Sympathetic.—You have traced out the branches of the abdominal aorta, and I trust you now understand the principle underlying the distribution of visceral nerve-branches, from one abdominal prevertebral plexus—the solar plexus. We shall find in the pelvis, likewise, organs receiving arteries ; these arteries will be accompanied by sympathetic nerves ; these nerves

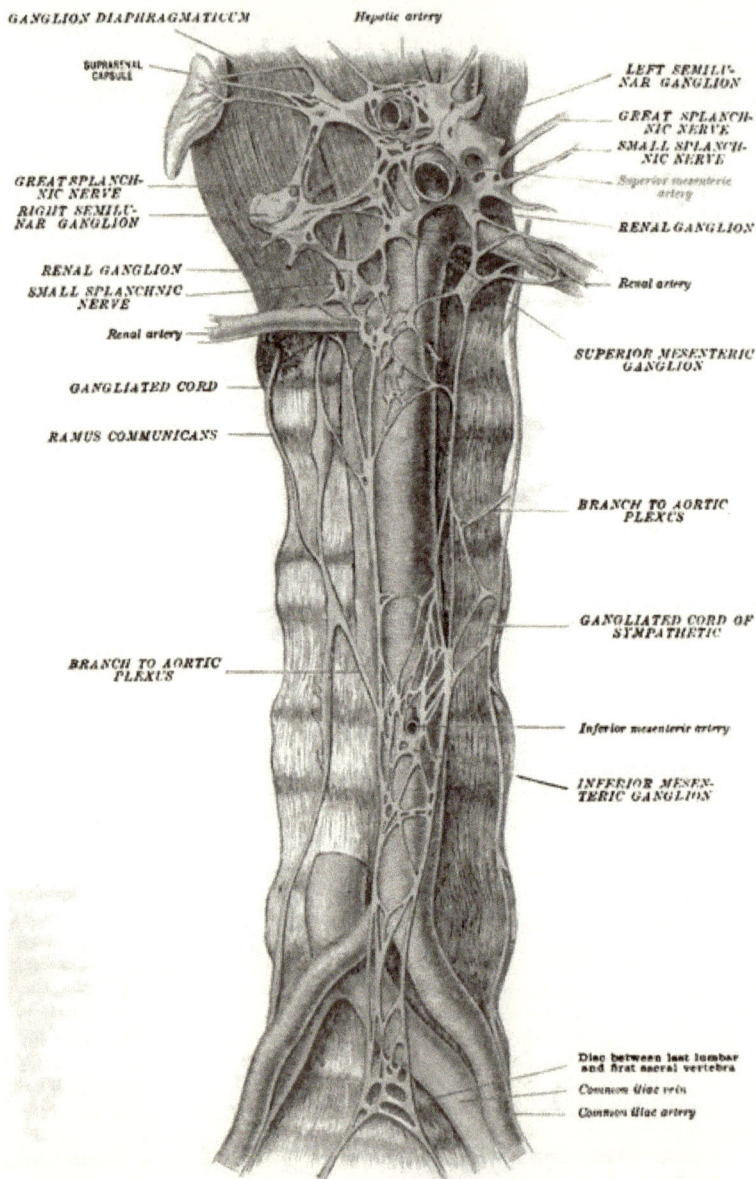

FIG. 189.—LUMBAR PORTION OF THE GANGLIATED CORD, WITH THE SOLAR AND HYPOGASTRIC PLEXUSES. (Henle.)

will come from plexuses on the artery; these plexuses will originate in ganglia near the origin of the artery; they will take the name of the artery and trace their own origin to the hypogastric plexus. We shall find, in the thorax, the heart and lung concerned in the circulation of the blood and its purification respectively. These organs receive their nourishment through arteries, and their arteries are attended by nerves, which nerves have their origin in plexuses; these plexuses spring from ganglia near the origin of the artery, and these ganglia trace their own origin to the cardiac plexus.

There is, then, one great sympathetic depot for the organs in the thorax, called *cardiac plexus;* one in the abdomen, called *solar plexus;* one in the pelvis, called *hypogastric plexus.* This arrangement is simply to make possible a physiological division of labor. Remember, then: the manner of distribution of sympathetic nerves is always the same in every region of the body; branches, plexuses, and ganglia take the name of the artery; there are three large plexuses—(1) *cardiac,* (2) *solar,* (3) *hypogastric;* there are as many small plexuses as there are arteries to supply viscera—and these small plexuses all draw their influence from the three large plexuses.

In view of what we have found, and what we shall find, let us gain a general idea of the sympathetic system in the following questions and answers:

1. *Is the sympathetic a system, separate and entire, and independent of the cerebro-spinal?*

No; sympathetic nerves are only nerves, branches of the spinals, set aside to supply organs.

2. *Of what does the sympathetic system consist?*

(1) Of gangliated cord, in the form of a horse-collar, communicating on the anterior communicating artery above and on the coccyx below; (2) of spinal branches, called rami communicantes—two from each spinal nerve; (3) of three prevertebral plexuses—*cardiac, solar,* and *hypogastric;* (4) of branches of distribution; (5) of plexuses and ganglia that take the name of the arteries on which found.

3. *Can we see the nerves and ganglia?*

Yes; in the thorax behind the pleura, and located opposite the heads of the ribs, you can see the ganglia; you can also find the two communicating branches from the spinal nerves; also the cord extending from one nerve to another.

The Hypogastric Plexus.—In the lumbar region you will find the gangliated cord on the front of the bodies of the vertebræ, along the inner margin of the psoas magnus. You can trace branches to the lumbar nerves behind the psoas magnus. You will see numerous branches from the lumbar cords, uniting with branches coming down from above to form the hypogastric plexus. This is between the common iliac arteries and passes down to form the pelvic plexus.

THE PELVIS.

DISSECTION OF THE PELVIS.

Inspect and become familiar with the following .

1. *The pelvic contents* are covered by peritoneum.
2. *The rectum* is on the posterior wall of the pelvis.
3. *The bladder* is on the anterior wall of the pelvis.
4. *The obturator nerve and vessels* are on the outer wall.
5. *The ureter* crosses into the pelvis and can be seen.
6. *The uterus and adnexa* are between bladder and rectum.
7. *The broad ligament* contains the uterine adnexa.
8. *The recto-vesical pouch* is between the rectum and bladder.
9. *The peritoneum* does not cover the anterior surface of bladder.
10. *Retzius' space* is between bladder and pubes.
11. *The recto-vaginal, recto-uterine,* and *cul-de-sac of Douglas.*
12. *Douglas' cul-de-sac* is between the rectum and vagina and uterus.
13. *Ileum and great omentum* gravitate into the pelvis.

Follow this order in your dissection :

1. The peritoneal ligaments of the bladder, rectum, and uterus, and the sympathetic nerves from the pelvic part of the hypogastric plexus to the same viscera, noticing they invariably follow the arteries to the organ, and take the name of the artery.

2. The pelvic fascia and its subdivisions and modifications incident to use, remembering, as you must, that growth is the correlative of function. Notice, too, that pelvic fascia is condensed subperitoneal connective tissue. The upper or uncondensed portions of the pelvic fascia contain fat.

3. Study the relations of the psoas carefully ; as nearly as may be, follow the steps just as given in the sequel, for this is the order in which you will meet all these various structures in your dissection.

4. The lumbar plexus ; its location in the substance of the psoas muscle ; its formation by the four upper lumbar nerves and a communicating branch from the twelfth thoracic ; and its immense distribution.

5. The internal iliac artery and all its branches, where the same escape from the pelvis, the spinal nerves they accompany, and the source of the sympathetic nerves that encircle them to supply the viscera.

6. A view of the ischio-rectal fossa from above, by cutting the origin of the levator ani muscle at the white line and looking in on the obturator internus muscle that forms its outer wall.

7. The sacral plexus, its location, relations, communications, and branches of distribution, and where these large branches escape from the pelvis.

8. In addition to the above, make dissection and study of the iliacus (origin of), the pyriformis, the levator ani, the obturator internus, the white line of the pelvis, and the delamination in the obturator fascia, called Alcock's canal.

In order to estimate the peritoneal ligaments in any peritoneal region, ask yourself how many viscera of considérable size grew up behind this membrane and pushed the same before themselves. In the pelvis, the *bladder, rectum,* and *uterus* grew up and developed behind the peritoneum; hence, not only are these three structures behind the peritoneum, but all their adnexal parts as well; they are held in position more or less by the peritoneum, and those processes having one end on the organ and the other fixed to a wall are called the peritoneal ligaments of these organs. Inflate the bladder, and see, posteriorly, two folds extending from the rectum to the bladder; also a fold on each side from the iliac fossa to the side of the bladder. The plica urachi is the superior ligament of the bladder. The peritoneal ligaments of the bladder are *two posterior, two lateral,* and *one superior.*

The Rectum.—The rectum begins at the pelvic brim, opposite the left sacro-iliac synchondrosis. You will ligate and cut the same at this point, leaving the sigmoid *in situ.* Notice that the upper part of the rectum has a fold of peritoneum—the meso-rectum. The next stage of the rectum is partially covered by peritoneum, like the descending colon. The lower part of the rectum, one and one-half inches in length, and extending from the tip of the coccyx to the anus, has no peritoneum, as you will see on the cadaver. Now search beside the rectum and bladder and you will find the pelvic plexus of sympathetic nerves, which supplies all the viscera in the pelvis. Here, as elsewhere, the nerves accompany the arteries to the parts and take the same name.

The pelvic fascia is the lower condensed layer of the subperitoneal connective tissue. It takes the following different names: iliac, obturator, recto-vesical, pubo-prostatic, anal, white line, Alcock's canal—according to its location. The white line you will see extending from the ischial spine to the pubic bone, one inch below the subpubic arch. It is a split in the obturator fascia, from which split arises the *levator ani muscle.* (Fig. 192.) The *anal fascia* covers the under surface of the levator ani muscle. The *recto-vesical fascia* extends across the upper surface of the levator ani muscle, investing the rectum and bladder. It also extends from the pubes to the neck of the bladder and prostate, under the name of *pubo-prostatic.* Demonstrate all these on the cadaver.

The iliac fascia covers the iliacus muscle, passes out of the pelvis behind the femoral vessels, and unites with the transversalis fascia by the side of the femoral vessels to form the femoral sheath, as described on page 235.

The obturator fascia covers the obturator internus muscle and delaminates to form the *white line.* (Fig. 192.)

Relations of the Psoas Magnus Muscle.—(1) Lying on the muscle you see the psoas parvus; (2) to the outer side, see the *iliacus muscle* and the *anterior crural nerve;* (3) to the inner side, skirting the pelvic brim, the *common and external iliac arteries* are seen; (4) to the inner side, and one-half of an inch below the pelvic brim, see the *obturator nerve and vessels;* (5) emerging from the front surface of the middle third of the muscle, see the *external cutaneous nerve;* (6) to the inner side, between the muscle and the common iliac artery, see the *genito-crural nerve;* (7) to the outer side of the muscle, above, see the *last thoracic,* the *ilio-hypogastric,* and the *ilio-inguinal nerves.* (8) in the substance of the muscle, this having been detached and removed piecemeal with the forceps from within outward, see the *lumbar plexus,* giving off the following branches.

Branches of the Lumbar Plexus (Fig. 190).—*Dissection:*

1. *Ilio-hypogastric,* to the skin over the gluteus maximus and over the hypogastrium; lies between the internal oblique and transversalis; communicates with the last dorsal. The hypogastric branch pierces the aponeurosis of the external oblique muscle one inch above the external abdominal ring, and is distributed to the skin over the region of the bladder.

2. *Ilio-inguinal,* crosses the quadratus and iliac muscles, pierces the transversalis, lies between the internal oblique and transversalis, supplies the internal

oblique, lies in front of the spermatic cord in the inguinal canal, escapes by the external abdominal ring, and supplies the scrotum and inner side of the thigh.

3. *Genito-crural*, descends on the front surface of the psoas magnus, divides into a genital and a crural branch; the genital branch enters the internal abdominal ring, lies behind the spermatic cord, and is distributed to the cremaster muscle. In the female it goes with the round ligament, supplies the same with motion, and is lost in the labia majora.

4. *The crural branch* of this nerve descends on the external iliac artery, lies in the femoral sheath, and is distributed to the skin covering the insertion of the psoas—according to Hilton's law.

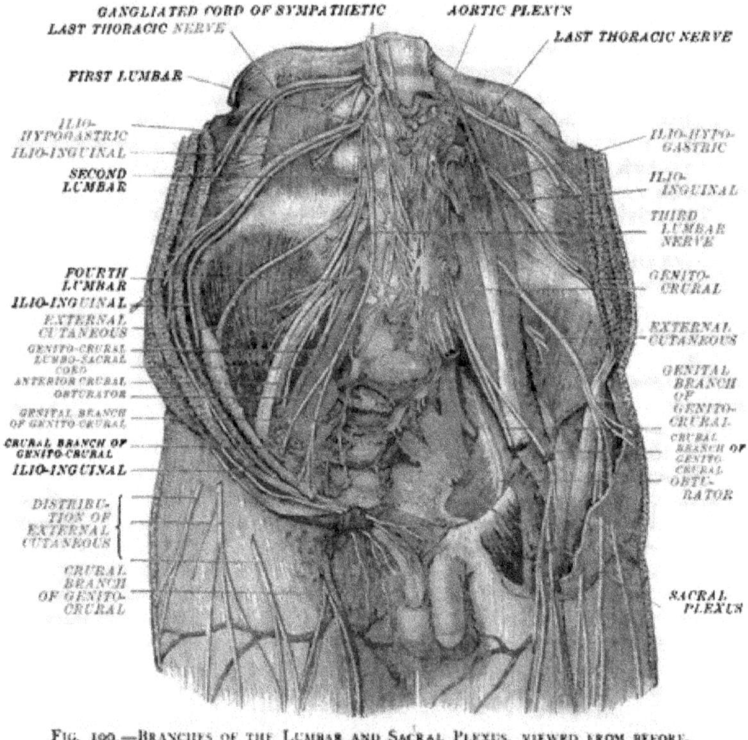

FIG. 190.—BRANCHES OF THE LUMBAR AND SACRAL PLEXUS, VIEWED FROM BEFORE.
(After Hirschfeld and Leveillé.)

5. *External cutaneous*, emerges from the outer side of psoas, crosses iliacus under the iliac fascia, leaves pelvis under crural arch, below anterior superior spine of ilium, pierces deep fascia of thigh, and supplies skin over outer side of thigh, being a dismembered branch of the anterior crural nerve, in reality.

6. *Obturator*, emerges from psoas on inner side, lies on outer wall of pelvis, leaves pelvis by obturator foramen, divides into anterior and posterior divisions, separated by the adductor brevis muscle; *the anterior division* communicates with the internal cutaneous and long saphenous nerves, forming the subsartorial plexus for the supply of the skin over the insertion of the adductors—*Hilton's*

late. It then supplies the gracilis, adductor longus, adductor brevis, and sartorius. *The posterior division* supplies the hip, knee, obturator externus, and adductor magnus.

7. *Accessory obturator* often absent; when present it lies to the inner side of the psoas, passes under outer border of pectineus, supplying this muscle and the hip-joint.

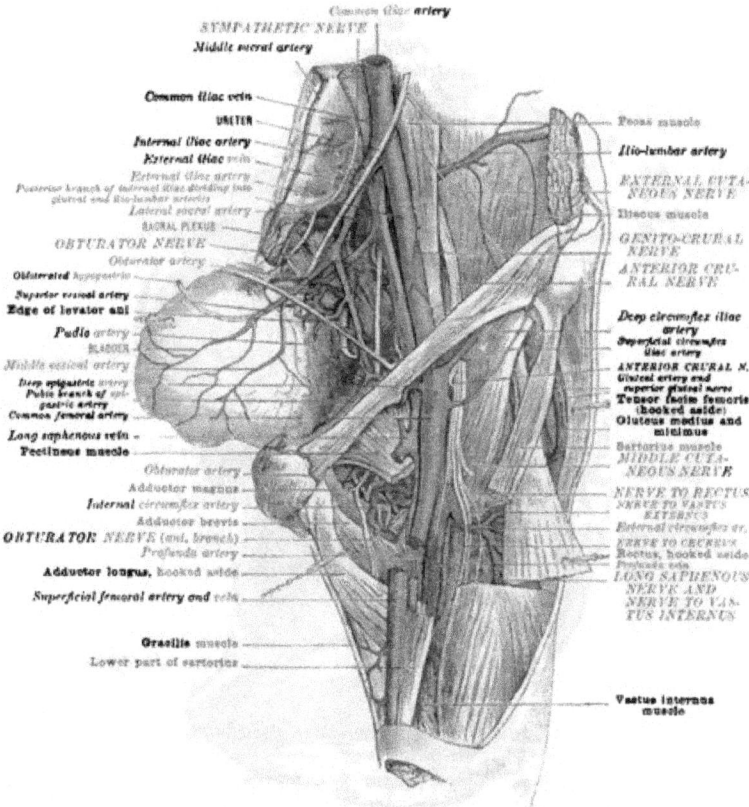

FIG. 191.—SIDE VIEW OF PELVIS AND UPPER THIRD OF THIGH, WITH THE EXTERNAL ILIAC, INTERNAL ILIAC, AND FEMORAL ARTERIES AND THEIR BRANCHES.
(From a dissection by W. J. Walsham in the Museum of St. Bartholomew's Hospital.)
The bladder is hooked over to expose back of pelvis.

8. *Anterior crural,* emerges between the iliacus and psoas muscles, under crural arch or Poupart's ligament, lies in a groove between these muscles under the iliac fascia; it gives off the internal cutaneous, middle cutaneous, and long saphenous branches. Its muscular branches are to the sartorius, rectus femoris, vastus internus, vastus externus, and crureus. The branches to the vasti also send filaments to the knee-joint; the branch to the rectus sends a filament to the hip-joint—*Hilton's law.*

9. *The lumbar nerves* send *rami communicantes* to the hypogastric plexus.

10. *The dorsi-lumbar cord* is the communication of the *first lumbar* with the *last thoracic nerve ;* the *lumbo-sacral cord* is the communication of the *lumbar plexus* with the *sacral plexus.* Find these two cords on the cadaver.

Dissection of the Internal Iliac Artery (Fig. 191).—The common iliac artery divides into the internal and external. Use only the forceps when you develop branches of vessels. Follow the direction of the artery and its branches as they are given off, and find :

1. The *ureter,* crossing into the pelvis at the bifurcation of the common iliac artery, and its small ureteric in its sheath.

2. The bladder being inflated, find the three vesical arteries—*superior, middle,*

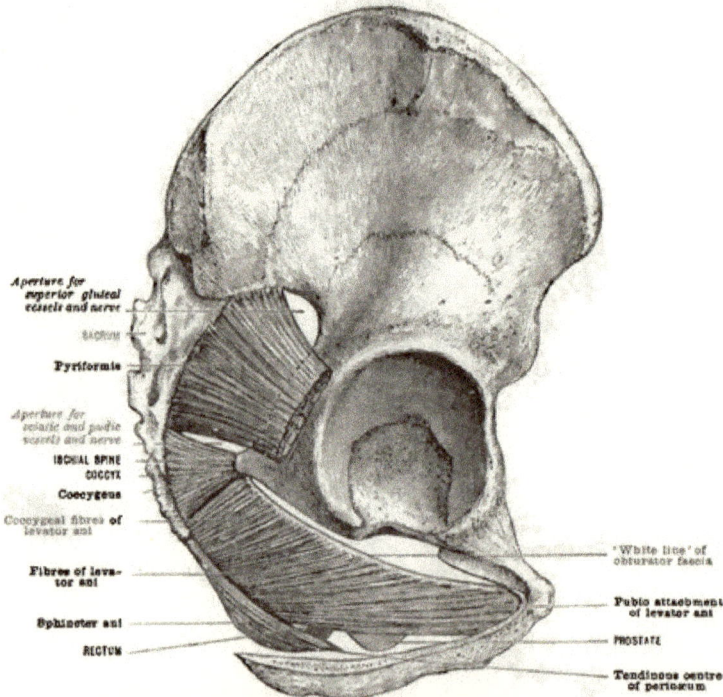

Fig. 192.—Muscles of the Floor of the Pelvis.
(A portion of the ischial and pubic bones sawn away.)

and *inferior.* In connection with the *superior vesical,* find the *obliterated hypogastric artery.*

3. Find also the *middle hæmorrhoidal,* to the rectum, in company with *sympathetic nerves* from the pelvic plexus.

4. Find the *uterine artery* (Fig. 204) passing between the folds of the broad ligament of the uterus.

5. Find the *vaginal arteries,* three or four in number, anastomosing with one another and with the uterine.

6. Trace the *ilio-lumbar* between the psoas magnus and vertebral column ; at this point see it divide into an iliac and a lumbar branch.

7. Trace out the *lateral sacrals* to the spine and front of the sacrum.

8. Follow the *superior gluteal* out of the pelvis above the pyriformis, through the greater sacro-sciatic foramen, with a nerve of the same name.

9. Follow the *obturator vessels* below the obturator nerve through the obturator foramen to the adductor muscles.

10. Observe the *internal pudic* leave the pelvis by the greater sacro-sciatic foramen below the pyriformis, cross the ischial spine, reenter the pelvis by the lesser sacro-sciatic foramen, and pass through Alcock's canal to supply the external organs of generation in both the male and female. This artery is attended in its course by the *internal pudic nerve.*

11. The *sciatic artery* is a large and important vessel. It crosses the pyriformis muscle and sacral plexus and leaves the pelvis by the greater sacro-sciatic foramen below the pyriformis muscle, in company with the internal pudic and sciatic nerves. The artery lies under the gluteus maximus muscle and on the external rotation of the thigh. It gives off the inferior gluteal artery, which anastomoses with the superior, and also branches to the pelvic floor and pelvic viscera; branches to the rotator muscles, and the arteria comes nervi ischiadici, a branch that accompanies the great sciatic nerve. This artery completes the crucial anastomosis, an important collateral circulation about the hip-joint, by inosculating *below* with the first perforating branch of the profunda, with the internal circumflex *internally*, and with the external circumflex externally.

The sympathetic distribution in this region will correspond to the visceral branches given off from the internal iliac artery. The bladder will receive its supply by the vesicals; the uterus by the uterine; the ovary by the ovarian; the rectum by the hemorrhoidals—superior, middle, and inferior; the urethra and erectile tissue of penis and clitoris by the internal pudic. Remember the order: (1) *nerve*; (2) *plexus*; (3) *ganglion*. All trace their sympathetic influence to that part of the sympathetic located in the pelvis, by the side of the bladder, rectum, vagina, and uterus—the *pelvic plexus.*

The Obturator Internus.—Cut through the levator ani muscle the whole length of the white line, from spine of ischium to pubic bone; you may now look down into the ischio-rectal fossa, and see the obturator internus muscle covered by the obturator fascia, forming the outer wall of this fossa; you can now appreciate the origin of the levator ani from the white line and the cut segment of the muscle, and study both the inner and outer walls of the ischio-rectal fossa. (Fig. 192.) You will also detach the *pyriformis*, at its origin, from the anterior surface of the sacrum. Observe it passing through the greater sacro-sciatic foramen. You will also see the *coccygeus muscle* extending from the spine of the ischium to the side of coccyx. Now find on the outer wall of the ischio-rectal fossa, in Alcock's canal, the *internal pudic structures*; trace the same forward and see the relation they bear to the tuber ischii and the triangular ligament, where they perforate the posterior layer of this latter, to enter the deep perineal space.

The ischio-rectal fossa is bounded by the obturator internus muscle externally, by the levator ani internally; its apex is at the white line; it extends from the pubic bone in front to the sacrum behind; the base is formed by the skin and fasciæ. When you dissect the pelvic outlet, you approach this fossa from the base. In the present dissection, however, you approach the fossa from above. When you cut through the white line, you opened into the fossa. You then pulled the levator ani muscle to the mid-line: this levator ani muscle with its investing anal fasciæ, above and below, is the inner wall of the fossa, and you are thus permitted to see the outer wall of the fossa—the obturator internus muscle—covered by the obturator fascia.

Iliacus Muscle.—Detach this muscle from its extensive origin in the iliac fossa, and find under the same the iliac branches of the ilio-lumbar artery—a

branch of the internal iliac artery.   Find the groove between this muscle and
the psoas, containing the anterior crural nerve.

Dissection of Sacral Plexus (Figs. 193 and 194).—Cut through the symphysis
pubis and pull outward on both pubic bones, and separate the ilium from the
sacrum, at the sacro-iliac synchondrosis.   Then study the following points :

    1. The plexus is located on the posterior pelvic wall.
    2. It rests on the pyriformis muscle.

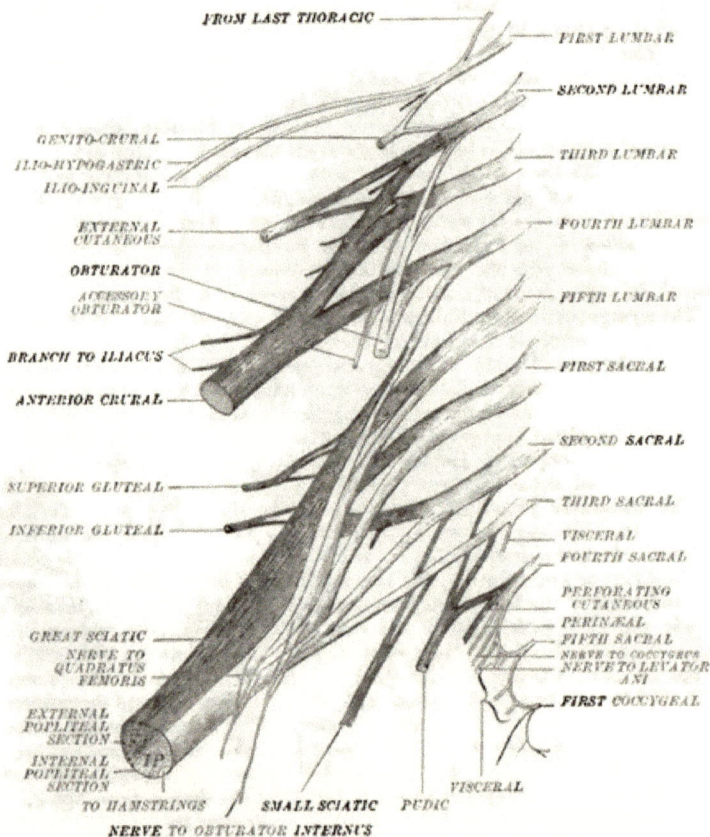

FIG. 193.—DIAGRAM OF THE LUMBAR AND SACRAL PLEXUSES.   Modified from Paterson.

    3. It is covered by the pelvic fascia and peritoneum.
    4. It is crossed by the internal iliac vessels.
    5. The gluteal and sciatic arteries transfix it.
    6. It receives the fifth and part of the fourth lumbar nerves.
    7. The part received from the lumbar is called the *lumbo-sacral cord.*
    8. The *sacral plexus* is formed by the union of the lumbo-sacral cord and
the anterior primary divisions of the first, second, third, and a part of the fourth
sacral nerves.

9. This union is called sacral plexus to the lower margin of the greater sacro-sciatic foramen, where it continues its course under the name of great sciatic nerve.

10. The *sacral nerves* all send rami communicantes to the pelvic sympathetic plexus, which you have already found, located to the inner side of the anterior sacral foramina, by the side of the rectum and bladder.

Dissection : (1) Trace the first, second, and third sacral nerves as far into the sacral foramina as possible. Notice! This must be done with a blunt forceps,

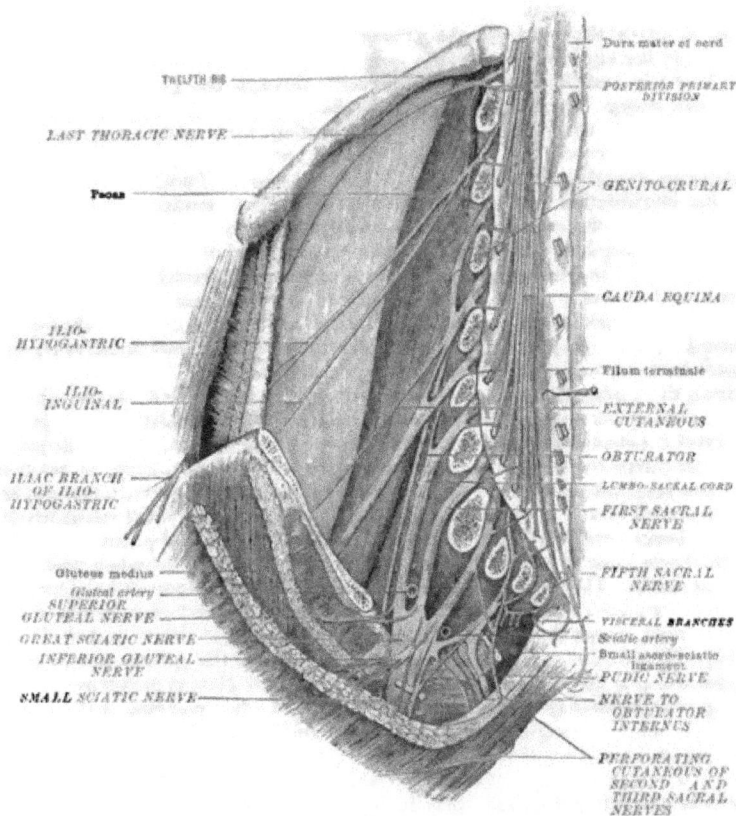

FIG. 194.—A DISSECTION OF THE LUMBAR AND SACRAL PLEXUSES, FROM BEHIND
(The anterior crural nerve is placed between the external cutaneous and obturator nerves.)

and with a very delicate touch, to avoid injuring the small branches given off to the sympathetic—pelvic part of the gangliated cord. After this is done, all the fat must be dissolved by ether, benzene, or gasoline. (2) Now dissect in the direction of the branches given off, and find the following :

BRANCHES OF THE SACRAL PLEXUS.

1. *The lumbo-sacral cord* is the communication between the sacral and lumbar plexuses. It is formed by the fifth lumbar and part of the fourth lumbar—

10

anterior primary divisions. This must now be located on the ala of the sacrum as it passes down into the pelvis, and its union with the first, second, and third sacral nerves carefully shown.

2. *Muscular branches* to the pyriformis, the quadratus femoris, the gemellus superior, the gemellus inferior, and the obturator internus muscles. Notice! You will find the branch to the obturator internus, leaving the pelvis by the greater and reentering by the lesser sacro-sciatic foramen. When you dissect the structures in the lesser sacro-sciatic foramen, you will find this nerve with the internal pudic structures. The nerve to the quadratus femoris and gemellus inferior leaves the pelvis by the greater sacro-sciatic foramen, and supplies a branch to the hip-joint as well.

You will trace the *superior gluteal nerve* through the greater sacro-sciatic foramen, above the pyriformis muscle, to the gluteus medius, gluteus minimus, and tensor vaginæ femoris. The skin covering these muscles is not supplied by this nerve-trunk apparently, but by the ilio-hypogastric and lumbar cutaneous branches from posterior primary divisions. Now, let me impress on the student this fact: A nerve-trunk supplying a muscle must supply the skin covering the muscle, or the serous membrane under the muscle,—if any serous membrane be present,—and the joint moved by the muscle. The skin branches, in the majority of cases, are given off directly from the nerve-trunk, and are easily found. In a minority of cases, however, the skin branch is not given off directly, but comes from *apparently* another source; to such cases as these I see no impropriety in applying the terms *divorced, dismembered,* or *erratic.* Confirmatory instances of divorced skin branches are to be found in the small sciatic, the dismembered cutaneous branch of the great sciatic, supplying the skin covering the muscles supplied with motion by the great sciatic nerve; in the external cutaneous nerve, on the outer part of the thigh, this nerve being divorced from its parent stem, the anterior crural; in the long pudendal nerve, or nerve of Soemmering, the dismembered branch of the internal pudic nerve, following, indeed, a most erratic course, ensconced in the sheath of the small sciatic nerve; in the descendens hypoglossi, a divorced branch of the cervical plexus. In view of the occurrence of anomalies, then, you are to ask yourselves, when dissecting mixed nerves Have I found (1) muscular branches, (2) cutaneous branches, (3) articular branches? and (4) are the cutaneous branches direct descendants of the nerve-trunk, or are they divorced therefrom?

3. *The inferior gluteal nerve* leaves the pelvis by the greater sacro-sciatic foramen below the pyriformis muscle and supplies the gluteus maximus.

4. *The perforating cutaneous nerve* will be found when you dissect the gluteal region, coming through the greater sacro-sciatic ligament to supply the skin in this region.

5. *The internal pudic nerve* is remarkable for the great number of important areas in which it and its branches are found. It leaves the pelvis by the greater sacro-sciatic foramen below the pyriformis muscle, crosses the ischial spine, reenters the pelvis by the lesser sacro-sciatic foramen, here enters Alcock's canal, in the outer wall of the ischio-rectal fossa, then pierces the triangular ligament, passes through the deep perineal space, again pierces the triangular ligament, passes through the suspensory ligament of the penis to gain the dorsum of this organ. In its course it throws off the following branches: As the nerve is entering the canal—Alcock's—it gives off the inferior hemorrhoidal and the perineal branches. These pass through the fat-bearing ischio-rectal fossa. The inferior hemorrhoidal nerve is distributed to the external sphincter ani muscle and the skin covering the same. The perineal branch is distributed to the accelerator urinæ, transversus perinæi, erector penis, levator ani, compressor urethra, and to the skin covering the scrotum and perineum. The internal pudic nerve

now continues its course in Alcock's canal, and through the deep triangular perineal space, under the name of dorsal nerve of the penis, to which, with the sympathetic from the pelvic plexus, it is distributed.

6. *The small sciatic* leaves the pelvis by the greater sacro-sciatic foramen below the pyriformis, and breaks up into four principal cutaneous branches—ascending, descending, external, and internal. The largest of the internal branches crosses the tuberosity of the ischium, and is distributed to the labia majora under the name of inferior pudendal, or nerve of Soemmering.

7. *The great sciatic nerve* takes its name at the lower border of the greater sacro-sciatic foramen, through which you see it leaving the pelvis below the pyriformis muscle. Examine closely and you can see some very small branches. These are articular to the hip-joint. Lower down are given off muscular branches to the biceps semitendinosus, semimembranosus, and adductor magnus muscles. As a rule, the nerve divides six inches above the knee into the internal and external popliteal nerves.

Branches of the Internal Popliteal Nerve of Great Sciatic :

1. *The communicans tibialis* to the communicans fibularis, a branch of the external popliteal to form by this union the short saphenous nerve. This nerve supplies the skin over the superficial group of muscles of the leg.

2. *Articular branches* to the knee, with the articular arteries, superior and inferior internal, and the azygos.

3. *Muscular branches* to the gastrocnemius, soleus, plantaris, and popliteus.

4. *Posterior tibial*, which gives off an articular branch to the ankle ; a plantar cutaneous branch to the heel and sole of foot on the inner side ; muscular branches to the flexor longus digitorum, tibialis posticus, flexor longus hallucis ; a communicating branch to the soleus muscle.

5. *The internal plantar*, which gives off articular branches to the tarsus and metatarsus, digital cutaneous branches to the four lesser toes, muscular branches to the abductor hallucis, flexor brevis digitorum, flexor brevis hallucis, and the two tibial lumbricales.

6. *External plantar*, that supplies one and one-half toes cutaneously ; muscular branches to flexor accessorius, abductor minimi digiti, the two outer lumbricales, the adductor hallucis obliquus, the adductor hallucis transversus, the flexor brevis minimi digiti, and all the interossei muscles.

Branches of the External Popliteal of Great Sciatic :

1. The *communicans fibularis* to the communicans tibialis to form the short saphenous nerve.

2. *Three articular branches* to the knee, proceeding with the superior and inferior external articular arteries and the azygos.

3. *Cutaneous branches* to the skin covering the peronei muscles, being given off with the communicans fibularis.

4. The *anterior tibial nerve*, which gives off an articular branch to the ankle, to the metatarso-phalangeal joints of all the toes. *Muscular branches* to the tibialis anticus, extensor proprius hallucis, extensor longus digitorum, extensor brevis digitorum, and a muscular branch to the first interosseous muscle.

5. The *musculo-cutaneous*, which gives off its muscular branches to the peroneus longus and peroneus brevis muscles. Cutaneous branches, internal and external, to the dorsum of the foot, communicating with the internal and external saphenous nerves.

1. *Where is the lumbar plexus located ?*

It is behind the peritoneum in the deep substance of the psoas magnus muscle.

2. *How is the lumbar plexus formed ?*

By the union of the anterior primary divisions of the first, second, third, and a part of the fourth lumbar nerves, and the dorsi-lumbar cord.

3. *With what does the lumbar plexus communicate?*

(1) With the twelfth thoracic nerve, by a branch named the dorsi lumbar; (2) with the sacral plexus, by a branch called the lumbo-sacral cord; (3) with the lumbar part of the gangliated cord, by branches called rami communicantes.

4. *Describe the sympathetic connection, and tell how to conduct a dissection of the same in this region.*

The lumbar part of the gangliated cord is quite easily found; the communications with the anterior primary divisions of the lumbar nerves are not difficult to trace out. The double chain of the sympathetic cord consists of four ganglia on each side; these ganglia are: (1) Right, which lie behind the ascending vena cava; (2) left, which lie behind and slightly external to the aorta. Find the cord communicating above with the thoracic, and below with the pelvic portion, by continuity. The rami communicantes are quite long in this region; as a rule, they are two in number, and they accompany the lumbar arteries. The lumbar arteries, then, are your guide in tracing the relation between the lumbar plexus of somatic nerves, and the lumbar part of the sympathetic gangliated cord.

5. *Name the branches of the lumbar plexus.*

The ilio-hypogastric, ilio-inguinal, genito-crural, external cutaneous, anterior crural, obturator, accessory obturator, muscular branches to the psoas magnus and quadratus lumborum, and the lumbar element of the lumbo-sacral cord.

6. *Give formation and location of the lumbo-sacral cord.*

It is formed by the union of the fifth lumbar nerve and a part of the fourth. In practical anatomy you find it buried, in a considerable quantity of fatty connective tissue behind the peritoneum, on the ala of the sacrum. This cord gives origin to the superior gluteal nerve, described previously.

7. *Describe the genito-crural nerve, and give its practical importance in diagnosis.*

The nerve arises from the first and second lumbar nerves. It lies on the psoas magnus muscle in the lower part of its course. It divides into: (1) A genital branch, which follows the spermatic vessels and supplies the cremaster muscle with motion. In practical anatomy you find this branch behind the spermatic cord. (2) A cutaneous branch, the crural, which is distributed to the skin of the upper inner part of the thigh. The cremasteric reflex depends on this nerve. Irritate the skin of the thigh corresponding to the sensory distribution of this, the genito-crural nerve, and the testicle of the corresponding side will be elevated by the contraction of the cremaster muscle. In diagnosis, the movement of the testicle thus produced argues in favor of the integrity of the spinal cord between the first and second lumbar nerves. In the female this nerve is rudimentary, and, there being no cremaster in this sex, the genital branch of the nerve is distributed to the round ligament of the uterus; the crural branch is distributed as in the male. The reflex phenomena in the female are recorded as a twitching of the external oblique; still, I think careful observation would prove such record erroneous. The twitching must be in the round ligament of the uterus, since this is the homologue of the spermatic cord.

8. *Describe a patellar reflex circuit and give its importance in diagnosis.*

The anterior crural nerve is concerned in this reflex. Sensory branches from this nerve, distributed over the insertion of the quadriceps, forming the plexus patellæ, convey sensation to a transfer centre; and motor branches from the anterior crural induce contraction of the extensor group of muscles on the front of the thigh. This reflex movement is normal in health. It is absent in locomotor ataxia and in the case of lesions and diseases affecting the anterior gray cornua of the spinal cord. It is increased after epileptic seizures, in spinal irritability, tumors of the brain and diseases of the lateral tracts of the cord, and in lateral and cerebro-spinal sclerosis.

9. *Explain the technique of obtaining knee-jerk, or patellar reflex movement.*

(1) Place the patient in the sitting posture on the table, with his legs hanging at right angles to the thighs ; (2) blindfold the patient, and do not acquaint him with the procedure ; otherwise, through nerve-influence from the obturator to the sartorious muscle, phenomena of a voluntary nature on the part of the patient might be observed which would cloud your diagnosis ; (3) tap the ligamentum patellæ with a ferrule, or, better still, gently prick the skin over the same with a sharp instrument. Students should be encouraged to practice obtaining these reflexes on one another.

DIAPHRAGM. (Fig. 195.)

Function.—Partition between thorax and abdomen.
Superior Serous Relations.—Pericardium and pleuræ.
Inferior Serous Relations.—The diaphragmatic peritoneum.
Structure.—A musculo-membranous sheet.
Apertures.—Œsophageal, aortic, and caval.
Ligamentum arcuatum internum, part of iliac fascia.
Ligamentum arcuatum externum, part of lumbar fascia.
Central tendon forms the summit of the dome.
Œsophageal opening transmits œsophagus and vagus nerves.
Aortic opening transmits aorta and thoracic duct.
Caval opening transmits the ascending vena cava.
Physiological Action.—Deepens the chest.
Nerve-supply.—The two phrenics and the sympathetic.
Source of Sympathetic Nerve-supply.—Solar plexus.
Blood-supply.—The phrenic arteries.
Source of Phrenic Arteries.—Aorta, renals, and intercostals.
How many origins has the diaphragm?

1. *Anterior or Sternal Portion.*—The lower border and back of the ensiform cartilage and the adjacent part of the back of the anterior aponeurosis of the transversalis abdominis.

2. *Lateral or Costal Portion.*—The lower border and inner surface of the cartilages of the six lower ribs, and sometimes also from the adjacent part of the ribs.

3. *Posterior or Vertebral Portion.*—(1) The ligamentum arcuatum externum, a fibrous thickening of the anterior layer of the lumbar fascia, which stretches from the tip of the transverse process of the second lumbar vertebra to the tip of the last rib ; (2) the ligamentum arcuatum internum—a fibrous thickening of the iliac fascia, which arches over the upper part of the psoas from the side of the body of the second lumbar vertebra to the tip of its transverse process ; (3) the crus of the diaphragm—a strong vertical band, fleshy externally, tendinous internally, arising on the right side from the front of the bodies of the first to the third or fourth lumbar vertebræ, from the intervening vertebral discs, and the anterior common ligament ; on the left side, from the bodies of the first to the second or third vertebræ only, as well as the discs and anterior common ligament.

Insertion.—The front, sides, and back of the central tendon.

The crura are perforated, and transmit structures as follows : The right, the sympathetic and the splanchnics of the right side ; the left, the splanchnics of the left side and the vena azygos minor.

Visceral Relations.—On careful dissection you will see the diaphragm bears very important relations to the following structures : (1) As previously indicated, the superior surface of the diaphragm is occupied by the three serous membranes. The *pulmonary* surface of the diaphragm corresponds to the base of the lung. The cardiac surface corresponds to the base of the pericardium. (2) The *liver, stomach, spleen, kidneys,* and *suprarenal capsules* are in relation with its under surface. (3) The *aorta, œsophagus, ascending vena cava,* the *vena azygos minor,* the *pneumogastric nerves,* the *splanchnics,* the *sympathetic*—all pass through the diaphragm. (4) Posteriorly, the diaphragm arches over the psoas magnus and the quadratus lumborum. A fibrous arch, formed by the iliac fascia, arches over the psoas ; one formed by the anterior layer of the lumbar fascia arches over the quadratus lumborum muscle. Consult carefully figure 195. Lastly, remove the peritoneal covering of the diaphragm, and see the beautiful manner in which the fibers from the three different sources of origin approach for insertion the

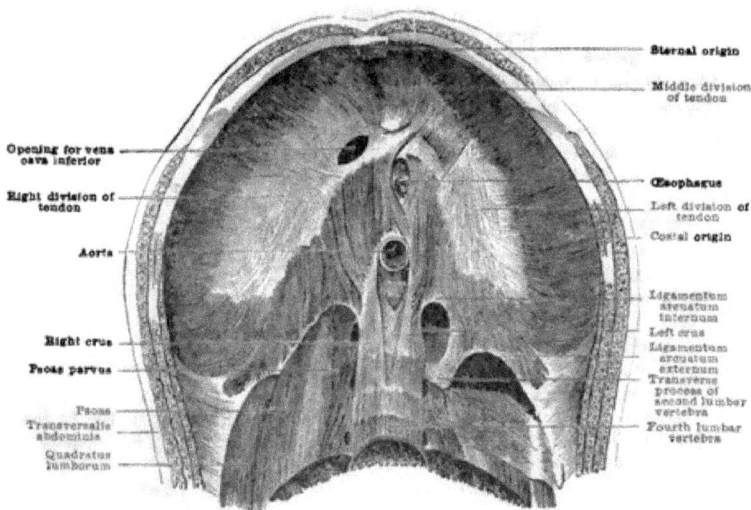

FIG. 195.—DIAPHRAGM.

central tendon. The central tendon corresponds to the base of the pericardium ; the fibrous part of the pericardium is the downward prolongation and expansion of the third layer of the deep cervical fascia, having firm bony attachments to the base of the skull ; hence, in deep inspiration it is not the central tendon that becomes *depressed,* but the *muscular part of the diaphragm.*

Dissection of Deep Muscles and other Structures of the Pelvis.—(1) The quadratus lumborum ; (2) psoas parvus ; (3) psoas magnus ; (4) iliacus ; (5) crura of diaphragm ; (6) anterior common ligament of vertebral column ; (7) obturator internus ; (8) obturator externus ; (9) obturator membrane ; (10) pyriformis ; (11) coccygeus ; (12) levator ani ; (13) white line of the obturator fascia.

Locate on your cadaver: (1) The iliac fossa; (2) the auricular surface of the ilium; (3) the tuberosity of the ilium; (4) the tuberosity of the ischium; (5) the spine of the ischium; (6) the spine of the pubes; (7) the ilio-pubal ridge; (8) the ilio-pectineal line; (9) the anterior and posterior surfaces of the pubes; (10) the ischio-pubic ramus; (11) the thyroid or obturator foramen; (12) the symphysis pubis; (13) the anterior superior and anterior inferior iliac spines;

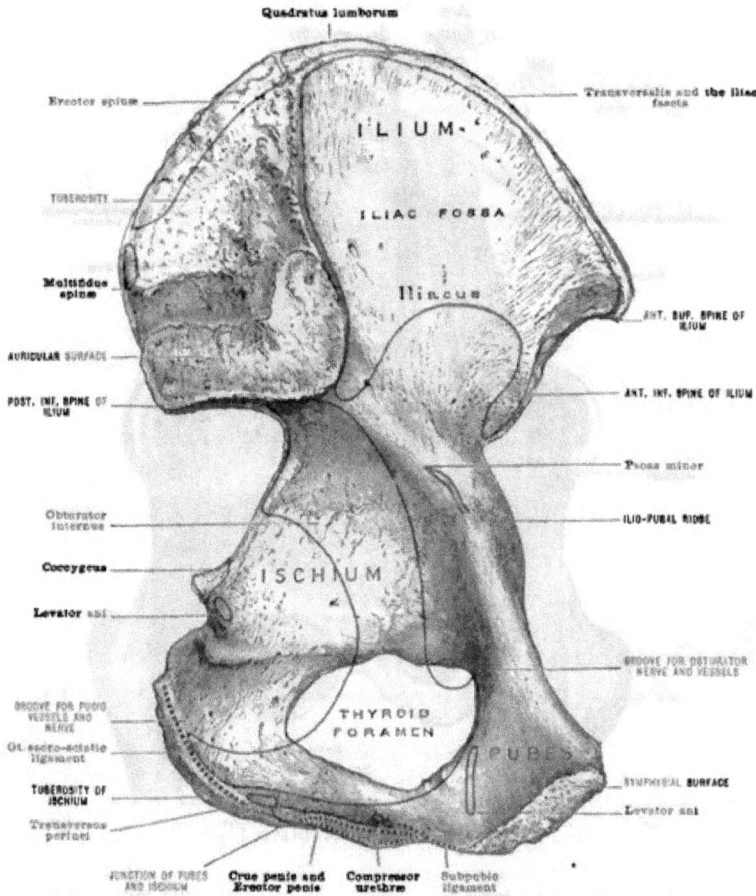

FIG. 196.—THE LEFT HIP-BONE. (Internal surface.)

(14) the posterior superior and posterior inferior iliac spines; (15) the greater and lesser sacro-sciatic foramina; (16) the greater and lesser sacro-sciatic ligaments; (17) the obturator groove. Locate the above points on the cadaver, and in the progress of your dissection find and dissect the soft structures attached to them or in any way bearing important relations thereto. Study thoroughly the os innominatum.

Dissection.—**Quadratus Lumborum.**—Find the origin on the ilio-lumbar ligament and the iliac crest for two inches. Trace the muscle to its insertion into the lower border of the twelfth rib and into the transverse processes of the four upper lumbar vertebræ. Now place the kidney in its original position on this muscle. This muscle lies in front of the erector spinæ muscle. In fact, these two muscles form a guide to operations on the kidney. See whether you can demonstrate on your work these structures, in the following relation from behind forward : (1) The *quadratus lumborum ;* (2) the *lumbar fascia ;* (3) the *last thoracic nerve*, the *ilio-hypogastric* and *ilio-inguinal nerves ;* (4) the *fatty capsule of the kidney ;* (5) the *kidney.* In operations on the kidney from behind,

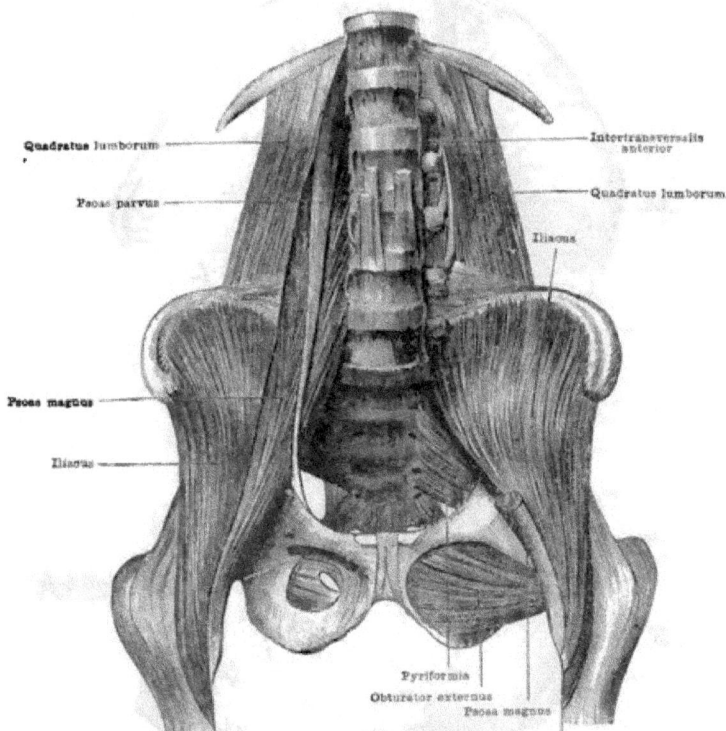

FIG. 197.—PSOAS, ILIACUS, AND QUADRATUS LUMBORUM.

you would find the above relations. Study the above relations until you have a perfect picture of all these structures in your mind. The lower end of the kidney is about one and one-half inches above the iliac crest.

The psoas magnus muscle in its relation to vessels and nerves has been previously described. The muscle has a strong synergist in the iliacus. These two are sometimes spoken of as one muscle, with two heads or parts—the *iliopsoas.* The two muscles are inserted into the lesser trochanter of the femur. By this action, when the thigh is the fixed point, they draw the pelvis forward, rotating the same on the heads of the femur. They thus antagonize the gluteus maximus. The psoas magnus *arises* from the sides of the bodies of the last thor-

acic and all the lumbar vertebræ, and from their intercentral cartilages ; from
the lower border of the transverse processes of the lumbar vertebræ. The
nerve-supply is from the lumbar plexus. The *nerve-supply* of the iliacus is from
the anterior crural part of the lumbar plexus.

The Iliacus.—Detach from its origin and turn it down to its insertion into
the lesser trochanter of the femur with the psoas magnus. This is a flexor of
the thigh on the abdomen. Its synergist is the psoas. It also pulls the body
forward, when the thigh is made the fixed point, antagonizing the gluteus maximus. Its nerve-supply is from the anterior crural of the lumbar plexus.

Origin.—(1) The upper surface of the ala of the sacrum ; (2) the front of the
ilio-lumbar, lumbo-sacral, and anterior sacro-iliac ligaments ; (3) the upper and
outer half of the venter of the ilium ; (4) the origin of the upper tendon of the
rectus femoris and the ilio-femoral ligament near the anterior inferior spine of
the ilium.

The Psoas Parvus.—Usually absent. Inserted into the ilio-pectineal line.
Origin : Side of twelfth thoracic and first lumbar and disc of cartilage between

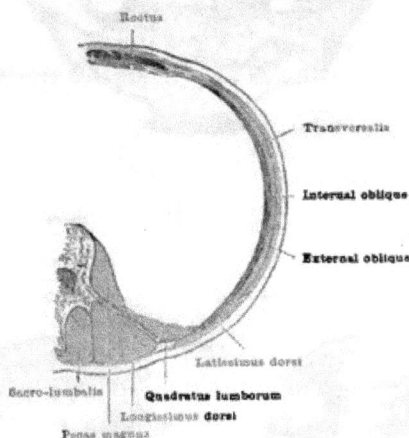

FIG. 198.—ARRANGEMENTS OF LUMBAR APONEUROSIS AT LEVEL OF THIRD LUMBAR VERTEBRA.

the two. Synergistic to psoas magnus. Its *nerve-supply* is from the first nerve
of the lumbar plexus.

The obturator internus may now be removed from its origin. (Fig. 196.)
Trace the same through the lesser sacro-sciatic foramen. Study the origin from
the margin of the obturator foramen, the obturator membrane, and the inner
surface of the body of the ischium. Its *nerve-supply* is from the first and second
nerves of the sacral plexus ; it passes through the lesser sacro-sciatic foramen.

Obturator Externus.—Detach the muscle from the bony margin of the
obturator foramen ; from the obturator membrane. It is inserted into the digital
or trochanteric fossa. It is supplied by the obturator nerve. Now you can see
and study the obturator membrane. See the obturator nerve coming through a
notch on under surface of the horizontal ramus of the os pubis.

Locate on the sacrum (Fig. 199) : (1) The bodies ; (2) the anterior sacral foramina ; (3) the transverse lines ; (4) the ala or wing ; (5) the sacro-coccygeal articulation ; (6) the origin of the pyriformis ; (7) the grooves leading to the foramina.

The pyriformis arises from the anterior surface of the sacrum by three

digitations. (Fig. 199.) It leaves the pelvis by the greater sacro-sciatic foramen, dividing this into an upper and a lower compartment, and is inserted into

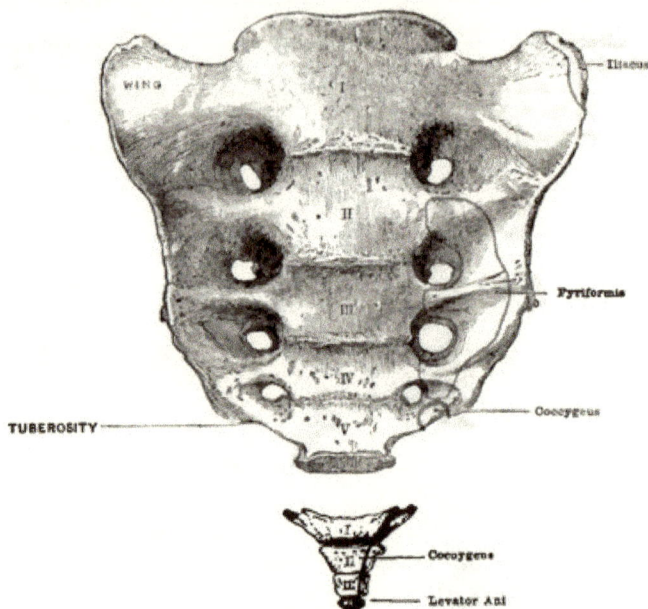

FIG. 199.—THE SACRUM AND COCCYX. (Anterior view.)

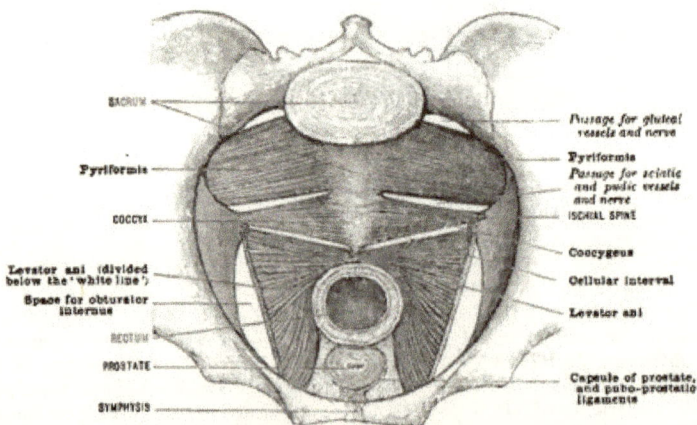

FIG. 200.—MUSCLES OF THE FLOOR OF THE PELVIS.

the upper border of the *greater trochanter.* It is an external rotator of the
thigh. Its *nerve-supply* is from the sacral plexus.

The coccygeus extends from the spine of the ischium to the anterior surface and side of the coccyx. It separates two intermuscular cellular spaces—(1) one between itself and the pyriformis, and (2) one between itself and the levator ani. (Figs. 199 and 200.) Its *nerve-supply* comes from the coccygeal plexus.

The levator ani (Fig. 200) arises from the back of the pubic bone and from the inner surface of the ischial spine. Between these two points it arises from the white line. The muscle has a triple insertion : (1) Into the tip of the coccyx ; (2) into the central point of the perineum ; (3) into the rectum. In the female the anterior fibres are inserted into the side of the vagina. *Function :* It draws the coccyx forward, elevates the floor of the pelvis, and aids in compressing the pelvic viscera.

The pelvic white line is formed by a delamination of the obturator fascia. It extends from the spine of the ischium to the pubic bone. (Fig. 200.) It gives partial origin to the levator ani muscle.

— · —

FEMALE GENERATIVE ORGANS.

Now thoroughly cleanse the pelvis and study the internal organs of generation of the female as follows :

1. Observe in front of the uterus, the bladder, partially covered by peritoneum. In front of the bladder you will see a cellular space called Retzius' space.

2. Pull the uterus to one side and study the relation of the following struc-

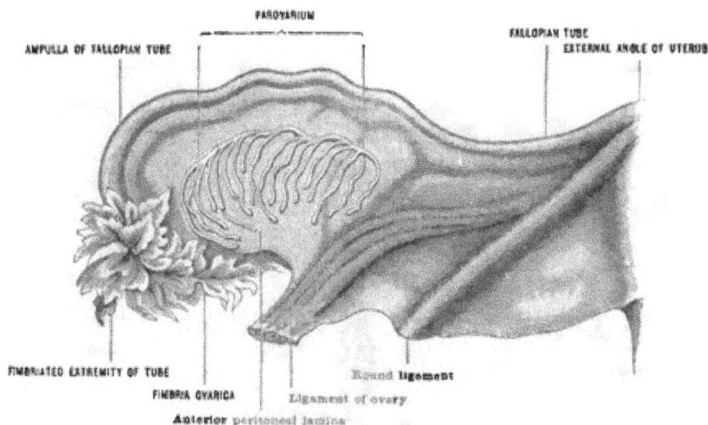

FIG. 201.—THE BROAD LIGAMENT AND ITS CONTENTS, SEEN FROM THE FRONT. (After Sappey).

tures : (1) The *round ligament of the uterus ;* (2) the *ligament of the ovary ;* (3) the *Fallopian tube,* or oviduct.

As you pull the uterus to the opposite side, you will notice these three structures dispersing to different parts from the angle of the uterus. Trace them out, remembering they are all between the two layers of the broad ligament (Fig. 201), as follows : (1) The round ligament to the internal abdominal ring, on the outer side of the deep epigastric artery, down through the inguinal canal, out through the external abdominal ring, to the labia majora ; (2) the ovarian ligament terminates in the capsule of the inner end of the ovary ; (3) trace the oviduct—*i. e.*, the Fallopian tube—to the outer end of the capsule of the ovary, to which it is attached by one of the fimbriæ, called the ovarian fimbria.

3. *Locate the ureter* as it passes down over the brim of the pelvis ; trace it

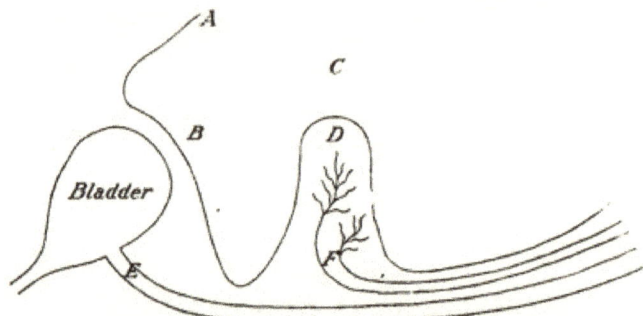

FIG. 202.—REPRESENTING SCHEMATICALLY THE RELATION OF THE URETER TO THE BROAD LIGAMENT OF THE UTERUS.

A. Peritoneum. *B.* Peritoneum in relation to bladder. *C.* Peritoneum pushed ahead of the uterus—called broad ligament. *D.* Space between the two layers of broad ligament containing uterus and adnexa. *E.* The ureter. *F.* Uterine and ovarian vessels.

carefully as it approaches the posterior surface of the broad ligament. (Fig. 202.) Now, as it passes under, cut a hole in the posterior surface of the broad ligament and follow it across the base and to the bladder. Notice, too, that as you take hold of the broad ligament and lift it up, the ureter remains in place. There would seem to be little danger of including this structure in the ligature in opera-

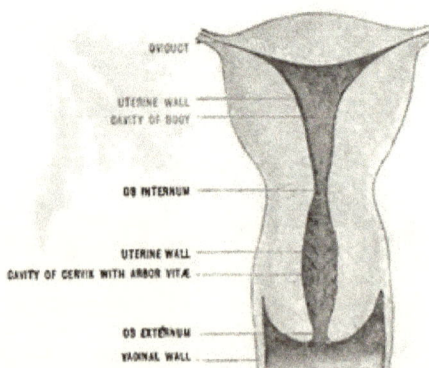

FIG. 203.—FRONTAL SECTION OF THE VIRGIN UTERUS. (After Sappey.)

tions on the ovary and tube. As the ureter lies on the levator ani, there would be more danger of wounding it in operations on the ischio-rectal fossa than in the operation before mentioned. The ureter remains firmly embedded in the deep layer of subperitoneal connective tissue, and can not be raised by any ordinary amount of traction exerted on the ovary, tube, and broad ligament.

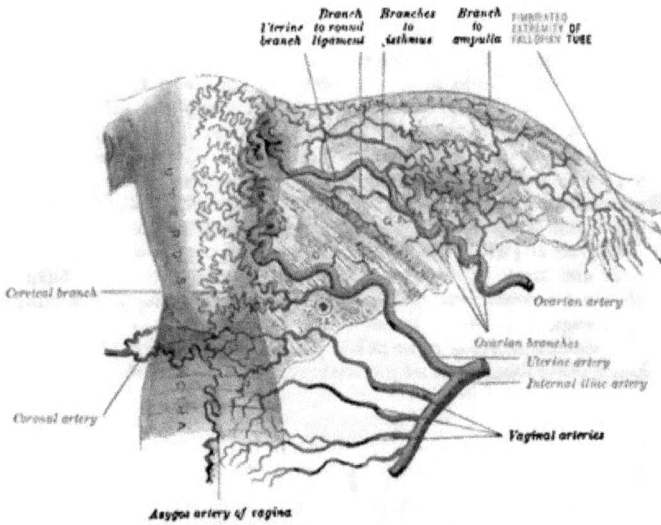

FIG. 204.—SCHEME OF THE OVARIAN AND UTERINE AND VAGINAL ARTERIES.

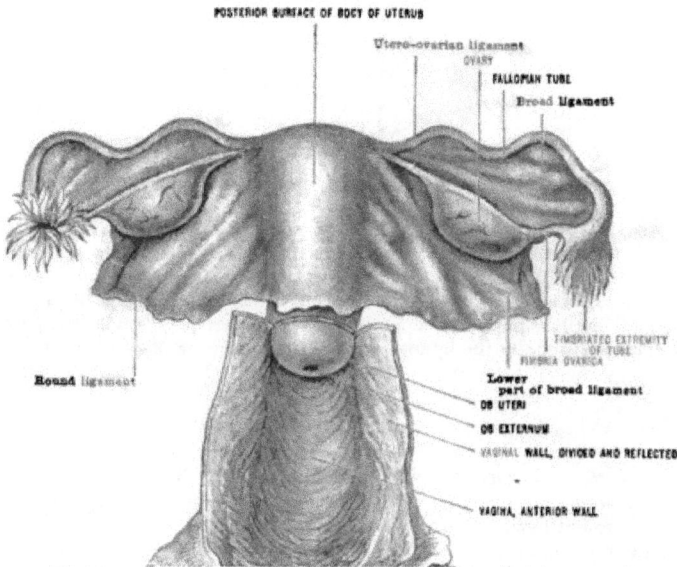

FIG. 205.—THE FEMALE ORGANS OF GENERATION. (Modified from Sappey.)
(Vagina divided and laid open behind.)

The Ovarian Artery and the Sympathetic Nerve.—This artery you will see shining through the peritoneum, passing behind the ureter, opposite the bifurcation of the aorta. You will remember its origin from the aorta, below the renal, and its homologue, the spermatic. Trace it down in this manner : Lift the peritoneum up, over the artery, and cut the same with the scissors. Make every effort not to disturb the artery in its bed. Trace the artery now to the lateral space, between the two folds of broad ligament (Fig. 202), to its distribution and anastomosis, as shown in figure 204. You will find the uterine artery embedded in the side of the uterus, between the two folds of the broad ligament. Find the artery, and trace it out to its origin from the internal iliac artery. (Fig. 204.) You will find an abundant supply of sympathetic nerves forming the uterine plexus. Figure 203 gives a frontal section of the uterus, which you may now imitate. In the angle of the uterus find the opening for the Fallopian tube. Locate the os internum and externum. Figure 206 is a sagittal section. Cut through the vagina and study the utero-vaginal junction.

Remember the analysis of the pelvic plexus. Formed : (1) by prolongation of the hypogastric plexus ; (2) by branches from the second, third, and fourth

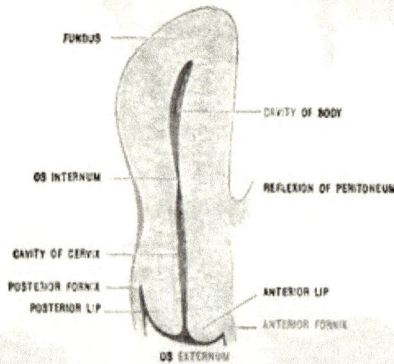

FIG. 206.—SAGITTAL SECTION OF THE VIRGIN UTERUS. (After Sappey.)

sacral nerves ; (3) by branches from the two upper sacral ganglia. Located : By the side of the rectum, vagina, and uterus in the female. Distribution : To all the viscera in the pelvis. The nerves accompany the branches of the internal iliac artery. The nerves proceed from a plexus on the artery ; the plexus proceeds from a ganglion ; the ganglion is supported by the pelvic plexus ; the pelvic plexus is in communication with the hypogastric plexus through a prolongation of this latter.

---

1. *Name the ligaments of the uterus you have thus far found, and explain their derivation.*

(1) The round ligament ; (2) one in front, called the anterior or vesico-uterine ; (3) one posterior, the recto-uterine ; (4) two sacro-uterine ; (5) two broad ligaments. The round ligament that in your dissection you traced through the inguinal canal, is lost in the labia majora , it is homologous to the spermatic cord. It is composed of fibrous and muscular tissue from the uterus. The cord is attended by a hood of peritoneum called the canal of Nuck. To see

this canal before it becomes obliterated, you must examine female fœtuses. The other uterine ligaments are of peritoneal origin, and can and must be demonstrated on your work.

2. *From what source does the uterus receive its blood?*

Through the uterine branch of the internal iliac artery. It anastomoses above with the ovarian and below with the vaginal arteries. It is surrounded by the uterine plexus of nerves, from the pelvic plexus. (Fig. 204.)

3. *Explain the rationale of contraction of adhesions in retroflexions.*

In the dissection of the peritoneum it was shown that an organ previously invested by peritoneum, may lose its peritoneum, *in parte* or *in toto*, by loss of the blood-supply at a given point, by pressure. (Fig. 170.) The uterus bears the same relation to its investing peritoneum that the colon does to its own. Things equal to the same thing are equal to each other; the two results are identically the same. The colon, by loss of its specific epithelial element, becomes physiologically adherent to the parietes, because this adhesion will cause no ulterior hindrance to the function of the colon. The uterus, however, contracts pathological adhesions, because the uterus can not, in the position these adhesions entail, discharge its function physiologically.

4. *Does the uterus undergo important structural changes?*

Anatomical structure is always the correlative of function. Not only the uterus, but all its appendages, obey this law of philosophy—correlation of structure and function. In the pre-puberty state, the female genitalia are products of growth, not of development; in the post-puberty state they are the products of both growth and development. During menstruation and pregnancy, the uterus and its appendages become relatively changed in size according to function. At the menopause, with cessation of functional activity, these parts become metamorphosed, in a retrograde manner.

5. *Explain the anatomical factors in the production of ovarian, uterine, rectal, and urethro-vesical reflexes—in short, of pelvic reflexes in general and particular.*

This belongs to physiology, to which chair I refer a detailed answer; but in order to make you familiar with the anatomical elements involved in reflex phenomena, I know you will pardon just enough physiological digression to keep up an interest in the anatomical regions involved.

Reflexes are of two general kinds: (1) simple; (2) complex. The simplicity of the one depends on the positive conductivity of somatic nerves and the simplicity of the factors involved.

The complexity of the other depends on the negative conductivity of sympathetic nerves and the complexity of the factors involved.

The factors involved are: (1) A nerve-center capable of receiving sensory impressions from sentient areas, and capable of originating motor impulses; (2) connecting media between the center and the sentient area capable of transmitting impressions of both sensation and motion.

Somatic nerves are those that supply the body-wall; they are usually called spinal and cranial. Those nerves are called sympathetic that are set aside principally for the regulation and movement of the viscera. They are the visceral branches of the spinal nerves. The former transmit pain and motion violently, rapidly, positively; the latter slowly, slight in degree, negatively, compared to the same in somatic areas. (See page 264, *et seq.*)

Law of Projectiles.—In addition to the central and peripheral factors, and their motor and sensory communicating media, we must be mindful of the law of

projectiles. The nerves to and from the brain are media by which the impressions travel. These impressions, then, and the nerves by which they travel, are projectiles, and are amenable to the law of philosophy governing all projectiles : (1) A projectile follows the line of least resistance ; (2) a projectile follows the point of greatest traction ; (3) a projectile may follow the resultant of these two —i. e., the line of greatest traction and the point of least resistance.

Anatomical factors determining the above are simply the amount and kind of nerve distribution at the locality whence the reflex proceeds. On this hypothesis alone do we account for the reference of both pain and motion, in their logical places, in simplex reflexes. In figure 187, a simple reflex, the logical place for the pain is the conjunctiva ; the logical place for the motion is in the orbicularis palpebrarum. Here the influences, sensory and motor, follow the greatest direct nerve distribution—over cranial nerves having maximal ability to transmit both sensation and motion. On this hypothesis do we account for pain, the result of an irritant,—for example, in the ovary, rectum, or uterus,—not in its logical place in the organ itself, but in its illogical place, far away in some other part of the body, where both reflex pain and reflex motion may be the symptoms. These are complex reflexes ; still, just as logical in every way and just as amenable to the law of projectiles as simple reflexes. (Fig. 185.)

Pain is diagnostically one of the most important subjective symptoms. It is reported peripherally. You must, then, have a thorough knowledge of the cutaneous and membranous distribution of somatic nerves, since it is through these nerves that pain from every organ in the body may make its demurrer. The steps, then, anatomically, in tracing pain for diagnostic purposes are :

1. What cutaneous or sensory nerves supply the part ?
2. Of what mixed nerve or plexus are they a part ?
3. Where do these mixed nerves leave the spinal canal ?
4. Do they correspond to the six upper intercostals ?
5. Do they correspond to the six lower intercostals ?
6. Do they correspond to the cervical plexus ?
7. Do they correspond to the brachial plexus ?
8. Do they correspond to the lumbar plexus ?
9. Do they correspond to the sacral plexus ?
10. What territory of organs does the hypogastric plexus supply ?
11. What territory does the solar plexus supply ?
12. What territory does the cardiac plexus supply ?

Any considerable pain, then, as previously stated, is reported by a somatic nerve. If this somatic nerve is the innervation pure and simple of the painful part under consideration, then the pain is a simple direct pain. If, however, as frequently happens, the somatic nerve is the medium through which vast territories are made to suffer, while the exciting cause is in some remote organ, then this pain is a reflex pain. The somatic nerves may also produce reflex muscular movements in the muscles corresponding to the region of the pain. (See page 264.)

The student can not become too well grounded in the philosophical distribution of a mixed nerve : The nerve-trunk that supplies a group of muscles supplies also the skin covering those muscles ; the articulation that these muscles move ; the serous or synovial membrane in contact with which these muscles may lie. The six upper intercostal nerves supply, then, not only intercostal muscles, but the pleura as well ; the six lower supply not only the muscles of the abdominal walls, but the peritoneum also. An organ, then, having a sympathetic nerve-supply, may manifest its pain either in skin, membranes, or articulations, and still not transcend the logic of the law of reflex phenomena.

THIGH AND LEG.

The following review of the practical osteological points on the bones of the lower extremity must be thoroughly mastered before the student can do intelligent work on the cadaver. These points the student must study with the bones in hand. Simply committing to memory a few pages of technical names, with no knowledge of a practical application of the subject, would be as useless here as in other departments of science.

The Tibia :

Name the bony points on the tibia of practical importance in dissection.

The internal tuberosity has one articular surface.

The external tuberosity has two articular surfaces.

The tubercle, into which the ligamentum patellæ is inserted.

The crest, or anterior border, is subcutaneous.

The internal surface is three-fourths subcutaneous.

The upper two-thirds of the external surface is occupied by tibialis anticus.

The popliteal notch transmits the popliteal artery and vein.

The oblique line is on the posterior surface.

The femoral articular surfaces of the head.

The internal malleolus of the tibia.

The astragaloid articular surfaces of the tibia.

The posterior surface and its subdivisions.

The nutrient canal, directed from the knee.

How many articular surfaces has the tibia ?

It has six articular surfaces—two femoral, two fibular, and two astragaloid. The student must find these surfaces and name technically the articulations they assist in forming. (See rule for forming compound words in introductory chapter.)

What can you say of the internal surface of the tibia ?

It is subcutaneous in its lower three-fourths. (Fig. 207.) Its upper one-fourth is occupied by the insertions of the semitendinosus and gracilis muscles, which are inserted by tendons, and the sartorius, inserted by aponeurosis.

What important flexor muscle is inserted into the posterior part of the internal tuberosity ?

The tendon of the semimembranosus muscle. It will be seen presently that this muscle has also three aponeurotic insertions that have very important relations and functions in the vicinity of the knee-joint.

Describe the posterior surface of the tibia.

It is divided by two lines—an oblique and a vertical one—into three subdivisions. (Fig. 208.) One of the three surfaces is occupied by the popliteus muscle; a second by the flexor longus digitorum; a third by the tibialis posticus.

Describe fully the oblique line of the tibia. (Fig. 208.)

It extends from the fibular articular surface to the middle one-third of the bone, on the inner border. It may be considered as possessing three lips—an upper, a middle, and a lower. Into the upper is inserted the popliteal fascia ; from the middle arises the tibial head of the soleus; from the lower, the flexor longus digitorum and tibialis posticus.

Describe the vertical line.

It extends in a curved line downward and outward to the middle one-third of the outer or interosseous border of the tibia. It separates the surfaces occupied by the tibialis posticus and flexor longus digitorum muscles. It also marks the location of the nutrient foramen of the tibia, proceeding, according to rule, *from the knee.*

Name the important parts of the internal malleolus of the tibia.

Posteriorly are two grooves for the tibialis posticus and flexor longus digitorum ; one surface articulates with the astragalus ; to the apex is attached the internal lateral ligament of the ankle-joint.

The Fibula :

Name the points on the fibula of importance in practical anatomy.

The head, for insertion of the biceps tendon.

The styloid process of the head, for external lateral ligament.

The tibial articular surface.

The external malleolus of the fibula.

The anterior surface of the fibula.

The outer surface of the fibula.

The internal surface of the fibula.

The posterior surface of the fibula.

What is the importance of the anterior surface of the fibula ? (Fig. 207.)

Its width is one-fourth of an inch. It gives origin to the extensor proprius or extensor longus hallucis, the extensor longus digitorum, the peroneus tertius— *i. e.,* all the muscles on the front of the leg, except the tibialis anticus.

What can you say about the external surface of the fibula ? (Fig. 207.)

It gives origin to and is occupied by the peroneus longus and peroneus brevis ; the former occupies the upper two-thirds ; the latter, the lower two-thirds.

What is the importance of the posterior surface of the fibula ?

It gives origin in its upper one-third to the fibular head of the soleus ; in its lower two-thirds to the flexor longus hallucis. (Fig. 208.)

Give the importance of the internal surface of the fibula.

It gives attachment to the tibialis posticus—*i. e.,* to the fibular head of this muscle.

What can you say of the patella?

It is the largest sesamoid bone in the body. It is developed in the common tendon of insertion for the group of muscles that extend the leg on the thigh—the *quadriceps extensor femoris.* It has two articular surfaces, corresponding to the trochlear articular surface of the femur. The ligamentum patellæ is inserted into the tubercle of the tibia. A pad of fat and a bursa intervene between the patella and tibia, except at the insertion.

The Foot and Ankle :

Name the osteological points of the foot and ankle of importance in dissection.

The posterior surface of the calcaneum.

The sustentaculum tali of the calcaneum.

The peroneal grooves of the calcaneum.

The inner and outer tuberosities of the calcaneum.

The cuboid surface of the calcaneum.

The astragaloid surface of the calcaneum.

The tuberosity of the scaphoid bone.

The cuneiform articular surfaces of the scaphoid.

The medio-tarsal articulation.

The base of the first metatarsal bone.

The base of the fifth metatarsal bone.

What two tendons do you expect to find on the outer surface of the calcaneum?

The tendons of the peroneus longus and brevis. They are separated by a little spine of bone—the peroneal intertendinous spine. Each tendon is in a separate canal here, but higher up, behind the outer malleolus, they are in the same canal.

How would you describe the peroneus muscle so as to give the bony parts of the limb due prominence?

The peroneus longus arises from the outer surface of the fibula in its upper two-thirds; its tendon passes behind the external malleolus of the fibula, with the tendon of the peroneus brevis; it passes on the outer surface of the calcaneum, in the inferior peroneal groove; it passes through a groove on the under surface of the **cuboid** bone; it is tendinously inserted into the base of the first metatarsal bone, having traversed an osseo-aponeurotic canal, from the cuboid bone to its insertion.

Describe the peroneus brevis.

It arises from the outer surface of the fibula, **middle one-third** ; passes behind the external malleolus of the fibula ; passes through the **superior peroneal** groove, on the **outer surface of the** calcaneum ; and **is inserted into the base of the** fifth **metatarsal bone.**

Name the grooves **through** *which the tendon of the flexor longus hallucis passes.*

This muscle arises from the posterior surface of the fibula, in its middle two-thirds. It passes through a groove in the posterior part of the tibia (Fig. 208), through a second groove, in the posterior **surface** of the astragalus, one inch below the first groove ; through a third groove, on the under surface of the sustentaculum tali of the os calcis or calcaneum ; through a fourth groove, **between** the two sesamoid bones, in the tendons of the **flexor** brevis hallucis, at the metatarso-phalangeal articulation of the great toe.

Name the muscles attached to the inferior surface of the os calcis.

The abductor hallucis, abductor minimi digiti, flexor brevis digitorum, and musculus accessorius.

Describe the insertion of the antagonistic muscles—tibialis anticus and tibialis posticus.

The *tibialis anticus* is inserted into the inner surface of the internal cuneiform bone and adjacent part of the base of the first metatarsal bone. (Figs. 213-215.) The *tibialis posticus* is inserted into the tuberosity of the scaphoid bone, into the sustentaculum tali, into the bases of the second, third, and fourth metatarsal bones, and into all the tarsal bones except the astragalus.

How shall we recognize and know these numerous small insertions of the tibialis posticus muscle?

You can easily trace them as small, tendinous bands from the tuberosity of the scaphoid bone to all other insertional points. (Fig. 215.)

Describe the importance of the sustentaculum tali.

A superior surface articulates with the inner articular facet of the astragalus ; its inferior surface is grooved and lined by synovial membrane, for the transmission of the tendon of the flexor longus hallucis.

Give the importance of the superior surface of the calcaneum.

(1) It gives origin to the extensor **brevis** digitorum ; (2) it **articulates** with the astragalus by two surfaces.

What is the importance of the posterior surface of the os calcis?

Into this surface is inserted the tendo Achillis.

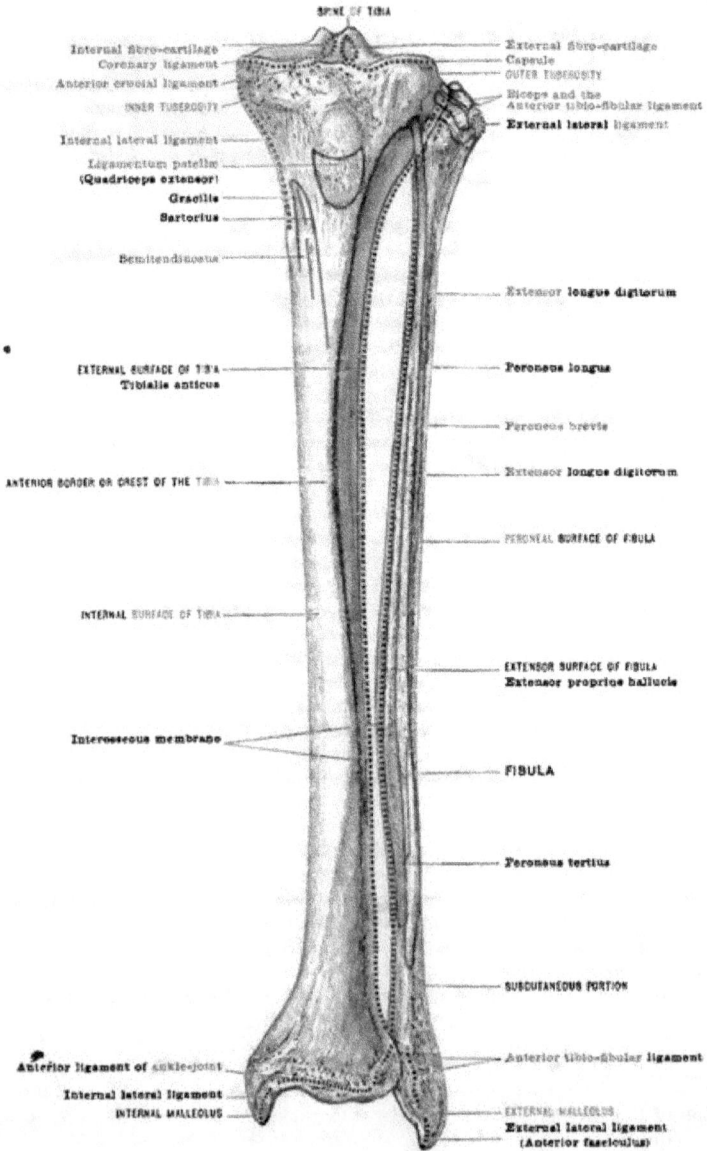

FIG. 207.—THE LEFT TIBIA AND FIBULA. (Anterior view.)

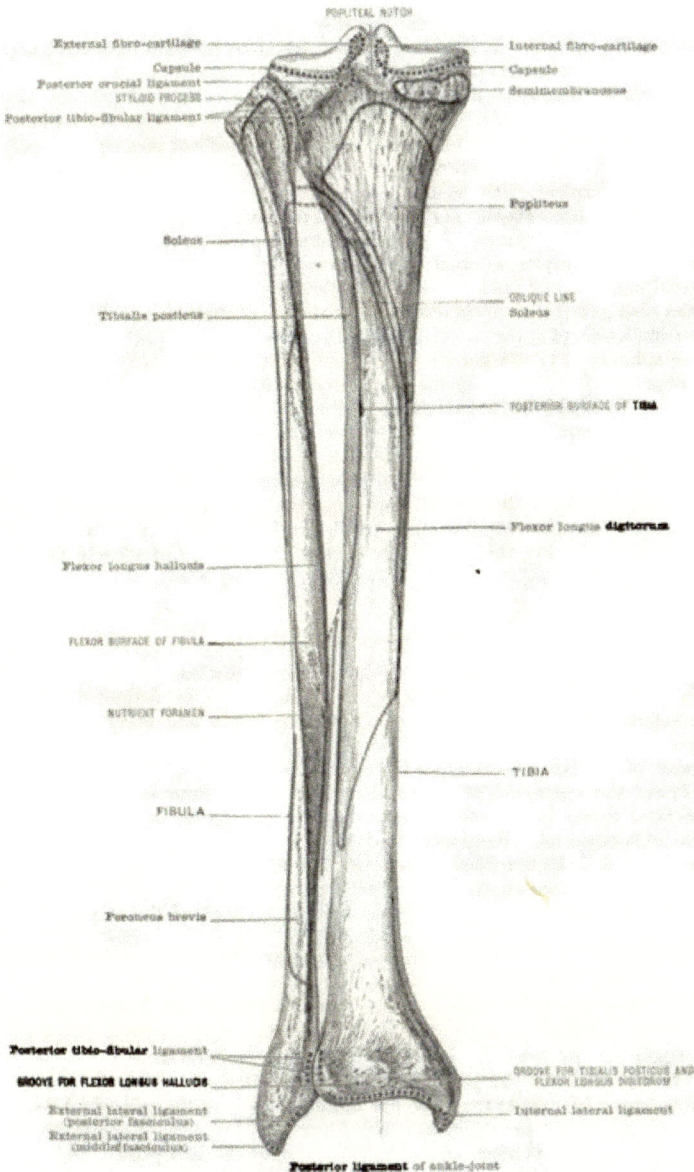

POPLITEAL NOTCH

External fibro-cartilage
Capsule
Posterior crucial ligament
STYLOID PROCESS
Posterior tibio-fibular ligament

Internal fibro-cartilage
Capsule
Semimembranosus

Popliteus

Soleus

OBLIQUE LINE
Soleus

Tibialis posticus

POSTERIOR SURFACE OF Tibia

Flexor longus digitorum

Flexor longus hallucis

FLEXOR SURFACE OF FIBULA

NUTRIENT FORAMEN

TIBIA

FIBULA

Peroneus brevis

Posterior tibio-fibular ligament
GROOVE FOR FLEXOR LONGUS HALLUCIS
External lateral ligament
(posterior fasciculus)
External lateral ligament
(middle fasciculus)

GROOVE FOR TIBIALIS POSTICUS AND
FLEXOR LONGUS DIGITORUM

Internal lateral ligament

Posterior ligament of ankle-joint

FIG. 208.—THE LEFT TIBIA AND FIBULA. (Posterior view.)

Locate by limitation the medio-tarsal joint, and give the surgical importance thereof.

Posteriorly, it is limited by the calcaneum and astragalus; anteriorly, by the cuboid and scaphoid. It is through this articulation the knife passes in Chopart's amputation at the medio-tarsal articulation.

Locate by limitation the tarso-metatarsal articulation, and give its surgical importance.

It is limited posteriorly by the cuboid and cuneiform bones; anteriorly, by the bases of the metatarsal bones. It is through this articulation Lisfranc's amputation is made.

Locate on the cadaver: (1) The crest of the ilium, (2) the anterior iliac spine; (3) the spine of the pubes; (4) the symphysis pubis; (5) the pubic crest; (6) the femoral condyles, internal and external; (7) the tuberosities of the tibia, internal and external; (8) the head, neck, and styloids of the fibula; (9) the crest of the tibia; (10) the subcutaneous inner surface of the tibia; (11) the inner and outer malleoli; (12) the os calcis; (13) the cuboid bone; (14) the tuberosity of the scaphoid; (15) the patella and ligamentum patellæ; (16) the tubercle of the tibia; (17) Poupart's ligament or crural arch.

How to Make Skin Incisions. — First cut through the skin from the centre of Poupart's ligament to the centre of the second toe, the incision passing through the mid-line of the patella. The second cut extends along Poupart's ligament, from the anterior superior spine of the ilium to the symphysis pubis. The third cut extends from one femoral condyle to the other. The fourth cut extends from the inner to the outer malleolus. Now remove the skin. Always follow this rule in removing the skin: Cut closely enough to the skin to permit light to shine through.

THE SUPERFICIAL FASCIA.

The superficial fascia is the second covering of the body. It contains a variable amount of fat. When, as a result of starvation or malnutrition, this fat disappears, the skin lies closely upon the deep fascia, and bony eminences are numerous—in other words, the individual is emaciated. This fascia always consists of two layers—an upper layer, containing the fat; a deep layer, in which are found the cutaneous vessels and nerves. The immense fatty mass in the superficial fascia is recorded in the surgical description of operations as the panniculus adiposus. Remember that there are numerous arteries in the superficial fascia for the nutrition of the skin, but few of them have special names. Collectively they are known as superficial, dermal, or cutaneous arteries. There are also numerous veins in this fascia. On account of their large size and surgical importance, quite a number have received special names—as long and short saphenous; still here, too, the rank and file of veins are collectively designated, as are the arteries—viz.: superficial, dermal, or cutaneous. By common consent, the word superficial is used by anatomists to designate anything in the superficial fascia; hence all arteries, veins, nerves, muscles, and lymphatics, in this fascia may properly be collectively designated superficial.

Having removed the skin, according to directions previously given, consult figures 209, 210, and 211, and find in the deep layer of the superficial fascia the following cutaneous or superficial structures:

1. The dorsal venous arch of the foot—arcus dorsalis pedis.
2. The long saphenous vein and its tributaries.
3. The short saphenous vein and its tributaries.
4. The internal and external femoral cutaneous veins.
5. The superficial epigastric vein and its tributaries.

6. The superficial circumflex iliac vein and tributaries.
7. The superficial external pudic vein.
8. The spermatic cord, just below the pubic spine. (Fig. 211.)
9. Superficial inguinal and femoral lymphatic glands. (Fig. 210.)
10. The long saphenous nerve, with a vein of like name. (Figs. 209 and 210.)
11. The short saphenous nerve and vein (posteriorly).
12. The cutaneous branch of the musculo-cutaneous nerve. (Fig. 209.)
13. The internal, middle, and external cutaneous nerves. (Fig. 209.)
14. The ilio-hypogastric and genito-crural nerves. (Fig. 209.)
15. The patellar plexus, formed by the union of cutaneous branches, from

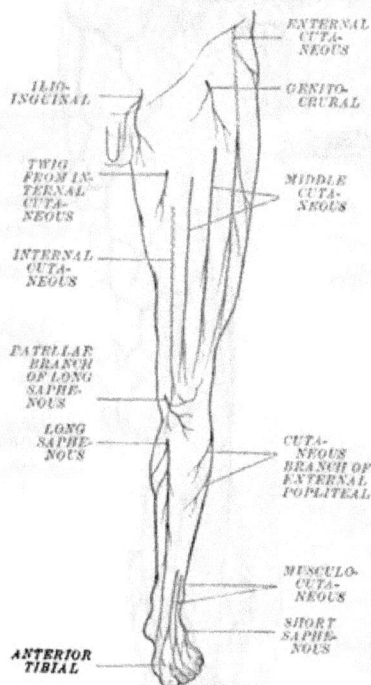

FIG. 209.—DISTRIBUTION OF CUTANEOUS NERVES ON THE ANTERIOR ASPECT OF THE INFERIOR EXTREMITY.

the internal, middle, external cutaneous nerves, and a branch from the long saphenous and obturator to supply the skin over the insertion of the extensor muscles of the leg, according to *Hilton's law*. (Fig. 211.)

16. The subsartorial plexus, formed by branches from the obturator, long saphenous, and internal cutaneous nerves, to supply the skin over the insertion of the adductors, according to *Hilton's law*.

17. The prepatellar bursa, found in the superficial fascia in front of the patella.

Dorsal Arch (Fig. 210.)—*How Formed.*—This arch is seen on the dorsum of the foot about one inch behind the clefts of the toes. It is formed by the confluence of the veins from the skin of the toes. It lies upon the cutaneous nerves

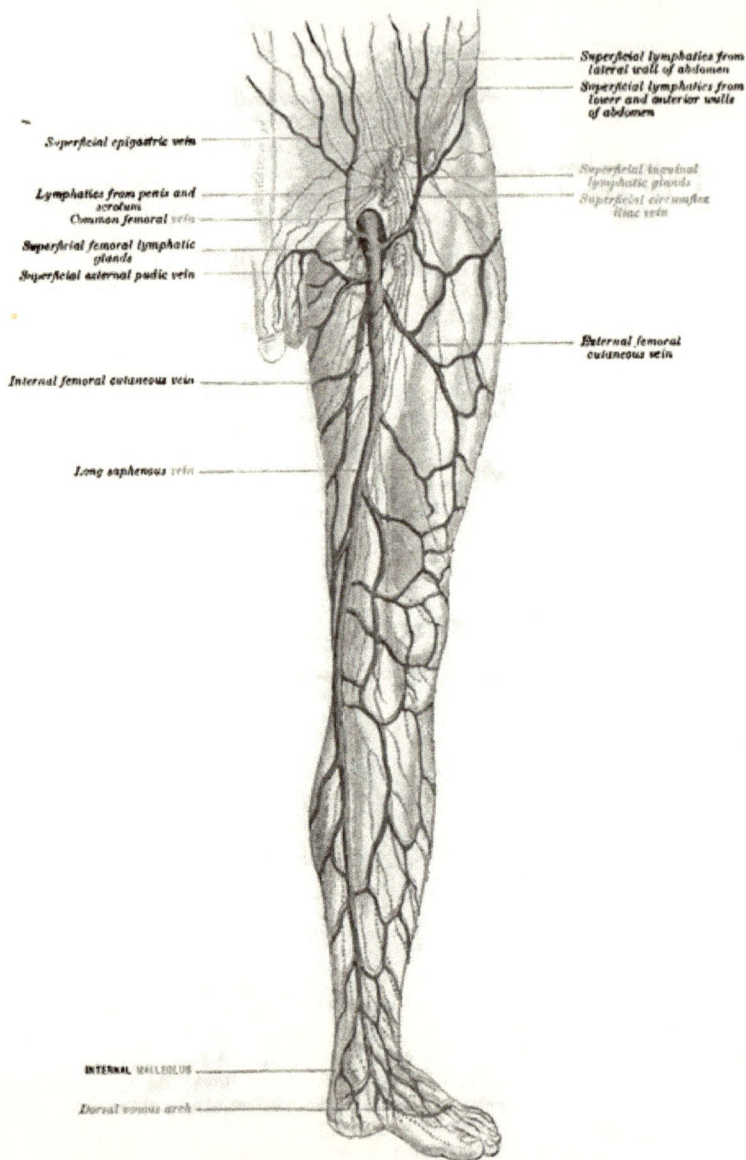

FIG. 210.—THE SUPERFICIAL VEINS AND LYMPHATICS OF THE LOWER LIMB.

of the dorsum of the foot. It has two ends, called inner and outer. This vein should be studied on your own feet in this simple way : Tie a bandage very tightly around your leg eight inches below the knee. The pressure will retard the return circulation and make all the superficial veins stand out.

Long Saphenous Vein.—This is the largest and longest superficial vein in the body. It is in the deep layer of the superficial fascia. It is attended by the long saphenous nerve below the knee. It begins at the inner end of the dorsal arch of the foot (*arcus dorsalis pedis*), passes in front of the internal malleolus, behind the inner tuberosity of the tibia and the inner condyle of the femur, and

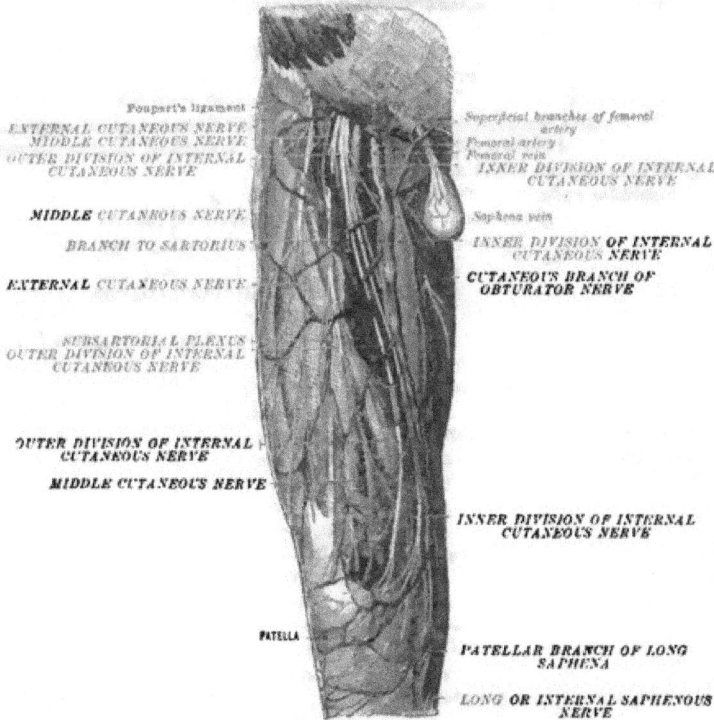

FIG. 211.—SUPERFICIAL DISSECTION OF THE FRONT OF THE THIGH.
(Hirschfeld and Leveillé.)

opens into the common femoral vein, passing through the saphenous opening in the fascia lata. This vein is ligated or obliterated in the radical operation for varicose veins of the lower extremity. (Fig. 210.)

The Short Saphenous Vein.—This begins at the outer end of the dorsal arch of the foot, passes behind the external malleolus of the fibula, soon gains the posterior mid-line of the leg, passes between the two heads of the gastrocnemius, and opens into the popliteal vein. This vein is attended by the short saphenous nerve.

The Long Saphenous Nerve.—This is the longest cutaneous nerve in the body. It is a branch of the anterior crural or femoral nerve. It is in Scarpa's triangle and Hunter's canal. It becomes cutaneous, by piercing the **deep fascia**, two inches below the knee, between the gracilis and sartorius muscles. **It joins** company at this place with the vein of like name, and is distributed **to the skin** of the *inner* side of the leg, foot, and great toe. This nerve has its **extensive** distribution below the knee, to supply the fullest insertion of the **sartorius** muscle and thereby verify Hilton's law.

The Short Saphenous Nerve.—This nerve has a double formation: a branch from the internal popliteal, called the *communicans tibialis*, joins a branch from the external popliteal, called the *communicans fibularis*, to form this nerve. This union may take **place high or low**. The nerve accompanies a vein of like name, and is distributed **to the** skin on the posterior part of the leg and the **outer** part of the foot and little **toe**.

The Cutaneous Branch of the Musculo-cutaneous Nerve (Fig. 209).— This is **a** branch of the external popliteal nerve. The cutaneous **branches are** distributed to the dorsum of the foot. The muscular branches **of this nerve** supply the peronei muscles. Anastomosis takes place between these cutaneous branches and also between the plantar nerves on the bottom of the foot, forming the general cutaneous pedal anastomosis.

The middle cutaneous nerve, a branch of the anterior crural, is usually found as two parallel branches, one of which crosses the sartorius, the other of which pierces the muscle. They end in the patellar plexus, having supplied the skin on the front of the thigh. (Fig. 211.)

The internal cutaneous nerve, a branch of the anterior crural, supplies the skin of the inner part of the thigh; supplies the skin over the primary insertion of the **sartorius**; assists the obturator in forming the subsartorial plexus, and ends in the **patellar** plexus. (Fig. 211.)

What is to be understood by the primary insertion of the sartorius?

The insertion of the muscle into the upper and inner one-third of the tibia. (Fig. 207.) The secondary insertion of the muscle is coextensive with the periosteum of the inner surface of the tibia; hence to carry out Hilton's law, the long saphenous branch of the anterior crural nerve has its distribution below the knee.

How do structures become cutaneous?

Cutaneous nerves become cutaneous by piercing the deep fascia. The hole through which the nerve comes is called an opening, and takes the technical name of the structure transmitted. Where several structures pass through one opening, the largest structure lends its name to the opening. Cutaneous **structures** piercing the deep fascia are centrifugal and centripetal. The former, **as nerves** and arteries, have their origin below the deep fascia and their explosion in the skin beyond; the latter, as veins and lymphatics, are made up in the skin, and, having collected all their tributaries, pass through the deep fascia, to become themselves tributary to deep vessels or glands.

How dissect the cutaneous nerves of the thigh?

There are two procedures, both of which **must be** closely followed: (1) Locate in figure 209 the **opening** where a **given** nerve comes through; then, with your forceps, plow through the fat in **the** long axis of the limb, always downward—never from side to side—until you find the emergence of the nerve. The location of these openings is never constant; they may vary several inches in any two cases. After having found the nerve, then trace the same out to its several branches. (2) We trace out the cutaneous nerves of the thigh by locating and lifting up, on the finger, the main trunk of the anterior crural nerve. This nerve is found under Poupart's ligament, beneath the **deep fascia, in a** groove, between the **iliacus and psoas** magnus muscles. **The nerve once on** the finger, and on the stretch, now plow downward with the forceps—never with the scalpel—in the direction of the branches given off. **In** this way you can find every branch of this nerve. Now, as the anterior crural gives off the internal and middle cutaneous nerves, it yet remains for us to find the external cutaneous nerve.

The **External Cutaneous Nerve.**—This is a branch of the lumbar plexus. You find its main trunk by cutting through the deep fascia just below the anterior superior iliac spine. Its branches are to be traced out in a manner similar to the others. The cutaneous nerves of the anterior region of the thigh, above referred to, become cutaneous, in average, at the junction of the upper and middle one-third of the thigh. They divide into many branches each, and collectively supply with sensation the front, outer, and inner parts of the thigh, as far as the knee. At this place they all communicate to form the plexus patellæ, above described.

The student should remember that while the external cutaneous nerve is usually given in the text-books as a separate nerve of the lumbar plexus, still, physiologically and rationally this nerve must be considered as a divorced or dismembered branch of the anterior crural or femoral nerve, for this reason : The nerve-trunk that supplies muscles, supplies the skin over the muscle, even to the fullest insertion. Now, the anterior crural nerve supplies the extensor group of muscles on the anterior and outer part of the thigh, hence the external cutaneous belongs philosophically to the anterior crural nerve, being divorced therefrom. Such cases will frequently be met. The small sciatic is the divorced cutaneous part of the great sciatic nerve.

The spermatic cord must be noticed in this section, on account of its important relation to two nerves and two arteries. You will find the spermatic cord emerging from the inguinal canal, by the external abdominal ring (Fig. 163). From this time on it is a cutaneous structure, because it is in the superficial fascia. With your scissors trace the cord into the scrotum, by cutting through the skin and superficial fascia. On the front of the cord you will find a little nerve, the ilio-inguinal ; behind the cord you find the genital branch of the genito-crural nerve. These nerves come from the lumbar plexus. The ilio-inguinal nerve supplies the side of the thigh and the scrotum ; the genital branch of the genito-crural supplies the cremaster in the male, and is lost on the round ligament in the female. The arteries in relation with the spermatic cord are the superficial and deep external pudics, branches of the common femoral ; the former is in front of the cord, the latter behind the cord.

THE DEEP FASCIA OF THE THIGH AND LEG.

In figure 210 the glistening structure, upon which the long saphenous vein and its tributaries rest, is the outer or circumferential part of the deep fascia of the lower extremity. The internal and central part of the deep fascia is, consequently, surrounded by the external. In looking at a well-filled comb of honey, you see only a small part of the honey-comb—the outside limiting part. If, however, you extract the honey, then you can see and examine the interior cellular arrangements ; in other words, you can examine the internal divisions of the connective tissue or deep fascia of the honey-comb. It is just so in examining the internal parts of the deep fascia of the thigh or of any organ of the body. The deep part of the deep fascia corresponds to the cellular-tissue spaces of the honey-comb. The spaces are occupied, not by one substance, but by many. Muscles, vessels, bone, nerves, fat, glands, and all the compounds which make up the limb occupy large spaces, just as effectually shut off from one another as is the honey in one cell separated from that in another in the comb.

Physiological division of labor and differential assimilation cause grouping of like compounds. This grouping of muscles takes place on the physiological

basis of antagonism. Flexors oppose extensors; adductors oppose abductors; **pronators** oppose supinators; levators oppose depressors. Nerves **that innervate** and vessels that feed them are as truly antagonistic as are the muscles.

The general law of antagonism is this: equality in length, strength, blood, nerve, and fascial environment. *Antagonistic groups* of muscles are separated from each other by deep fascia, called septa. A group of muscles acting in unison, as the extensors or flexors of the leg, is called a musculature. Adjacent musculatures are separated from each other by deep fascia, called septa. The individual muscles, of which groups are composed, are also separated by deep fascia. Bone is surrounded by deep fascia, called periosteum; articulations are held in place by deep fascia, called capsular ligament. Vessels and nerves receive their sheath, and glands their capsules from deep fascia—*i. e.*, from the internal division of the same.

Minor Details.—With this understanding of the general distribution of the deep fascia of the thigh, consider some of the minor details:

1. One characteristic of deep fascia is *the presence of openings* in the outer or circumferential portion, for the transmission of both centripetal and centrifugal cutaneous vessels and nerves; in the central or septal portions for the transmission of those branches of antagonistic vessels and nerves, called communicating and anastomotic, which preserve a physiological balance of power. Instance, the perforating branches of the profunda reaching the hamstrings, and the communications between the anterior crural, obturator, and great sciatic nerves.

2. Deep fascia is *not uniform in thickness.* As structure is the correlative of function, those parts subjected to the greatest degree of tension and use must become thick and strong in comparison with less used parts.

3. Deep fascia of the leg and thigh receives in some localities *special names*, founded on no rational or logical ground, but which are perpetuated in anatomy and surgery, and revered, as Egypt points to her ruins. The special names for the deep fascia of the lower extremity are these: (1) The *fascia lata*, on the upper front part of the thigh; (2) the *ilio-tibial band*, on the outer side of the thigh; (3) the *popliteal fascia*, covering the popliteal space; (4) the *internal annular ligament*, between the os calcis and internal malleolus; (5) the *external annular ligament*, between the os calcis and external malleolus; (6) the *anterior annular ligament*, between the malleoli; (7) the *dorsal fascia*, on the back of the **foot**; (8) the *plantar fascia* on the sole of the foot; (9) the *ligamenta vaginales*, as sheaths for the flexor tendons of the toes; (10) the *vincula*, as intertendinous slips; (11) the *various intermuscular septa.*

4. *Deep Fascia Gives Origin and Insertion to Muscles.*—This fact is often not fully appreciated by the student. Instance, the *vastus internus* and *vastus externus*, the muscles on the *anterior part of the tibia* and *fibula.* Be ever ready to see deep fascia giving origin or insertion to muscles. The gluteus maximus has one of its insertions into the deep fascia; the ilio-tibial band is in reality only the insertion of the tensor vaginæ femoris, by aponeurosis.

5. *Subdivisions of fascia lata* are: (1) the iliac and (2) pubic portions. These are separated from each other by a large opening—the saphenous. The iliac or anterior portion is attached to the crest of the ilium, to the whole of Poupart's ligament. The pubic or posterior part covers the pectineus muscle and gracilis; is continuous with the femoral sheath behind the femoral vessels.

Attachments of Deep Fascia.—Deep fascia is continuous with periosteum at subcutaneous areas. These are called the attachments. The deep fascial attachments of the lower limb are as follows. In the region of the hip: (1) To crest of ilium; (2) to pubic spine and body; (3) to ischio-pubic ramus; (4) to the tuberosity of ischium; (5) to sacrum and coccyx. In the knee region, demonstrate its attachment to : (1) The patella ; (2) the tuberosities ; (3) the condyles ; (4) the crest and inner surface of the tibia. In the ankle region, see its attachment : (1) To the os calcis ; (2) to the malleoli ; (3) to the tuberosity of the scaphoid bone and numerous other places.

Saphenous Opening.—This is the largest opening in the deep fascia. It is situated between the iliac and pubic portions of the fascia lata. Behind it, or on its floor, is the femoral sheath, containing the femoral artery, vein, and canal. In front of the opening, forming its roof or covering, is the superficial fascia, called in this locality, the cribriform fascia. The structures passing through the saphenous opening are numerous : the long saphenous vein and the small vessels given off from the common femoral artery and vein, and some lymphatic vessels.

Cribriform Fascia.—Just that part of the deep layer of the superficial fascia, that covers the saphenous opening, is called by this name. It was so called on account of the numerous perforations, transmitting the structures referred to in the previous paragraph. The word " cribriform " means sieve-like.

Importance.—Since fascia often determines the direction taken by burrowing pus, or the course of a bullet, its various superficial attachments should be care-

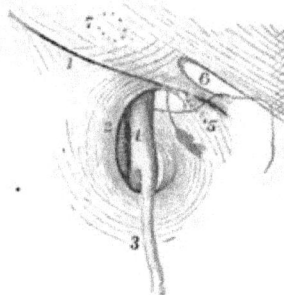

FIG. 212.—THE FEMORAL RING AND SAPHENOUS OPENING. (After Holden.)
(The arrow is introduced into the femoral ring.)
1. Crural arch. 2. Saphenous opening of the fascia lata. 3. Saphena vein. 4. Femoral vein. 5. Gimbernat's ligament. 6. External abdominal ring. 7. Position of internal ring.

fully examined, and its deep ones found by dissection. The guide to finding these is to remember that muscles are arranged in antagonistic groups, and that these groups are called musculatures, and that these musculatures are separated by septa, and these septa are attached to bone—i. e., continuous with its periosteum.

The Application.—Now apply the principle of musculatures to the lower extremity. On the thigh you will presently find : (1) An *extensor group* of muscles in front ; (2) a *flexor group* behind ; (3) an *adductor group* internally.

You will also find : (1) A septum between the *extensor* group and *flexor* group ; (2) between the *flexor* group and the *adductor* group ; (3) between the *adductor* group and the *extensor* group.

Again, below the knee you will find : (1) A group of muscles on the *anterior tibio-fibular* region ; (2) a group on the *outer surface of the fibula* ; (3) two groups on the *posterior tibio-fibular* region. Here, also, you may demonstrate the rule that adjacent musculatures are separated from one another by fascial septa, since here are strong bands of fascia separating the peronei muscles on the outer surface of the fibula, from different musculatures both in front and behind. Likewise, on the back of the leg, the superficial group is separated from the deep by the transverse fascia, and both these groups from lateral musculatures.

ANTERIOR REGION OF LEG.

Remove carefully the deep fascia. (Fig. 210.) You have now, by the removal of the skin and superficial fascia, exposed the deep fascia covering all the muscles on the anterior and outer regions of the leg (Fig. 213):

1. The cutaneous branch of the musculo-cutaneous nerve.
2. The tibialis anticus and the tuberosity of scaphoid bone.
3. The extensor proprius hallucis muscle.
4. The extensor longus digitorum muscle.
5. The peroneus tertius muscle.
6. The extensor brevis digitorum muscle.

Now dissect down between the tibialis anticus and the extensor proprius hallucis by cutting through the deep fascia with scissors and find the anterior tibial nerve and vessels. (Fig. 214.)

Anterior Tibial Artery.—It is one of the two terminal branches of the popliteal artery. It gains the front of the leg by passing between the two heads of the tibialis posticus and interosseous membrane. It lies in a deep groove, bounded internally by the tibialis anticus; externally, by the extensor longus digitorum and extensor proprius hallucis. It terminates on the dorsum of the foot, continuing its course under the name of dorsalis pedis. Its branches are: (1) The anterior tibial recurrent, which anastomoses with the articular branches of the popliteal and anastomotica magna; (2) posterior tibial recurrent, when present, is given off before the anterior tibial passes through the interosseous membrane; (3) muscular branches, which supply the muscles on the anterior region of the leg (Fig. 214); (4) malleolar branches, which supply the ankle-joint. On the back of the foot you cut through the deep fascia between the extensor proprius hallucis and extensor longus digitorum muscles, to find the dorsalis pedis artery.

Dorsalis Pedis Artery.—This is a continuation of the anterior tibial. It lies between the extensor proprius hallucis and extensor longus digitorum, on the dorsum of the foot. (Fig. 214.) It gives off: (1) A communicating branch to the external plantar, by which the plantar arch is completed; (2) a tarsal artery to the extensor brevis digitorum and the tarsus; (3) the dorsalis hallucis to the great toe; (4) interosseous arteries to the interosseous spaces and their contents. Trace the arteries out carefully with the forceps. Remember they are beneath the dorsal fascia.

Anterior Tibial Nerve.—It is a branch of the external popliteal nerve, being given off with the musculo-cutaneous nerve. It gives off articular branches, according to Hilton's law, to the (1) ankle-joint, (2) the tarsal, and (3) the metatarso-phalangeal joints. It gives muscular branches to the extensor brevis digitorum and to all the muscles on the anterior regions of the tibia and fibula.

Muscles on Anterior Surface of Fibula.—These are: (1) Extensor proprius hallucis; (2) extensor longus digitorum; (3) peroneus tertius, the fifth tendon of the extensor longus digitorum.

Insertion of Extensor Brevis Digitorum.—This muscle has four tendons. The first is inserted independently into the base of the first phalanx of the great toe; the remaining three are inserted, conjointly with the tendons of the extensor longus digitorum, into the bases of the second and third phalanges.

Give nerve-supply of the five muscles now exposed on the anterior tibio-fibular region.

The anterior tibial, a branch of the external popliteal.

Intermuscular Septum.—Examine the septum that separates the muscles on the anterior surface of the fibula from the peronei on the external surface.

How to Dissect these Muscles.—*Tibialis Anticus.*—Cut through the

anterior annular ligament on the tendon of the muscle. Now follow the
tendon to its insertion into the internal cuneiform bone and base of the

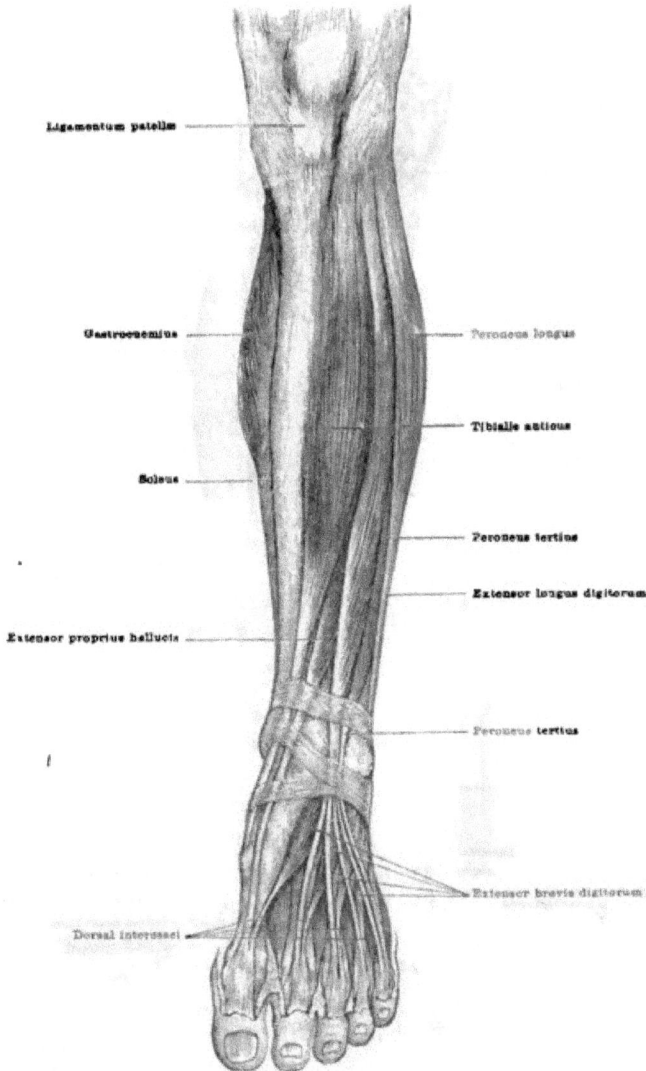

Ligamentum patellæ

Gastrocnemius

Peroneus longus

Tibialis anticus

Soleus

Peroneus tertius

Extensor longus digitorum

Extensor proprius hallucis

Peroneus tertius

Extensor brevis digitorum

Dorsal interossei

FIG. 213.—THE MUSCLES OF THE FRONT OF THE LEG.

first metatarsal. Then trace the muscle to its origin : (1) the outer surface of the
tibia, upper two-thirds ; (2) outer tuberosity. Notice the nerve-supply and blood-
supply.

Extensor Proprius Hallucis.—Begin at the end of the great toe and cut down on the tendon. Follow it up through a separate compartment under the

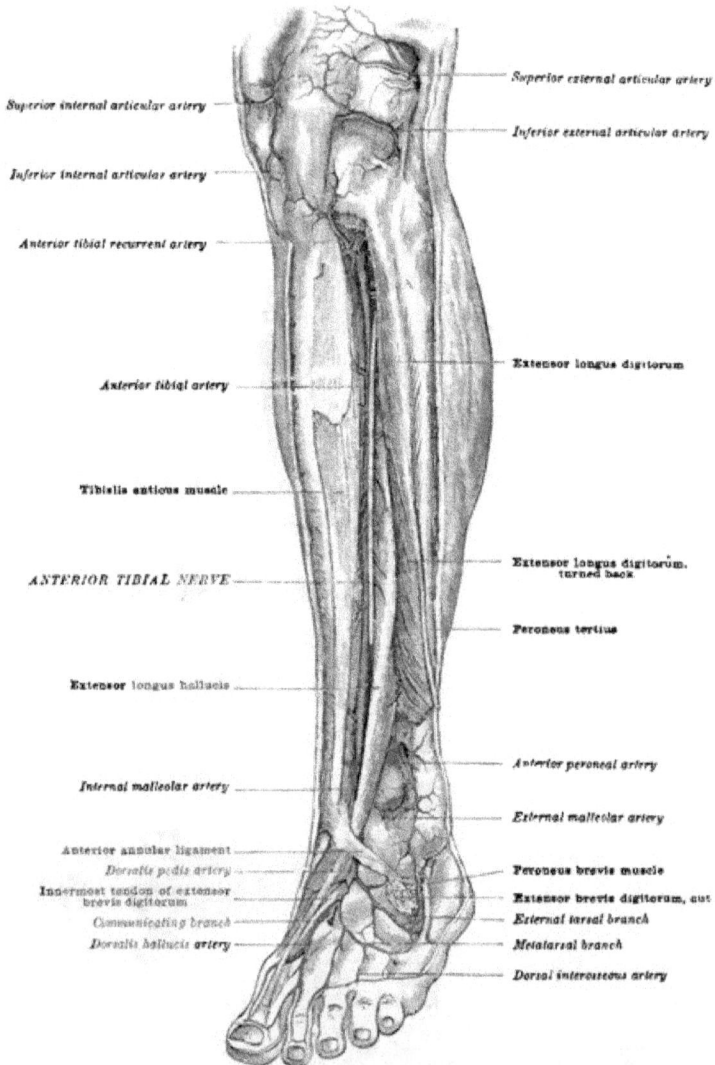

Superior internal articular artery

Inferior internal articular artery

Anterior tibial recurrent artery

Anterior tibial artery

Tibialis anticus muscle

ANTERIOR TIBIAL NERVE

Extensor longus hallucis

Internal malleolar artery

Anterior annular ligament
Dorsalis pedis artery
Innermost tendon of extensor brevis digitorum
Communicating branch
Dorsalis hallucis artery

Superior external articular artery

Inferior external articular artery

Extensor longus digitorum

Extensor longus digitorum, turned back

Peroneus tertius

Anterior peroneal artery

External malleolar artery

Peroneus brevis muscle
Extensor brevis digitorum, cut
External tarsal branch
Metatarsal branch
Dorsal interosseous artery

FIG. 214.—THE ANTERIOR TIBIAL ARTERY, DORSAL ARTERY OF THE FOOT, AND ANTERIOR PERONEAL ARTERY, AND THEIR BRANCHES.

annular ligament to its origin on the middle half of the anterior surface of the fibula and interosseous membrane.

The Extensor Longus Digitorum.—To dissect this muscle successfully, cut

down on the same through the anterior annular ligament. Then use your
forceps as a director, and carefully cut the dorsal fascia over each of the four

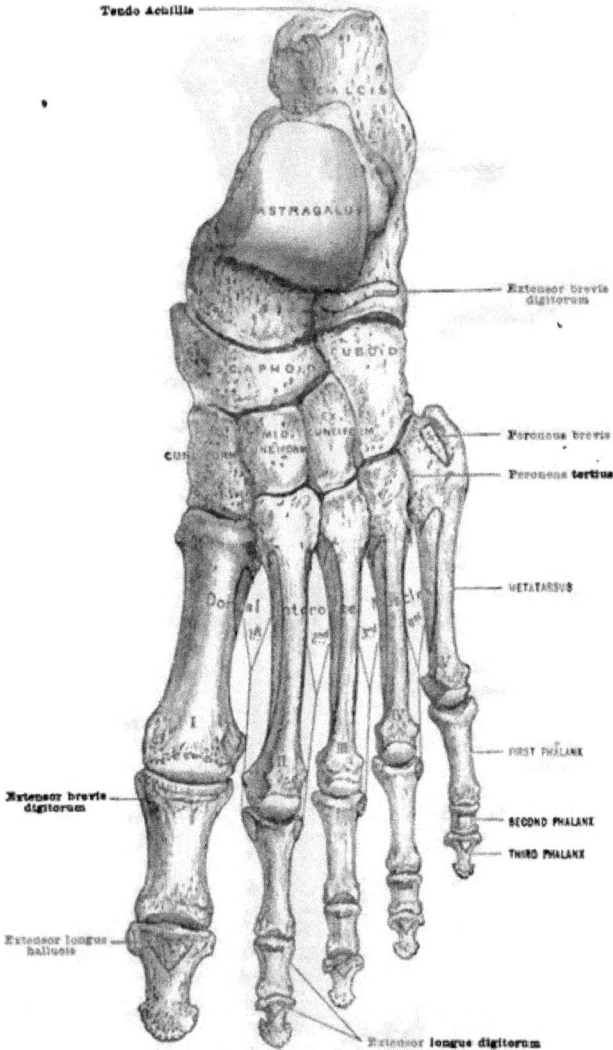

FIG. 215.—THE LEFT FOOT. (Dorsal surface.)
(Study origin and insertion of muscles on this figure and compare with your dissection.)

tendons. In like manner trace each tendon to the end of the toe. Notice that
each tendon divides in three slips, on the dorsal surface of the first phalanx;
that the middle slip is inserted into the base of the second phalanx; that the

21

SPINE OF TIBIA

Internal fibro-cartilage
Coronary ligament
Anterior crucial ligament
INNER TUBEROSITY

Internal lateral ligament

Ligamentum patellæ
(Quadriceps extensor)
Gracilis
Sartorius

Semitendinosus

EXTERNAL SURFACE OF TIBIA
Tibialis anticus

ANTERIOR BORDER OR CREST OF THE TIBIA

INTERNAL SURFACE OF TIBIA

Interosseous membrane

External fibro-cartilage
Capsule
OUTER TUBEROSITY
Biceps and the
Anterior tibio-fibular ligament
External lateral ligament

Extensor longus digitorum

Peroneus longus

Peroneus brevis

Extensor longus digitorum

PERONEAL SURFACE OF FIBULA

EXTENSOR SURFACE OF FIBULA
Extensor proprius hallucis

FIBULA

Peroneus tertius

SUBCUTANEOUS PORTION

Anterior ligament of ankle-joint
Internal lateral ligament
INTERNAL MALLEOLUS

Anterior tibio-fibular ligament

EXTERNAL MALLEOLUS
External lateral ligament
(Anterior fasciculus)

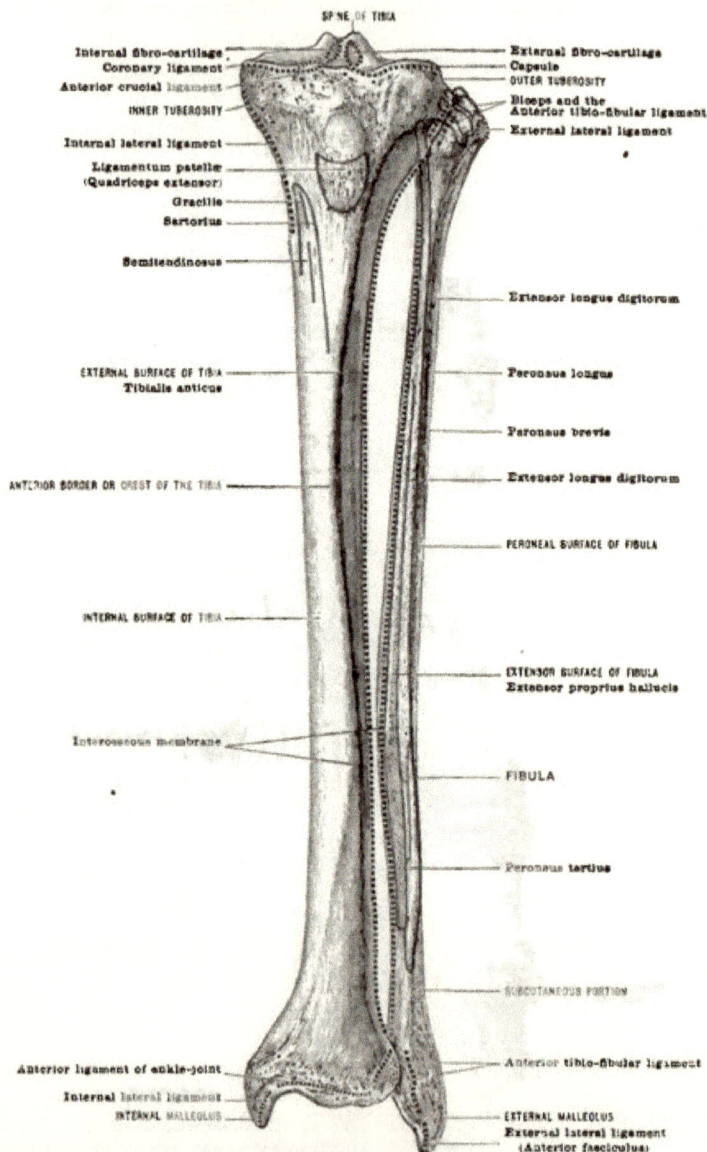

FIG. 216.—THE LEFT TIBIA AND FIBULA.   (Anterior view.)
(Study origin of muscles on this figure and compare with your dissection.)

two lateral slips are inserted into the base of the third phalanx. Also notice
the relation of the tendons of this muscle to those of the extensor brevis digitorum. Origin, outer tuberosity of the tibia and anterior surface of fibula.

The Peroneus Tertius.—Find the insertion of this muscle in the base of the
fifth metatarsal. (Fig. 215.) Trace it to the anterior surface of the fibula, and notice
that it blends with the preceding muscle, a part of which it really is. (Fig. 207.)

THIGH, ANTERIOR AND INTERNAL REGIONS.

Remove the deep fascia and expose the vessels, nerves, and muscles. (Figs.
218, 219, and 220.)

First locate the **sartorius**. (Fig. 218.) Begin at the anterior superior iliac spine
(Fig. 226) and expose the muscle to its insertion (Fig. 207) into the tibia, by gently
removing all deep fascia. Now lift up the muscle from its bed, taking care not to
damage the vessels and nerves that enter it. Notice that this muscle is crossed by
one and pierced by another branch of the middle cutaneous nerve. (Fig. 211.)

Locate the **gracilis** next, on the inner surface of the thigh. (Figs. 218 and
219.) Trace it from the descending ramus of the pubes (Fig. 226) to its insertion into the inner surface of the tibia (Fig. 207), taking care not to damage
the nerve branches of the obturator that enter it. This dissecting, by which
this muscle is separated from its fellow-muscles, must be done with the forceps.

Locate the **adductor longus**. (Fig. 218.) Trace it from the anterior surface
of the pubes to its insertion into the middle of the middle lip of the linea aspera.
(Fig. 227.) Divide the connective tissue between this muscle and the pectineus.
(Fig. 218.) Now cut the adductor longus, at its origin, and turn the same
aside. (Fig. 219.) Also cut the pectineus and turn it aside. (Fig. 218.)

Obturator Nerve, Anterior Division (Fig. 219).—This you will see lying on
the adductor brevis. (Fig. 219.) Take this nerve up gently, and follow out its
branches to: (1) The gracilis; (2) the adductor longus; (3) the adductor brevis; (4)
the sartorius; (5) a branch to the hip-joint; (6) a branch to the femoral artery; (7)
an occasional branch to the pectineus muscle; (8) a cutaneous branch to assist the
long saphenous and internal cutaneous in forming the subsartorial plexus.

Obturator Nerve, Posterior Division.—This lies on the adductor magnus.
The adductor brevis muscle then separates the anterior from the posterior division of the nerve. This division supplies: (1) The obturator externus; (2) the
adductor magnus; (3) an articular branch to the hip-joint; (4) an articular
branch to the knee.

Why does the obturator nerve send a branch to the knee?

Because this nerve supplies the sartorius muscle, a muscle that moves the
knee-joint. (Fig. 219.)

How does the articular branch from the obturator nerve enter the hip-joint?

It passes through the cotyloid notch.

When the accessory obturator nerve is present, where may it be found?

To the inner side of the psoas magnus muscle.

What is the function of the subsartorial plexus?

To supply the skin over the adductor muscles, and to exercise a sensory
balance of power between the obturator and anterior crural nerves.

The Anterior Crural Nerve (Fig. 219).—You will find the main trunk of this
nerve deeply buried in a space between the iliacus and psoas magnus. (Fig. 197.)
Take this nerve up on your finger and gently follow out its muscular branches to
all the muscles on the front of the thigh: (1) Sartorius; (2) rectus femoris; (3)

vastus internus ; (4) vastus externus ; (5) crureus ; (6) subcrureus ; (7) pectineus. In tracing these nerves, divide the connective tissue in the direction of the *long axis*

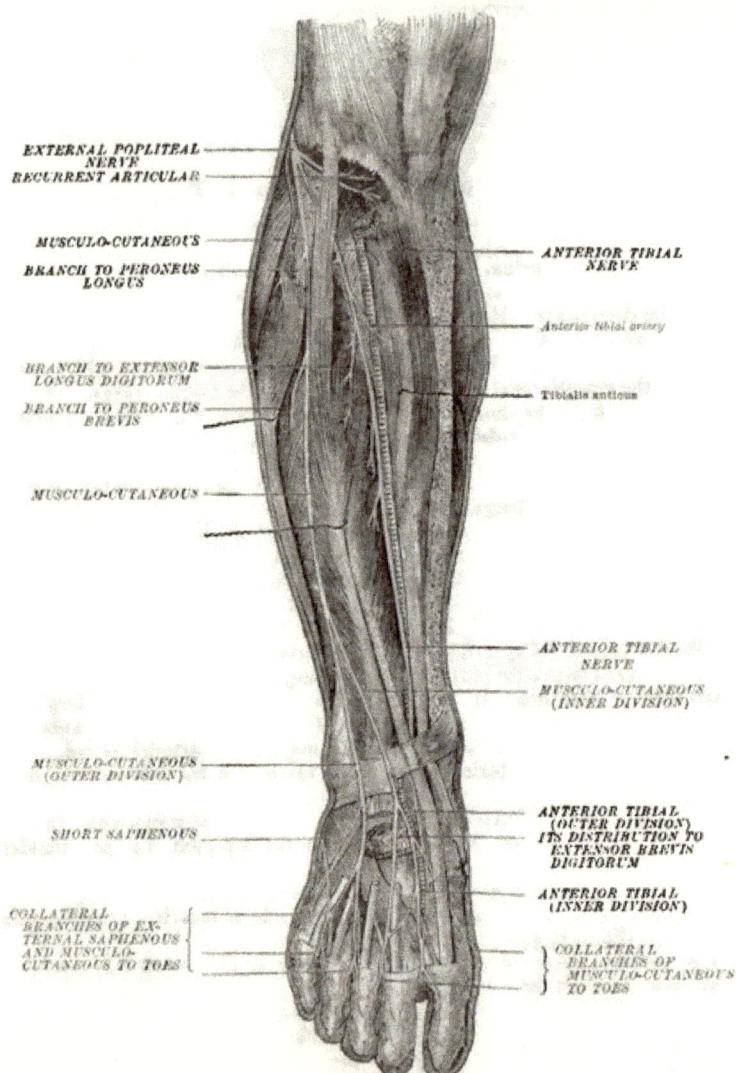

EXTERNAL POPLITEAL NERVE
RECURRENT ARTICULAR

MUSCULO-CUTANEOUS

BRANCH TO PERONEUS LONGUS

ANTERIOR TIBIAL NERVE

Anterior tibial artery

BRANCH TO EXTENSOR LONGUS DIGITORUM

BRANCH TO PERONEUS BREVIS

Tibialis anticus

MUSCULO-CUTANEOUS

ANTERIOR TIBIAL NERVE

MUSCULO-CUTANEOUS (INNER DIVISION)

MUSCULO-CUTANEOUS (OUTER DIVISION)

SHORT SAPHENOUS

ANTERIOR TIBIAL (OUTER DIVISION) ITS DISTRIBUTION TO EXTENSOR BREVIS DIGITORUM

ANTERIOR TIBIAL (INNER DIVISION)

COLLATERAL BRANCHES OF EXTERNAL SAPHENOUS AND MUSCULO-CUTANEOUS TO TOES

COLLATERAL BRANCHES OF MUSCULO-CUTANEOUS TO TOES

FIG. 217.—BRANCHES OF THE EXTERNAL POPLITEAL NERVE.

of the nerve—never crosswise. And trace it to the skin covering the same by : (1) The internal cutaneous ; (2) the middle cutaneous ; (3) the long saphenous. **The Femoral Artery** (Fig. 223).—Get your finger under this vessel and trace out its branches. Great care must be taken to divide the connective tissue binding the artery and vein together. To learn the branches of this artery, see page 320.

Dissection of the Rectus (Fig. 218).—You can easily lift this muscle from its bed. It lies on the *vastus internus.* (Fig. 220.) Joining the rectus on its outer

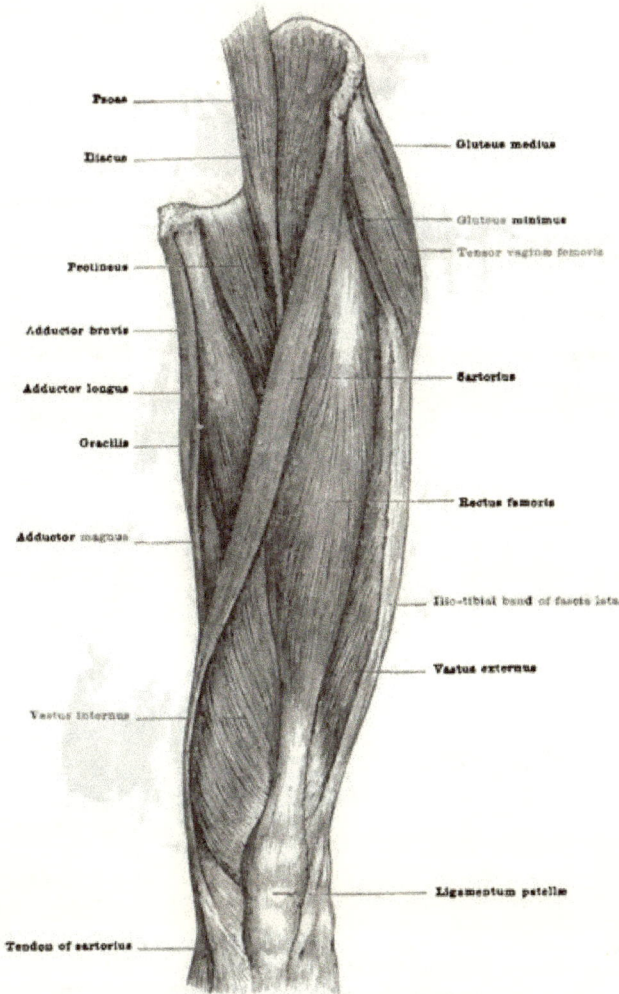

FIG. 218.—MUSCLES OF THE FRONT OF THE THIGH.

border is the *vastus externus.* (Fig. 218.) Trace the rectus upward and develop its two heads (Fig. 226): (1) The straight, from the anterior inferior iliac spine ; (2) the reflected, from the depression above the brim of the acetabulum. Notice the nerves and arteries enter the under surface, and trace them to their sources. (Fig. 223.)

The Vasti Internus and Externus (Fig. 220).—Cut the rectus four inches above the patella and turn it aside, guarding well the vessels and nerves. You will take particular notice of the form of the groove or bed in which the rectus lay. This bed is on the front surface of the vastus internus, and is called the *crureus muscle.* (Fig. 220.) It is not a separate muscle, but only that part of the vastus internus on which the rectus lay. Along the outer margin of the rectus-bed you will see the descending branch of the external circumflex artery (Fig. 223),

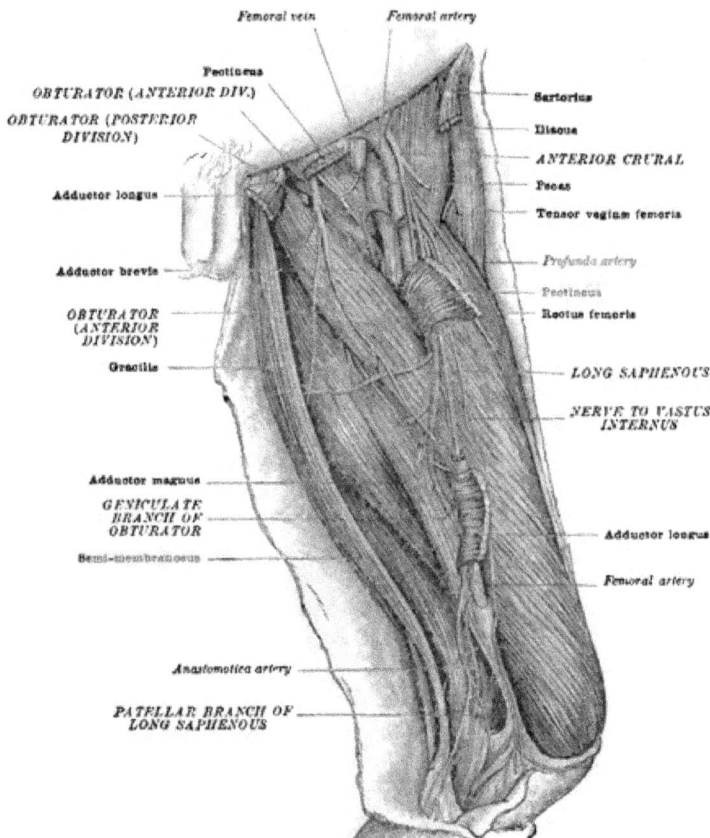

FIG. 219.—ANTERIOR CRURAL AND OBTURATOR NERVES. (Ellis.)

attended by the nerve to the vastus externus. This artery lies in a groove which separates the vastus internus from the vastus externus. Develop this groove, and you will be able to turn the vastus externus aside. This latter muscle, the vastus externus, overlaps the outer part of the vastus internus as far as the margin of the rectus. (Fig. 218.) Now trace the rectus and the two vasti downward to their conjoined insertion into the tubercle of the tibia, by the ligamentum patellæ. (Fig. 216.) Also note that the vastus internus receives its nerve-supply on its anterior surface. (Fig. 219.)

Obturator Nerve (Fig. 219).—This is a branch of the lumbar plexus. It passes below the brim of the pelvis, with the obturator artery and vein. (Fig. 223.) It escapes from the pelvis by the *obturator canal*, on the under surface of the horizontal ramus of the pubes. It divides into anterior and posterior branches. These are separated by the adductor brevis muscle. (Fig. 219.) The anterior branch

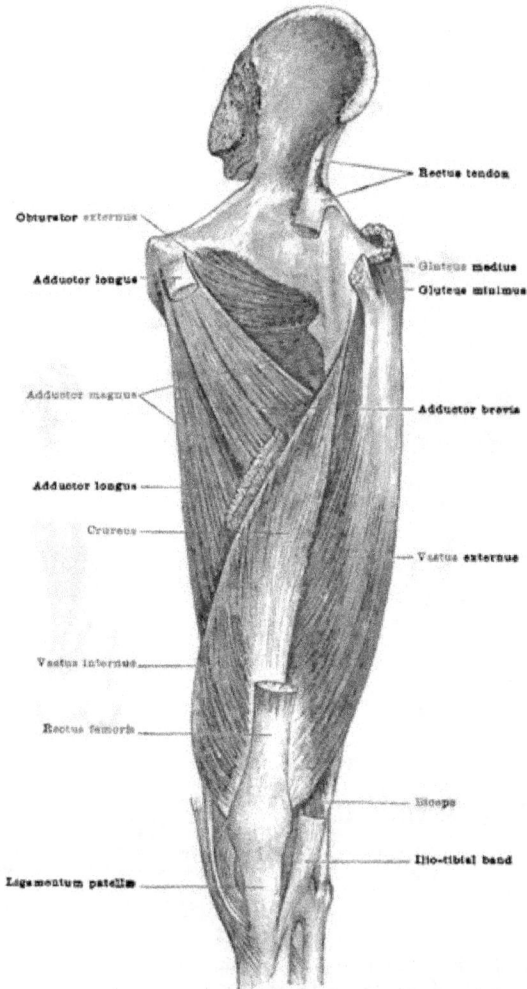

FIG. 220.—THE DEEP MUSCLES OF THE FRONT OF THE THIGH.

supplies the gracilis, adductor longus, adductor brevis, and sartorius. (Fig. 221.) The posterior branch supplies the adductor magnus, the knee, and the hip.

Branches of the Anterior Crural Nerve.—This nerve is from the lumbar plexus. (Fig. 193.) It is in a groove under the crural arch between the iliacus and psoas magnus. It gives off the following branches:

1. *A branch to the iliacus muscle.* This is given off within the pelvis, but is easily found distributed to the under surface of the muscle.

2. *Cutaneous branches*—the long saphenous, the internal and middle cutaneous.

3. *A small branch to the femoral artery.* This is given off within the pelvis.

4. *Muscular branches* to all the muscles on the front of the thigh. Note that the sartorius muscle is supplied either by the middle or internal cutaneous nerve.

5. *An articular branch to the hip,* given off from the branch to the rectus; also articular branches to the knee, from the nerves to the two vasti.

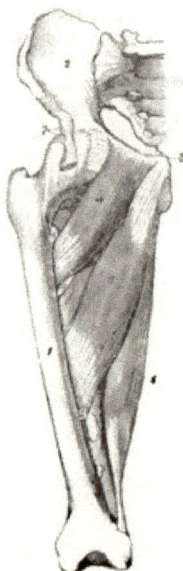

FIG. 221.—PECTINEUS AND ADDUCTOR LONGUS.
1. Femur. 2. Ilium. 3. Pubis. 4. Pectineus. 5. Adductor longus. 6. Lower portion of adductor magnus. 7. Tendon of rectus femoris. 8, 8. Orifices for vessels. 9. Orifices for femoral vessels.

FIG. 222.—ADDUCTOR BREVIS AND ADDUCTOR MAGNUS.
1. Femur. 2. Ilium. 3. Pubis. 4. Obturator externus. 5. Upper portion of adductor magnus. 6. Upper portion of adductor brevis. 7. Inferior portion of adductor brevis. 8. Middle portion of adductor magnus. 9. Inferior portion. 10. Tendon of insertion into internal condyle of femur. 11. Orifice for femoral vessels. 12. Orifice for internal circumflex artery and veins.

Branches of the Femoral Artery.—The common femoral is a continuation of the external iliac. (Fig. 223.) Usually, it is about one and one-half inches in length. It then divides into the superficial and deep femoral. Each has branches.

Branches of the Common Femoral Artery.—(1) Superficial circumflex iliac; (2) superficial epigastric; (3) superficial external pudic; (4) deep external pudic. These are all small and surgically insignificant branches. (Fig. 223.)

Branches of the Profunda.—(1) The internal circumflex; (2) the external circumflex; (3) the three perforating arteries.

Why are perforating arteries so called?

Because they perforate the adductor group, to reach the hamstring muscles on the back of the thigh, which they supply. (Fig. 222.)

The external circumflex supplies the muscles on the anterior part of the thigh. (Fig. 223.) Its ascending branch anastomoses with the gluteal and circumflex iliac; its descending branch with the superior external articular branch of the popliteal artery.

The Use of the Internal Circumflex (Fig. 223).—To assist the obturator artery in supplying the hip-joint and the adductor muscles, and to anastomose with the sciatic and external circumflex to complete the crucial anastomosis.

Where, on the posterior part of the thigh, will you find the terminal branch of the internal circumflex artery?

Between the quadratus femoris and the adductor magnus.

Branches of the Superficial Femoral Artery.—The superficial femoral artery lies in both Scarpa's triangle and Hunter's canal—surgical areas to be presently described. The branches given off by this artery are: muscular, to the sartorius and vastus internus; the anastomotica magna, given off just before the artery leaves Hunter's canal.

The anastomotica magna divides into two branches: (1) A superficial one, that accompanies the long saphenous nerve, and (2) a deep one, that anastomoses with the internal articular and the anterior recurrent tibial. This artery supplies branches to the knee-joint. It will be found on the bone, above the condyle, forming an arch with the external articular.

Review of the Work on the Thigh.—Thus far in the deep dissection of the anterior and lateral regions of the thigh, you have:

1. Traced the origin, insertion, and nerve-supply of the sartorius, and lifted this muscle from its bed.

2. You have found the exact origin and insertion of the gracilis, and liberated the same without doing violence to its nerves and blood-vessels.

3. You have traced the adductor longus from origin to insertion, preserved its nerves, cut the origin of the muscle, and seen the anterior division of the obturator nerve lying below, on the adductor brevis muscle; and the branches of this nerve you have followed to four muscles.

4. You have located the anterior crural nerve in a space between the psoas magnus and iliacus, taken this nerve up, and traced out its branches to the muscles on the front part of the thigh and to the skin covering these muscles and to the joints these muscles move.

5. You have removed the rectus from its bed, traced the same to its origins, and preserved the nerve- and blood-supply; finally, you have cut the rectus four inches above the patella.

6. You have seen that the crureus is not a separate muscle, but a part of the vastus internus.

7. You have traced the femoral artery well down and studied its profunda division.

8. You have traced the vasti and rectus to their common insertion into the tubercle of the tibia by the ligamentum patellæ.

What to Dissect Next.—(1) Cut the origin of the adductor brevis (Fig. 220), gently lift the same, and see the posterior division of the obturator nerve. Trace this nerve to the *adductor magnus* and *obturator externus* muscles. (2) Cut the pectineus (Fig. 219) and turn it down. Now, on the outer surface of the *obturator externus* muscle you will find the *obturator artery*, anastomosing with the *internal circumflex*, a branch of the profunda. (Fig. 223.) (3) Now you may replace in this order the muscles you have cut and turned aside: (*a*) Pectineus, (*b*) adductor brevis, (*c*) adductor longus, (*d*) rectus. Place the sartorius in its original oblique position.

Two Important Surgical Areas.—*Scarpa's triangle* and *Hunter's canal.*

Scarpa's triangle is located on the upper anterior one-third of the thigh. It is divided into an inner and an outer part by the femoral vessels, which bisect the space vertically. It is of surgical importance for the following reasons:

1. Inguinal and femoral herniæ are seen here.
2. Varicocele, dilatation of scrotal or spermatic veins.
3. Removal of inguinal lymphatic glands may become necessary.
4. Fracture of the femur and dislocations of its head.

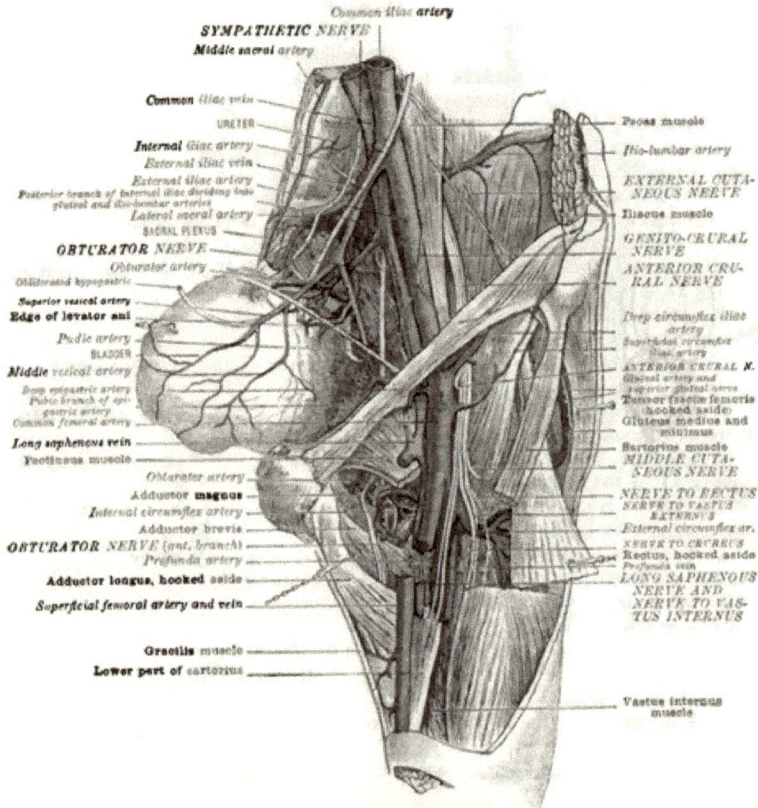

FIG. 223.—THE FEMORAL ARTERY IN SCARPA'S TRIANGLE.
(From a dissection by W. J. Walsham in St. Bartholomew's Hospital Museum.)

5. Burrowing of pus from regions above may point here.
6. Bloodless amputation at the hip involves this region.
7. Injuries to the vessels and nerves in the space may occur.

The triangle should be studied as having the following geometrical parts: (1) Roof: skin, superficial and deep fasciæ. (2) Floor: iliacus, psoas, pectineus, and adductor longus. (3) Superior boundary: Poupart's ligament. (4) Internal boundary: adductor longus muscle. (5) External boundary: the sartorius muscle. (6) The base: the superior boundary. (7) The apex: conjunction of adductor

longus and sartorius. *Contents :* (*a*) The *femoral sheath,* formed of the iliac fascia
behind the vessels, and the transversalis fascia in front of the same. This sheath
contains the femoral artery, vein, and femoral canal. (Fig. 212.) (*b*) The termination of the *long saphenous vein.* (*c*) The *external cutaneous nerve* of the thigh, which
you will find under the anterior superior iliac spine. (*d*) The *spermatic cord* in
the male, and its homologue, the round ligament, in the female, both of which
you will find under the spine of the pubes. (Fig. 224.) (*e*) The *obturator nerve*
and vessels. (*f*) The *lesser trochanter* of the femur, giving insertion to the iliacus
and psoas magnus muscles ; (*g*) the *anterior crural* or femoral nerve and its
branches ; (*h*) the *common femoral artery* and vein, bifurcating into the profunda
and superficial femoral arteries. In this, as in other regions, the student should
become so familiar with structures that he can, by the sense of feel alone, say
positively what he touches. I remember well a freshman examined the deep

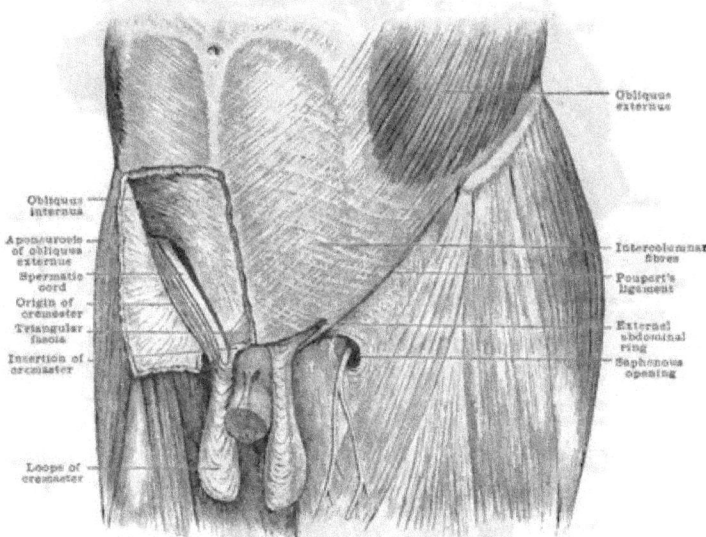

FIG. 224.—OBLIQUUS EXTERNUS AND FASCIA LATA.

part of Scarpa's triangle, and feeling the *lesser trochanter* of the femur, made a
diagnosis of osteo-sarcoma !

Poupart's ligament, or *crural arch,* is the lower part of the aponeurosis of the external oblique muscle of the abdominal wall. (Fig. 224.) It extends from the
anterior superior iliac spine to the pubic spine. It is continuous below with the
fascia lata. (Fig. 212.) It forms the floor of the inguinal canal, upon which
floor rest the spermatic cord and round ligament of the uterus, just before they
emerge through the external abdominal ring into Scarpa's triangle.

Hunter's canal (Fig. 219) (1) extends from the apex of Scarpa's triangle to
the aperture in the adductor magnus muscle. (Fig. 222.) In other words, it is a
groove linking together the popliteal space and Scarpa's triangle. (2) It is
bounded externally by the inner part of the vastus internus muscle ; internally by
the front surfaces of the adductor longus and adductor magnus muscles. (3) Its
roof is the deep fascia passing across from the adductors to the vastus internus.
It contains : (1) The superficial femoral artery and vein. The vein lies behind

the artery, according to the rule governing the relations of veins and arteries below the diaphragm. (2) The long saphenous nerve. Note that this nerve is on the outer side of the artery. (3) The communicating branch to the obturator nerve *via* the subsartorial plexus.

Explain the quadriceps femoris.

This is a collective noun indicating a four-headed muscle, on which depends

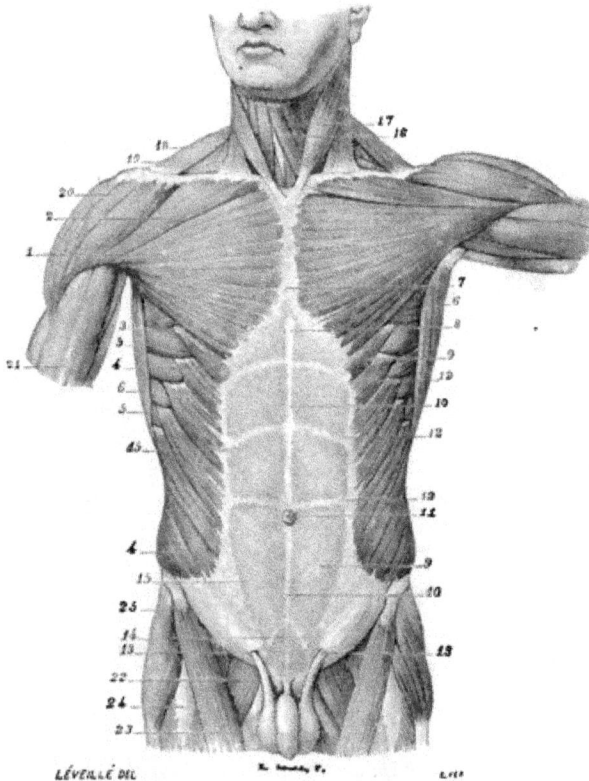

FIG. 225.—MUSCLES OF THE ANTERIOR ASPECT OF THE BODY.

1. Pectoralis major. 2. Its clavicular fasciculus. 3. Fasciculus attached to abdominal aponeurosis. 4, 4. External oblique. 5, 5. Serratus magnus. 6, 6. Anterior border of latissimus dorsi. 7. Decussation of tendinous fibers of pectorales majores. 8. Ensiform cartilage. 9, 9. Abdominal aponeurosis. 10, 10. Linea alba. 11. Umbilicus. 12, 12, 12. Tendinous intersections of rectus abdominis. 13, 13. External abdominal ring. 14. Pyramidalis. 15, 15. External border of rectus abdominis. 16. Sterno-hyoid. 17. Omo-hyoid. 18. Sterno-mastoid. 19. Cervical portion of trapezius. 20. Deltoid. 21. Biceps brachialis. 22. Pectineus. 23. Sartorius. 24. Rectus femoris. 25. Tensor vaginæ femoris.

extension of the leg. The individual muscles forming the extensor quadriceps femoris all converge to form one tendon,—the ligamentum patellæ,—which is inserted into the tubercle of the tibia. (Fig. 220.) These muscles, from their diverse origin and extensive attachments to the femur and ilium, represent both strength and celerity of motion—not an easy combination. The rectus, vastus externus, vastus internus, and crureus make up the extensor musculature. Trace these muscles to the following origins (from Morris):

The Rectus.—*Origin.*—Anterior head, from the front of the anterior inferior spine of the ilium; posterior head, from the upper surface of the rim of the acetabulum just external to the attachment of the capsular ligament.

Vastus Externus.—*Origin.*—(1) The upper half of the anterior intertro-

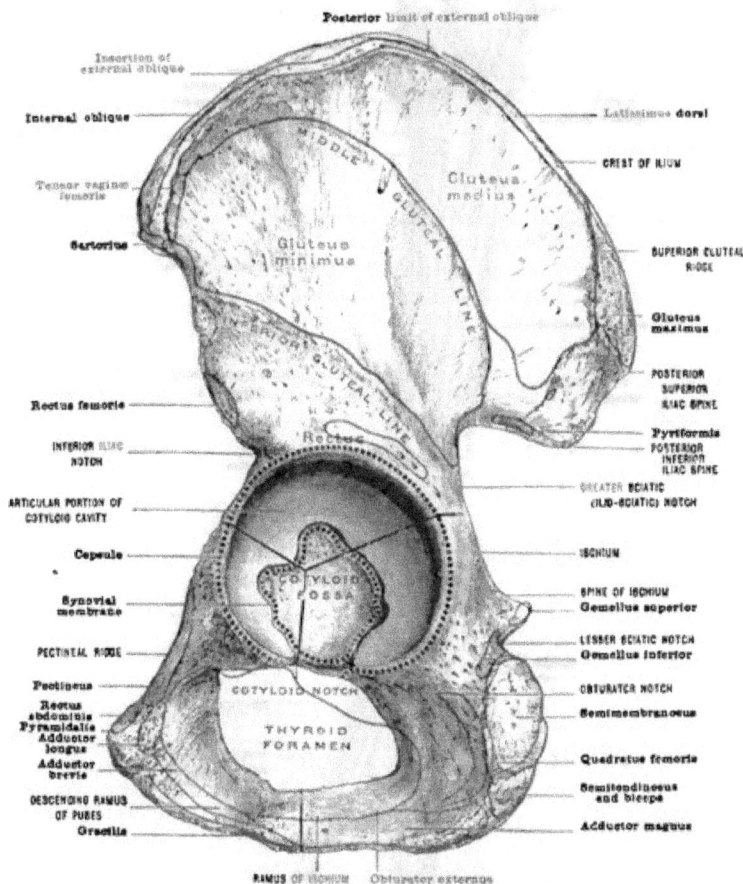

FIG. 226.—THE LEFT HIP-BONE. (Posterior view.)
(Trace out the origin and insertion of muscles and consult this figure.)

chanteric line and the front of the upper part of the femur along the anterior border of the great trochanter; (2) a horizontal line, which forms the lower border of the great trochanter; (3) the outer lip of the gluteal ridge; (4) the upper half of the outer lip of the linea aspera and the adjacent portion of the shaft of the femur for about one-sixth of an inch; (5) the external intermuscular septum in the neighbourhood of its attachment to the linea aspera.

Vastus Internus.—*Origin.*—(1) The outer lip of the lower half of the linea

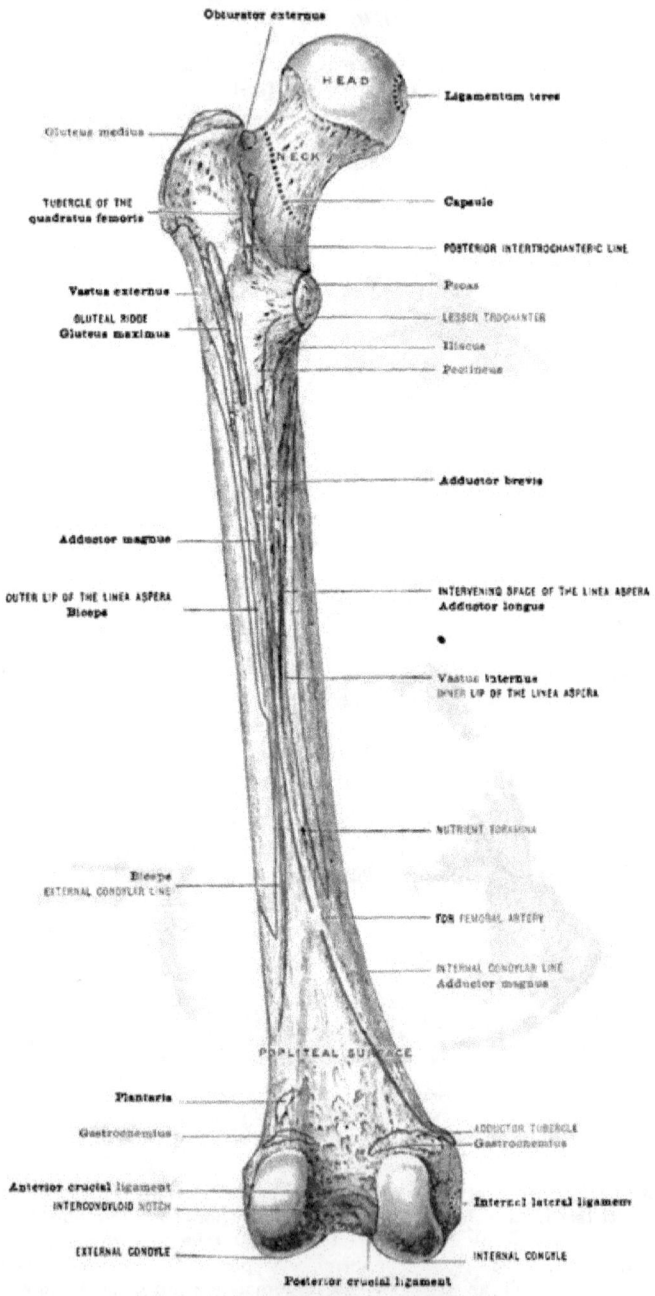

Obturator externus

HEAD

Ligamentum teres

Gluteus medius

NECK

TUBERCLE OF THE
quadratus femoris

Capsule

POSTERIOR INTERTROCHANTERIC LINE

Vastus externus

Psoas

GLUTEAL RIDGE
Gluteus maximus

LESSER TROCHANTER

Iliacus

Pectineus

Adductor brevis

Adductor magnus

OUTER LIP OF THE LINEA ASPERA
Biceps

INTERVENING SPACE OF THE LINEA ASPERA
Adductor longus

Vastus internus
INNER LIP OF THE LINEA ASPERA

NUTRIENT FORAMINA

Biceps
EXTERNAL CONDYLAR LINE

FOR FEMORAL ARTERY

INTERNAL CONDYLAR LINE
Adductor magnus

POPLITEAL SURFACE

Plantaris

ADDUCTOR TUBERCLE
Gastrocnemius

Gastrocnemius

Anterior crucial ligament
INTERCONDYLOID NOTCH

Internal lateral ligament

EXTERNAL CONDYLE

INTERNAL CONDYLE

Posterior crucial ligament

FIG. 227.—THE LEFT FEMUR. (Posterior view.)
(Study the insertion of the adductor on this figure.)

326

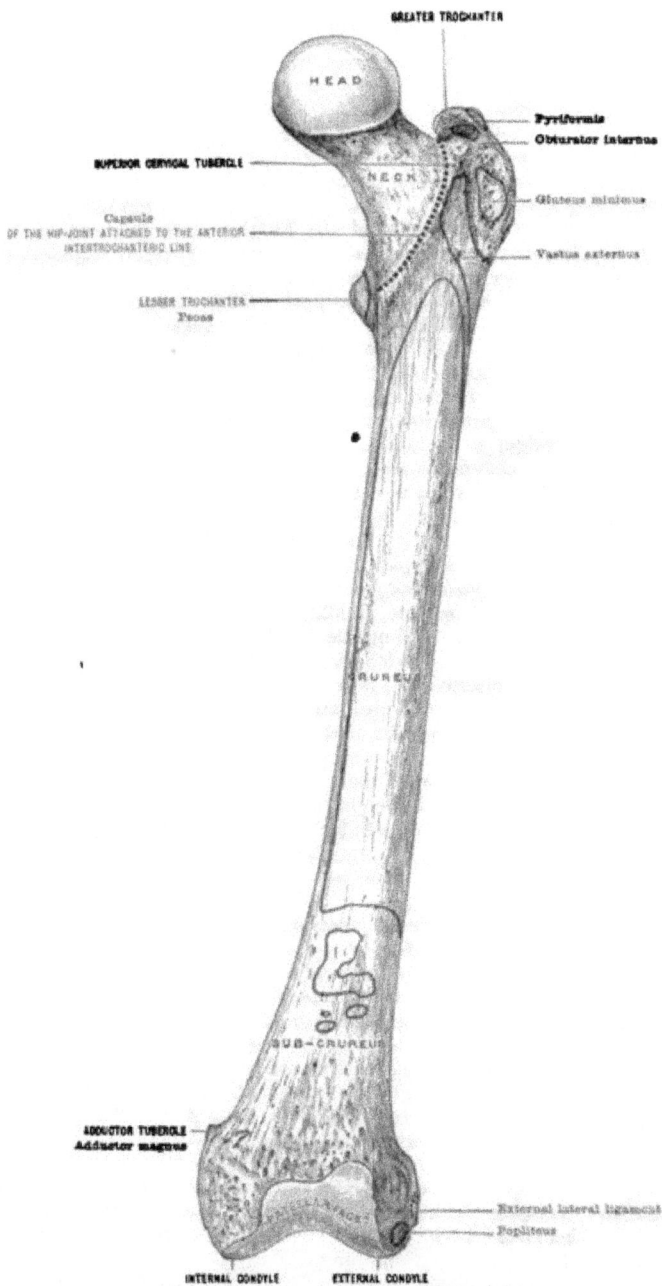

GREATER TROCHANTER

HEAD

Pyriformis
Obturator internus

SUPERIOR CERVICAL TUBERCLE

NECK

Gluteus minimus

Capsule
OF THE HIP-JOINT ATTACHED TO THE ANTERIOR
INTERTROCHANTERIC LINE

Vastus externus

LESSER TROCHANTER
Psoas

CRUREUS

SUB-CRUREUS

ADDUCTOR TUBERCLE
Adductor magnus

External lateral ligament
Popliteus

INTERNAL CONDYLE   EXTERNAL CONDYLE

FIG. 228.—THE LEFT FEMUR. (Anterior view.)
(Study the origin of the vasti on this figure.)

327

aspera and its external bifurcation, together with the adjacent external inter-muscular septum ; (2) the lower part of the anterior intertrochanteric line and the spiral line of the femur ; (3) the inner lip of the whole length of the linea aspera and its internal bifurcation, together with the adjacent part of the internal inter-muscular septum, and the front of the tendon of the adductor magnus ; (4) the greater part of the front and sides of the femur within the limits formed by the three preceding attachments and the origin of the vastus externus.

The Physiological Adductors of the Thigh.—In figure 222 this group of muscles may be seen. They are four in number. They are supplied by the obturator nerve. They are all ligamentous in action to the hip-joint ; hence the obturator nerve will send a nerve to this joint. They all figure conspicuously in the architecture of Scarpa's triangle and Hunter's canal. Study the origin of these muscles very carefully on figure **226**, and then on the **cadaver**, as follows :

The adductor longus arises from the anterior surface of the body of the pubis immediately below the crest and angle.

The adductor brevis arises from the **body** and ramus **of the pubis, below and** external to the origin of the adductor longus. If the student will study the divergence of the ischio-pubic rami, he will then understand why the origin of the adductor brevis is described as external to that of the adductor longus.

The adductor magnus arises from the tuber of the ischium, and from the ischio-pubic ramus. (Fig. 226.) Examine the perforations in this muscle for the superficial femoral artery to pass into the popliteal space.

Name the adductor muscles and give their insertions. (Fig. 227.)

Adductor longus, inserted into middle lip of linea aspera, middle one-third. Adductor brevis, inserted into middle lip of linea aspera, upper one-third. Adductor gracilis, inserted into inner surface of tibia, upper one-third. (Fig. 216.)

Adductor magnus, inserted : (1) Into the back of the femur, in a line beginning at the lower extremity of the linea quadrati, and extending along the inner border of the gluteal ridge and the middle of the linea aspera down to its bifurcation ; (2) into the adductor tubercle on the upper and posterior part of the internal condyle ; (3) into the lower part of the internal intermuscular septum.

Fascial Septa.—Having thoroughly dissected all the muscles on the inner part of the thigh and front part of both leg and thigh, I now desire you to answer these questions :

1. *How many musculatures have you found—that is, how many groups of muscles acting in unison?*

You say : (1) An *extensor group*, innervated by an extensor **nerve**, called an-terior crural or femoral ; (2) a *flexor group*, on the posterior part of the thigh, innervated by a flexor nerve, called the great sciatic ; (3) an *adductor group*, on the **inner** part of the thigh, innervated by an adductor nerve, called the obturator ; (4) an abductor group, on the outer side of the pelvis, innervated by the superior **gluteal**, an abductor nerve ; (5) a *group on the front of the leg*—flexors of the tarsus and extensors of the toes—supplied by the anterior tibial nerve ; (6) a *group on the outer surface of the fibula*, supplied by the muscular branches of the musculo-cuta-neous nerve ; (7) two *groups of muscles on the posterior part of the leg*, innervated by the internal popliteal nerve, extensors of the tarsus and flexors of the toes.

2. *Are the groups individually separated from each other by septa?*

You answer yes, for you have found a fascial septum between the adductors and extensors ; between the flexors and adductors ; between the anterior and lateral groups below the knee ; between the lateral and posterior groups ; be-tween the superficial and deep groups on the back of the leg.

3. *Are these intermuscular septa attached to bone? Do they in any manner give origin to muscle? Are they derived from deep fascia?*

You answer the above in the affirmative.

Anatomical transition is never abrupt, but gradual. Compact bone-tissue gradually merges into cancellous; muscle gradually becomes tendinous; at muco-cutaneous areas skin and mucous membrane gradually partake of the nature of each other. The transition from extension of the leg to adduction of the thigh would be abrupt. This is moderated by the sartorius muscle, which, in consequence of its dual function, has a double nerve-supply and an independent fascial sheath. The adductor magnus moderates between flexion and adduction; hence its dual nerve-supply.

The Femur (Figs. 227 and 228).—*Name the points of the femur of importance in practical anatomy.*

The head articulates with the acetabulum.

The neck joins the head of the bone to the shaft.

The greater trochanter facilitates axial rotation.

The lesser trochanter facilitates axial rotation.

The tereal depression is for the ligamentum teres.

The digital or trochanteric fossa is for the obturator externus.

The intertrochanteric lines are anterior and posterior.

The oblique line of the greater trochanter is for the gluteus medius.

The anterior intertrochanteric line is for the capsule.

The tip of the greater trochanter is for the gemelli, **pyriformis, obturator externus.**

The condyles articulate with the **tibia and** patella.

The condylar ridges limit laterally the popliteal surface.

The intercondylar notch lodges the crucial ligaments.

The patellar facet articulates with the patella.

The adductor tubercle is for the adductor magnus tendon.

How and where are the gluteal muscles inserted?

They are inserted principally by tendon as follows: The gluteus maximus into **the** gluteal ridge; the gluteus medius into the oblique line of the greater trochanter; the gluteus minimus in the anterior border of the greater trochanter.

What is the third trochanter?

A name given to the gluteal ridge when this is unusually prominent. **Growth** is the correlative of function; where extraordinary function is imposed on any group of muscles, or on any single muscle of a group, the bony insertion point of the muscle will be correspondingly increased in size. Trochanter means a bony eminence for muscular attachment, situated favorably for axial rotation of the limb; hence the name, third trochanter.

Describe the linea aspera.

This may be compared to a tree with (1) a trunk, (2) definite roots, (3) definite branches **above.** Now, **when we** speak **in** general terms of a tree, we have in mind the trunk; when we speak specifically of this object as a shade-tree, we have in mind the branches; when we speak specifically of its attachments to the earth, we have in mind the roots of the tree. The linea aspera has a trunk which is in the middle one-third of the femur; branches which ramify definitely in the upper one-third of the femur; roots which are definitely disposed in the lower one-third of the femur.

Describe the trunk of the linea aspera.

It is composed of three lips or ridges, produced by muscular traction; these lips are called outer, middle, inner. This part of the linea aspera occupies the middle one-third of the shaft of the femur, and forms the posterior border of the same.

How is the root of the linea aspera disposed?

The outer lip is continued downward and outward to the condyle of the

22

femur as the external condylar ridge. The inner lip is continued downward to the adductor tubercle as the internal condylar ridge. The middle lip is continued downward, and expands to form the popliteal surface of the femur.

How are the branches of the linea aspera disposed?

The inner lip continues upward, around and below the lesser trochanter, as the spiral line ; to the lower half of this spiral line is attached the vastus externus ; to the upper, the capsule of the hip-joint.

The outer lip of the linea aspera continues upward and outward, and ends at the base of the great trochanter in the gluteal ridge for the gluteus maximus.

The middle lip divides into an inner and an outer division. The inner breaks up into two branches,—one for the iliacus, the other for the pectineus muscle,—both of which continge the lesser trochanter. The outer division of the middle lip of the linea aspera terminates above as a vertical ridge, midway between the trochanters, as the linea quadrati, for the insertion of the quadratus femoris.

Describe the attachment of muscles to the three lips of the trunk of the linea *aspera.*

To the outer, the short head of the biceps femoris and the vastus externus ; to the inner lip, the vastus internus ; to the middle lip, the adductor brevis, adductor longus, and adductor magnus.

THE PELVIC OUTLET.

Locate : (1) The subpubic arch ; (2) the tuberosities of the ischium; (3) the tip of the coccyx ; (4) the ischio-pubic ramus on each side, and give its composition ; (5) the spine of the pubes ; (6) the symphysis pubis.

Place the subject in the position shown in figure 229—*i. e.*, flex the leg on the thigh, the thigh on the abdomen, and abduct the thighs. Now fill the rectum with cotton and sew the margins of the anus together. Do not take your stitches too deeply—one-fourth of an inch external to the junction of the skin and mucous membrane is enough.

Incisions.—(1) Cut from the subpubic arch through the skin to the front margin of the anus ; then from the posterior margin of the anus to the tip of the coccyx ; (2) cut from one tuberosity of the ischium to the other ; (3) make a circular incision around the anus just a little outside of the stitches employed in sewing together the margins of the anus ; (4) now begin, and dissect back the four flaps. The greatest care must be taken to keep the skin very thin—thin enough to see through.

You have now removed the skin from the outlet of the pelvis. This space is diamond in shape, and has the following boundaries : Two lines above and two below will outline the space called the pelvic outlet—(1) a line from the tuberosity of the ischium to the subpubic arch on each side above ; (2) a line from the tip of the coccyx to the tuberosity on each side below. This outlines two triangles with their bases together on an imaginary line from one tuberosity to the other.

The anterior triangle is called the perineal ; **the posterior,** the ischiorectal. Continue removing skin until you have exposed a space as large as that represented in figure 229. The next step is to introduce a sound into the bladder, and retain it there by means of string tied tightly around the penis near the end.

A dissection of this region really consists of only two stages : (1) Of the structures seen in figures 229 and 230; (2) of structures seen in figure 233. After having removed the skin, only a little developing is necessary to expose all the structures seen in figures 229 and 230. To give an adequate idea of the subject, study the following to learn what you are expected to find in the first stage of the dissection:

1. *The deep layer of the superficial fascia* that you see in figure 230 turned outward and marked Colles' fascia. Notice this is attached to the ischio-pubic ramus and to the triangular ligament, whose cut edge shows in figure 230. This is called the *fascia of Colles*.

2. Around the anus you will see a very delicate layer of muscular fibres that

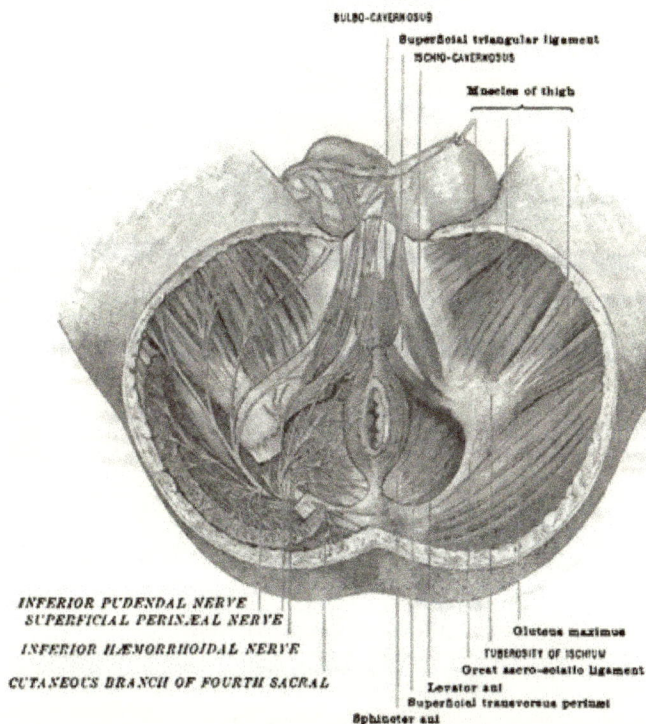

FIG. 229.—THE MALE PERINÆUM. (Modified from Hirschfeld and Leveillé.)

throws the margin of the anus into folds—the *corrugator cutis ani*, a dermal muscle.

3. In the anterior triangle you see the three large pyramids originating, which come together to form the penis: (1) The *bulb* in the centre, being the lower part of the corpus spongiosum; (2) the *crura*, parts of the corpora cavernosa on each side, in firm connection with the *ischio-pubic rami*.

4. The bulb, covered by a muscle called the *accelerator urine* in the male. Depress the sound and you can make this bulb stand out very prominently. (Fig. 230.)

5. *The crus penis*, covered by a muscle called the *erector penis*—ischio-cavern-

osus—in the male.   Pull the penis upward and from side to side gently, and this
will bring the crura and their investing muscles into view.   (Fig. 239.)

6. *The external sphincter ani* you will see reaching from the tip and sides of
the coccyx, surrounding the anus, to the central perineal point.   The outer
margins of this muscle seem to become lost in a quantity of fatty and connective
tissue.

7. Another muscle, the *superficial transversus perinæi*, lies between the
sphincter ani and the erector penis, extending in an oblique direction from the
tuberosity of the ischium to the central perineal point.

8. *The central perineal point, or tendinous centre* of the perineum, is in the
mid-line from side to side and about one inch in front of the anus.   As you will

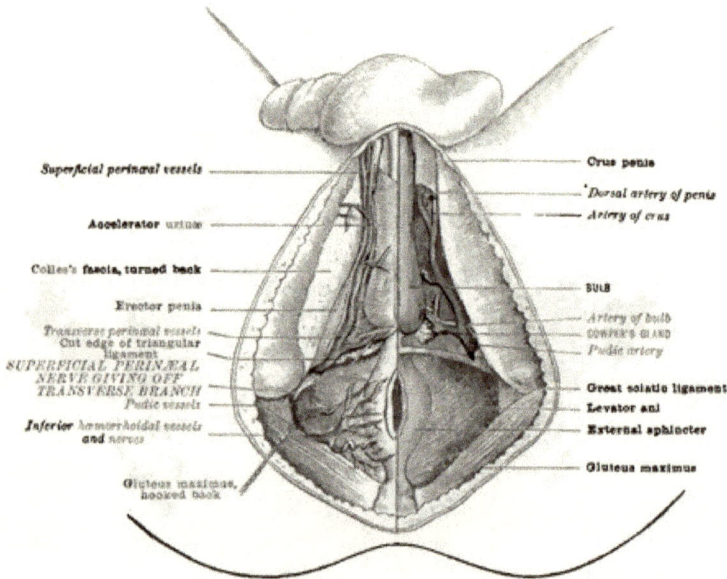

FIG. 230.—THE ARTERIES OF THE PERINEUM.

On the right side Colles' fascia has been turned back to show the superficial vessels.   On the left side the
superficial vessels have been cut away with the anterior layer of the triangular ligament to show the
deep vessels.

demonstrate in your dissection, and as you may see in figure 229, this point is
where the following muscles meet: (1) The sphincter ani ; (2) the superficial
transversus perinæi ; (3) the accelerator urinæ or bulbo-cavernosus.

9. *The levator ani* shows well in both figures, but at this stage of the dissection is obscured by (1) the sphincter ani and (2) the large mass of fat in the
ischio-rectal fossa.   The muscle is seen in figure 230 on each side of the anus.

10. The space on each side of the anus is the *ischio-rectal fossa.*   It is now
full of fat.   Figure 230 shows this fat removed and the levator ani muscle
divested of the anal fascia.

11. *The greater sacro-sciatic ligament* must now be located.   In the figures it
is shown cut on one side to expose the internal pudic structures.

12. Notice the *hæmorrhoidal branches*, and the perineal branches of the internal

pudic vessels and nerves; **also the inferior pudendal nerve, a branch of the small** sciatic, distributed to the scrotum and side of the thigh.

Dissection.—Locate the *greater sacro-sciatic ligament.* Cut this ligament and find under the same, in the *lesser sacro-sciatic foramen,* the main branches of the internal pudic nerve and vessels. You will find two fair-sized branches given off. They will both be found to break up into numerous branches. Now carefully trace these branches through the fat of the *ischio-rectal fossa*—the **one to the** anus, the other forward to the perineum. The branch to the anus is the *inferior hæmorrhoidal;* the other is the *superficial perineal.* Notice: (1) These branches are both from the internal pudic; (2) each divides into *cutaneous* and *muscular* branches, for the supply of the muscles and the skin in these regions. The arteries follow the nerves and take the same name. The internal pudic nerve now continues **in a** canal—called Alcock's. This nerve pierces the posterior layer of the triangular ligament (Fig. 233), and gives off the following branches: (1) The *dorsal nerve of the penis;* (2) *nerve to the bulb;* (3) *nerve to the corpus cavernosum;* (4) *superficial perineal.* This, the dorsal nerve of the penis, is the largest division of the internal pudic **nerve.** You will find it on the inner surface of the **ramus of** the ischium. **It** passes behind the superficial transversus

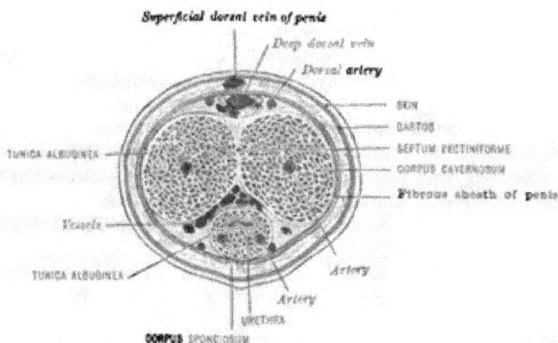

FIG. 231.—TRANSVERSE SECTION THROUGH THE BODY OF THE PENIS.

perinæi **muscle,** gains the **deep perineal** space, **pierces the** anterior layer of the deep perineal fascia, **and is distributed to the** dorsum of the penis.

Crus Penis (Dissection of) and Its Muscle—the Erector Penis, or Ischio-cavernosus.—The muscle covers the crus. Cut through the centre of the muscle in the long diameter and turn the two halves aside. Notice its extensive origin: (1) Tuberosity of ischium, (2) ramus of ischium, (3) surface of the crus. Trace the muscle up to its insertion into the sides and under surface of the crus; also notice its nerves and vessels. Now cut through the thick, strong capsule of the crus in its long diameter, and examine the *erectile tissue* of its interior. This tissue is the specific element of the crus, and is called, in common with the remainder of the lateral pyramid of the penis, the corpus cavernosum. **Observe the** strong origin of the crus from the ischio-pubic ramus.

Dissection of Accelerator Urinæ and the Bulb.—Depress the sound and make the bulb, covered by the *accelerator,* stand out full. This muscle, as you see, is in **the mid-line, in front of** the anus. The muscle being on the stretch, you will now **see in the middle a raphe separating the** two halves. Now cut, with a very sharp **knife, through the center of this** raphe and turn the muscle outward, and thus expose the bulb. **(Fig. 230.)** Trace the muscle and note its insertions: (1)

The lowest fibres into the triangular ligament; (2) *the middle,* surround the bulb; (3) *the upper* ones pass out across the corpus spongiosum and are inserted into the dorsum of the penis and pubic bone. Notice, also, the *nerve-supply* to this muscle from the internal pudic. Now cut through the bulb down to the **sound** and see (1) the bulbous part of the urethra and (2) the erectile **tissue; the latter** is **the specific** element of the bulb, and, taken with the remaining part **of the middle** cylinder of the penis, is called the *corpus spongiosum.* Notice the dense fibrous capsule of the bulb, derived from the outer layer of the triangular ligament, upon which the bulb rests. See the nerves and vessels from the internal pudics.

Dissection of Superficial Transversus Perinæi.—Trace this little muscle from the tuberosity of the ischium to the central point of the perineum, it being inserted between the sphincter ani and the accelerator. Notice the nerve-supply.

Dissection of the Penis.—With the scissors cut through the mid-line of the skin of the scrotum and under part of the penis. Now carefully remove the integument from the entire organ. As stated above, the penis is composed of three cylinders, which converge from three different sources. These three cylinders are (1) the two corpora cavernosa and (2) the urethal cylinder or corpus spongiosum, which unite to form a triangular body with three compartments. In transverse section. (Fig. 231.) See, now, that each cylinder has special characteristics. (1) The central cylinder, resting on the outer layer of the triangular ligament, begins in the *bulb* and terminates in an expansion, the *glans penis,* which receives the conical extremities of the lateral cylinders. (Fig. 232.) (2) The lateral cylinders, the *corpora cavernosa,* begin in blunt crura from the ischiopubic ramus, are covered by compressor muscles, erectores penis, and terminate in pointed extremities in the distal expansion of the *corpus spongiosum*—the head of the penis. Observe, also, that each cylinder has common characteristics: (1) a fibrous elastic envelope; (2) erectile tissue. Lastly, see the middle **cylinder** traversed from end to end by the urethra. Now cut down on the sound and expose the urethra. (Fig. 232.)

The *urethra* is the mucous canal extending from the bladder to the extremity of the glans penis. In its course it pierces the prostate from base to apex, the deep and superficial triangular ligaments with the intervening compressor urethræ muscle, and the whole length of the corpus spongiosum. It may hence be divided into three segments: (1) Prostatic; (2) membranous (the portion lying in the space between the two transverse triangular ligaments), and (3) spongy. (Fig. 232.)

Having completed the first stage of the **dissection of the pelvic** outlet, review your work carefully, and see if you have found the following structures:

1. You removed the skin, learned the boundaries of the pelvic outlet, and classified the same into an *anterior* or *perineal* portion, and a *posterior* or *ischio-rectal portion.*

2. You saw the deep layer of the superficial fascia,—Colles' fascia,—having two attachments: (1) To the *ischio-pubic ramus;* (2) to the *triangular ligament.*

3. You located the greater sacro-sciatic ligament, cut the same, and found the main trunk of the *internal pudic nerve* with its accompanying vessels. In this locality you found two nerves given off. One of these you traced through the fat in the ischio-rectal fossa, in many branches, to the corrugator cutis ani and external sphincter ani muscles, and also to the skin covering these muscles; this was the *inferior hæmorrhoidal nerve.* The other you traced, under the name of *superficial perineal nerve,* to (1) the accelerator, (2) the erector penis, (3) the superficial transversus perinæi, and to the skin covering these muscles.

4. Then you located the *central point of the perineum,* and learned the muscles meeting there and their nerve-supply as follows: (1) Accelerator urinæ; (2)

erector penis; (3) superficial transversus perinæi; (4) levator ani. Nerve-supply, the internal pudic.

5. You then located the origin of the three cylinders of which the penis is composed,—the central one on the triangular ligament as the bulb, the lateral ones on the ischo-pubic rami as the crura,—and you studied the muscles associated with these three cylinders: accelerator urinæ and erector penis.

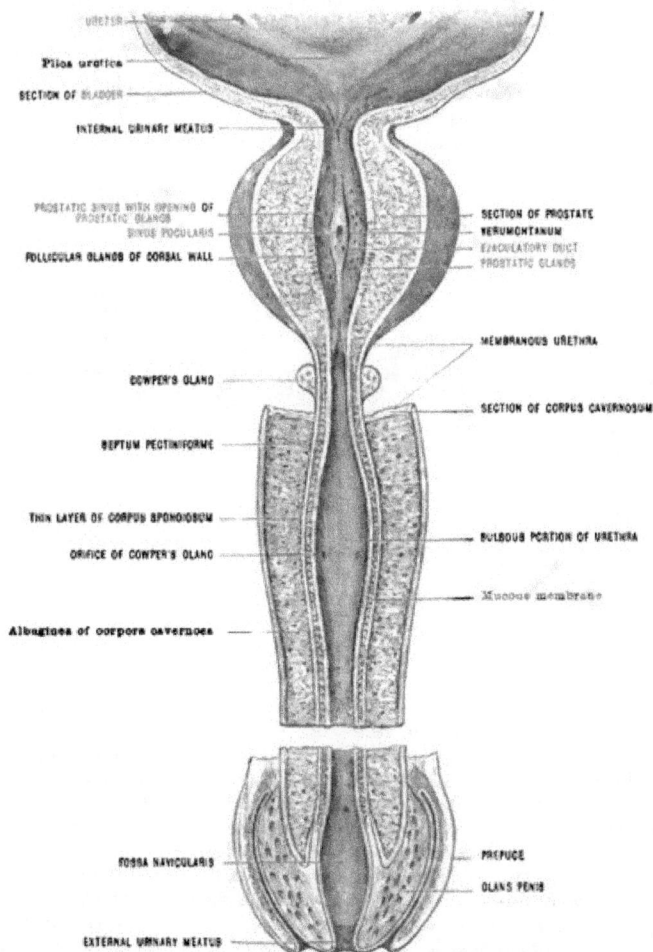

FIG. 232.—THE MALE URETHRA, CLEFT DORSALLY TO SHOW VENTRAL MUCOUS WALL.

6. You examined the three component cylinders of the penis; saw the *general* and *special* characters of each. You saw each cylinder surrounded by a dense tunica albuginea and composed internally of erectile tissue. You saw the central cylinder traversed by a conduit—the urethra; you opened this urethra and studied its interior. In the cuts and in your dissections you studied both the transverse and longitudinal sections of the penis.

Second Stage in Dissecting the Pelvic Outlet.

Dissect the crura from the rami and turn them upward over the pubic bone and fasten them out of your way. Cut the central cylinder—the corpus spongiosum—off even with the outer layer of the triangular ligament. Detach the external sphincter ani and the two superficial transverse perineal muscles at the central perineal point. Then your field will look like the left side of figure 233.

The triangular ligament (Fig. 233) occupies the space under the pubes. It consists of two layers, superior and inferior, also called superficial and deep triangular ligaments, between which are found: (1) The membranous urethra (Fig. 232); (2) the duct of Cowper's gland perforating the superficial triangular ligament (Fig. 233) and opening into the bulbous urethra (Fig 232); (3) the dorsal nerves and vessels of the penis perforating the suspensory ligament of the penis and supplying the back of the penis (Fig. 233); (4) the deep transversus perinæi or compressor urethræ, the muscle of Guthrie. (Fig. 234.)

The above structures are seen on removing the inferior layer of the triangular

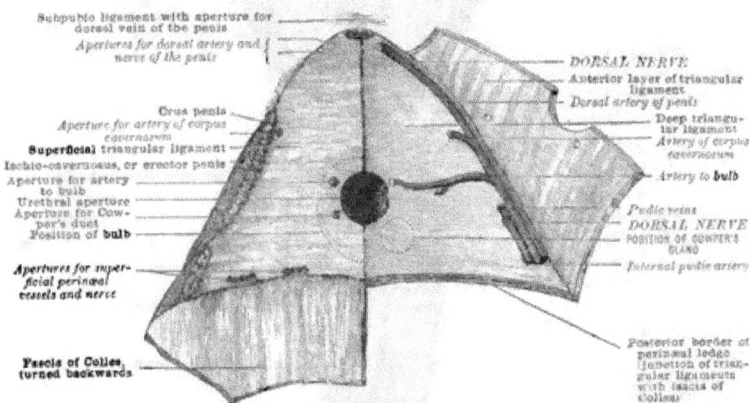

Fig. 233.—Diagram of the Superficial and Deep Triangular Ligaments.

ligament. (Fig. 233.) Now study the attachments of, and the foramina in, the triangular ligament and compare your work with figure 233.

You will now dissect the ischio-rectal fossa. You have traced the inferior hæmorrhoidal vessels through the fatty mass of this fossa to the rectum and anus. You will now remove the fat and study the geometrical parts of this region.

The outer wall is formed by the obturator internus muscle and the innominate bone. This muscle is covered by a dense fascia of the same name. This fascia splits to form a canal (Alcock's) in which are the internal pudic vessels and nerve. *The inner wall* of the ischio-rectal fossa is formed by the levator ani and coccygeus muscles. These, as you will see, embrace the rectum and give it support. The levator ani muscle arises from the white line. This line extends from the spine of the ischium to the pubic bone. The obturator fascia delaminates at the white line, and gives origin to the levator ani muscle in the split between the upper and lower branches into which it divides. The apex, then, of the fossa is at the white line. The boundaries, as you will now see them on your work, of the inlet of the fossa are: (1) The external sphincter ani; (2) the superficial transversus perinæi; (3) the greater sacro-sciatic ligament; (4) the gluteus maximus muscle; (5) the tubers of the ischium. The base of the fossa is the skin and fasciæ covering the inlet. The fossa extends in an upward direction to the pubic bone ; in a backward direction to the sacrum, forming the *anterior* and *posterior recesses* of Morris. In liberating pus in this fossa, cut through the base ; then lay your knife aside and continue the operation with a blunt instrument.

1. *Give the shape of, and tell what you mean by, the pelvic outlet.*

It is lozenge-shaped, and is the space bounded by the pubic arch above, the coccyx below, and the greater sacro-sciatic ligament, tuberosity of the ischium, and ischio-pubic ramus laterally.

2. *Give the subdivisions of this space, and tell where the boundary-line between the two is located.*

A line from tuberosity to tuberosity of the ischium divides the space into an anterior triangle, the perineum, and a posterior triangle, the ischio-rectal fossa.

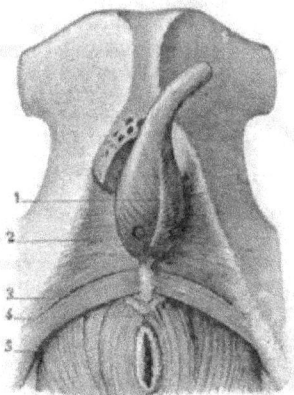

FIG. 234.—MUSCLE OF GUTHRIE.

1. Bulbo-cavernosus (accelerator urinæ) muscle. 2. Muscle of Guthrie (transversus perinæi profundus). 3. Superficial transverse muscle. 4. External sphincter ani. 5. Levator ani.

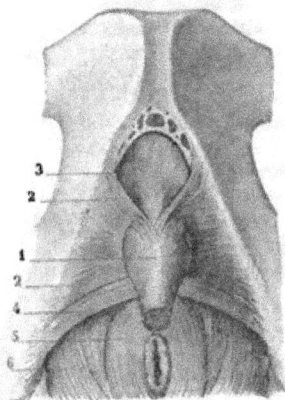

FIG. 235.—MUSCLE OF GUTHRIE AND WILSON.

1. Bulb of urethra. 2, 2. Muscle of Guthrie (transversus perinæi profundus). 3. Muscle of Wilson. 4. Transversus perinæi superficialis. 5. External sphincter ani. 6. Levator ani.

3. *Give the contents of the ischio-rectal fossa.*

It contains the rectum and the fossa on each side of the rectum, called ischio-rectal fossa.

4. *Give the geometrical parts of this fossa and the importance you attach to each part.*

(1) Apex, at the white line, the origin of the levator ani ; (2) base, the skin and fasciæ, between anus and tuber of ischium ; (3) outer wall, the obturator internus, Alcock's canal and contents ; (4) inner wall, the levator ani and coccygeus muscles covered by the anal fascia ; (5) an anterior or pubic recess ; (6) a posterior or sacral recess.

5. *Where is Alcock's canal, how is it found, and what are its contents ?*

It is in the outer wall of the ischio-rectal fossa, in a delamination of the obturator fascia ; it contains the internal pudic vessels and nerve and extends from the lesser sacro-sciatic foramen to the deep triangular ligament.

6. *Name the contents of the ischio-rectal fossa.*

The fossa contains fat and connective tissue, in which are embedded the inferior hæmorrhoidal nerves and vessels. On dissection you will see these structures crossing the fossa to supply the rectum.

338 *PRACTICAL ANATOMY.*

7. *Name the structures that meet at the central point of the perineum.*

The accelerator urinæ and bulb. The external sphincter ani. The superficial transversus perinæi. The rectum and labia majora. The triangular ligament and labia majora. The fascia of Colles. The superficial perineal space.

8. *What is the importance of the fascia of Colles in extravasation of urine through the bulb?*

This fascia limits its extension laterally and posteriorly, since it is attached both to the ischio-pubic ramus and lower part of the triangular ligament.

9. *Give the derivation of the superior layer of the triangular ligament.*

It is a continuation of the obturator fascia.

10. *Define the anal fascia.*

It is a division of the obturator fascia that covers the under surface of the levator ani muscle.

11. *Name the several modifications of the pelvic fascia.*

(1) Iliac ; (2) obturator ; (3) white line ; (4) Alcock's canal ; (5) anal fascia ; (6) recto-vesical fascia ; (7) triangular ligament ; (8) pubo-prostatic ligament.

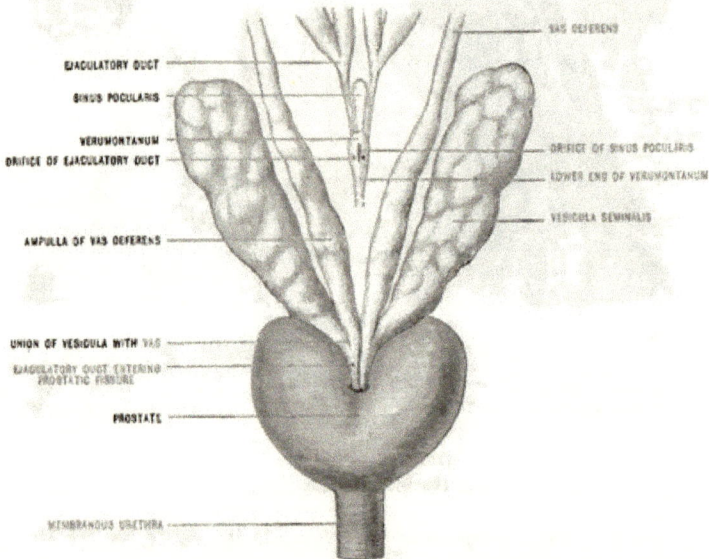

FIG. 236.—VASA DEFERENTIA AND VESICULÆ SEMINALES. (After Sappey.)

The Base of the Bladder and Seminal Vesicles.—Having thoroughly studied the perinèum and ischio-rectal fossa, make a dissection to show (1) the membranous urethra, (2) the prostate gland, (3) the lower part of the rectum, (4) the vesiculæ seminales, (5) the base of the bladder, in the following manner (turn to Fig. 229): Cut the external sphincter ani and pull the rectum backward and downward over the coccyx. This will put the levator ani on the stretch. Now divide the levator ani near the white line—its origin, as you will remember. Now you will see in the mid-line (1) the membranous urethra, with Cowper's gland on each side ; (2) the prostate gland, receiving at its central part the ejaculatory duct of the seminal vesicles (Fig. 236); (3) the base of the bladder, occupied by fatty connective tissue, in which you will find a remarkable plexus of veins—the vesico-prostatic. A small amount of work with the forceps will expose to view all the above structures.

1. *Describe the prostate gland.*

It is in front of the neck of the bladder, the part resting on the bladder being the base. (Fig. 232.) The posterior surface is on the rectum. The anterior surface is behind the symphysis. The apex reaches the triangular ligament. It has three lobes. You will see, between the lateral lobe and the middle lobe, the ejaculatory duct. The prostate is traversed by the urethra and ducts **just** mentioned. The gland is surrounded by a firm capsule. **Its** substance **is gland**ular and muscular, and not easily torn, as is the spleen.

2. *Describe the vesico-prostatic plexus.*

This is now easily shown. It consists of engorged veins which surround the neck and base of the bladder and the prostate, just mentioned. The dorsal vein of the penis passing close under the subpubic arcade discharges into this plexus. (Fig. 233.) Behind this plexus is the hæmorrhoidal plexus. The two plexuses communicate very freely.

3. *Give function of vesiculæ seminales.* (Fig. 236.)

They act as a reservoir for the semen, just as the gall-bladder is a receptacle for bile. The vas deferens brings semen from the testicle, as the hepatic duct brings bile from the liver. The seminal duct and vas come together to form the excretory duct, just as the cystic and hepatic unite to form the common bile **duct.** The vas deferens is two feet in length.

4. *Name the structures pierced by the male urethra.*

(1) The prostate from base to apex ; (2) the deep and superficial triangular ligaments ; (3) the compressor urethræ muscle (Guthrie's muscle) ; (4) the corpus spongiosum.

5. *What is the average length of the male urethra?*

It is about six and one-half inches long. Its length is increased in senile hypertrophy of the prostate.

6. *Give the length of the prostatic urethra and tell what you found in the* **same.**

Its length is one and one-fourth inches. In this part of the urethra we **found :** (1) The sinus pocularis, or uterus masculinus, believed to be homologous to the uterus of the female ; (2) the openings for the seminal ejaculatory ducts ; (3) the **orifices** for the prostatic glands.

7. *Name and locate the dilated parts of the spongy portion of the urethra.*

(1) The pars bulbosa in the bulb, one inch in length ; (2) the fossa navicularis, **one** inch in length, situated in the glans penis.

8. *Locate the penile angle.*

It is about two inches in front of the external layer **of the triangular ligament.**

9. *Describe the blood-supply to and from the penis.*

A distinction must be made between the superficial circulation and the deep : (1) The circulation in the retractile or loose **covering, or** the superficial circulation, reaches the penis by (*a*) the external pudic artery, a branch of the common femoral artery ; (*b*) **the** superficial perineal, a branch of the internal pudic ; (*c*) **the** superficial branch of the dorsal artery of the penis. The blood from this superficial circulation is collected in one or two rather large superficial veins, and conveyed by them to the femoral vein and its large tributary, the long saphenous. (2) Blood reaches the penis proper by arteries to the bulb and corpus cavernosum. **These vessels are branches of** the internal pudic artery. They are given off from **this artery in the space** between the superficial and deep triangular ligaments. **(Fig. 233.) They** pierce the superficial triangular ligament preparatory to supplying these parts. The dorsal artery of the penis (Fig. 233) also supplies the **deep parts** of the penis. Blood from this region is returned to the venous circulation **by** two routes : (*a*) By the dorsal vein of the penis ; (*b*) by the veins that accompany the arteries to the corpora **cavernosa** and **corpus** spongiosum. The **dorsal vein** passes under the pubic arch and is tributary to the vesico-prostatic **plexus, at** the base and sides of the **bladder.** The **other veins accompany their** arteries and become tributary to the internal iliac vein.

10. *Describe the nerve-supply of the penis.*

The erectile bodies receive nerves from the dorsal nerve of the penis ; the superficial perineal and sympathetic filaments from the hypogastric plexus. The skin receives sensory branches from the genito-crural and superficial perineal.

THE FEMALE PELVIC OUTLET.

The surgical importance **of the female** perineum will warrant a thorough consideration of its *dissection* in **this book.** The homologies are so striking that some of our best text-books on anatomy seem content to refer the student to the male **perineum to** gain his conception of **the** structure of the female **peri-** neum. To my mind this is a most neglectful practice, since the desideratum is a practical one, involving acquisition of knowledge pending operative procedure, instead of a knowledge in the direct line of morphological research. Would the student be competent to treat the male urethra, who had studied only the clitoris from an anatomical standpoint ? Still, the penis is homologous to the clitoris, and in the male the urethra occupies the entire length of the central cylinder of the **penis.** Would the student be competent to remove a diseased testicle, whose **anatomical** knowledge was all in homologies ? Again, my experience teaches **me that** medical students look upon the outlet of the female pelvis not from the standpoint **of the** morphologist. It requires study and practice to see in the fore- skin, the labia minora ; in the scrotum, the labia majora ; in the accelerator urinæ, the sphincter vaginæ ; in the bulb of the urethra, **the** bulb of the vagina ; in the spermatic cord, the round ligament of the uterus ; in the uterus mascu- linus of the prostatic urethra, the uterus of the female ; in the testicle of the male, the ovary of the female.

The repair of the female perineum is seconded in point of frequency only by extirpation of diseased ovaries and tubes ; in comparison to these operations on the female, seen thousands of times by our students, I would ask our seniors, How many operations on the male perineum have you witnessed ? How many castrations have you been party to ? A student may be expert in introducing the catheter into the male urethra, but without special knowledge he will be an ignominious failure when he attempts to perform this operation on the female. A knowledge of the specific anatomy of the female generative organs must be acquired by dissection ; homological studies are most useful and most necessary adjuncts.

Examine before making any incisions :

1. **The mons veneris,** or **mons pubis,** the fatty elevation covered with crisp hair, surmounting the pubes.

2. **The vulva** is a collective noun by which all the external genitals of the female, except the mons pubis, are designated.

3. **The clitoris** is an erectile structure analogous to the penis. Its free extremity is the glans. It has a suspensory ligament and erector muscles. On each side it has a crus originating from the ischio-pubic ramus.

4. **The major labia** are homologous to the **scrotum.** They are cleft along the mid-line (rima pudendi). They extend from the mons to the tendinous centre

of the perineum. They flatten out posteriorly into the smooth covering of the perineum. Their junction here is called **posterior commissure.**

5. **The minor labia** are homologous to the prepuce. They are cutaneous folds. They have neither hair nor fat. In the adult they are concealed by the major labia; not so in the fœtus. They are connected posteriorly by the **fourchette.**

6. Examine carefully the **vestibule.** This is a guide to the meatus urinarius in using the catheter. It opens into the meatus and extends from this opening to the clitoris.

7. **The meatus urinarius externus** is the distal end of the urethra. It is nearly an inch from the clitoris. Introduce a catheter into the bladder and feel the urethra on the anterior vaginal wall, in the mid-line and 1.5 inch in length.

8. **The** vaginal orifice lies in the mid-line. It is narrowed by the hymen in the virgin, and surrounded by the remains of the hymen in females who have borne children. Its homologue would be artificially produced by a vertical slit

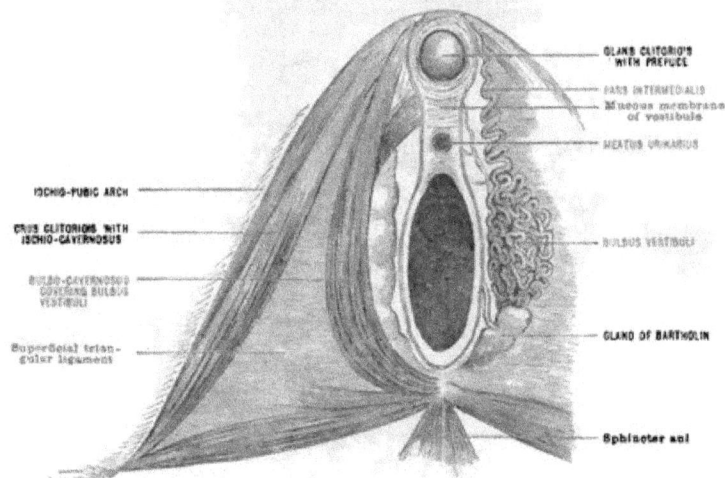

FIG. 237.—DIAGRAMMATIC REPRESENTATION OF THE PERINEAL STRUCTURES IN THE FEMALE.

through the bulb of the male urethra; the two parts of the accelerator urinæ would then be homologous to the sphincter vaginæ, and the erectile tissue of the bulb to the bulbi vestibuli.

9. **"The Hymen"** (Fig. 238), says Morris, "has been a subject of much speculation among the learned and unlearned of all ages. Its very existence was at one time denied by many great authorities." Failure to find this in careful examination of several hundred female fœtuses and young babes has caused me to ask if it be possible that the hymen is absent in females until the age when other pubal changes occur. The remains of the hymen are called *carunculæ myrtiformes.* They surround the entrance to the vagina.

Having located the foregoing nine structures, and thoroughly learned both their *function* and *homologies,* where the latter exist, you are now ready to begin a careful dissection of the *female pelvic outlet.* Remember, there is no difference in the posterior triangular region—the ischio-rectal fossa—in the sexes. The differences will all be found in the anterior, or perineal region.

Dissection.—Place the subject in the proper lithotomy position, as described previously for dissecting the male perineum. Thoroughly remove all pubic hair. Find the following structures and dissect them in this order:

1. *The clitoris,* homologous to the penis.
2. *The labia majora,* homologous to the scrotum.
3. *The bulbo-cavernosus,* homologous to the accelerator.
4. *The bulbi vestibuli,* homologous to the bulb of the urethra.
5. *The gland of Bartholin,* homologous to Cowper's gland.
6. *The sphincter ani,* the superficial transversus perinæi, and superficial part of the triangular ligament.

The Clitoris.—The clitoris is homologous to the penis. Its size is often underestimated, because the foreskin—the labia minora—binds it down so you are

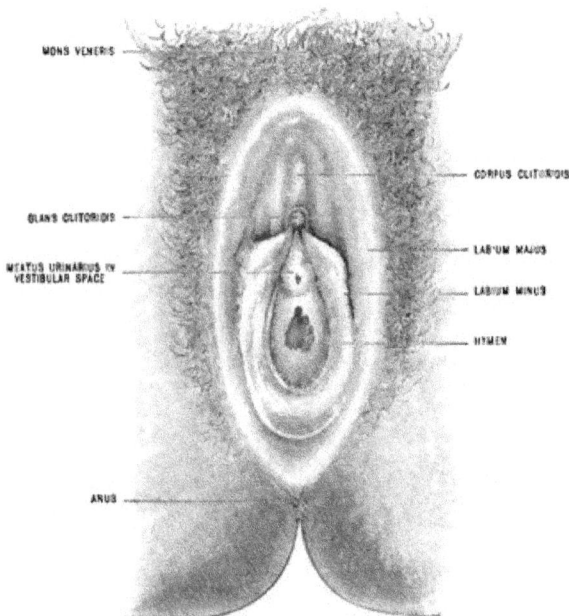

FIG. 238.—EXTERNAL GENITALS OF THE VIRGIN, WITH DIAPHRAGMATIC HYMEN. (Sappey.)

able to see only the distal end, called the glans. In the penis the prepuce covers the glans and terminates below in a frenum; in the clitoris the prepuce terminates in the two elongated labia minora, owing to the vertical cleft of the vaginal orifice. To dissect properly this organ, grasp the glans in the forceps of your left hand, and, with the scissors in your right hand, cut the skin along the dorsum of the clitoris for an inch upward. Then you can see the corona glandis and the glans clitoridis in full view. Next remove carefully the skin from the two crura along the ischio-pubic ramus. You will find these crura covered by a muscle, the erector clitoridis, homologous to the erector penis.

Remove the skin from the labia majora and trace the crescentic muscular fibres of the bulbo-cavernosus muscle around the orifice of the vagina. This is the *sphincter muscle* of the vagina. The transversus perinæi and sphincter ani are the same as in the male. The nerve- and blood-supply are *precisely* the same. The deep perineal fascia transmits the urethra. It is smaller than in the male, on account of the vagina. (See blood-supply on page 339.)

The perineal body (Fig. 239) occupies the space between the lower part of the vagina and the rectum. It has a base, covered by skin, and an apex between the termini of its anterior and posterior surfaces. From side to side it extends from tuber to tuber of the ischium. Its posterior surface lies in front of the anterior wall of the rectum; its anterior surface lies behind the posterior vaginal wall. Into it are attached all the muscles of the central perineal point or tendon: (1) The external sphincter ani; (2) the superficial transversus perinæi; (3) the

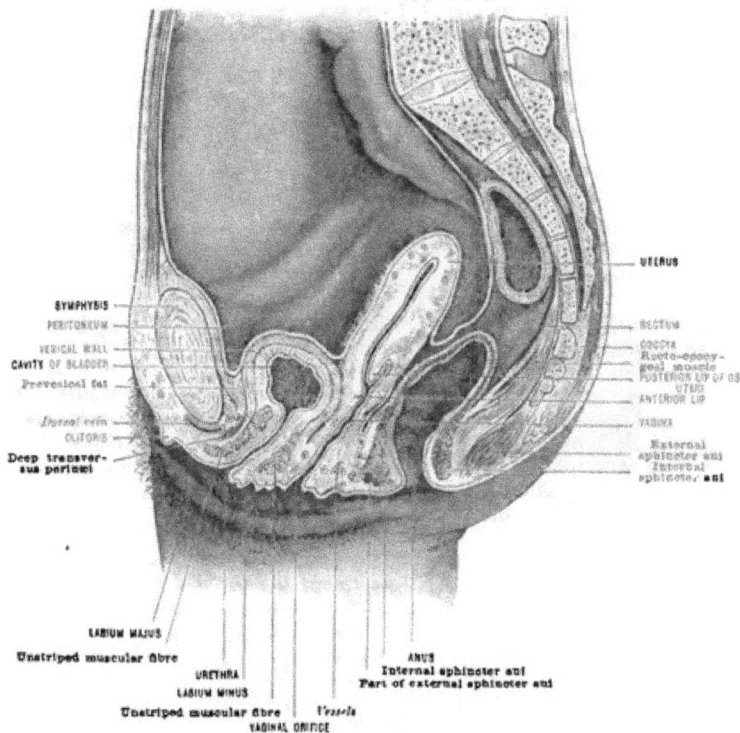

FIG. 239.—SECTION OF THE FEMALE PELVIS. (After Henle.)

sphincter vaginæ. This perineal body is strong; it owes its strength to elastic connective tissue. The perineal body is about an inch and a half in length from before backward.

The glands of Bartholin probably represent Cowper's glands in the male, but are more superficially placed. They are two little racemose glands, about a third of an inch long, situated one on each side beneath the lateral wall of the vestibule and behind the bulbi vestibuli. The duct, about three-quarters of an inch in length, opens immediately in front of the vaginal orifice opposite its meridian.

Locate the vagina by superior and inferior limitations.

It is limited inferiorly by the introitus vaginæ at the vestibule, and superiorly by its junction with the cervix uteri.

Define and locate the utero-vaginal fornix.

It is the highest part of the vagina; a space surrounding that part of the cervix uteri within the vagina, and formed by the reflexion of the vaginal mucous membrane on to the cervix. The vaginal passage extends upward and backward. Its long axis extended would pass through the upper segment of the sacrum.

What are the anterior relations of the vagina?

It is in close relation with the urethra and bladder. The ureters open into the posterior wall of the bladder, anterior to the vagina, and about an inch below the lowest part of the cervix uteri. There is a large amount of connective tissue between the anterior wall of the vagina and the bladder and urethra.

Describe the posterior relations of the vagina.

The vagina is in relation posteriorly with the rectum. Intervening between the rectum and posterior wall of the vagina from above downward are (1) the peritoneal cul-de-sac of Douglas; (2) subperitoneal connective tissue; (3) the perineal body.

Describe the lateral relations of the vagina.

(1) The vaginal branch of the uterine artery—the uterine being a branch of the internal iliac; (2) a plexus of veins, lying at the base of the broad ligament of the uterus, communicating with the hæmorrhoidal, and tributary to the internal iliac vein.

How is the vaginal inlet narrowed?

By the action of the sphincter vaginæ, the levator ani, and engorgement of the erectile tissue of the inlet, called the bulbi vestibuli.

Describe the structure of the vaginal wall.

It is composed of three coats: (1) An outer fibrous coat, a derivative of the recto-vesical fascia. In this coat you will find in some cases a beautiful plexus of veins. (2) A muscular coat, consisting of outer longitudinal fibres and an inner layer of circular fibres. The muscular coat consists of unstriped fibres. (3) A mucous coat, highly elastic, and continuous above with the mucous membrane of the uterus, and below with that of the vulva.

Describe the blood-supply of the vagina.

The arteries, derived from the internal iliac, from the inferior branch of the uterine, and from the external pudic branch of the common femoral artery, lie along the lateral part of the vagina and divide into anterior and posterior branches for the supply of the anterior and posterior surfaces respectively. The veins follow the course of the arteries, form a rich plexus between the inner and middle coats, and finally become tributary to the internal iliac vein. The nerve-supply of the vagina is from the hypogastric plexus, internal pudic nerve, and the fourth sacral.

POSTERIOR PART OF SHOULDER, ARM, FOREARM, AND HAND.

Dissection.—Locate—(1) the *acromion process* and *spine of the scapula ;* (2) the *inner and outer humeral condyles ;* (3) the *olecranon process of the ulna ;* (4) the *head of the radius ;* (5) the *inferior outer border of the clavicle ;* (6) the *axil-*

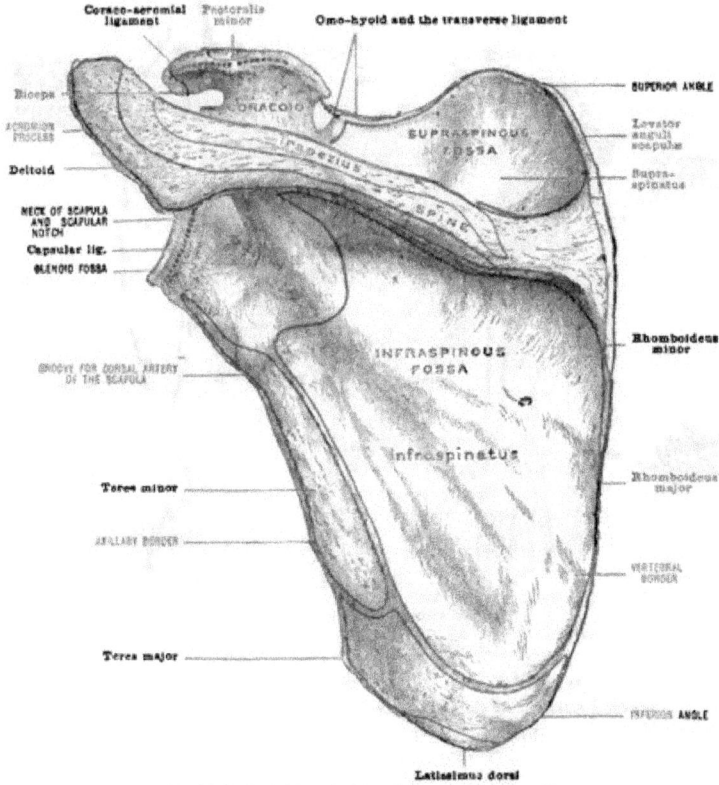

Fig. 240.—THE LEFT SCAPULA. (Dorsal surface.)

lary border of the scapula ; (7) the *superior border of the scapula ;* (8) the *greater and lesser tuberosities of the humerus ;* (9) the *olecranon fossa ;* (10) the *ulnar and radial styloid processes ;* (11) the *radio-carpal articulation ;* (12) the *metacarpo-*

phalangeal articulation ; (13) the *phalangeal* **articulations ;** (14) the *clefts of the fingers ;* (15) the *thenar and hypothenar* **eminences.**

Incisions.—(1) From the acromion **process to the end of the** middle finger ; (2) from **the** outer one-third of the clavicle to the acromion process, thence along the scapular spine ; (3) from one condyle to the other.

Dissection (Figs. 241 and 242).—Carefully remove the skin and find the ·

1. Radial nerve distributed to three and one-half fingers dorsally.
2. Ulnar dorsal branch to one and one-half fingers dorsally.
3. External cutaneous, a branch of the musculo-spiral nerve.
4. Internal cutaneous nerve from the inner cord of the brachial plexus.

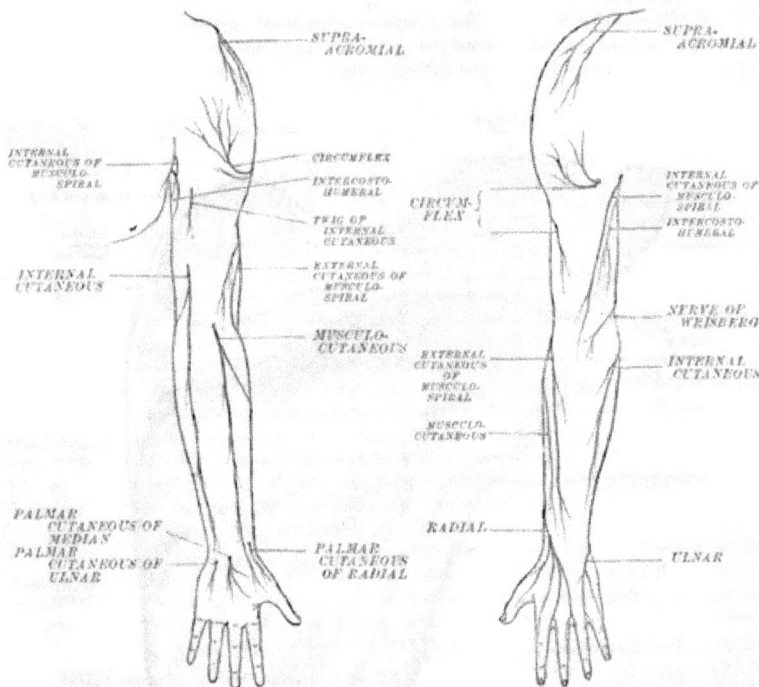

FIG. 241.—DISTRIBUTION OF CUTANEOUS NERVES ON THE **ANTERIOR** AND POSTERIOR ASPECTS OF THE SUPERIOR EXTREMITY.

5. Lesser internal cutaneous or the nerve of Wrisberg.
6. Internal cutaneous nerve, a branch of musculo-spiral.
7. Cutaneous branches of the circumflex nerve to skin of deltoid.
8. Supraacromial, a descending branch of the cervical plexus.
9. Intercosto-humeral nerve—the lateral cutaneous branch of the second intercostal nerve.

General **directions** only can be given for doing this cutaneous dissection : (1) Keep close to the skin when you remove it. (2) Consult the figures. Marginally, you will see the name of the nerve, which, having found, take your forceps and search in the superficial fascia of the region until you find the nerve.

(3) Then lift the nerve on your finger and follow the same by dividing the connective tissue in the direction of the branches given off.

Remember that the circumflex nerve sends its cutaneous branches to the skin over the insertion of the deltoid, according to Hilton's law (Fig. 241.) The musculo-cutaneous nerve sends a cutaneous branch to supply the skin over the insertion of the supinator longus, according to the same law, since this muscle is one of the forearm flexors and is supplied by the musculo-cutaneous The

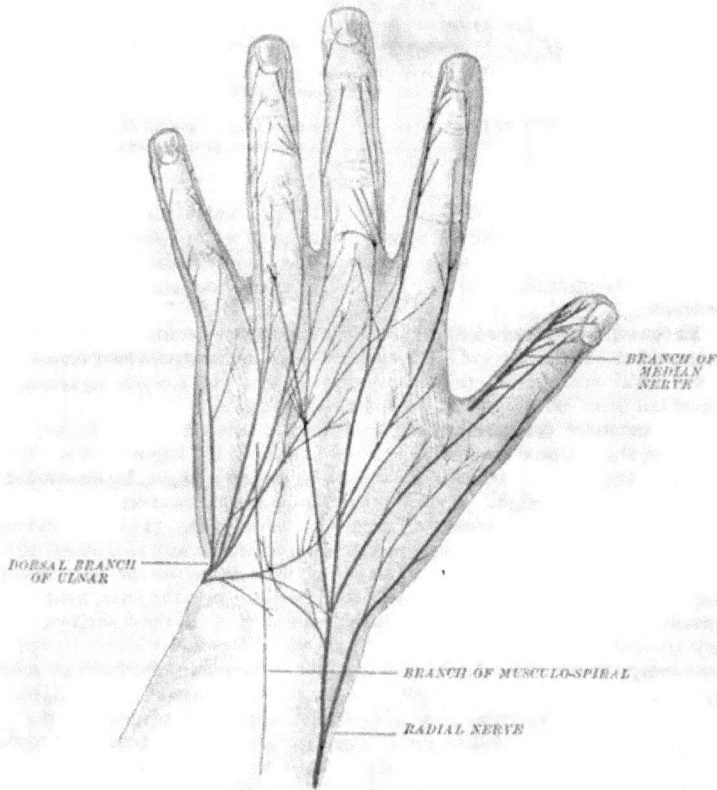

FIG. 242.—A DISSECTION OF THE CUTANEOUS NERVES ON THE DORSAL ASPECT OF THE
HAND AND FINGERS. (H. St. J. B.)
The branches of the median nerve are shown in black.

distribution of the radial nerve is explained in the same way. You will note that the cutaneous nerves divide into numerous small branches. (Fig. 242.)

The deep fascia is quite heavy below the elbow, but thin above the elbow. Opposite the wrist it takes the name *posterior annular ligament.* Notice in figure 243—that the muscles all become tendinous before they reach this ligament. Remove the deep fascia as in the figure. "Dorsal fascia" is the name for the deep fascia on the back of the hand.

CLASSIFICATION OF MUSCLES ACCORDING TO FUNCTION.

Extensors of the pollux
1. Extensor longus pollicis ; insertion, base of second phalanx.
2. Extensor brevis pollicis ; insertion, base of first phalanx.
3. Extensor ossis metacarpi pollicis ; insertion, base of metacarpal.

Extensors of the carpus
1. Extensor carpi radialis longior ; insertion, base of second metacarpal.
2. Extensor carpi radialis brevior; insertion, base of third metacarpal.
3. Extensor carpi ulnaris ; insertion, base of fifth metacarpal.

Extensors of the digits
1. Extensor communis digitorum ; insertion, base of second and third phalanges.
2. Extensor minimi digiti ; insertion, with extensor communis.
3. Extensor indicis digiti ; insertion, with extensor communis.

Extensors of the forearm
1. The triceps ; insertion, into the olecranon process of the ulna.
2. The anconeus ; insertion into the oblique line of ulna.

Supinator of the radius
1. Brachio-radialis ; insertion, base of styloid of radius.
2. Supinator radii brevis ; insertion, posterior and external surfaces of radius.

Dissection.—Trace out the tendons of the three extensors of the thumb by cutting through the deep digital fascia. Take each to its insertion according to insertions given in the classification above. Do not cut the annular ligament. See if two compartments in the annular ligament transmit the three thumb extensors.

Extensor Carpi Radialis Longior and Brevier.—Locate the insertions of these tendons, in the bases of the second and third metacarpal bones respectively on the radial side. Trace the same upward, under the annular ligament, and behind the three extensors of the thumb. (Fig. 243.)

The extensor communis digitorum sends a tendon to each finger. The tendon of the extensor communis digitorum to the index finger is joined by the tendon of the extensor indicis ; the tendon to the little finger, by the tendon of the extensor minimi digiti. Demonstrate this on your dissection.

The insertion of the extensor tendons is as follows (Fig. 243) The extensor tendon of each finger divides into three parts, opposite the first phalangeal articulation (Fig. 244) ; the middle portion is inserted, just across the joint, into the base of the second phalanx ; the two lateral portions pass the joint, unite on the dorsum of the second phalanx, and pass to the base of the third phalanx, where they are inserted. See whether this muscle and the extensor indicis occupy the same compartment under the annular ligament. Now trace the extensor minimi digiti through a separate compartment in the annular ligament. (Fig. 244.)

Insertion of the extensor carpi ulnaris : Cut down on its tendon near the base of the fifth metacarpal ; follow the tendon through a groove behind the styloid process of the ulna, and trace it through a separate compartment in the posterior annular ligament.

Compartments of the Posterior Annular Ligament (Fig. 244).—Cut through the annular ligament and find the following compartments and their contents, and see whether the same are lined by synovial membrane.

1. For extensor ossis metacarpi pollicis and extensor brevis pollicis.
2. For extensor carpi radialis longior and brevior.
3. For extensor longus pollicis.
4. For extensor communis digitorum and extensor indicis.
5. For the extensor minimi digiti.
6. For the extensor carpi ulnaris.

The Brachio-radialis, or Supinator Radii Longus.—Supinate the radius and you will find this muscle, on the radial side, inserted into the base of the styloid process of the radius. (Fig. 246.)

Trace to their origins the **radial group** of muscles :

1. Supinator longus, to upper two-thirds of external condylar ridge of the humerus. (Fig. 246.)

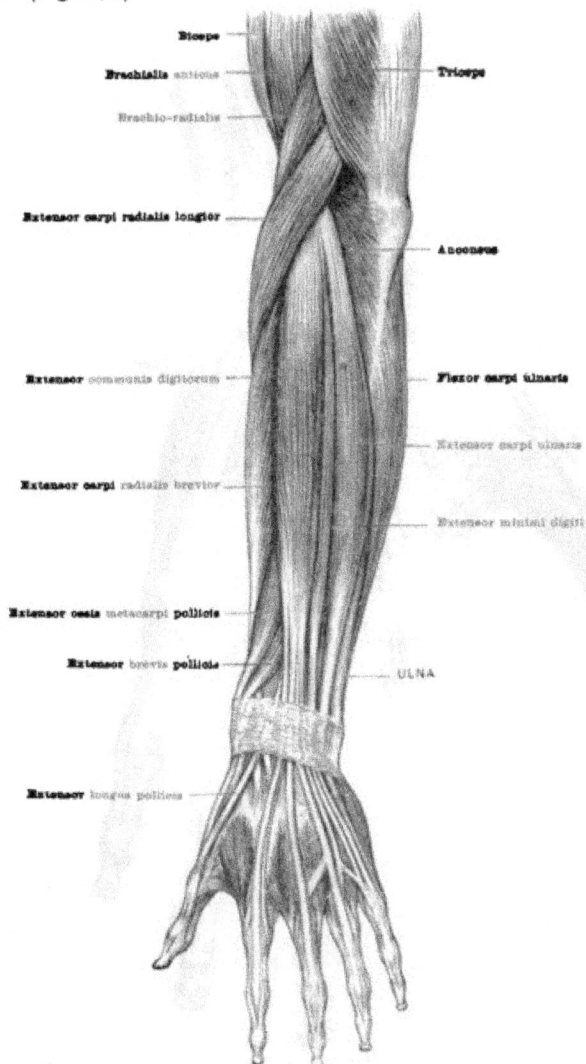

FIG. 243.—MUSCLES OF THE RADIAL SIDE AND THE BACK OF THE FOREARM.

2. Extensor carpi radialis longior, trace to the lower one-third of external condylar ridge of the humerus.

3. Extensor carpi radialis brevior, to the outer condyle of the humerus.

To do this properly, begin with the tendon and follow it upward, dividing the

connective tissue between the muscle and the ones on each side of it.   Be careful,
in doing this, not to destroy the vessels and nerves ; these structures will become
numerous in the region of the elbow.   Trace these adjacent muscles also up to
their origin.

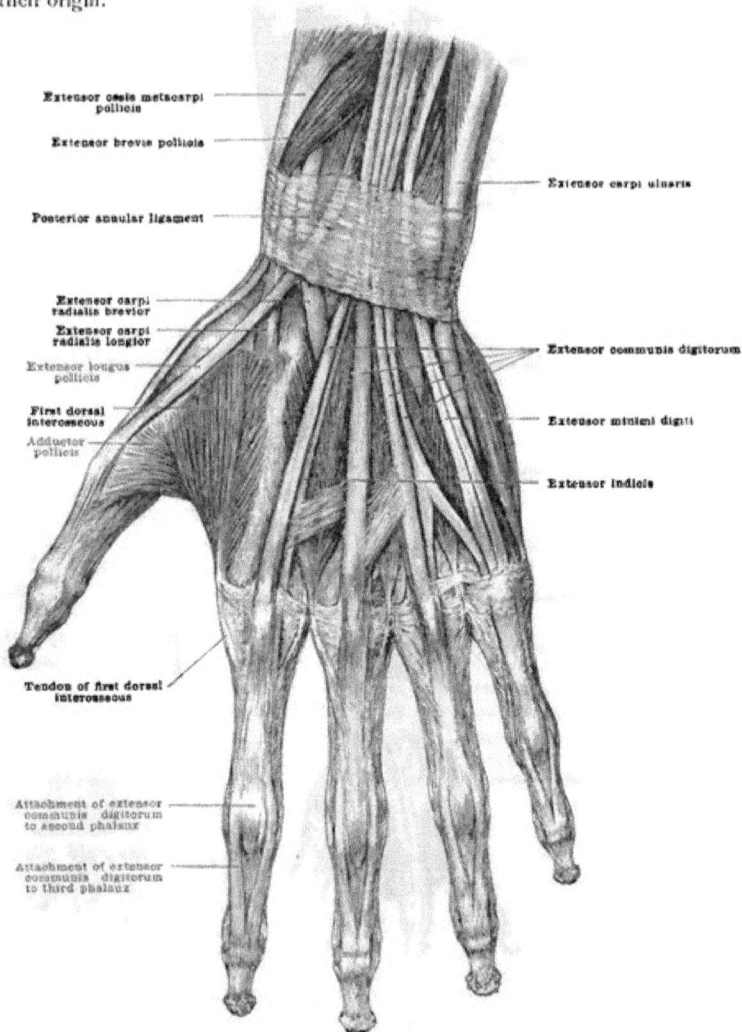

Extensor ossis metacarpi
pollicis

Extensor brevis pollicis

Posterior annular ligament

Extensor carpi
radialis brevior

Extensor carpi
radialis longior

Extensor longus
pollicis

First dorsal
interosseous

Adductor
pollicis

Tendon of first dorsal
interosseous

Attachment of extensor
communis digitorum
to second phalanx

Attachment of extensor
communis digitorum
to third phalanx

Extensor carpi ulnaris

Extensor communis digitorum

Extensor minimi digiti

Extensor indicis

FIG. 244.—TENDONS UPON THE DORSUM OF THE HAND.

Trace the superficial group to origin as follows (Fig. 243):

1. Extensor communis digitorum to the outer condyle of the humerus.
2. Extensor minimi digiti to the outer condyle of the humerus.
3. Extensor carpi ulnaris to the outer condyle of the humerus.

Notice the triple origin of the extensor carpi ulnaris : (1) The outer condyle ; (2) the posterior surface of the ulna ; (3) posterior border of the ulna, in common with the flexor carpi ulnaris. Dissection of this latter muscle requires care.

The deep group of muscles and their origin (Figs. 245 and 246) :

1. The extensor longus pollicis—ulna and interosseous membrane.

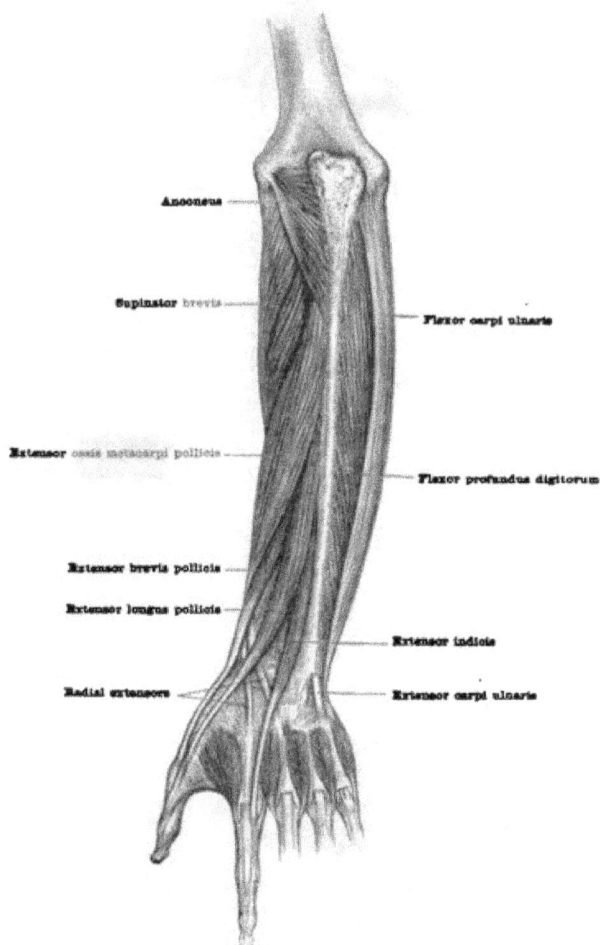

FIG. 245.—THE DEEP LAYER OF THE BACK OF THE FOREARM.

2. The extensor brevis pollicis—radius and interosseous membrane.

3. The extensor ossis metacarpi pollicis—radius interosseous membrane and ulna.

4. The extensor indicis from the ulna.

In view of the special importance of the three extensors of the thumb and the index finger, it is necessary for the student to trace each muscle to its specific origin and insertion according to the following scheme for aiding the memory :

The extensor indicis (Fig. 245) arises practically from the posterior surface of the lower one-third of the ulna and interosseous membrane. Demonstrate on the cadaver the amount of ulnar surface in the lower third of the bone which

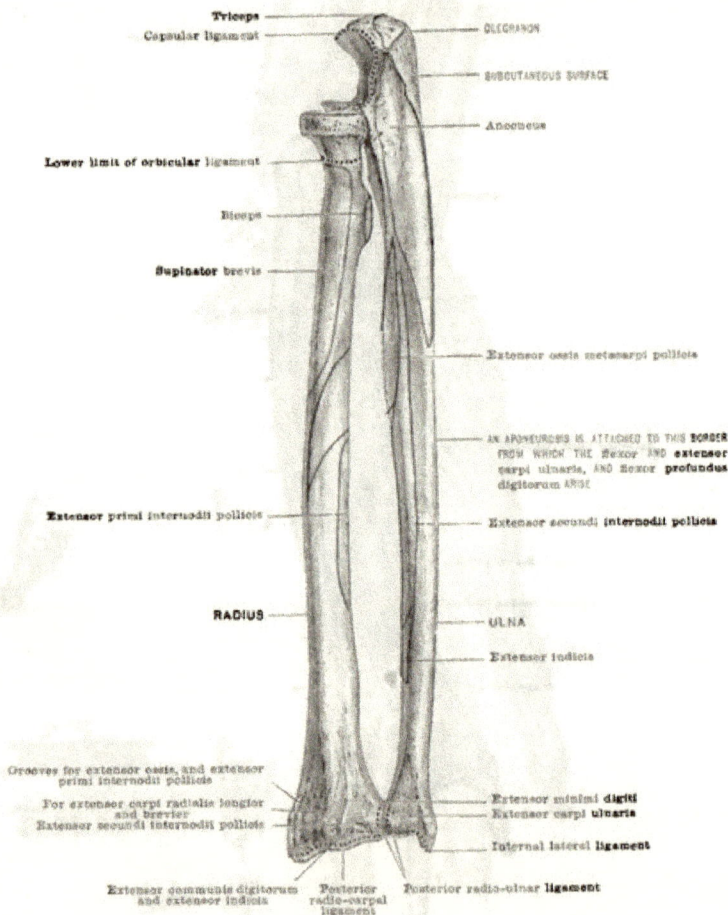

Triceps
Capsular ligament
OLECRANON
SUBCUTANEOUS SURFACE
Anconeus
Lower limit of orbicular ligament
Biceps
Supinator brevis
Extensor ossis metacarpi pollicis
AN APONEUROSIS IS ATTACHED TO THIS BORDER FROM WHICH THE flexor AND extensor carpi ulnaris, AND flexor profundus digitorum ARISE
Extensor primi internodii pollicis
Extensor secundi internodii pollicis
RADIUS
ULNA
Extensor indicis
Grooves for extensor ossis, and extensor primi internodii pollicis
For extensor carpi radialis longior and brevior
Extensor secundi internodii pollicis
Extensor minimi digiti
Extensor carpi ulnaris
Internal lateral ligament
Extensor communis digitorum and extensor indicis
Posterior radio-carpal ligament
Posterior radio-ulnar ligament

FIG. 246.—THE LEFT ULNA AND RADIUS. (Postero-external view.)

does not give origin to this muscle. Observe, too, the septal origin of the muscle, between this and the extensor longus pollicis.

Now consider the origins of the three extensors of the thumb in reference to : (1) The posterior surface of the radius ; (2) the posterior surface of the ulna ; (3) the posterior surface of the interosseous membrane. Notice, too, you have to deal with a long bone, the ulna ; a short bone, the radius ; and an interosseous membrane associated with both bones. For purposes of convenience call the extensor ossis metacarpi pollicis the strong muscle. Then say :

1. The long muscle (the extensor longus pollicis) arises from the posterior surface of the long bone (the ulna) and the interosseous membrane.

2. The short muscle (the extensor brevis pollicis) arises from the posterior

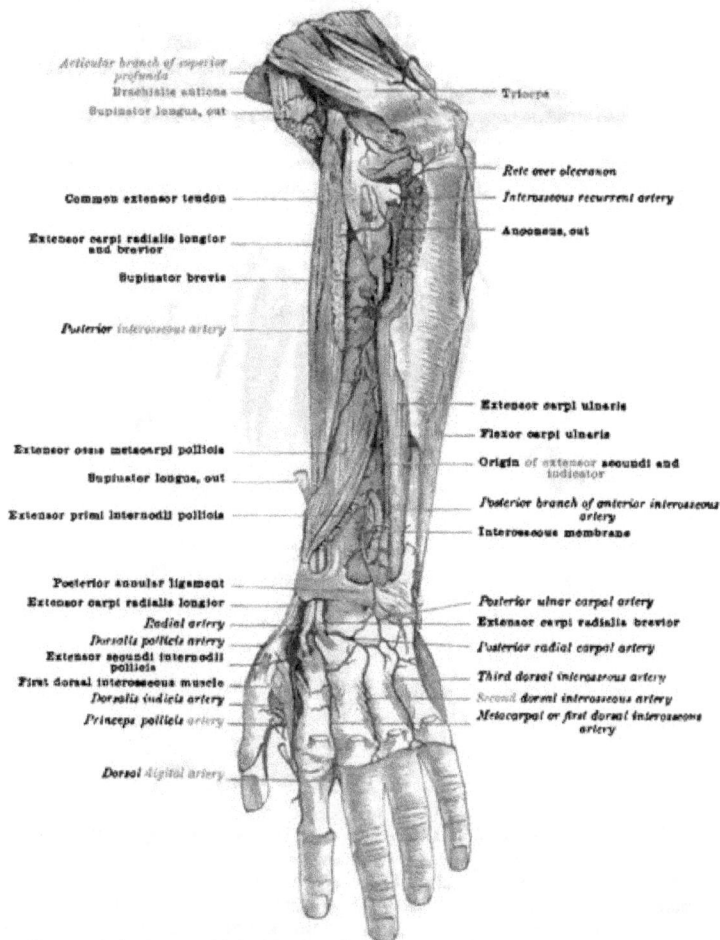

FIG. 247.—THE BACK OF THE FOREARM, WITH THE POSTERIOR INTEROSSEOUS ARTERY AND BRANCHES OF THE RADIAL AT THE BACK OF THE WRIST.
(From a dissection in the Hunterian Museum.)

surface of the short bone (the radius) and the interosseous membrane. Insertion into the base of the first phalanx.

3. The strong muscle (the extensor ossis metacarpi pollicis) arises from the posterior surfaces of both the long and short bones (ulna and radius) and the interosseous membrane. Insertion into base of metacarpal bone.

Observe that the collective origin of the three thumb extensors practically occupies the middle posterior third of ulna, radius, and interosseous membrane. Compare the origins of these muscles as you find them on the cadaver, with the origins as indicated in figure 246.

Now locate (1) the posterior interosseous artery. (Fig. 247.) It lies between the superficial and deep groups of muscles. It is a branch of the common interosseous branch of the ulnar. Find the space between the oblique ligament and the interosseous membrane, where this vessel appears. Also find where it anastomoses with the anterior interosseous artery below. Notice that in its course this artery crosses the deep group of muscles, and is attended in a part of its course only by the posterior interosseous nerve. Trace branches from this artery to all the muscles in this locality. This artery is attended by the posterior interosseous nerve in the upper part of its course. Show the exact relation of artery and nerve on your work. (2) Locate the *radial artery* in the pulse region of the wrist. You

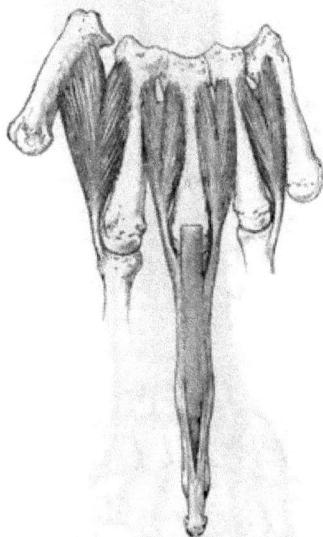

FIG. 248.—THE DORSAL INTEROSSEI.

will find it passing under the three extensor muscles of the thumb, where it pierces the first interosseous muscle and disappears into the deep palm.

The Dorsal Interosseous Muscles (Figs. 247 and 248).—Study the muscles and their blood-supply. Find four of these muscles. Notice that they arise by two heads from the contiguous surfaces of the metacarpal bones. Trace their tendons of insertion to the tendon of the extensor communis digitorum. These muscles (Fig. 248),—as you might infer from their insertion,—acting alone, are abductors of the fingers. Acting with the palmar interossei and lumbricales, they produce *flexion* at the metacarpo-phalangeal and extension at the phalangeal articulations. Acting in this way, they place the fingers in position for holding the pen. These muscles are all supplied by the ulnar nerve.

The Supinator Radii Brevis.—Detach all the muscles that arise from the outer condyle: The extensor carpi radialis brevior, the extensor communis digitorum, the extensor minimi digiti, and the extensor carpi ulnaris. Do this in such a way as not to injure the nerve-supply from the musculo-spiral. Now pull these muscles slightly aside and see the supinator brevis. (Fig. 247.) Trace it to the following origins and insertions

Origins.—(1) Lower and back part of external condyle ; (2) the external lateral ligament of the elbow-joint ; (3) the orbicular ligament ; (4) the triangular depression below the lesser sigmoid cavity of the ulna, especially along its posterior margin, which forms the upper part of the external border of the ulna.

Insertion.—(1) The back of the neck of the radius ; (2) the anterior and outer surfaces of the radius above and at the upper border of the oblique line.

You will now find the posterior interosseous nerve issuing from the lower border of the muscle. This nerve supplies all the muscles in this region except the radial group, which were supplied by the musculo-spiral, before this nerve divided into the radial and posterior interosseous. The radial nerve is a branch of the musculo-spiral.

The radial nerve perforates the deep fascia, between the extensor carpi radialis and supinator longus about three inches above the carpus. Find where the nerve comes through the deep fascia and trace the same to the dorsum of the hand, and to three and one-half fingers.

The anconeus (Fig. 245) arises from the back part of the outer condyle. It is inserted into the olecranon and posterior surface of the ulna. Remove this muscle from its attachment, turn the same aside, and see the orbicular ligament of the radius and the elbow-joint. Cut through the orbicular ligament and dislocate the head of the radius.

The region of the shoulder (Fig. 249) shows :

1. The *deltoid muscle* and its nerve-supply, the circumflex.
2. The *supraspinatus muscle* (Fig. 251) under the trapezius. (Fig. 250.)
3. The *infraspinatus muscle* under the deltoid aponeurosis.
4. The *teres minor muscle*, from the axillary border of scapula.
5. The *triceps muscle* and its three heads.
6. The *teres major muscle*, with the latissimus dorsi.
7. The *triangular space* and the dorsalis scapulæ artery.
8. The *quadrangular space* and its circumflex structures.
9. The *circumflex nerve* and vessels in the quadrangular space.
10. The *anterior* and *posterior circumflex* arteries and veins.
11. The *musculo-spiral nerve* and the superior profunda artery.
12. The *scapular head of the triceps muscle* between the tereals.

The circumflex quadrangular space (Fig. 249) is bounded : (1) Externally by the surgical neck of the humerus ; (2) internally by the long or scapular head of the triceps ; (3) superiorly by the teres minor ; (4) inferiorly by the latissimus dorsi and teres major muscles ; (5) contents posterior circumflex artery, a branch of the axillary, and the circumflex nerve, from the posterior cord of the brachial plexus.

The triangular space (Fig. 249) is bounded : (1) Externally by the scapular head of the triceps ; (2) superiorly by the teres minor ; (3) inferiorly by the teres major and latissimus dorsi ; (4) contents (Fig. 249) : arteria dorsalis scapulæ, a branch of the subscapular artery.

A second triangular space is bounded (Fig. 249) : (1) Externally by the humerus and outer humeral head of triceps ; (2) internally by the scapular head of the triceps ; (3) superiorly by the teres major and latissimus dorsi ; (4) it contains the musculo-spiral nerve, from the posterior cord of the brachial plexus, and the superior profunda artery, a branch of the brachial ; (5) it transmits the external cutaneous branch of musculo-spiral nerve.

Dissection.—The Deltoid Muscle (Fig. 250).—Develop the posterior border of this muscle by cutting freely with the scissors the deltoid aponeurosis between this muscle and the triceps, teres minor, infraspinatus, and trapezius below. Now, with a sharp scalpel cut the scapular origin of the deltoid, along the whole length of the inferior border of the spine and acromion process of the scapula. (Fig. 249.) Turn the muscle forward and see, in a mass of connective tissue and fat, the posterior circumflex artery and circumflex nerve emerging from the *quadrangular space.* (Fig. 249.) Trace the circumflex nerve to the deltoid and teres minor muscles. Trace the deltoid to its insertion into the deltoid depression on the outer surface of the humerus, middle one-third. Take note of the triangular shape of a number of muscles about the shoulder.

The Teres Major and Latissimus Dorsi (Figs. 250–253).—Divide the connective tissue between these two muscles. Trace them under the scapular

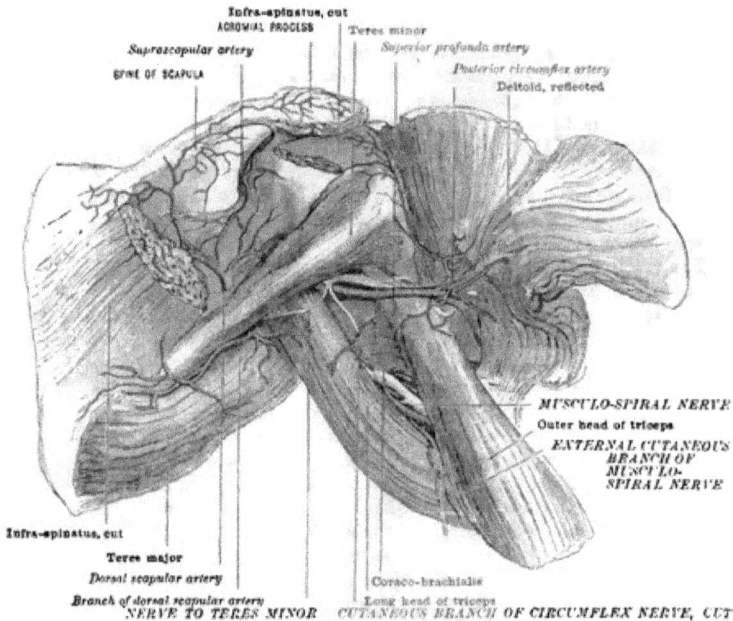

FIG. 249.—THE POSTERIOR CIRCUMFLEX ARTERY.
(From a dissection by Mr. Horner in the Museum of St. Bartholomew's Hospital.)

head of the triceps to the humerus. These muscles form the lower boundary of both the quadrangular and triangular spaces. They are inserted as follows: The latissimus dorsi into the bottom of the bicipital groove as high as the lesser tuberosity of the humerus; the teres major into the inner lip of the bicipital groove for two inches in length. Note that both these tendons are inclosed in a rather tough sheath of connective tissue.

The teres minor (Figs. 251 and 253) is separated near the humerus from the teres major by the scapular head of the triceps. Divide the connective tissue between the teres minor and the infraspinatus above and the teres major below. (Fig. 250.) The teres minor arises from the axillary border of the scapula. It is synergistic in its action to the deltoid and derives its nerve-supply from the same circumflex nerve. It is inserted into the lowest facet of the

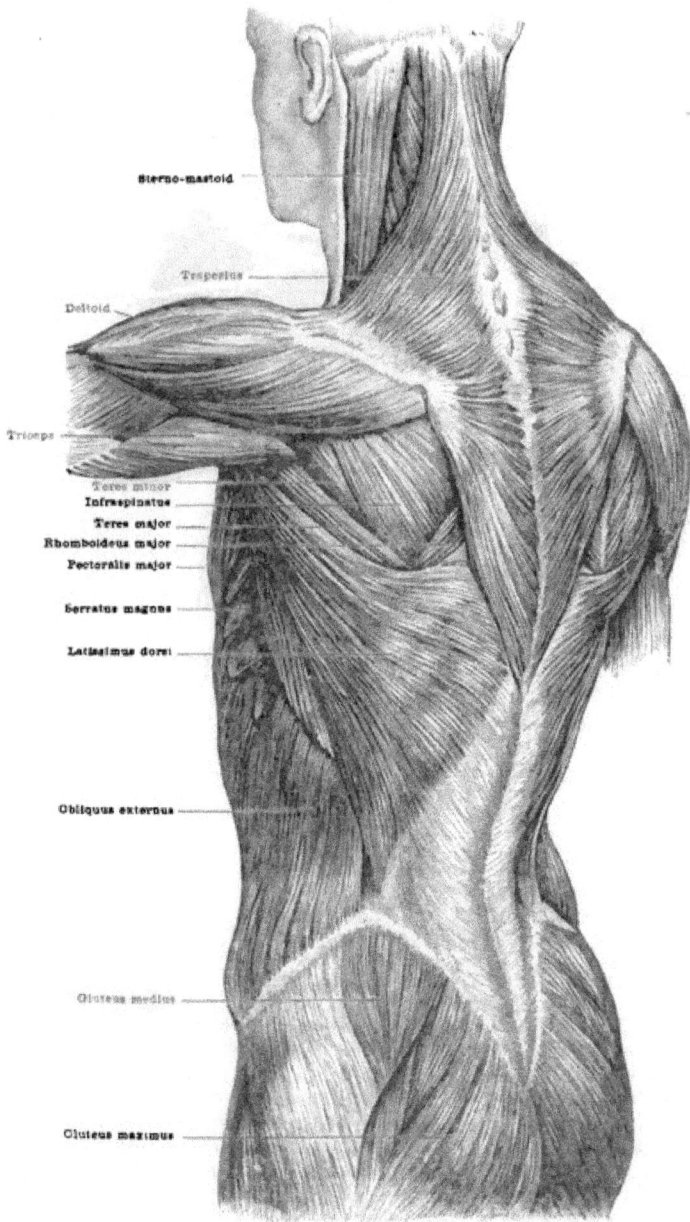

Sterno-mastoid

Trapezius

Deltoid

Triceps

Teres minor
Infraspinatus
Teres major
Rhomboideus major
Pectoralis major

Serratus magnus

Latissimus dorsi

Obliquus externus

Gluteus medius

Gluteus maximus

FIG. 250.—FIRST LAYER OF THE MUSCLES OF THE BACK.

greater tuberosity of the humerus. (Fig. 253.) Demonstrate the quadrangular and triangular spaces according to the outline given in the preceding pages. (Fig. 251.)

The Infraspinatus (Fig. 251).—Now cut this muscle at its insertion into the middle facet of the greater tuberosity; turn the same backward, taking care not to injure the capsule of the shoulder-joint and the vessels and nerves that enter under the surface of the muscle. This muscle arises from the outer two-thirds of the infraspinous fossa. (Fig. 252.) It is covered by a dense layer of

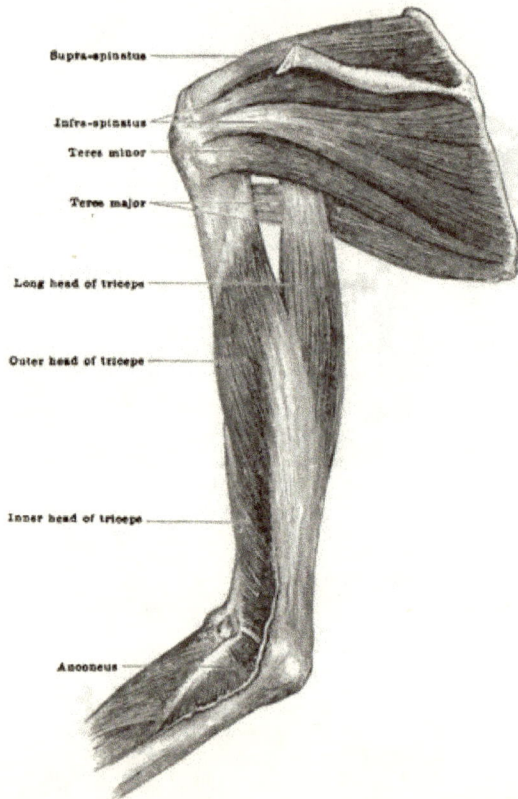

Supra-spinatus

Infra-spinatus

Teres minor

Teres major

Long head of triceps

Outer head of triceps

Inner head of triceps

Anconeus

FIG. 251.—BACK VIEW OF THE SCAPULAR MUSCLES AND TRICEPS.

deep fascia. The muscle is triangular in shape, and derives its nerve-supply from the infraspinous branch of the suprascapular nerve.

The supraspinatus muscle (Fig. 251) arises from the outer two-thirds of the fossa of the same name and also from the dense aponeurosis covering the muscle. It is inserted into the upper facet of the greater tuberosity of the humerus. It is covered (1) by a dense fascia of the same name; (2) by the trapezius muscle, which is inserted into the upper lip of the scapular spine. (Fig. 250.) Expose this supraspinatus muscle by cutting the insertion of the trapezius. Then you may trace the supraspinatus under the acromion process to its insertion. Cut this muscle at its insertion, turn it back (Fig. 254), and find the vessels and nerves—the suprascapular—that supply this region.

The suprascapular artery and nerve will come into view as you turn aside the supraspinatus muscle. (Figs. 249–254.) The nerve you will trace through the suprascapular foramen under the transverse ligament. (Fig. 252.) The artery crosses the transverse ligament. (Fig. 252.) Each structure divides into supra-spinous and infraspinous branches. Trace these vessels and nerves out to their respective localities of distribution. (Fig. 254.)

The *anastomosis* between the suprascapular artery and the dorsal branch of the subscapular takes place in the infraspinous fossa, on the bone. Find this. (Fig. 249.) Anastomosis about the shoulder includes the following arteries, which you may now demonstrate. (Fig. 254): Circumflex, subscapular, dorsalis

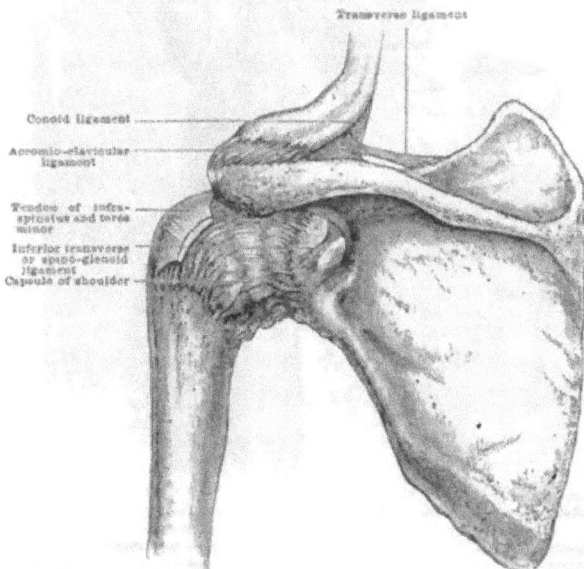

FIG. 252.—POSTERIOR VIEW OF THE SHOULDER-JOINT, SHOWING ALSO THE ACROMIO-CLAVICULAR JOINT AND THE PROPER LIGAMENTS OF THE SCAPULA.

scapulæ, suprascapular, posterior scapular artery, a continuation of the trans-versalis colli, a branch of the thyroid axis of the subclavian.

The triceps muscle (Fig 253) has three heads. The longer scapular head you will now very carefully trace to the axillary border of the scapula, immediately below the glenoid fossa. As you have already observed, this head assists in forming boundaries for the quadrangular space and also for the two triangular spaces. The whole posterior surface of the humerus is occupied by the inner and outer heads of the triceps muscle, and the musculo-spiral groove.

Dissection.—Locate the musculo-spiral nerve and the superior profunda artery in the second triangle. (Fig. 249.) Now the nerve will pass between the bone and muscle. Cut the muscle in a line corresponding to the course of the nerve, until you come to a point two inches above the outer condyle. The mus-cular mass attached to the posterior surface of the humerus above the groove, in which you now see the musculo-spiral nerve, is called the external, or long humeral head of the triceps; the mass below the musculo-spiral groove is called the internal or short humeral head. Trace the triceps muscle to its insertion, into the olecranon process of the ulna. Cut the insertion of the triceps and study the capsule of the elbow-joint and the olecranon fossa.

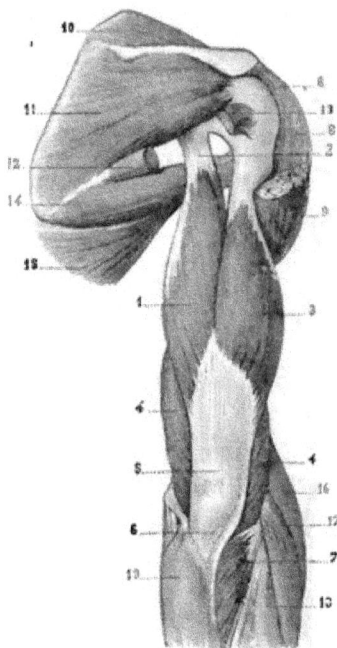

FIG. 253.—TRICEPS BRACHIALIS, POSTERIOR ASPECT.

1. Long, middle, or scapular head. 2. Its tendon of origin from scapula. 3. External, or long humeral head. 4. Internal, or short humeral head. 5. Common tendon. 6. Attachment to olecranon. 7. Anconeus. 8, 8. Upper portion of deltoid, posterior half removed. 9. Lower portion. 10. Supraspinatus. 11. Infraspinatus. 12. Teres minor, middle portion removed. 13. Insertion of teres minor into humerus. 14. Teres major. 15. Upper extremity of latissimus dorsi. 16. Supinator longus. 17. Extensor carpi radialis longior. 18. Extensor carpi ulnaris. 19. Flexor carpi ulnaris.

Nerve-supply to the Triceps.—As you dissect between the two humeral heads of the triceps, notice the large number of branches given off to the muscle from the musculo-spiral nerve, a branch of the posterior cord of the brachial plexus.

1. *How many origins has the deltoid muscle?*

Two; one from the lower lip of the scapular spine and acromion, another from the outer one-third of the clavicle. (Fig. 252.)

2. *The origin of the deltoid, then, may be said to correspond to the insertion of what muscle?*

To the insertion of the trapezius, since this muscle is inserted into the upper lip of the scapular spine and acromion, and into the posterior outer one-third of the clavicle. (Fig. 252.)

3. *Describe the circumflex nerve.*

It is a branch of the posterior cord of the brachial plexus. (Fig. 42.) It passes through the circumflex quadrangular space, to the back of the shoulder. It gives muscular branches to the deltoid and teres minor muscles, and cutaneous branches to the skin covering the insertion of these muscles and articular branches to the shoulder-joint. (Hilton's law.)

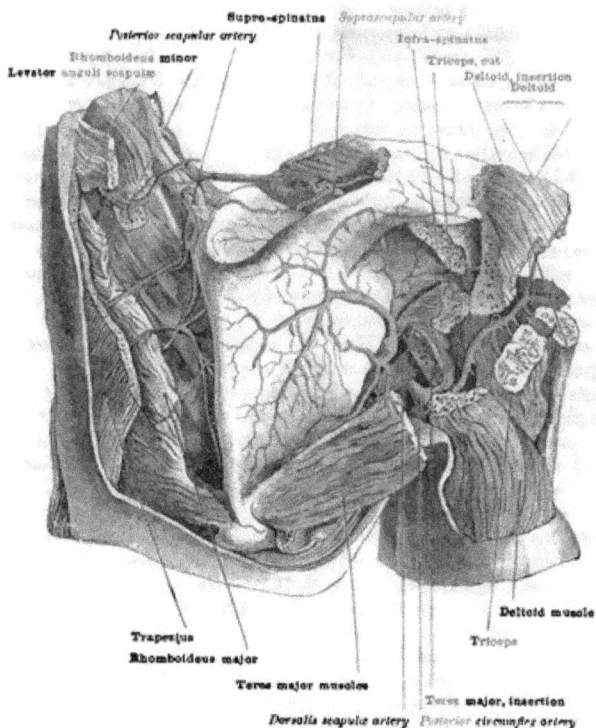

FIG. 254.—THE DORSAL SCAPULAR ARTERY.
(From a dissection in the Museum of the Royal College of Surgeons.)

4. *Describe the suprascapular nerve.*

A branch of the upper trunk of the brachial plexus. (Fig. 42.) It passes through the constant suprascapular foramen. It gives muscular branches to (1) the supraspinatus, (2) the infraspinatus, and articular branches from each muscular branch, to the shoulder-joint. (Hilton's law.)

5. *Describe the musculo-spiral nerve.*

It is a branch of the posterior cord of the brachial plexus. (Fig. 42.) It may be found on the floor of the second triangular space. (Fig. 249.) It passes between the two humeral heads of the triceps in the musculo-spiral groove of the humerus. It lies between the brachialis anticus and brachio-radialis muscles in the lower third of the arm. It gives muscular branches to the triceps and anconeus

24

and **to all** the muscles in the radial region, and also to all the muscles on the posterior part of the foramen. Its cutaneous branches are as follows : (1) Radial, to the integument of the ball of the thumb **and to** three and one-half fingers dorsally, as far as the base of the distal phalanx (Fig. 242) ; (2) internal cutaneous branches to the skin on the inner and posterior part of the arm ; (3) external cutaneous branches to the skin on the outer and front part of the arm. (Fig. 241.) The articular branch to the shoulder is sometimes taken with the circumflex. The **posterior** nerve-cord is composed of both the circumflex and musculo-spiral and other branches.

6. *Name the muscles arising from the external condylar ridge.*

The brachio-radialis and extensor carpi radialis longior.

7. *Give the origin, course, and distribution of the posterior interosseous nerve.*

A branch of the musculo-spiral ; it passes between the upper and deep layers of the supinator brevis muscle. It lies between the deep and superficial groups of posterior forearm muscles, which it supplies. Note that the musculo-spiral nerve gives off branches **to the** radial group of muscles prior to its division into the radial and posterior interosseous. Demonstrate this on your dissection.

If you have not already **done** so, review you dissection of the musculo-spiral nerve and show that the branches that supply the humeral heads of the triceps muscle are given off at the back of the humerus ; that branches are given off in the axilla to the scapular head. This nerve sometimes accompanies the **ulnar nerve** as a long slim branch, known as the ulnar collateral.

8. *Describe the dorsal interossei muscles.*

They are four in number. They are supplied by the ulnar **nerve.** They have **origin** from the contiguous sides of the metacarpal bones. (Fig. 248.) Their single independent action is abduction of the fingers, which is always associated physiologically with supination of the radius. Their synergistic action is with the palmar interossei and lumbricales to produce metacarpo-phalangeal flexion, and extension of the second and third phalanges. They are inserted into the bases of three fingers, as shown in figure 248.

1. *Describe the median nerve.*

It arises by two heads—an outer and an inner—from the outer and inner **cords** of the brachial plexus respectively. (Fig. 42.) It accompanies the axillary artery and **its** prolongation, the brachial, to the elbow. It gets into the forearm by passing between the condylar and coronoid heads of the pronator radii teres muscle. It gives off no branches to the flexor muscles of the forearm. It is, in rare cases, fused with the musculo-cutaneous nerve. It is said to give off two articular branches to the elbow-joint. It supplies all the muscles on the anterior surface of the forearm, except the flexor carpi ulnaris, and one-half of the flexor profundus digitorum. It supplies three and one-half fingers with sensation. In the hand it supplies the two radial lumbricales and all the muscles of the thenar eminence, except the adductor pollicis and one head of the flexor brevis pollicis. It is accompanied by an artery below the elbow. Sometimes this artery is as large as the radial. It passes under the anterior annular ligament. In the forearm the median nerve lies in the sheath of the flexor sublimis digitorum, on the posterior surface of this muscle. The digital branches of this nerve supply the finger-joints.

2. *Describe the ulnar nerve.*

It **is a** branch of the inner cord of the brachial plexus. It is attended by the inferior profunda artery above the elbow, and by the ulnar below. It gains the forearm by passing between the two heads—olecranon and condylar—of the flexor carpi ulnaris. It supplies the flexor carpi ulnaris, and the ulnar half of the flexor profundus digitorum. It supplies one and one-half fingers dorsally and palmarly. It supplies in the hand all the muscles not innervated by the median **nerve.** It gives articular branches to the elbow and wrist, and all subsequent joints moved by the muscles which it supplies. Demonstrate the dorsal ulnar cutaneous nerve. (Fig. 242.)

3. *Describe the musculo-spiral nerve.*

It is a branch of the posterior cord of the brachial plexus. (Fig. 42.) It is found between the two humeral heads of the triceps muscle in the musculo-spiral groove with the superior profunda artery. In the lower one-third of the arm it lies between the brachialis anticus and supinator longus. In this place it gives off branches to the radial group of muscles—the supinator radii longus, the extensor carpi radialis longior and brevior. It divides then into the radial and posterior interosseous. The posterior interosseous passes through the supinator brevis and then supplies the muscles on the posterior part of the forearm. The radial branch becomes cutaneous by piercing the deep fascia in the lower one-third of the forearm, between the supinator longus and the extensor carpi radialis. In the hand the nerve supplies the ball of the thumb and adjacent three and one-half fingers dorsally. (Fig. 242.) The musculo-spiral nerve gives off two cutaneous branches to the skin over the insertion of the triceps and anconeus.

4. *Describe the musculo-cutaneous nerve.*

This is a branch of the outer cord of the brachial plexus. (Fig. 42.) It perforates, as a rule, the coraco-brachial muscle, and gains the intermuscular space between the biceps and brachialis anticus. It supplies the coraco-brachial and the flexor muscles of the forearm, the biceps, brachialis anticus, and supinator longus ; **the** cutaneous part is distributed to the skin over the insertion of these muscles.

5. *Describe the circumflex nerve.*

It is a branch of the posterior cord of the brachial plexus. (Fig. 42.) It passes through the quadrangular space to the back of the shoulder with the posterior circumflex artery. (Fig. 249.) It supplies the deltoid and teres minor muscles. It gives two articular branches to the shoulder-joint and cutaneous branches to the skin over the insertion of the deltoid and teres minor muscles.

6. *Name the important regions of geometrical parts in the upper extremity.*

(1) The radial groove ; (2) the ulnar groove ; (3) the cubital fossa ; (4) the brachial groove ; (5) the axilla ; (6) the quadrangular space ; (7) the first triangular space ; (8) the second triangular space.

7 *Give the special names for deep fascia in the upper extremity.*

(1) Anterior annular ligament ; (2) posterior annular ligament ; (3) dorsal fascia ; (4) palmar fascia ; (5) ligamenta vaginales ; (6) bicipital fascia ; (7) axillary fascia ; (8) clavi-pectoral fascia ; (9) costo-coracoid ligament ; (10) pectoral aponeurosis ; (11) deltoid aponeurosis ; (12) supraspinous fascia ; (13) infraspinous fascia ; (14) superior transverse scapular ligament ; (15) inferior transverse scapular ligament. (Fig. 252.)

8. *What structures did you **find** under the anterior annular **ligament at the** carpus?*

(1) The flexor sublimis digitorum ; (2) the flexor profundus digitorum ; (3) the flexor longus pollicis ; (4) the median nerve **and its artery.**

9. *Give boundaries of the radial groove, and **its** contents.*

On the radial side is the supinator longus muscle ; on the ulnar side, in the upper third, **the** pronator radii teres ; in the remainder of its course, the flexor **carpi** radialis. In front of the artery the skin and fascia form the roof of the radial groove. **Behind the** artery, from above, are : (1) The biceps ; (2) supinator **brevis** ; (3) pronator **radii** teres ; (4) flexor sublimis digitorum ; (5) flexor longus **pollicis** ; (6) pronator quadratus ; (7) a small part of the lower end of the radius. The radial groove contains the radial artery and its venæ comites.

10. *Give boundaries and contents of the ulnar groove.*

The ulnar groove contains the ulnar artery and its venæ comites. The artery is a branch of the brachial artery, being given off from this vessel in the cubital fossa. The groove is limited on the ulnar side by the flexor carpi ulnaris, this muscle being the guide to the artery in the lower two-thirds of its forearm course ; on the radial side by the flexor sublimis digitorum ; behind by the brachialis anticus, the flexor profundus digitorum, and the pronator quadratus. The ulnar artery is covered in the lower two-thirds by the skin and fasciæ. In its course from the cubital fossa to the ulnar groove proper, the artery passes behind all the muscles that originate from the inner condyle of the humerus, except the flexor carpi ulnaris. On entering the groove proper the artery is joined by the ulnar nerve.

11. *Give boundaries of the cubital fossa, and tell what you found in this fossa.*

Externally, is the supinator longus muscle ; internally, the pronator radii teres ; above, an imaginary line passing through the humeral condyles. The roof is formed by the skin and fasciæ. The floor is formed by the brachialis anticus and supinator brevis, containing between its superficial layers the posterior interosseous nerve. The fossa contains the tendon of the biceps and bicipital tuberosity ; the median nerve ; the brachial artery, dividing into the radial and ulnar arteries ; the musculo-spiral nerve, dividing into the radial and posterior interosseous nerves ; the cutaneous branch of the musculo-cutaneous nerve ; the median cephalic and median basilic veins.

The student should now review the scheme of the brachial plexus on page 77, *et seq.*

POSTERIOR PART OF LOWER EXTREMITY.

Dissection.—Locate greater trochanter, sacrum, coccyx, tuber of ischium, crest of ilium, inner and outer hamstrings, head of fibula, neck of fibula, os calcis, inner and outer malleoli.

Incisions.—(1) From centre of crest of ilium to centre of the os calcis ; (2) from greater trochanter **to** coccyx ; (3) from internal to external condyle ; (4) from inner malleolus **to** outer malleolus. Cut deeply enough to permit the edges of the skin to separate one-half of an inch with ease. In removing the skin from any area, cut closely enough that light may shine through the skin. The thing you see now having removed the skin is the superficial fascia. **It** contains the following structures in a variable amount of fat (Fig. 255):

Cutaneous or sensory nerves in figure 255.

1. *The twelfth or last thoracic nerve below the twelfth rib.*
2. *The iliac branch of the ilio-hypogastric nerve.*
3. *Posterior branches of the lumbar and sacral nerves.*
4. *The external cutaneous nerve of the lumbar plexus.*
5. *The small sciatic of the sacral plexus.*
6. *The cutaneous branch of the obturator nerve.*
7. *Twigs from the long or internal saphenous.*

8. *The short saphenous*, formed, as you see in figure 255, by the communicans **tibialis** and communicans fibularis, branches of internal and external popliteal **respectively.**

9. **The short** *saphenous vein*, with the nerve described in the preceding paragraph. This vein comes from the outer end of the dorsal arch of the foot, runs behind the **outer** malleolus of the fibula, passes up the mid-line of the leg, between the two heads of the gastrocnemius muscle, pierces the deep popliteal fascia, and opens into the *popliteal* vein.

Describe the twelfth thoracic nerve.

It is seen giving (1) a cutaneous branch to the anterior part of the gluteal region ; (2) a cutaneous branch to the anterior part of the abdominal walls as low as the hypogastric region. The nerve lies below the last rib, with the first lumbar artery. It crosses the quadratus lumborum muscle, and its muscular branches are distributed like the other abdominal intercostals.

Describe the ilio-hypogastric nerve.

This is a branch of the lumbar plexus, being given off from the first lumbar with the ilio-inguinal nerve. The iliac branch supplies the integument of the front part of the gluteal region. The hypogastric branch pierces the aponeurosis of the external oblique muscle one inch above the external abdominal ring, and supplies the skin in this region.

Observe that the posterior branches of the lumbar and sacral nerves supply the skin over the gluteus maximus muscle.

Remember, the external cutaneous branch of the lumbar plexus supplies the skin covering the vastus externus, and may be considered a dismembered branch of the anterior crural nerve, according to Hilton's law.

The small sciatic nerve supplies the skin covering the flexor **muscles of the** leg. It may be regarded as the dismembered branch of the great **sciatic nerve.** The long pudendal branch of this nerve **crosses** the tuber ischii **and is** distributed to the scrotum of **the** male and its homologue, the labia majora **of the** female.

The long saphenous nerve is a branch of the anterior crural, but to carry out the law governing cutaneous distribution, it extends to the inner side of the ankle, to supply the skin over the fullest insertion (periosteal) **of the** sartorius.

If you will now clean off all the fat, you will see a rather dense fascia covering the muscles. This is the deep fascia. That part of this fascia that covers **the** popliteal **space is** called the popliteal fascia. Now remove the deep fascia and make your **muscles** look like figure 256, taking pains not to destroy vessels and nerves.

You have now exposed to view the gluteus medius and maximus, biceps, semitendinosus, semi-membranosus, gracilis, sartorius, plantaris, gastrocnemius, and tendo Achillis.

THE POPLITEAL SPACE

is a surgical area occupying the lower posterior one-third of the thigh and the upper posterior one-sixth of the tibia. It is diamond in shape. Its floor is formed by **the** femur, ligament of Winslow, and popliteus muscle. Its roof is formed by the skin, superficial fascia, and deep popliteal fascia. (See above.) The space is bounded (Fig. 256), above the joint, externally by the biceps muscle, **internally by the** semitendinosus, semi-membranosus, gracilis, and sartorius; below the joint, externally by the outer head of the gastrocnemius and plantaris muscle, **internally by the** inner **head** of the gastrocnemius. The space contains, in a variable amount of fat and connective tissue:

1. The terminus **of the** short saphenous vein in the popliteal.
2. The descending branch of the small sciatic nerve. (Fig. 260.)
3. The communicating branches to the short saphenous nerve.
4. The internal popliteal nerve in the centre of the popliteal space. (Fig. 266.)
5. The external popliteal nerve with the biceps tendon.
6. The articular branch of the obturator nerve to the knee.
7. The popliteal artery and vein and their branches.

In figure 255 note in particular the formation of the short or external saphenous nerve, by the union of two communicating branches, from the internal and external popliteal nerves. There is no constant level at which this union takes place between the communicans tibialis and the communicans fibularis.

Locate the popliteal space.

It occupies the lower posterior one-third of the thigh and the upper posterior one-sixth of the leg; it extends from the aperture in the adductor magnus muscle to the lower border of the popliteus muscle.

Give the relation, in the centre of the space, of the internal popliteal nerve, and the popliteal vein and artery. (Fig. 261.)

The nerve is above, the vein in the middle, and the artery below. (Fig. 269.)

Remember, when we dissect the popliteal space, the subject is face downward, hence the order referred to is as follows: having cut through the skin, superficial fascia, **and deep fascia, called** in this region popliteal fascia, the most superficial structure **seen in the space** under the deep fascia is the internal popliteal nerve; the next deep structure immediately under the nerve is the popliteal vein; the third deep structure is the popliteal artery. This relation of the vein to its artery is in harmony with the rule governing the relation of arteries to their veins, above and below the diaphragm.

Why this relation of artery and vein?

The **rule** is: Veins are behind their arteries **below the diaphragm, and in** front **of them** above the diaphragm—where they are not on the same plane. Exception to rule: the renal vessels.

What can you say about the internal popliteal nerve? (Fig. 266.)

It begins at the bifurcation of the great sciatic, and ends at the lower border of the popliteal space, where its continuation through the leg is called posterior tibial. In its course it throws off these branches:

1. The communicans **tibialis** or communicans poplitei.

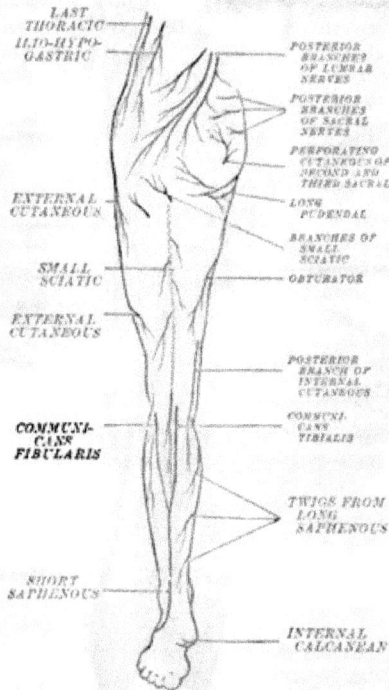

FIG. 255.—DISTRIBUTION OF CUTANEOUS NERVES ON THE POSTERIOR ASPECT OF THE INFERIOR EXTREMITY.

2. Three articular branches to the knee-joint.

3. Muscular branches to the gastrocnemius, soleus, plantaris, popliteus.

The articular branches reach the knee-joint how?

One goes with the azygos artery, **and** two accompany the internal articular arteries, superior and inferior.

Describe the external popliteal nerve.

It begins at the **bifurcation** of the great sciatic, is found in the outer side of the popliteal space **in the** sheath of the biceps muscle. It leaves the space between the biceps and the outer head of the gastrocnemius muscle. It crosses the neck of **the** fibula, having passed behind the head of this bone, and terminates in the **anterior tibial and musculo-cutaneous nerves.** In its course it gives off these

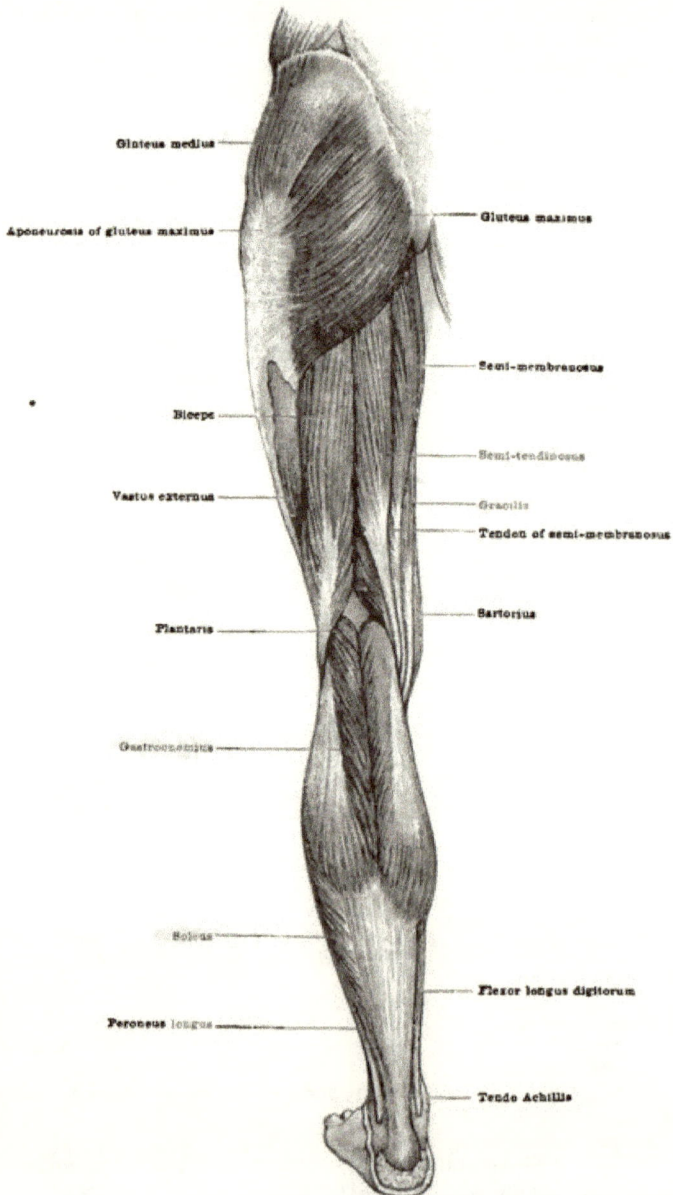

Gluteus medius

Aponeurosis of gluteus maximus

Biceps

Vastus externus

Plantaris

Gastrocnemius

Soleus

Peroneus longus

Gluteus maximus

Semi-membranosus

Semi-tendinosus

Gracilis

Tendon of semi-membranosus

Sartorius

Flexor longus digitorum

Tendo Achillis

FIG. 256.—SUPERFICIAL MUSCLES OF THE BACK OF THE THIGH AND LEG.

branches : (1) The communicans fibularis or **communicans peronei** ; (2) three articular branches to the knee. (Fig. 264.)

The articular branches reach the joint how?

One goes with the anterior tibial recurrent artery. Two accompany the external articular arteries, superior and inferior. Remember, nerve trunks that supply muscles, supply the skin covering the muscles, and the joints the muscles move.

Describe the popliteal artery.

It extends from the aperture in the adductor magnus muscle, to the lower border of the popliteus muscle. It is a continuation of the superficial femoral artery. In its course it gives off these branches : (1) Muscular to the borders of the popliteal space ; (2) articular branches, five **in** number, called superior two (internal and external), **inferior two** (internal **and** external), and one called the azygos. These arteries, **as above indicated, are** all accompanied by articular nerves from the popliteal **branches of the** great sciatic **nerve.**

Short saphenous nerve **is** *how formed, and where distributed?*

It **is formed by** the union of the communicans tibialis **and communicans** fibularis **branches of the** internal and external popliteal **nerves respectively, and** distributed **to the** posterior part of the leg and to the **outer border of the foot.**

In how **many** *ways* **is** *the semimembranosus muscle* **inserted!**

In two ways : (1) Tendinously, into the inner tuberosity **of** the tibia , (2) aponeurotically, **in** (a) ligament of Winslow, (b) into internal lateral ligament of the knee, (c) **into** the oblique line of the tibia, as the popliteal aponeurosis. Notice, aponeuroses are named according to the name of the muscle they cover **or** the surgical area they invest—*e. g.*, axillary, masseter, parotid, etc.

Explain the insertion of the gracilis, semitendinosus, and sartorius.

They are inserted into the upper one-third of **the** inner surface of the **tibia,** below the tuberosity ; the gracilis and semitendinosus *by tendon,* the former **one-**half inch the higher. The sartorius is inserted by a broad aponeurosis, which completely covers in the insertions of the other two muscles.

The heads of the gastrocnemius differ in origin how?

The inner head arises above the condyle ; the outer, from the side of the condyle. The latter seems to have been crowded out of place by the plantaris muscle—at least let this suggestion aid the memory.

Geometrical Inventory of the Popliteal Space :

Roof—skin, superficial and deep fasciæ.

Floor—femur, ligament of Winslow, popliteus muscle.

Superior end—aperture in the adductor magnus muscle.

Inferior end—oblique line of the posterior surface of the tibia.

Superior border—externally, the biceps muscle.

Superior border—internally, semitendinosus, semimembranosus, gracilis, and sartorius muscles.

Inferior border—internally, inner head of the gastrocnemius.

Inferior border—externally, outer head of the gastrocnemius and plantaris.

Shape—diamond ; resembling two triangles with bases together.

In the center are seen the internal popliteal nerve and popliteal vessels.

Outer border—the external popliteal nerve is in the sheath of biceps.

THE GLUTEAL REGION.

The gluteus maximus muscle (Fig. 256) must now be put on the stretch
by rotating the entire limb inward, and at the same time drawing the subject to
the edge of the table and slightly lowering the leg. Now, with a sharp scalpel
remove the deep fascia from the muscle itself (Fig. 257), and locate the line of
differentiation between the gluteus maximus and medius. Theoretically, this is
easily done, but practically some experience is necessary to obtain perfect results,
since the gluteal aponeurosis continues forward and completely covers the gluteus
medius muscle. The gluteus maximus originates : (1) From the aponeurosis of
the gluteus medius ; (2) from the lumbar aponeurosis and the greater sacro-sciatic
ligament ; (3) from the side of the coccyx ; (4) from the back of the lateral part

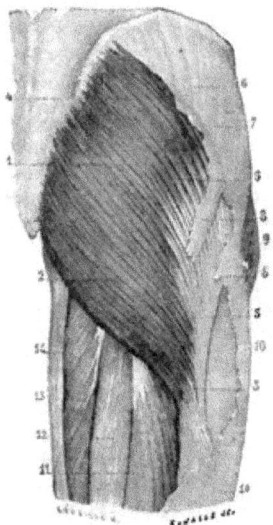

FIG. 257.—GLUTEUS MAXIMUS MUSCLE.

1. Gluteus maximus. 2. Its inferior portion. 3. Fibres of attachment to linea aspera. 4. Superior
portion. 5, 5. Tendinous fibres of insertion into linea aspera. 6. Upper portion of femoral aponeurosis. 7. Duplicature of this aponeurosis at superior level of gluteus maximus. 8. Portion of
its superficial layer attached to tendinous bands. 9. Lower extremity of tensor vaginæ femoris.
10, 10. Portion of femoral aponeurosis continuous with tendinous fibres of gluteus maximus.
11. Upper portion of biceps femoris. 12. Upper portion of semitendinosus. 13. Upper portion of
semimembranosus. 14. Gracilis.

of the lower two segments of the sacrum ; (5) from the posterior one-fifth of the
outer lip of the iliac crest ; (6) from the surface of the bone between the crest of
the ilium and the superior gluteal ridge. (Fig. 226.) .The muscle is inserted
into : (1) The ilio-tibial band ; (2) into the gluteal ridge of the femur. (Fig.
227.) The inferior gluteal nerve, a branch of the sacral plexus, supplies this
muscle.

Cut at its insertion into the deep fascia of the thigh, and into the gluteal
ridge of the femur, the **gluteus maximus** ; turn the same back, as seen in figure
258. Also cut the insertion of the **gluteus medius** at the oblique line of the
greater trochanter (Fig. 227), turn same up, and develop your work like
figures 258 and 259. Caution ! Take the greatest care, in turning the

above muscles back, not to injure the vessels and nerves, upon which both
the beauty and benefit of your dissection depend. You have exposed to view
now these structures :

1. The greater sacro-sciatic foramen. (Fig. 226.)
2. The lesser sacro-sciatic foramen. (Fig. 226.)
3. Sacro-sciatic ligaments—greater and lesser. (Fig. 258.)
4. The three gluteal muscles.
5. The pyriformis muscle.
6. The obturator externus muscle.
7. The obturator internus muscle.
8. The gemelli—superior and inferior.
9. The quadratus femoris muscle.
10. The superior gluteal nerve and artery.
11. The small sciatic nerve and its branches.

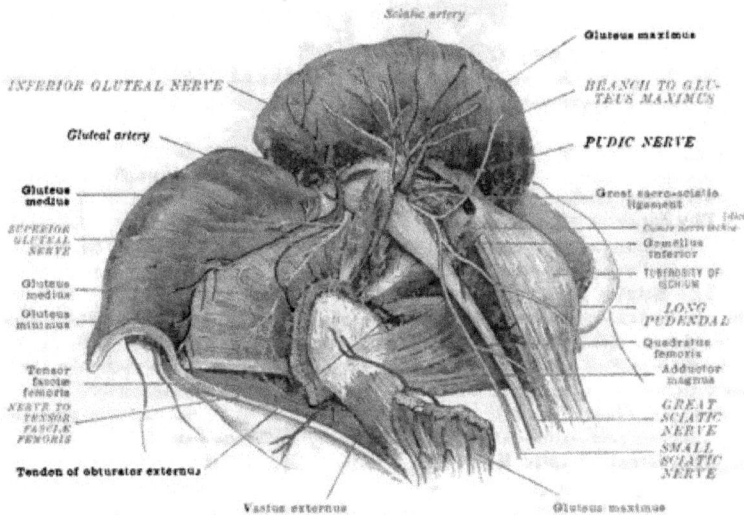

FIG. 258.—A DISSECTION OF THE NERVES IN THE GLUTEAL REGION.
(The gluteus maximus and gluteus medius have been divided near their insertions, and thrown upward.)

12. The sciatic artery and its branches.
13. The great sciatic nerve and its relations.
14. The internal pudic nerve and artery.
15. The origin of the biceps, semitendinosus, and semimembranosus.
16. The gluteus medius muscle.
17. The gluteus minimus muscle.
18. The subgluteal bursa over the greater trochanter.

The gluteus medius arises from the outer lip of the iliac crest for the anterior
four-fifths ; from the outer surface of the ilium, between the superior and middle
gluteal curved lines (Fig. 226) ; from the dense aponeurosis covering the muscle
and separating the same from the gluteus maximus. The muscle is inserted into
the oblique line of the greater trochanter of the femur. (Fig. 227.) Nerve-supply
is the superior gluteal from the lumbo-sacral cord of the sacral plexus. (Fig. 193.)

The gluteus minimus arises from the outer surface of the ilium, between the middle and the inferior curved gluteal lines (Fig. 226), from the capsule of the hip, and is inserted into the anterior border of the greater trochanter. (Fig. 228.) The nerve-supply is from the superior gluteal. (See page 281, Superior Gluteal Nerve.)

Notice the common origin of the **long head of the biceps** and **semitendinosus** from the tuber of the ischium. (Fig. 226.) See where the **semimembranosus**, in its origin from the tuber, crosses the two preceding muscles since it arises from the upper and outer part of the tuberosity.

Greater Sacro-sciatic Foramen.—The notch of the same name (Fig. 226) is converted into this foramen by the lesser sacro-sciatic ligament. It transmits the pyriformis muscle, which divides the foramen into an upper and a lower compartment. (Fig. 258.) Above the muscle emerge the *superior gluteal artery* and *nerve ;* below the muscle, the *greater sciatic nerve*, the *lesser sciatic nerve*, the *sciatic artery*, the *internal pudic nerve* and *artery*, and the *inferior gluteal nerve*.

The pyriformis muscle arises from the anterior surface of the sacrum, between the four upper **anterior sacral** foramina. The **muscle leaves** the pelvis by the greater **sacro-sciatic foramen, and is** inserted **into the upper** border of the greater trochanter.

The Gluteal Artery (Fig. 258).—The largest **branch of the** internal iliac ; **leaves** the pelvis by the greater sacro-sciatic foramen, above the pyriformis, divides into a superficial and a deep branch. The former **lies between** the gluteus **maximus and medius**, and anastomoses with the *circumflex iliac branch* of the **external iliac artery ;** the latter—the deep branch—**lies between** the gluteus **medius and minimus**, and anastomoses with the *external circumflex artery*,—a branch of the profunda.

The superior gluteal nerve—a branch of the lumbo-sacral cord of the sacral plexus—accompanies the artery of like name (Fig. 258) and supplies the gluteus medius and minimus and the tensor vaginæ femoris.

The Sciatic Artery (Fig. 258).—A branch of the internal iliac. It leaves the pelvis below the pyriformis muscle. It gives branches to the rectum, base of bladder, seminal vesicles, prostate, and all the muscles of the pelvic floor. It sends branches to the muscles on the back of the hip ; one branch, the *comes nervi ischiadici*, to the great sciatic nerve. (Fig. 258.) It anastomoses with the gluteal, the obturator, both circumflex arteries, internal and external, and with the superior perforating artery, a branch of the profunda **femoris.**

The Great Sciatic Nerve (Fig. 193).—The **largest** branch of the sacral plexus. Leaves **the pelvis** below the pyriformis. **Lies upon** successively the superior gemellus, the **internal obturator, the inferior** gemellus, the external obturator, the quadratus femoris, **and the adductor** magnus. (Fig. 258.) It ends in the popliteal space, in the *internal* and *external popliteal nerves*. It gives motor branches to the biceps, semitendinosus, semimembranosus, and adductor magnus muscles. It gives to the hip-joint *articular branches*, which perforate **the** posterior part of the capsule.

The Small Sciatic Nerve (Fig. 260).—A branch of the sacral plexus. (Fig. 193.) It leaves the pelvis below the pyriformis muscle, lying behind the greater sciatic nerve. Its branches are cutaneous : *ascending, internal, external* and *descending*. The latter passes through the popliteal space to supply the skin over the insertion of the flexors of the leg. (Fig. 255.) The small sciatic is then the great divorced sensory branch of the great sciatic, carrying out Hilton's law. The internal branches of this nerve are numerous. (Fig. 260.) The short ones supply the skin on the inner side of the thigh ; the longest one of the internal branches is called the inferior pudendal, or nerve of Soemmering, and supplies the scrotum and labia majora.

How to find the **long pudendal** in practical anatomy : It crosses the tendons of the biceps and semitendinosus, on the tuber of the ischium. (Fig. 260.)

The Inferior Gluteal Nerve.—A branch of the sacral plexus. Leaves the pelvis by the greater sacro-sciatic foramen, below the pyriformis, and supplies the gluteus maximus. (Fig. 258.)

The Lesser Sacro-sciatic Foramen.—The notch of the same name is con-

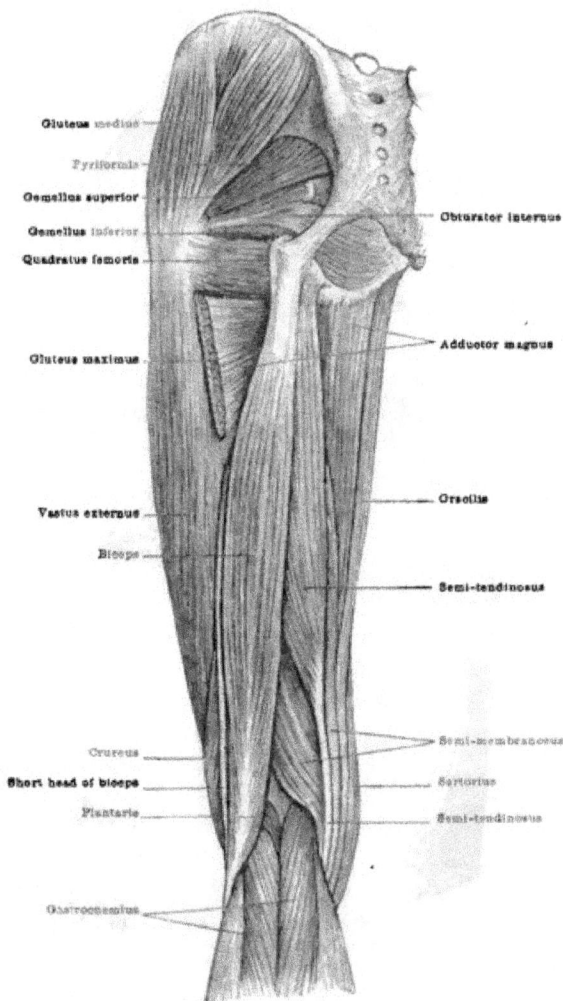

Fig. 259.—The External Rotators and the Hamstring Muscles.

verted into this foramen by the *greater sacro-sciatic ligament.* (Fig. 259.) It transmits the internal obturator tendon and its nerve, the internal pudic nerve, artery, and vein. (Fig. 258.) The vessels and nerve lie above the tendon. To trace these structures properly, it will be necessary to remove the gluteus maximus from the greater sacro-sciatic ligament, and then cut the latter. You will then see the internal pudics all entering Alcock's canal. (Fig. 258.)

Branches and Course of Internal Pudic Nerve.—A branch of the sacral plexus. (Fig. 194.) The nerve escapes from the pelvis by the greater sacro-sciatic foramen, crosses the spine of the ischium, enters the pelvis by the lesser sacro-sciatic foramen (Fig. 226), above the tendon of the obturator internus muscle, traverses a delamination of the obturator fascia, called Alcock's canal, in

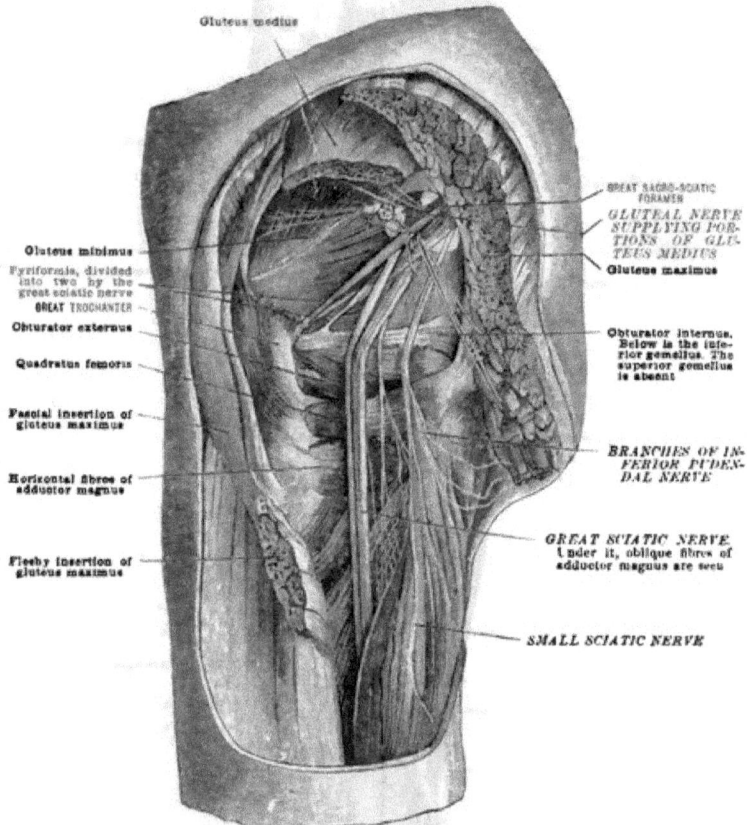

Gluteus medius

Gluteus minimus
Pyriformis, divided into two by the great sciatic nerve
GREAT TROCHANTER
Obturator externus
Quadratus femoris

Fascial insertion of gluteus maximus

Horizontal fibres of adductor magnus

Fleshy insertion of gluteus maximus

GREAT SACRO-SCIATIC FORAMEN
GLUTEAL NERVE SUPPLYING POR-TIONS OF GLU-TEUS MEDIUS
Gluteus maximus

Obturator internus. Below is the inferior gemellus. The superior gemellus is absent

BRANCHES OF IN-FERIOR PUDEN-DAL NERVE

GREAT SCIATIC NERVE Under it, oblique fibres of adductor magnus are seen

SMALL SCIATIC NERVE

FIG. 260.—DEEP DISSECTION OF THE GLUTEAL REGION.
(From a preparation in the Hunterian Museum.)

the outer wall of the ischio-rectal fossa. It gives off the following branches : (1) The *inferior hæmorrhoidal*, which crosses the ischio-rectal fossa and supplies the skin around the anus ; (2) the *perineal*, which supplies the integument of the scrotum and the labia majora, the transversus perinæi, the accelerator urinæ, the erector penis, and the compressor urethræ, or deep internal transversus perinæi muscle ; (3) the *dorsal nerve of the penis*, which gives off branches to the skin of the penis, and to the corpus cavernosum.

Greater Trochanter (Figs. 227 and 228).—*Muscles inserted into:* (1) The *gluteus medius* into the oblique line. (2) The *gluteus minimus* into the anterior surface. (3) The *obturator externus* into the digital fossa. (4) Into the superior border, in the following order, from before backward : the *pyriformis*, the *gemellus superior*, the *obturator internus*, and the *gemellus inferior*. This is the logical order of insertion, since they leave the pelvis in this order. (5) Into the base you will see the *vastus externus*. (Fig. 228.) Now turn to figures 227 and 228 and study these insertions on the bone. In your dissection trace each tendon to its complete insertion.

Gluteus Maximus Insertion.—This muscle is inserted into the gluteal ridge of the femur. This is called the third trochanter when it is very large. It is one of the constituent parts of the linea aspera. Turn to figure 227, and study the *linea aspera*. (Read page 329 on linea aspera.)

External Rotators of the Thigh.—From the nature of the origin and insertion of the pyriformis, gemelli, obturators, and quadratus femoris, they will produce external rotation of the limb, when this is straight—that is, when its long axis parallels the long axis of the trunk, as in the standing or recumbent posture. In the sitting posture, however, when the limb is at right angles to the trunk, they make traction parallel to the long axis of the femur, and thereby

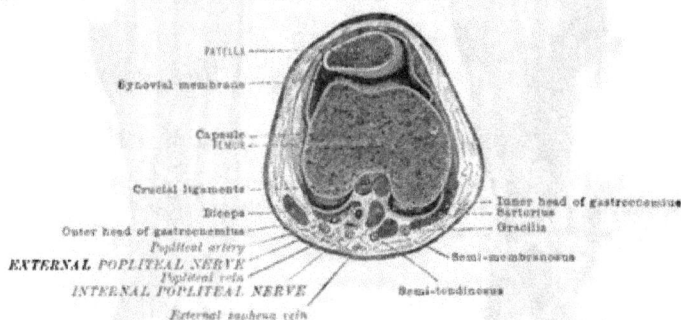

FIG. 261.—HORIZONTAL SECTION OF THE KNEE-JOINT. (One-half.)

become physiological abductors and antagonize the adductor group. The *obturator externus* remains an external rotator even in the sitting position. Pull the cadaver to the edge of the table, and make this simple demonstration at this stage of your dissection.

Demonstrate on yourselves external rotation (1) in the standing position ; (2) in the sitting position. In the first case notice the ease with which you may turn the whole limb outward. This act was done with ease because the six external rotators participated in the act. It would be very much more difficult to do if five of these refused to act. This is exactly what occurs when you sit and do external rotation, since five of these muscles cease to be external rotators when exerting traction in the long axis of the limb, as they do in the sitting posture. In the standing position six muscles are exerting traction in the region of the greater trochanter almost at right angles to the long axis of the femur. This makes external rotation easy. In sitting, only one muscle is left to do this work—*i. e.*, the external obturator muscle.

Internal Rotators of the Thigh.—The anterior segments of the *gluteus medius* and *minimus* rotate the thigh inward.

Tendon of Obturator Internus.—To show this tendon, carefully separate the

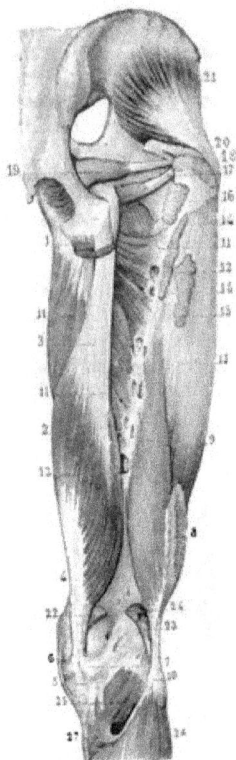

FIG. 262.—SHORT HEAD OF BICEPS AND SEMI-MEMBRANOSUS.

1. Attachment to ischium of long head of biceps and semitendinosus. 2. Semimembranosus. 3. Its superior tendon. 4. Its inferior tendon. 5. Middle portion of tendon. 6. Its anterior portion. 7. Its posterior portion. 8. Section of long head of biceps. 9. Its short head. 10. Its attachment to head of fibula. 11, 11, 11. Adductor magnus. 12, 12. Orifices for passage of perforating arteries and veins. 13. Vastus externus. 14, 14. Insertion of gluteus maximus. 15. Divided expansion of tendon of this muscle, continuous with the aponeurosis of the vastus externus. 16. Attachment of quadratus femoris. 17. Tendon of obturator externus. 18. Attachment of gluteus medius. 19. Obturator internus. 20. Tendon of pyramidalis. 21. Gluteus minimus. 22. Divided inner head of gastrocnemius. 23. Outer head. 24. Plantaris. 25. Popliteus. 26. Soleus. 27. Fibrous ring for artery, vein, and nerve.

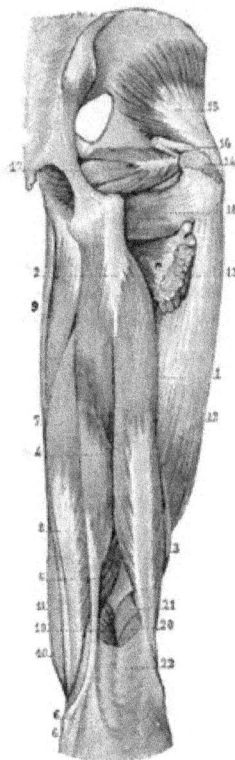

FIG. 263.—LONG HEAD OF BICEPS AND SEMI-TENDINOSUS.

1. Long head of biceps. 2. Common tendon of head of biceps and semitendinosus. 3. Inferior tendon of biceps. 4. Semitendinosus. 5. Its tendon. 6, 6. Its tendinous expansions, continuous with aponeurosis of leg. 7. Semimembranosus. 8. Its inferior tendon. 9. Gracilis. 10. Its tendon. 11. Sartorius. 12. Vastus externus. 13. Femoral attachment of gluteus maximus. 14. Insertion of gluteus medius. 15. Gluteus minimus. 16. Tendon of pyriformis. 17. Obturator internus. 18. Quadratus femoris. 19. Inner head of gastrocnemius. 20. Outer head of gastrocnemius. 21. Plantaris. 22. Popliteal aponeurosis.

muscle above and below from the gemelli. (Fig. 259.) Cut the muscle near its
insertion into the greater trochanter, and then trace it back into the pelvis
through the lesser sacro-sciatic foramen. Notice the strong converging tendinous bands on the under surface of the muscle. Also see the origins of the
gemelli from the ischial tuber and spine. (Fig. 226.)

*Reduced to its simplest terms, what is a rational physiological grouping of the
muscles primarily concerned in the movements of the hip-joint?*

(1) Flexors of the thigh on the abdomen ; (2) extensors of the thigh on the
abdomen ; (3) adductors of the thigh on the pelvis ; (4) abductors of the thigh
on the pelvis ; (5) internal rotators of the thigh; (6) external rotators of the
thigh.

Name the flexor muscles of the thigh.

The iliacus and psoas magnus, inserted into the lesser trochanter of the femur,
are, as Morris points out, almost pure flexor muscles. When, however, they
deviate from flexion, they are to some extent internal rotators, and not, as is
sometimes suggested, external rotators. Nerve-supply, second and third lumbar
nerves.

How is the thigh extended on the pelvis?

By the action of the gluteus maximus, which is the direct antagonist of the
psoas and iliacus. Nerve-supply, the inferior gluteal nerve.

Name the abductors of the hip-joint.

The tensor vaginæ femoris, the gluteus medius, the gluteus minimus.
Nerve-supply, superior gluteal.

Name the adductors of the thigh or hip-joint?

The obturator externus, the adductor longus, the adductor brevis, the adductor magnus, and the adductor gracilis. Nerve-supply, the obturator.

How is the thigh or hip-joint rotated inward?

By the anterior segments of the members of the abductor group of muscles—
viz., the glutei minimus and medius and the tensor vaginæ femoris ; and, in addition,
according to Morris, by the ilio-psoas muscle. Nerve-supply, superior gluteal.

Name the muscles that rotate the thigh outward.

The pyriformis, the obturator internus, the two gemelli muscles, and
the quadratus femoris. These muscles all derive their nerve-supply from the
sacral plexus. The obturator externus, although strictly speaking an adductor
muscle, becomes an external rotator when the patient is in the sitting posture.

Now, on figure 263 study the relation of the quadratus femoris, obturator
internus, and the gemelli to each other. On figure 262, on which the quadratus
and gemelli are removed, study the relation of the obturators to each other

BACK PART OF THE LEG.

Locate the fibrous arch connecting the heads of the gastrocnemius muscle.
(Fig. 259.) Notice passing downward into the deep intermuscular space the
popliteal artery and vein, in their proper relation to each other, and the internal
popliteal nerve. (Fig. 264.) Carefully isolate these structures from each other
and from the arch. Then lift the arch upward and cut down between the two
heads of the gastrocnemius to the point where this muscle joins the soleus. Also
notice between these two muscles the plantaris, with its very long tendon. (Fig.
264.) Now cut the gastrocnemius (Fig. 264), and trace each head across the
joint to its exact origin on the femur. (Fig. 227.) Notice the difference in origin

on the two sides. Also trace the plantaris up to its origin—and be sure you find
its nerve, a branch of the internal popliteal. (Fig. 264.)

Locate the fibrous arch of the soleus. (Fig. 264.) See passing under the
same the popliteal artery, nerve, and vein, now called posterior tibial, as far down
as the ankle, having passed under the arch of the soleus. Pull the arch up and
cut down between the two heads of the soleus. By cutting in the mid-line of
these two muscles you preserve the nerve-supply intact, which nerves it is desired
you trace out carefully. You will remember the gastrocnemius, plantaris, soleus,
and popliteus are supplied by the *internal popliteal nerve*. (Fig. 264.) Now
let your dissection verify the complete distribution of the internal popliteal nerve.
Preserve every branch of this nerve.

On each side of the fibrous arch of the soleus, on the *tibia* and *fibula*, you

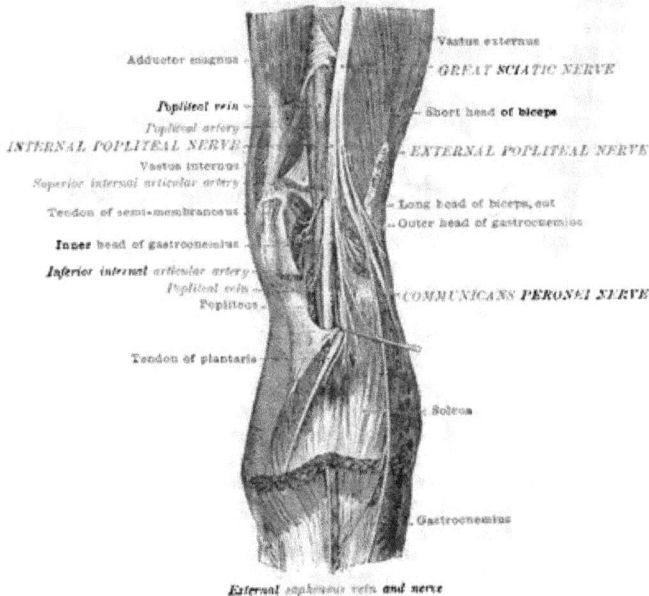

Fig. 264.—Deep View of the Popliteal Space. (Hirschfeld and Leveillé.)

will see the inner and outer heads of this muscle. (Fig. 208.) Divide them
carefully one-half of an inch from their origin, and turn the three muscles of the
superficial group aside, and figure 266 will represent what you should have.
You now have in view these structures :

1. The posterior tibial nerve and its branches. (Fig. 266.)
2. The posterior tibial vessels and their branches. (Fig. 266.)
3. The remaining origins of the plantaris, soleus, and gastrocnemius, and the
popliteus muscle, as seen in figure 265.
4. The tibialis posticus, an extensor muscle of the tarsus.
5. The flexor longus digitorum muscle—a flexor of the four outer toes.
6. The flexor longus hallucis muscle—a flexor of the great toe.
7. The thin, deep transverse fascia, or deep intermuscular fascia of the leg,
which you will now see covering in all the structures below the popliteus muscle.

If you examine this fascia, you will find it attached internally to the *tibia*, externally to the *fibula*, and above to the oblique line of the tibia. (Fig. 208.)

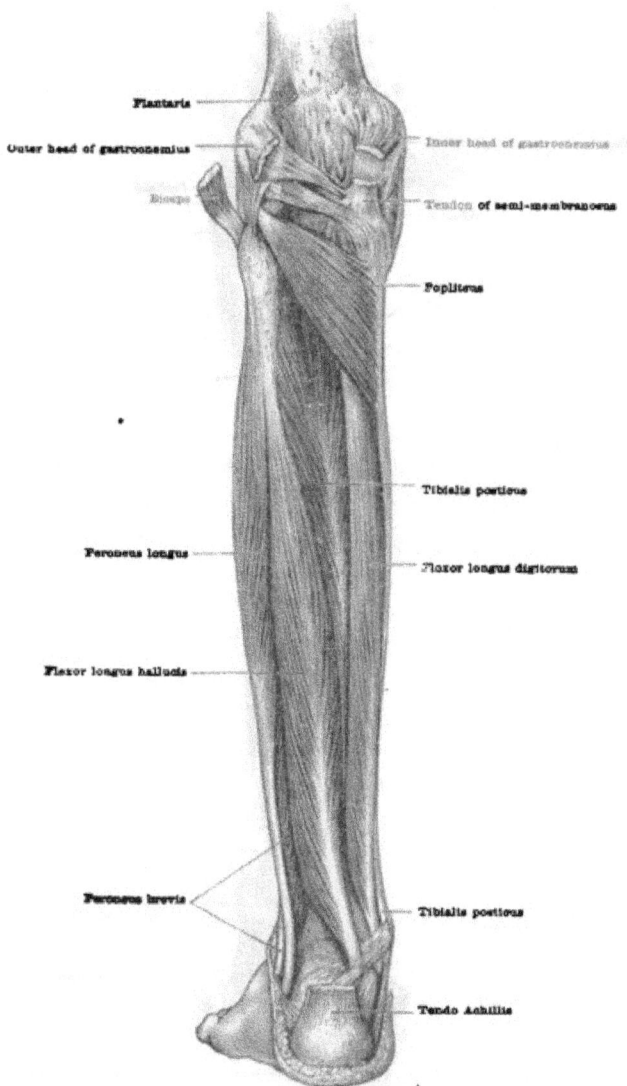

FIG. 265.—THE DEEP MUSCLES OF THE BACK OF THE LEG.

It is quite thin, yet strong, and permits the posterior tibial vessels and nerves to be seen through it.

The Posterior Tibial Nerve (Fig. 266).—It begins at the lower border of the popliteal muscle; it is a continuation of the internal popliteal nerve downward. It ends under the internal annular ligament (Fig. 266) of the ankle,

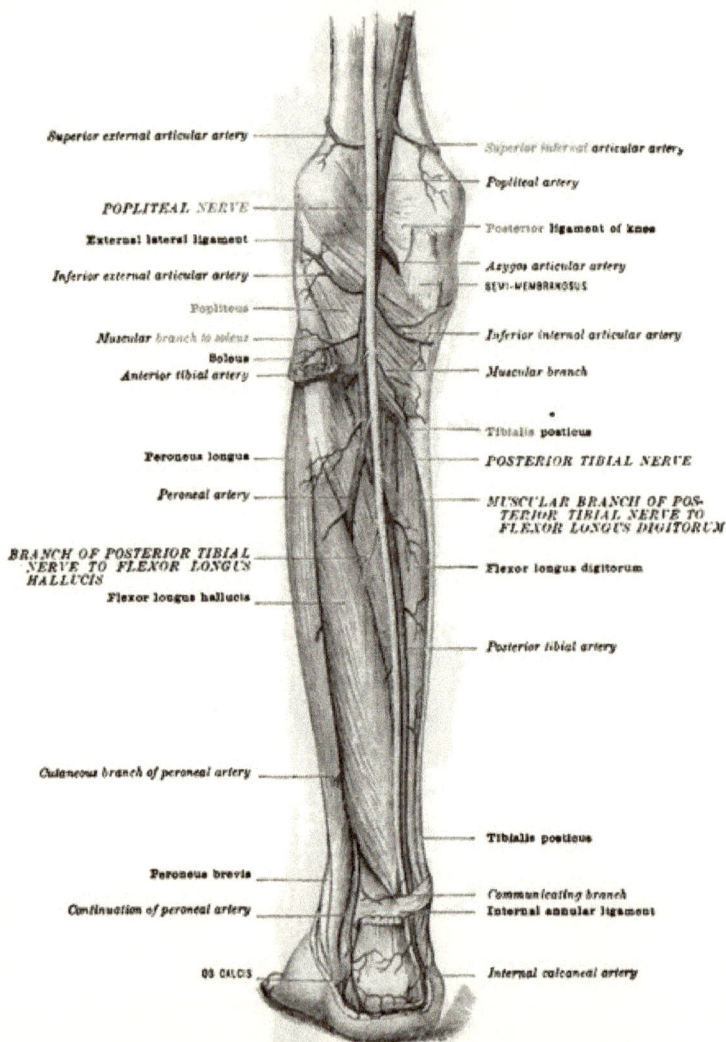

Superior external articular artery

POPLITEAL NERVE
External lateral ligament
Inferior external articular artery
Popliteus
Muscular branch to soleus
Soleus
Anterior tibial artery

Peroneus longus

Peroneal artery

BRANCH OF POSTERIOR TIBIAL
NERVE TO FLEXOR LONGUS
HALLUCIS
Flexor longus hallucis

Cutaneous branch of peroneal artery

Peroneus brevis

Continuation of peroneal artery

OS CALCIS

Superior internal articular artery
Popliteal artery
Posterior ligament of knee
Azygos articular artery
SEMI-MEMBRANOSUS
Inferior internal articular artery
Muscular branch

Tibialis posticus
POSTERIOR TIBIAL NERVE
MUSCULAR BRANCH OF POS-
TERIOR TIBIAL NERVE TO
FLEXOR LONGUS DIGITORUM
Flexor longus digitorum

Posterior tibial artery

Tibialis posticus

Communicating branch
Internal annular ligament

Internal calcaneal artery

FIG. 266.—RELATIONS OF THE POPLITEAL ARTERY TO BONES AND MUSCLES.
(The structures seen in this figure are covered by the deep transverse fascia. See page 378.)

between the internal malleolus and os calcis, by dividing into the internal and external plantar nerves. (Fig. 272.) It accompanies the posterior tibial vessels, being first to the inner and later to the outer side of these vessels.

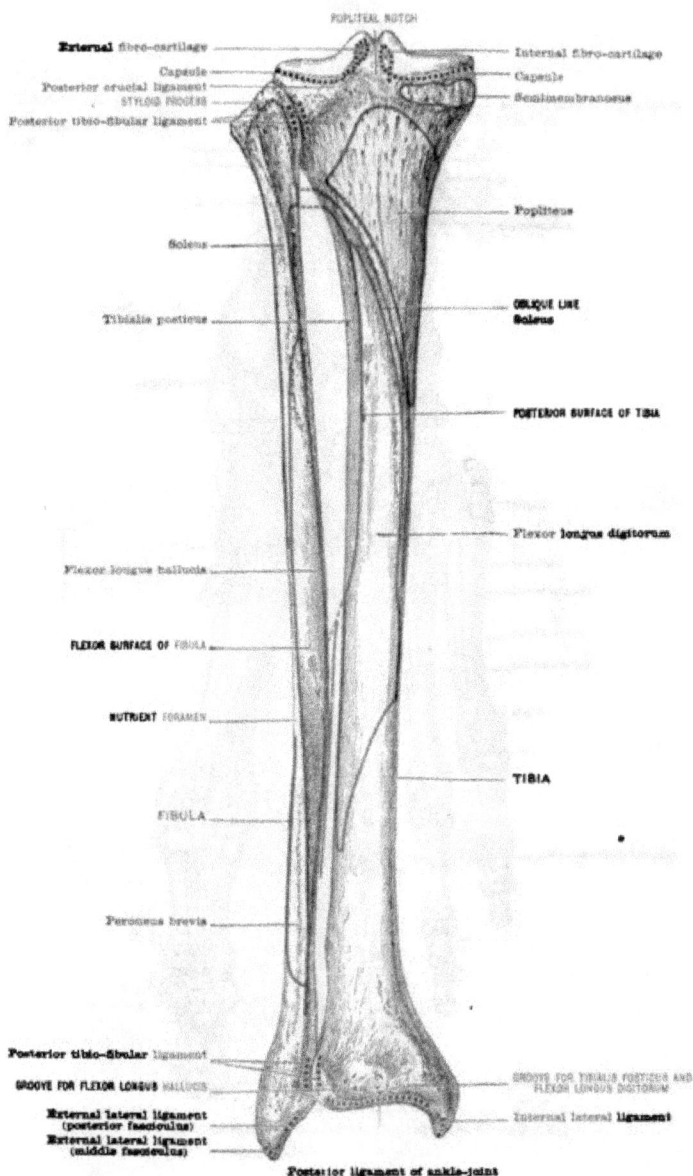

POPLITEAL NOTCH

External fibro-cartilage
Capsule
Posterior crucial ligament
STYLOID PROCESS
Posterior tibio-fibular ligament

Internal fibro-cartilage
Capsule
Semimembranosus

Popliteus

Soleus

OBLIQUE LINE
Soleus

Tibialis posticus

POSTERIOR SURFACE OF TIBIA

Flexor longus digitorum

Flexor longus hallucis

FLEXOR SURFACE OF FIBULA

NUTRIENT FORAMEN

TIBIA

FIBULA

Peroneus brevis

Posterior tibio-fibular ligament
GROOVE FOR FLEXOR LONGUS HALLUCIS
External lateral ligament
(posterior fasciculus)
External lateral ligament
(middle fasciculus)

GROOVE FOR TIBIALIS POSTICUS AND
FLEXOR LONGUS DIGITORUM
Internal lateral ligament

Posterior ligament of ankle-joint

FIG. 267.—THE LEFT TIBIA AND FIBULA. (Posterior view.)

Branches of Posterior Tibial Nerve.—(1) An articular branch to the ankle, given off just above the bifurcation of the nerve into the internal and external

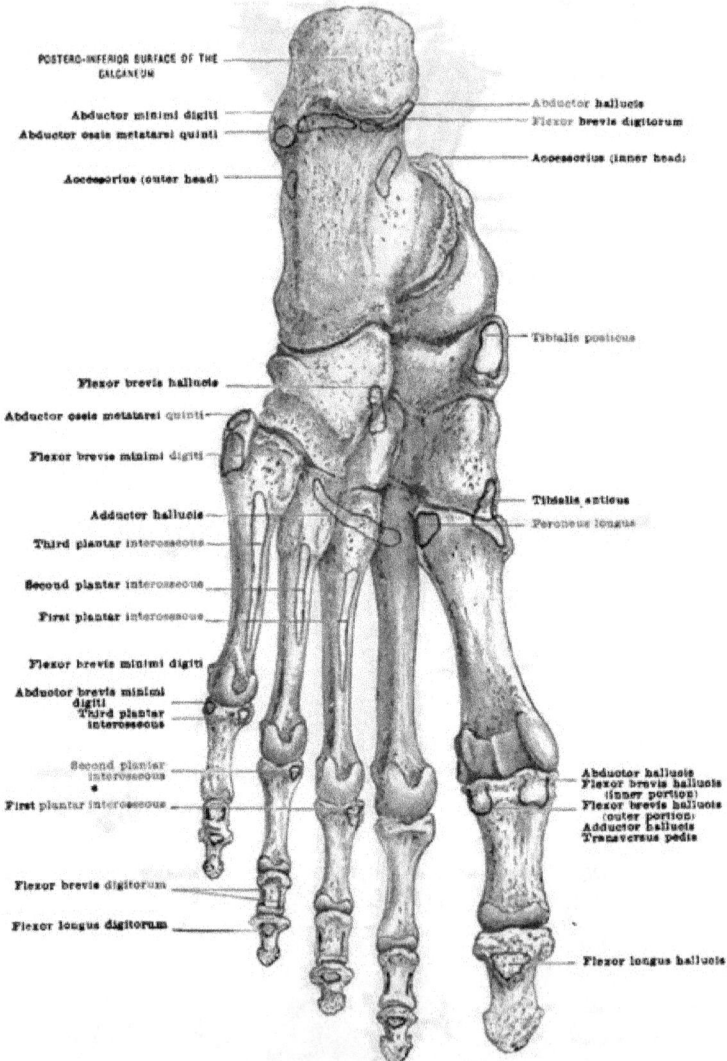

POSTERO-INFERIOR SURFACE OF THE
CALCANEUM

Abductor minimi digiti

Abductor ossis metatarsi quinti

Accessorius (outer head)

Abductor hallucis

Flexor brevis digitorum

Accessorius (inner head)

Tibialis posticus

Flexor brevis hallucis

Abductor ossis metatarsi quinti

Flexor brevis minimi digiti

Adductor hallucis

Third plantar interosseous

Second plantar interosseous

First plantar interosseous

Tibialis anticus

Peroneus longus

Flexor brevis minimi digiti

Abductor brevis minimi
digiti
Third plantar
interosseous

Second plantar
interosseous

First plantar interosseous

Abductor hallucis
Flexor brevis hallucis
(inner portion)
Flexor brevis hallucis
(outer portion)
Adductor hallucis
Transversus pedis

Flexor brevis digitorum

Flexor longus digitorum

Flexor longus hallucis

FIG. 268.—THE LEFT FOOT. (Plantar surface.)

(Study the insertion and location of tendons on this and compare your dissection therewith.)

plantar nerves. (2) A plantar cutaneous nerve, to the heel and inner side of the sole of the foot. (Fig. 272.) (3) Muscular branches to the tibialis posticus, flexor longus hallucis, and flexor longus digitorum muscles. (Fig. 266.)

(4) Two terminal branches—the internal and external plantar nerves. (Fig. 272.)

Branches of the Posterior Tibial Artery.—(1) The peroneal artery to the outer side of the leg. (Fig. 266.) (2) A nutrient artery to the tibia, which you will find entering the foramen as far down as the lower end of the oblique line of the tibia, on the ridge separating the origin of the tibialis posticus from the flexor longus digitorum. (Fig. 208.) Find this artery and recall the rule for the direction of these foramina in the long bones of the extremities : from the knee and

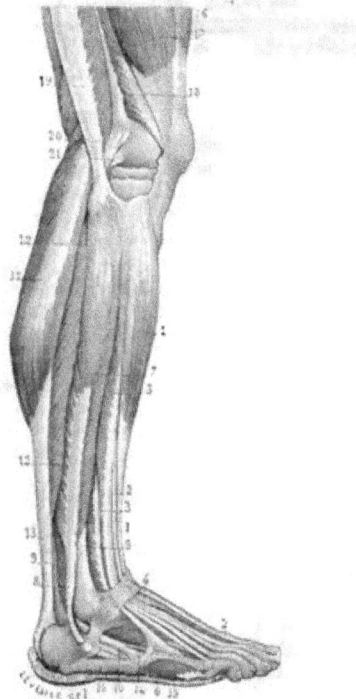

FIG. 269.—MUSCLES OF THE LEG, EXTERNAL ASPECT.

1. Tibialis anticus. 2, 2. Tendon of extensor proprius pollicis. 3, 3. Extensor longus digitorum. 4. Its tendons for four last toes. 5. Peroneus tertius. 6. Its attachment to last two metacarpal bones. 7. Peroneus longus. 8. Its tendon. 9. Peroneus brevis. 10. Its tendon. 11. Outer head of gastrocnemius. 12, 12. Soleus. 13. Tendo Achillis. 14. Extensor brevis digitorum. 15, 15. Abductor minimi digiti. 16. Rectus femoris. 17. Vastus externus. 18. Its inferior fibres. 19. Tendon of biceps femoris. 20. External lateral ligament of knee. 21. Tendon of popliteus.

toward the elbow. (3) A communicating branch to the peroneal, on the back of the tibia, under the flexor longus hallucis, two inches above the joint. (4) Internal calcanean, which communicates with the external calcanean branch of the peroneal artery behind the tendo Achillis. (Fig. 266.) (5) Muscular branches, small and numerous, to the soleus and deep muscles on the back of the leg. (Fig. 266.)

The Posterior Tibial Canal.—For practical purposes we may locate the *posterior tibial nerve* and its accompanying *vessels* in a canal, bounded as follows :

Externally, by the *flexor longus hallucis*; internally, by the *flexor longus digitorum*; its floor is the *tibialis posticus, flexor longus digitorum, tibia*, and *internal lateral ligament*; its roof, the *deep transverse fascia*, described in a previous paragraph.

Deep Muscles of the Leg.—(1) Popliteus; (2) flexor longus hallucis; (3) **flexor** longus digitorum; (4) tibialis posticus. It is necessary now to study the posterior surfaces of the tibia. (Fig. 208.) Notice that the oblique line of the tibia has three lips: (1) An upper one, occupied by the insertion of the *popliteus muscle*; (2) a middle one, by the *soleus*; (3) a lower one, by two muscles— the *flexor longus digitorum* on the outer side of the vertical line, and the *tibialis posticus* on the inner side. Notice, too, that the posterior surface of the fibula (Fig. 208) is occupied by the origin of two muscles: the outer head of the *soleus*, the upper one-third, and the *flexor longus hallucis*, the middle one-third, of this posterior surface.

What can be said regarding the rather peculiar origin and insertion of the flexor longus digitorum and flexor longus hallucis?

They arise from surfaces of bone above, opposite their insertions, into the phalanges below: The *great toe* is on the side opposite the *fibula*; the *four lesser toes* are on the side opposite the *tibia*. Now, the muscle that bends the great toe—flexor longus hallucis—arises from the posterior surface of the fibula, middle one-third; likewise, the muscle that bends the four lesser toes—the flexor longus digitorum—arises from the posterior surface of the tibia, middle one-third. To gain their **insertions** they cross in the sole of the foot in the second layer of muscles **of this** region. (Fig. 270.)

The flexor longus hallucis traverses four grooves. (Fig. 268.) The flexor longus hallucis needs special mention. It passes through (1) a little groove on the posterior part of the tibia (Fig. 208); (2) through a pronounced one on the narrow posterior surface of the astragalus; (3) through a well-marked groove on the under part of the sustentaculum tali of the os calcis (Fig. 268); (4) through a groove between the two sesamoid bones, in the tendon of the flexor brevis hallucis at the metatarso-phalangeal articulation. This muscle is inserted into the base of the distal phalanx of the great toe. (Fig. 268.)

Trace the popliteus muscle to its origin. (Fig. 266.) Detach it at the oblique line of the tibia, its insertion, and trace it between the outer tuberosity and head of fibula, under the external lateral ligament of the knee and the biceps, to the outer side of the external condyle of the femur.

Structures under Internal Annular Ligament.—From the internal malleolus to the os calcis: (1) Tendon of tibialis posticus, next the malleolus; (2) tendon of flexor longus digitorum; (3) a sheath of connective tissue containing the posterior tibial nerve and artery, with a vein on each side of the artery; (4) behind this sheath, and deeply buried in its three upper grooves, above referred to, is the tendon of the flexor longus pollicis. In the dissection of the sole of the foot these several structures will be traced to their various destinations.

The Peronei Longus and Brevis (Figs. 266 and 269).—These two muscles occupy almost the entire **outer** surface of the fibula. Commonly, the origin is as follows: the longus from the upper one-third and the brevis from the middle one-third. (Fig. 207.) The muscles are separated in front and behind from adjacent musculatures by intermuscular septa. They pass under the external annular ligament, immediately behind the malleolus, and are inserted: (1) the brevis into the base of the fifth metatarsal (Fig. 269); (2) the longus passes through a groove on the under surface of the cuboid bone, then through an osseo-aponeurotic canal, to the base of the first metatarsal bone. (Fig. 275.) These muscles are supplied by muscular branches of the musculo-cutaneous nerve, a branch of the external popliteal. (Fig. 264.) These muscles, when acting together, assist the tibialis posticus in extending the ankle. They are antagonized by the tibialis anticus and peroneus tertius.

THE SOLE OF THE FOOT. (Fig. 272.)

The importance of a knowledge of this region can not be overestimated. The large number of important structures in a comparatively small area, combined with thick skin, dense plantar fascia, and delicate lumbrical muscles, all taken

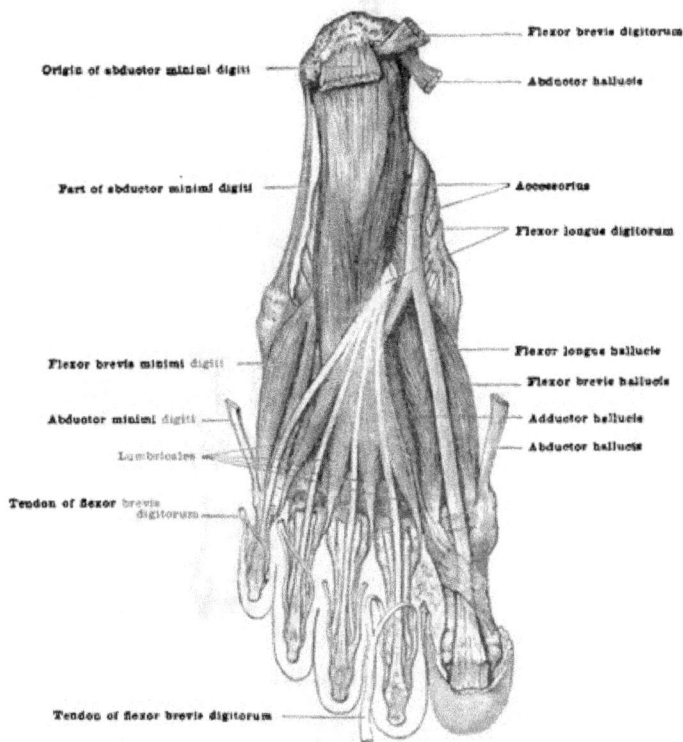

FIG. 270.—SECOND LAYER OF THE MUSCLES OF THE SOLE.

together make dissection of this area somewhat difficult. A dissection of the four layers of muscles alone would be an easy task, but here, as in larger areas of the body, nerves and vessels must be saved. The only general rule I can give you is this: Follow painstakingly as a guide the dissections given in the text, and learn from the figures what you expect to find. Having done this, follow each nerve and vessel out carefully to its distribution. No number of written pages will make an awkward man improve his touch; hence it is impossible to

do more than suggest to the student that he should early learn to cultivate a
delicate touch. It requires but little effort with the forceps to divide the delicate
connective tissue that surrounds all vessels and nerves, and intervenes between
muscles. In the sole of the foot you will find, in this order, from without inward, the following structures:

1. The thickest skin of the body—on the sole of the foot.
2. Superficial fascia, very thick, containing granular fat.
3. The deep or plantar fascia, with its three divisions.

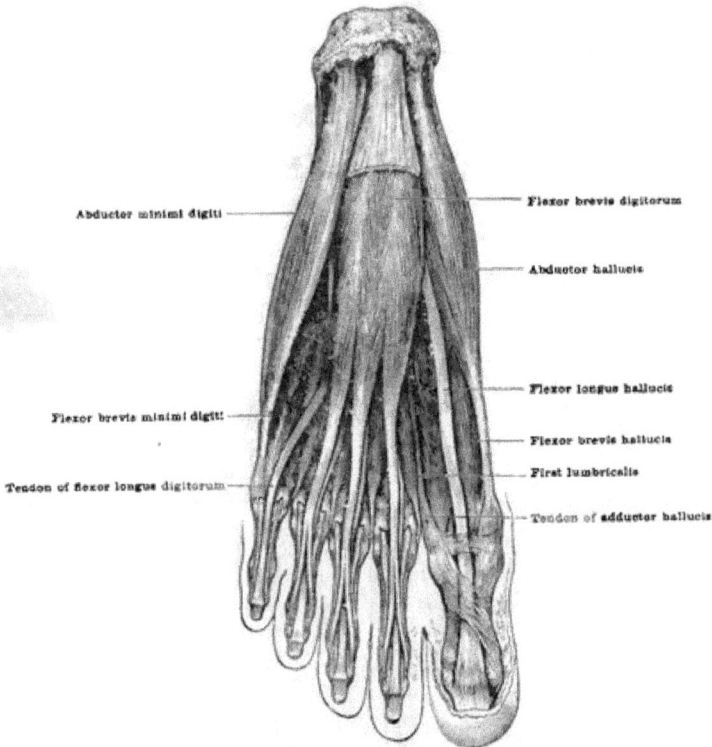

FIG. 271.—FIRST LAYER OF THE MUSCLES OF THE SOLE.

4. *The first layer of muscles:* The abductor hallucis, the flexor brevis digitorum, and the abductor minimi digiti.
5. *The internal and external plantar nerves and vessels.* (Fig. 272.)
6. *The second layer of muscles:* The tendon of flexor longus hallucis and a
slip of contribution to the flexor longus digitorum; tendons of the flexor longus
digitorum; the musculus accessorius and lumbricales; the nerve- and blood-
supply to these muscles.
7. *The third layer of muscles:* Flexor brevis hallucis and its sesamoid bones;
abductor hallucis muscle; flexor brevis minimi digiti muscle; transversus pedis
muscle; nerve- and blood-supply to these muscles.

8. *The fourth layer of muscles :* Dorsal interossei, four in number ; plantar interossei, three in number ; the nerve- and blood-supply to these. (Fig. 275.)

9. *The tendon of the peroneus longus* and its sesamoid bone. (Fig. 275.)

Incisions.—Start at the middle of the os calcis and make one cut through the skin to the middle of the great toe ; the other to the middle of the little toe. This V-incision will permit you to find :

1. **The plantar cutaneous nerve,** a branch of the posterior tibial nerve.

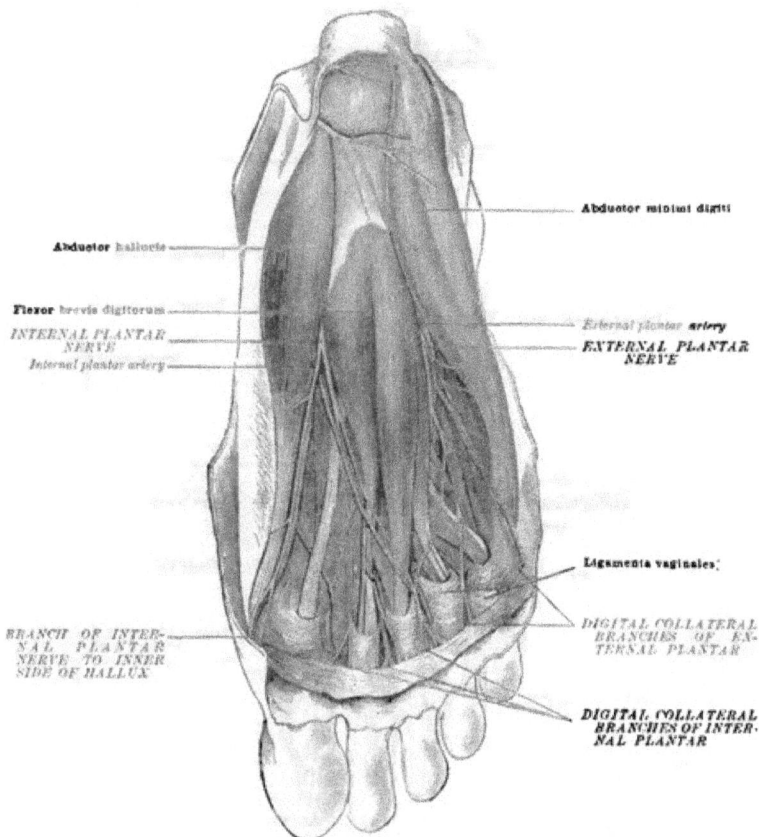

FIG. 272.—SUPERFICIAL NERVES IN THE SOLE OF THE FOOT. (Ellis.)
(In this dissection the greater part of the plantar fascia has been removed.)

(Fig. 272.) This is a sensory nerve to the heel, and you will find it in the thick pad of fat in the superficial fascia. Sometimes the granular fat here is three-fourths of an inch thick.

2. **The cutaneous branches of the internal and external plantar vessels and nerves.** Exercise now your common sense in removing the skin.

The superficial fascia contains much fat. The fat found in the palms of the hands, in the soles of the feet of the adult, and all over the body of the child at term, is called granular. Remove all this fatty tissue, and notice it becomes less abundant in the hollow of the foot and toward the toes.

The deep fascia occurs in the sole of the foot, under two names : (1) From the heel to the clefts of the toes it is called plantar fascia ; (2) from the clefts of the toes onward it forms very dense sheaths for the flexor tendons of the toes. These sheaths are called the ligamenta vaginales (ligamentum vaginalis, in the singular).

Dissection of Plantar Fascia and Ligamenta Vaginales.—Having now cut through the dense, deep fascia of the toes,—the ligamenta vaginales,—you

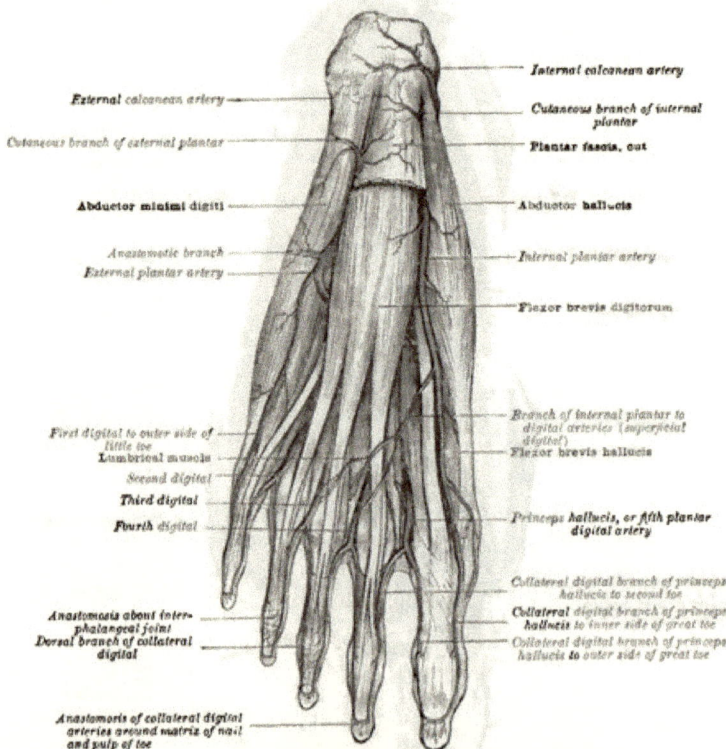

External calcanean artery

Cutaneous branch of external plantar

Abductor minimi digiti

Anastomotic branch
External plantar artery

First digital to outer side of little toe
Lumbrical muscle
Second digital
Third digital
Fourth digital

Anastomosis about inter-phalangeal joint
Dorsal branch of collateral digital

Anastomosis of collateral digital arteries around matrix of nail and pulp of toe

Internal calcanean artery

Cutaneous branch of internal plantar

Plantar fascia, cut

Abductor hallucis

Internal plantar artery

Flexor brevis digitorum

Branch of internal plantar to digital arteries (superficial digital)
Flexor brevis hallucis

Princeps hallucis, or fifth plantar digital artery

Collateral digital branch of princeps hallucis to second toe
Collateral digital branch of princeps hallucis to inner side of great toe
Collateral digital branch of princeps hallucis to outer side of great toe

FIG. 273.—THE PLANTAR ARTERIES.
(From a dissection in the Museum of St. Bartholomew's Hospital.)

must find : (1) The theca and the thecal culs-de-sac, the synovial sheath of the tendons ; (2) the vincula, delicate thread-like bands of deep fascia ; (3) the tendons of the flexor longus digitorum ; (4) the tendons of the flexor brevis digitorum ; (5) the slits, in tendons of the flexor brevis digitorum ; (6) the grooves on the plantar surfaces of the pedal phalanges ; (7) the outer division of the plantar fascia, quite thin, covering the abductor minimi digiti muscle ; (8) the inner division of the plantar fascia, quite thin, covering the abductor hallucis ; (9) the middle division of the plantar fascia, very thick, strong, and glistening, covering the flexor brevis digitorum ; (10) two intermuscular grooves almost under the diverging branches of the V-incision you made in the skin. You will notice that the three divisions of the plantar fascia correspond to the three muscles in the first layer, cover them, and in great part are almost inseparably connected therewith. (Fig. 273.) The plantar fascia is part of the origin of these muscles. Remove this fascia (Figs. 281 and 273) and trace out with the forceps the branches of the internal and external plantar nerves and vessels. (Fig. 272.)

The plantar fascia has a central part that covers the flexor brevis digitorum; an outer portion that covers the abductor minimi digiti, and is continuous with the dorsal fascia of the foot externally; an inner portion that covers the abductor

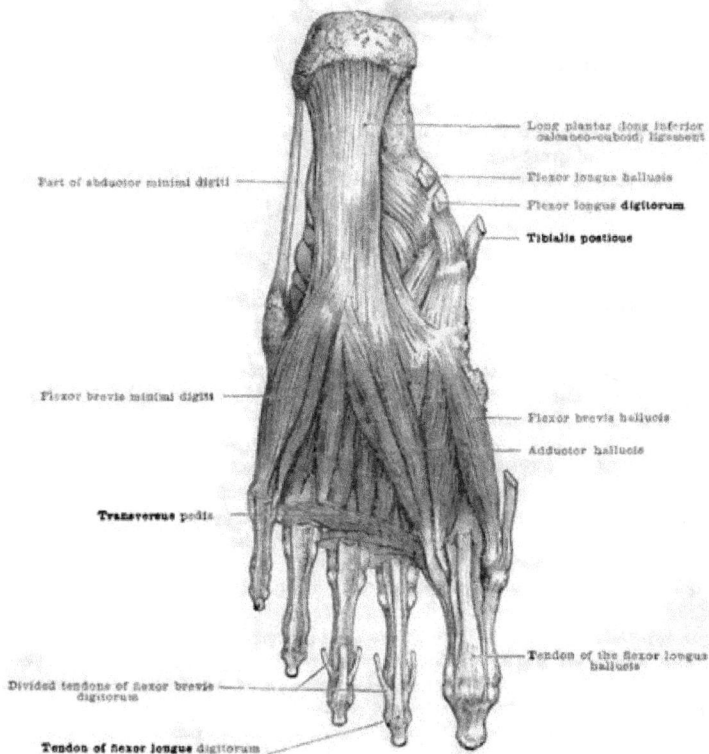

Part of abductor minimi digiti

Flexor brevis minimi digiti

Transversus pedis

Divided tendons of flexor brevis digitorum

Tendon of flexor longus digitorum

Long plantar (long inferior calcaneo-cuboid) ligament

Flexor longus hallucis

Flexor longus digitorum

Tibialis posticus

Flexor brevis hallucis

Adductor hallucis

Tendon of the flexor longus hallucis

FIG. 274.—THIRD LAYER OF THE MUSCLES OF THE SOLE.

hallucis, and is continuous internally with the dorsal fascia of the foot. Posteriorly, the plantar fascia is attached to the os calcis; anteriorly, it is continuous with the ligamenta vaginales.

Muscles of the First Layer (Fig. 273).—Dissect carefully in the grooves between the middle muscle and the muscles on each side of it. Trace all three muscles back to their origins on the os calcis. (Fig. 268.) Likewise trace each forward to its insertion: the *abductor hallucis* to the base of the first phalanx of the great toe (Fig. 268); the *abductor minimi digiti* to the base of the first phalanx of the little toe; the *flexor brevis digitorum* to the second phalanges of the four lesser toes, by cutting through the ligamenta vaginales the long

way of the tendon. (Fig. 268.) Now cut all three of these muscles and turn
them forward, and expose the second layer of muscles. (Fig. 270.) Having
cut through the ligamenta vaginales of the toes (Fig. 272), observe that the
four tendons of the flexor longus digitorum pass through slits in the four
corresponding tendons of the flexor brevis digitorum, in the same manner the
deep flexors of the fingers pass through the superficial ones.

Second Layer (Fig. 270).—In the figure of the flexor longus digitorum
notice **three** accessories: (1) The *muscular accessorius* (Fig. 268); (2) the four
lumbricales (Fig. 273); (3) a slip of contribution from the *tendon of the flexor
longus hallucis.* (Fig. 270.) Having studied the origin and insertion of these
muscles, **divide** the flexor longus hallucis and flexor longus digitorum near the
os calcis and **turn** them all forward, as you did the previous layer.

Third Layer (Fig. 274).—Trace out to its origin each one of the three muscles inserted into the base of the great toe. Cut down between the two sesamoid
bones in the tendon of the flexor brevis hallucis. Trace the tendon of the
peroneus longus across the sole of the foot to its insertion into the internal cuneiform and base of great toe metatarsal. (Fig. 275.)

In the **fourth layer** we have the dorsal interossei and the plantar interossei.
In number they are seven, four being dorsal and three plantar. In action they
are analogous to the interossei of the hand. In **the foot** they are all supplied
by the external plantar nerve; in the hand by the ulnar nerve.

Function of the Interossei and Lumbricales.—It is comprehensive to consider
the action of these **muscles** in this manner in the hand:

1. **All the interossei** and lumbricales acting together, flex **at** the metacarpo-
phalangeal articulation **and** extend at the first and **second** phalangeal articulations. This gives about the position for holding the **pen.**

2. The four dorsal interossei acting alone, abduct **the** digits. Notice that
supination of the forearm is involuntarily associated with abduction **of the**
digits.

3. The palmar interossei acting alone, adduct the digits. Notice that prona-
tion of the forearm is associated involuntarily with adduction of the fingers. In
the above I have described the action of these muscles in the hand; the action
in the foot is the **same,** only less under control of the will, since the foot of man
is not a prehensile member.

4. The adductor line of fingers and toes. The adductor finger is the
middle; the adductor toe is the second. In adduction and abduction the other
digits approach and recede from this adductor digit.

The Adductor Digit.—(1) The adductor digit of the hand has no palmar
interossei inserted into it; (2) the adductor digit of hand has two dorsal interossei inserted into it; the same is true of the foot.

Difference in Origin.—Notice that the dorsal interossei are very large, and
arise from the contiguous sides of the metatarsals or metacarpals; that the
palmar are small, and arise from one side only of the metatarsal or metacarpal
corresponding to the phalanx into which they are inserted. Practise the
movements of the lumbricales and interossei on your own fingers until you
master this subject thoroughly.

Granular Fat.—Found in the *superficial fascia* of the hand and foot—palmar
and plantar surfaces—and in the general superficial fascia of the fœtus and infant.
The toughness of this kind of fatty tissue depends on a larger amount of connective tissue.

Divisions of Plantar Fascia.—(1) An outer thin one that covers the *abductor minimi digiti.* (2) An inner one that covers the *abductor hallucis.* (3) A
middle one that **covers** the *flexor brevis digitorum.* This middle one is very
strong.

Ligamenta Vaginales.—These are the fibrous sheaths of deep fascia that cover the flexor tendons. They are formed by the deep fascia, in a modified form. Above, they are continuous with, the plantar fascia. They are attached on the sides to the margins of the grooves on the plantar surface of the phalanges.

The Theca and Thecal Culs-de-sac.—The vaginal synovial membrane, consisting of a *visceral* and a *parietal* layer, is called theca. The point in conjunction between the two layers—about the middle of the metatarsal bones—forms a cul-

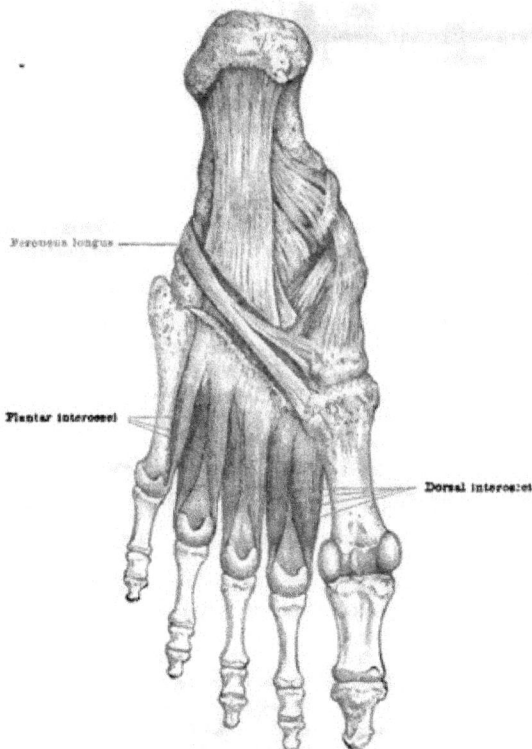

FIG. 275.—FOURTH LAYER OF THE MUSCLES OF THE SOLE.

de-sac. Pus forming below in the *cavity of the theca* could burrow to this point. Note the difference in the fingers.

Normally, the thecal cul-de-sac is found as above described in each toe, thus making five in number. The homologous structures of the fingers are described as three, the thumb and little finger having none, since the thecal synovial membrane corresponding to these two members is continuous above with the two bursæ under the annular ligament. My experience is, however, that in the majority of cases no communication will be found between the thecal sacs of the thumb and little finger, and the bursæ under the anterior annular ligament at the wrist.

The Vincula.—These are small silvery cords that extend from the bottom of

the groove to the tendons, and from tendon to tendon, in the ligamenta vaginales
of the digits of the lower as well as the upper extremity. By means of these
vincula capillary vessels reach the synovial membrane of the tendon.

The Internal Plantar Nerve.—Begins under the internal annular ligament
at the bifurcation of the posterior tibial nerve. It lies between the first and second
layers of muscles in the sole of the foot. (Fig. 272.) It gives off: (1) Articular
branches to the tarsus and metatarsus and phalangeal articulations ; (2) a cutaneous branch to the skin of the heel (Fig. 272) ; (3) digital branches to the
inner three **and** one-half toes ; **(4) muscular** branches to the abductor hallucis,
flexor brevis hallucis, flexor brevis digitorum, and two lumbricals.

The External Plantar Nerve.—Begins at the bifurcation of the posterior
tibial ; it accompanies the external plantar artery into the sole of the foot. This
nerve is similar in its distribution to the ulnar nerve. It supplies a larger number
of muscles, still it is smaller than the internal plantar. The muscles innervated
by this nerve are : (1) All the interossei, seven in number, four dorsal and
three plantar ; (2) two lumbricals on the outer side ; (3) adductor hallucis ;
(4) the transversus pedis. This nerve also supplies one and one-half toes—
the flexor accessorius and abductor minimi digiti.

The Internal Plantar Artery (Fig. 273).—This is a small artery. It
begins with the external plantar ; at the bifurcation of the posterior tibial artery
under the internal annular ligament. It accompanies the internal plantar nerve.

The External Plantar Artery (Fig. 273).—This is the larger plantar
artery. It crosses the sole of the foot obliquely to the head of the little toe metatarsal, lying, of course, in common with all the other plantar structures, between
the first and second layers of muscles ; here it turns abruptly inward and forward
to the first intermetatarsal space, where it anastomoses with the communicating
branch of the *dorsalis pedis artery*, to complete the *plantar arterial arch*.

The Plantar Arterial Arch.—This arch is formed between the external
plantar artery and the dorsalis pedis. This arch lies deep on the interossei
muscles. It sends off digital branches to the outer three and one-half toes.
At the clefts of the toes these digital branches send off perforating branches,
which, passing up through the interosseous spaces, communicate with the metatarsal artery on the dorsum of the foot.

Study the origin, insertion, and function of muscles on your dissection, and
use the tabulated list of muscles that each student is expected to prepare for
himself for reference.

*Name the muscles of the first layer of the sole of the foot, giving their origin,
insertion, nerve-supply and blood-supply, and action.*

The Abductor Hallucis.—Origin, from the plantar fascia ; greater tuberosity of
os calcis ; intermuscular fascia between itself and the flexor brevis digitorum.
Insertion, under surface of base of first phalanx of the great toe. This muscle
abducts and flexes the great toe. Nerve-supply, the internal plantar nerve.

The Abductor Minimi Digiti.—Origin, plantar fascia ; lesser tuberosity of the
os calcis ; the intermuscular septum between itself and the flexor brevis digitorum ; the long plantar ligament. Insertion, under surface of base of first phalanx
of little toe. This muscle abducts the little toe from the mid-line. Nerve-supply,
the external plantar.

The Flexor Brevis Digitorum.—Origin, the plantar fascia ; the greater tuberosity of the os calcis ; the lateral intermuscular fascia. Insertion, into the sides
of the middle phalanges. These tendons are perforated by the tendons of the
flexor longus digitorum. Nerve-supply, the internal plantar.

*Name the muscles of the second layer of the sole of the foot, giving origin, inser-
tion, function, and nerve-supply.*

The Musculus Accessorius.—Origin, the concave inner surface of the os calcis ;

the under surface of the os calcis; the long plantar ligament. Insertion, the upper surface and outer oblique border of the flexor longus digitorum. Demonstrate carefully on your dissection the specific double origin and double insertion of this muscle. To do this, hold the heel firmly and extend the toes vigorously. Nerve-supply, the external plantar, a branch of the posterior tibial.

The lumbrical muscles are four in number. They arise in connection with the tendons of the flexor longus digitorum, as follows: The first from the great-toe side of the first tendon, beginning at the point where this tendon leaves the main tendon of the muscle. The other three arise from the adjacent tendon, as seen in figure 270. Insertion, into the inner border of the expanded tendons of the extensor communis digitorum. The nerve-supply is from the external plantar for the two outer; the internal plantar for the two inner. Acting with the interossei, these muscles flex the metatarso-phalangeal articulation and extend the first and second interphalangeal joints. Note, then, the homology, both in nerve-supply and function, of these muscles and the lumbricales manus.

The intimate association of the accessorius and lumbricales pedis with the flexor longus pollicis and flexor longus digitorum justifies some authors in classifying these latter also in the second group. For details of these muscles, which are not here considered as belonging to the second, the reader is referred to page 389.

Name the muscles of the third layer of the sole of the foot, giving origin, insertion, function, and nerve-supply.

The Flexor Brevis Hallucis.—Origin, from the under surface of the cuboid bone; the long plantar ligaments; the expansion of the tibialis posticus in the middle of the sole of the foot. Insertion, base of the first phalanx of the great toe, both on the inner and outer borders. (Fig. 268.) The tendon of the flexor longus hallucis passes between the two parts of the muscle, playing over a pulley formed by the two sesamoid bones developed in the tendon of flexor brevis hallucis. (Fig. 274.) This muscle flexes and slightly adducts the first phalanx.

The Adductor Hallucis (Fig. 274).—Origin, the sheath of the peroneus longus formed by the long plantar ligament; the under surface of the second, third, and fourth metatarsals. Insertion, into the outer part of the base of the first phalanx of the great toe. Nerve-supply, the external plantar.

The Transversus Pedis (Fig. 274).—Origin, the deep transverse metatarsal ligaments and the plantar ligaments of the three outer metatarso-phalangeal articulations. Insertion, into the base of the first phalanx of great toe on the outer side. Nerve-supply, the external plantar. Action, to adduct the great toe.

The Flexor Brevis Minimi Digiti.—Origin, base of fifth metatarsal bone, under surface (Fig. 268); sheath of peroneus longus. (Fig. 274.) Insertion, into the under surface of base of first phalanx of the little toe. Action, to flex the little toe. Nerve-supply, the external plantar.

The interossei have been fully considered. The student is urged to study carefully the action of these muscles, both separately and in conjunction with the lumbricales

THE BACK.

Dissection.—Dissection of the back is often almost wholly ignored; and why ignored? I would ask. Simply because (1) it is considered as belonging to a region of the body called *not practical;* because (2) its structures are somewhat *difficult to expose on dissection.* Says one: "There are only two incisions in the whole category of diseases of this region—(1) for renal operations; (2) for carbuncles. Why, then, should one learn to dissect seven layers of muscles?" I would answer, "If the former incision is made in the proper place, and the latter deep enough, these alone are enough to justify careful study and dissection of this area." I desire to accentuate the following:

1. **The thick superficial fascia,** and its very dense connective-tissue framework, since this is a favorite locality for boils and carbuncles, offering greater resistance to pus than do all the other structures combined, between this and the suboccipital triangle. (Law of projectiles.)

2. **The erector spinæ** and the **quadratus lumborum,** since these muscles are guides to lumbar colotomy and lumbar nephrectomy and nephrotomy.

3. **The last rib,** since in the above operations, nephrectomy and nephrotomy, the pleura must not be cut or wounded, and the last rib is the guide.

4. **The complexus muscle,** since this forms the roof of the *suboccipital triangle.* On the floor of the triangle are the vertebral artery and suboccipital nerve; crossing the triangle, immediately under the complexus muscle, is the great occipital nerve.

5. **The vertebral aponeurosis,** since this separates the muscles which act upon the shoulder girdle—the first and second layers—from the proper muscles of the back—viz., those that move the vertebral column by acting on the spinous and transverse processes and the parts of the skull serially continuous therewith.

6. **The erector spinæ,** since this is the location of lumbago, the so-called *muscular rheumatism* of the back. I wish you to note particularly in this connection the anatomical reasons why the pain of lumbago need not be mistaken for pain in diseases of the kidneys and surrounding viscera, of the rectum, and of the uterus. The pain in lumbago (in any of the proper muscles of the back) is logically located in the small of the back, because these muscles are supplied by somatic nerves. It could not be mistaken for the above pains, because those organs—uterus, kidney, rectum—have a sympathetic nerve-supply, and this pain is reflected in the distribution of the *anterior primary divisions* of the somatic nerves; the proper muscles of the back are supplied by the *posterior primary divisions* of the spinal nerves, and these latter have no sympathetic connections. (See Fundamental Principles of Anatomy: Application of Law of Projectiles in Cases of Pain Remote from Place of Injury.)

MUSCLES OF THE BACK, GROUPED ACCORDING TO LAYERS, AND THEIR NERVE-SUPPLY.

First layer, . . Trapezius. Nerve-supply, spinal accessory, cervical plexus.

Second layer.
- Levator anguli scapulæ. Nerve-supply, cervical plexus.
- Rhomboideus minor. Nerve-supply, brachial plexus.
- Rhomboideus major. Nerve-supply, brachial plexus.
- Latissimus dorsi. Nerve-supply, the long subscapular.

Third layer, . Serratus posticus superior. } Nerve-supply, posterior divisions **of spinals,** Serratus posticus inferior. } external branches.

Fourth layer, . Splenius capitis. } Nerve-supply, posterior divisions of spinals, external Splenius colli. } branches.

Fifth layer, . (Erector spinæ.)
Outer division (nerve-supply, all by posterior divisions, spinal nerves):
1. Ileo-costalis.
2. Musculus accessorius ad ileo-costalem.
3. Musculus cervicalis ascendens.
Middle division (nerve-supply, posterior divisions **of spinals**):
1. Musculus longissimus dorsi.
2. Musculus transversalis colli.
3. Musculus trachelo-mastoideus.
Inner division (nerve-supply, posterior divisions of spinals):
Musculus spinalis dorsi.

Sixth layer, . . .
- Musculus complexus. Nerve-supply, suboccipital, great occipital, posterior division.
- Semispinalis colli. Nerve-supply, posterior divisions of spinal.
- Semispinalis dorsi. Nerve-supply, posterior divisions of spinal.
- Rotatores spinæ. Nerve-supply, posterior divisions of spinal.
- Multifidus spinæ. Nerve-supply, posterior divisions of spinal.

Seventh layer.
- Interspinales. Nerve-supply, posterior divisions of spinal.
- Intertransversales. Nerve-supply, posterior divisions of spinal

Occipital Group (nerve-supply, suboccipital nerve).—Rectus capitis posticus minor, rectus capitis posticus major, rectus capitis lateralis, obliquus superior, obliquus inferior. Nerve-supply, great occipital and suboccipital nerves.

ANALYSIS OF THE SIMPLE MOVEMENTS OF THE SPINAL COLUMN AND THE MUSCLES WHICH PRODUCE THEM.

View the skeleton from behind, and note the juxtaposition of spines, laminæ, transverse processes, articular processes, and bodies or centra of vertebræ. Note the head of the rib articulating with the body, and the tubercle of the rib articulating with the transverse process of a vertebra. (Fig. 276.) Each is a movable articulation, because it has a synovial sac. Note the central pulpy nature of the intervertebral disc. Note the manner in which the head of the rib articulates with two bodies and the cartilaginous disc, and recall the exceptions to this rule in the first, ninth, tenth, eleventh, and twelfth thoracic vertebræ. (Fig. 277.)

SIMPLE MOVEMENTS OF THE SPINE.

1. *Lateral flexion of the column* by the intertransversales.
2. *Extension of the column* by the interspinales and spinalis dorsi.
3. *Lateral rotation* by the rotatores spinæ, extending from the transverse process of the vertebra below to the lamina of the bone above.
4. *Extension and lateral rotation of the column* by the multifidus spinæ, in a typical region extending from the transverse processes of the vertebræ below, to the lower border of the spines above, from the last lumbar to the second cervical vertebræ.

Note that extension and lateral rotation are the two most common movements of the human vertebral column; that there are three regions where this movement is specially pronounced: (1) the *thoracic region;* (2) the *cervical region;* (3) the *region between the head and neck.*

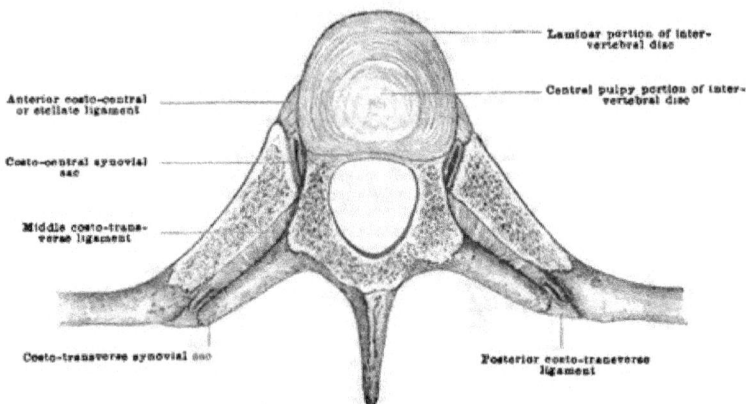

Lamioar portion of inter-vertebral disc

Central pulpy portion of inter-vertebral disc

Anterior costo-central or stellate ligament

Costo-central synovial sac

Middle costo-trans-verse ligament

Costo-transverse synovial sac

Posterior costo-transverse ligament

FIG. 276.—HORIZONTAL SECTION THROUGH THE INTERVERTEBRAL DISC AND RIBS.

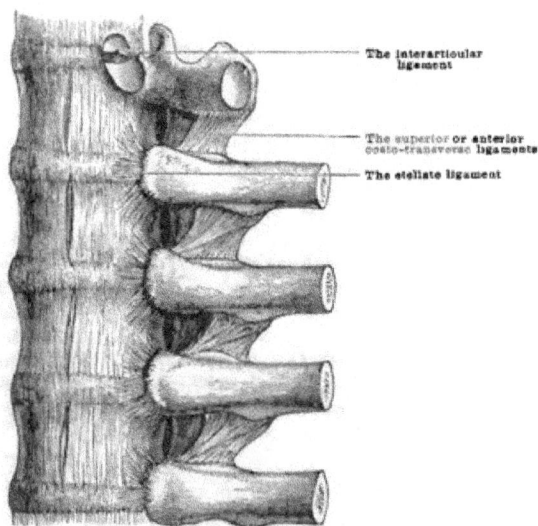

The interarticular ligament

The superior or anterior costo-transverse ligaments

The stellate ligament

FIG. 277.—SHOWING THE ANTERIOR COMMON LIGAMENT OF THE SPINE, AND THE CONNECTION OF THE RIBS WITH THE VERTEBRÆ.

The **semispinalis dorsi, semispinalis colli,** and **complexus** are accessory to the multifidus spinæ.

Note the means to an end, then, in (1) the *semispinalis dorsi* (Fig. 285) extending from transverse processes to the spines in such a manner as to strengthen the dorso-cervical junction. The *semispinalis colli* is likewise disposed to strengthen the same region. So we will see that from the transverse processes of all the dorsal vertebræ to the spines of all the dorsals and cervicals there is a continuous, uninterrupted plane of oblique muscular fibres. (Fig. 285.) They are called semispinales dorsi and colli. They are synergistic physiologically to the multifidus spinæ—only a larger edition of the same. Their function is extension and lateral rotation.

Note, again, there is an upward continuation of this same muscular arrangement to the head—the *complexus*, a semispinalis capitis. (Fig. 283.) It arises from the transverse processes of the upper six thoracic and last cervical, and is

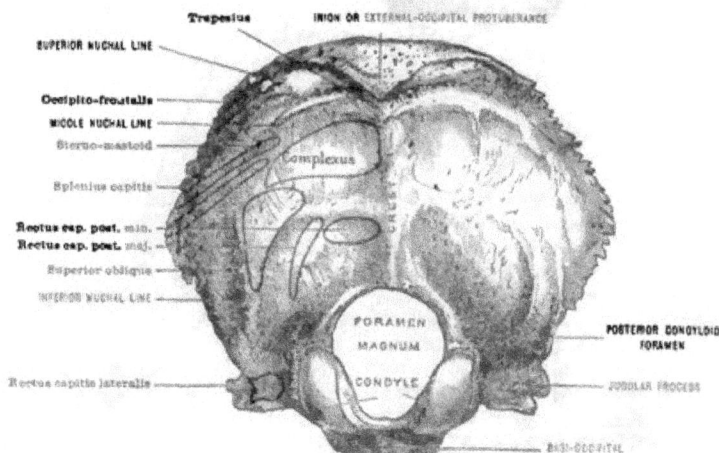

FIG. 278.—THE OCCIPITAL. (External view.)

inserted into the occipital bone. Its action is extension and lateral rotation of the head.

The above are the simple movements of the vertebral column. Complex movements may be had by combination of different simple movements. I have spoken of the mechanism in advance of the dissection, to inspire the student to greater care in his work.

NERVES THAT SUPPLY THE BACK.

1. *The long subscapular nerve* to the latissimus dorsi (brachial plexus).
2. *The spinal accessory nerve* to the trapezius (twelfth cranial nerve).
3. *The suboccipital nerve* to the complexus, recti, and oblique muscles.
4. *Muscular branches from the cervical plexus* to the rhomboids and levator anguli scapulæ.
5. *Posterior divisions of the spinal nerves,* thirty-one pairs to other muscles.
6. *The great occipital nerve,* to the complexus muscle and to the scalp.

Scheme for the Posterior Divisions of the Spinals.—All except the first cervical divide into internal and external branches; the internal branches supply the sixth and seventh layers, the external branches the remainder; they all supply the skin of the back; there are twelve dorsal cutaneous nerves; the suboccipital is the first cervical posterior division; the great occipital is the posterior division of the second cervical; the third occipital is a branch of the third cervical, posterior division.

You will find the (1) suboccipital nerve in the suboccipital triangle; (2) the great occipital under the complexus muscle crossing the suboccipital triangle; (3) the spinal accessory nerve between the trapezius and sterno-mastoid muscles; (4) the third occipital internal to the great occipital; (5) the suboccipital emerges between the occipital bone and atlas behind the vertebral artery; (6) the great occipital nerve emerges between the atlas and axis and passes under cover of the complexus muscle through the lower and inner part of the suboccipital triangle; (7) the great occipital nerve joins the occipital artery under the trapezius and lies on the complexus muscle; (8) other posterior divisions emerge between the transverse processes.

Arteries of the back are: (1) the dorsal branches of all the intercostals; (2) the deep cervical, a branch of the superior intercostal; (3) the arteria princeps cervicis, a branch of the occipital; (4) these arteries, represented by numbers 2 and 3, anastomose under the complexus.

The student who studies in advance of his dissection the foregoing pages will have no difficulty in making a thorough dissection of the back.

Dissection.—Locate: (1) The *spinous processes* of the vertebra in the dorsal and lumbar regions; (2) the *iliac crest;* (3) the *crest* of the *scapular spine;* (4) the *three lips* of the crest of the scapular spine; (5) the *acromion process* of the scapula; (6) the *external occipital protuberance;* (7) the *vertebra prominens;* (8) the *angles* of the ribs, and note the distance they are from the spinous processes; (9) the *iliac junction* with the sacrum; (10) the rudimentary *sacral spines;* (11) the *mastoid processes;* (12) the *clavicle;* (13) the *twelfth rib.*

Incisions.—(1) From external occipital protuberance to the rudimentary sacral spines; (2) from one acromion process of the scapula to the other. Begin to remove the skin at the intersectional point of the vertical and horizontal incisions just made.

Observe the cutaneous nerves piercing the trapezius and latissimus dorsi, on removing the skin. These are the cutaneous branches of the posterior divisions of the spinal nerves. (Fig. 279.)

Note the very heavy variety of the superficial fascia. This, you will observe, is tightly bound to the deep fascia by fibrous trabeculæ that make removal of the skin difficult. This fascia, especially in the region of the neck, is a common place for boils and carbuncles. Observe the great depth you will have to cut to get through this fascia.

Clean all the superficial fascia off and make your work look like figure 227. Here you have exposed two muscles: (1) The trapezius; (2) the latissimus dorsi.

The Trapezius.—Trace its origin to the external occipital protuberance, the inner third of the superior nuchal line, the ligamentum nuchæ, the seventh cervical spine, and all the dorsal spines. Now pull the arm outward (Fig. 280), and put the muscle on the stretch. Study its descending fibres to the outer third of the clavicle and acromion; its ascending and horizontal ones to the upper lip of the spine of the scapula.

Develop the lower margin of the trapezius muscle, and observe that it overlaps the latissimus dorsi. Find the space between the trapezius and sterno-cleido-mastoid muscle. Locate in this space the spinal accessory nerve, and trace the same to the trapezius. Also trace the nerve upward to where it comes through the sterno-cleido-mastoid muscle.

Latissimus Dorsi.—Trace the origin to the six lower thoracic spines, lum-

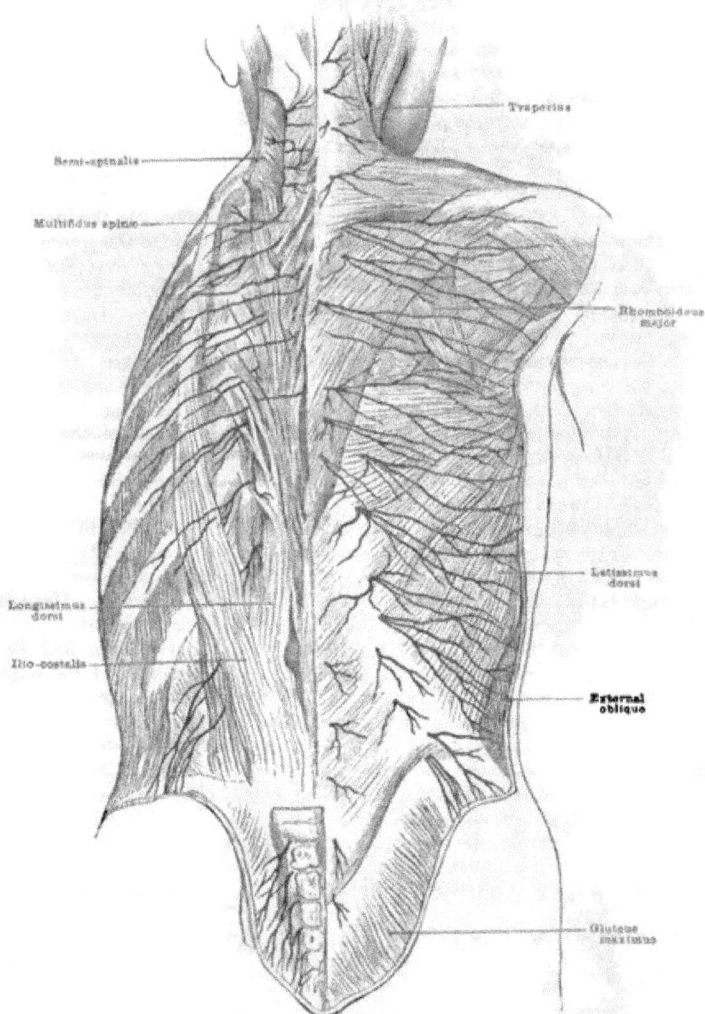

FIG. 279.—DISTRIBUTION OF THE POSTERIOR PRIMARY DIVISIONS OF THE SPINAL NERVES. (Henle.)

bar aponeurosis, and iliac crest, posterior one-third of outer lip. Develop the upper border and expose the infraspinatus and teres major in the triangle formed by this muscle, the deltoid, and the trapezius. (Fig. 280.) Develop with the scissors the lower border of the muscle. The latissimus dorsi will be inserted into the posterior lip of the bicipital groove of the humerus with the major tereai muscle. **These** two muscles, the trapezius and the latissimus dorsi, may now be removed by cutting through their origins, tracing the same carefully to their respective insertions, where they may be detached, when you will have exposed (Fig. 281)

1. The musculus levator anguli scapulæ.
2. The musculus rhomboideus minor.
3. The musculus rhomboideus major.
4. The musculus splenius capitis.
5. The musculus complexus.
6. The musculus serratus posticus inferior and superior.
7. Vertebral aponeurosis.
8. The occipital artery on the complexus muscle.
9. The great occipital nerve.

Now develop with scissors the four serrations of insertion into the lower four ribs of the serratus posticus inferior. This muscle arises from the spines ot the eleventh and twelfth thoracic, and the first and second dorsal vertebræ. Pull the arm outward, and put the rhomboids on the stretch. Study their insertion into the middle lip of the vertebral border of the scapula. Develop their spinous origin. (Fig. 281.)

Cut the **rhomboids** near their origins and turn them outward. (Fig. 281.) See the nerve-supply on the under surface near the centre. See also the posterior scapular artery, a continuation of the transversalis colli artery

The rhomboideus minor (Fig. 281) arises from the spine of the seventh cervical and first thoracic vertebræ, and from the ligamentum nuchæ. It is inserted into the vertebral border of the scapula opposite the spine. Its nerve-supply is from the brachial plexus.

The **rhomboideus major** (Fig. 281) arises from the five upper thoracic vertebræ and their supraspinous ligament, and is inserted into the middle lip of the vertebral border of the scapula from a point opposite the spine above to the inferior angle below. The nerve-supply is from the brachial plexus. The action of the rhomboids is to lift the scapula upward, backward, and inward.

Levator Anguli Scapulæ (Fig. 281).—Pull the arm outward and find the insertion of this muscle into the middle lip of the vertebral border of the scapula above the spine. Trace its three or four tendons of origin to the anterior tubercles of the transverse processes of the four upper cervical vertebræ. This muscle derives its nerve-supply from the cervical plexus. Its action is to elevate the scapula and, by producing rotation of the same, it depresses the *point* of the scapula.

Serratus Posticus Superior (descending fibres) (Fig. 282).—This muscle lies under the levator anguli scapulæ and rhomboids. It arises by aponeurosis from the seventh cervical spine and the upper three dorsal spines. It is inserted into the second, third, fourth, and fifth ribs, a little beyond the angle. The nerve-supply is from the posterior primary divisions of the second and third intercostals.

The serratus posticus inferior (ascending fibres) (**Fig.** 282) arises, by aponeurosis, from the two lower dorsal and three upper lumbar spines; it is inserted into the four lower ribs a little beyond the angle. Notice the difference between the two serrati muscles. The upper is inserted into the upper border, the lower into the lower border of ribs.

The Vertebral Aponeurosis (Fig. 282).—Internally you see this aponeurosis attached to the spines; externally it is attached to the angles of the ribs; below it blends with the serratus posticus inferior and latissimus dorsi; above it passes behind the superior serratus. It separates the proper muscles of the back from those that act on the shoulder girdle.

The Splenius Capitis and Colli.—Develop the long, pointed origin of this

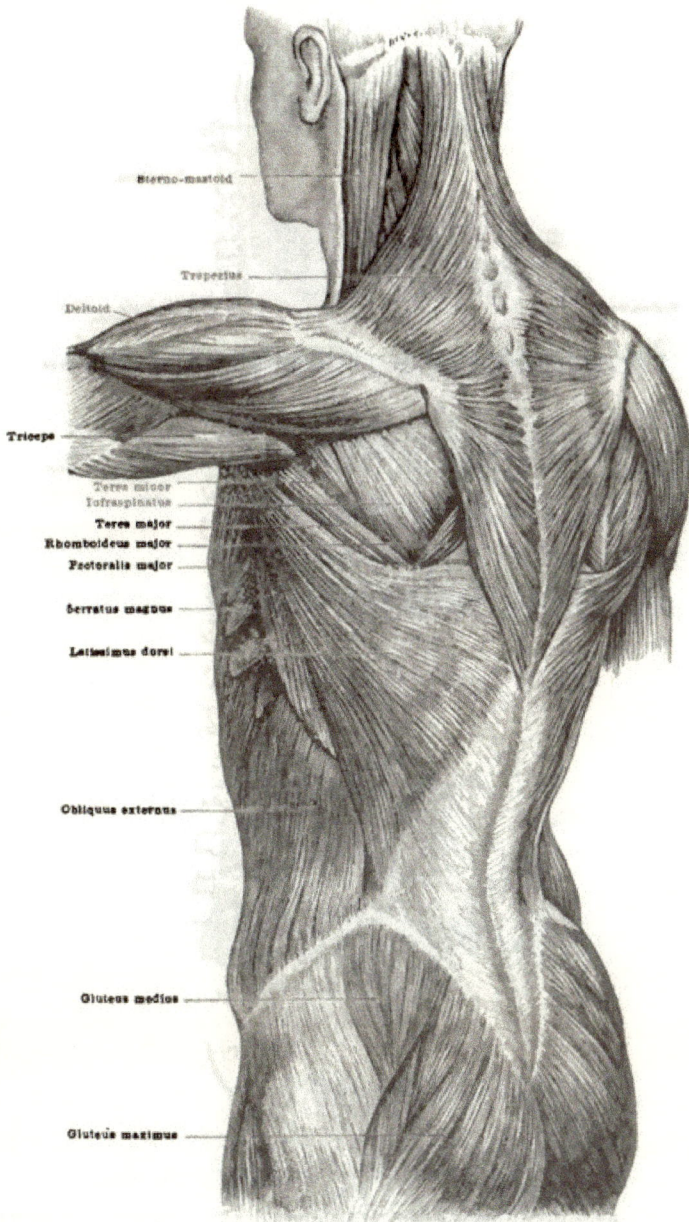

Sterno-mastoid

Trapezius

Deltoid

Triceps

Teres minor
Infraspinatus
Teres major
Rhomboideus major
Pectoralis major

Serratus magnus

Latissimus dorsi

Obliquus externus

Gluteus medius

Gluteus maximus

FIG. 280.—FIRST LAYER OF MUSCLES OF THE BACK.

muscle from the sides of the spines of the seventh cervical and six upper dorsal vertebræ. (Fig. 281.) The muscle has two insertions: (1) into the mastoid

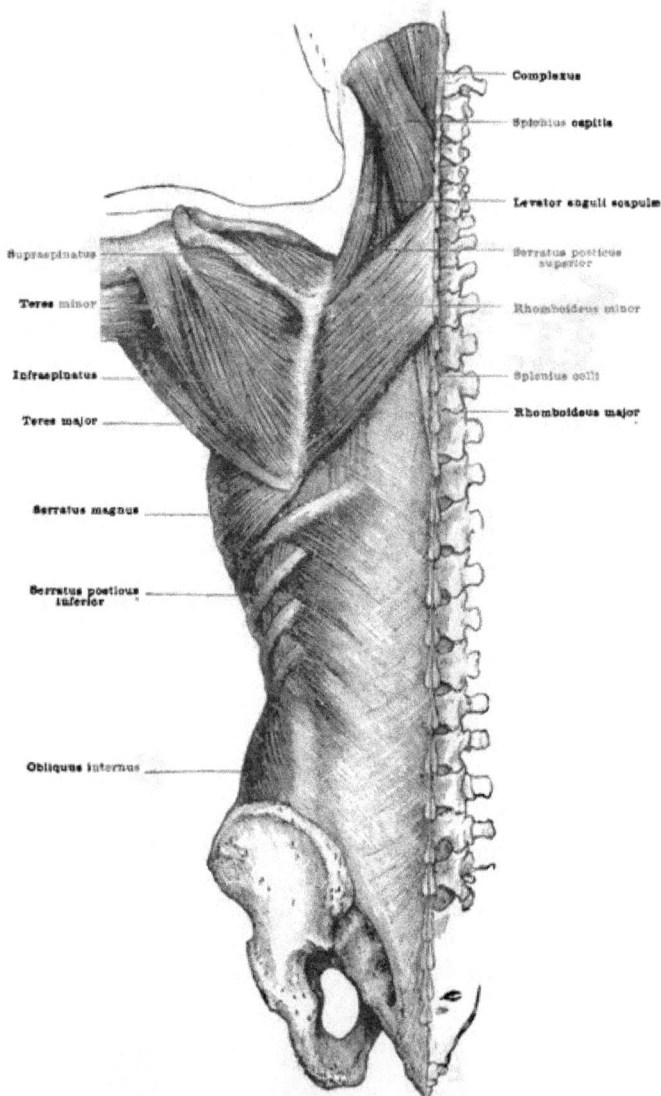

FIG. 281.—THE LEVATOR ANGULI SCAPULÆ AND RHOMBOIDEI.

process of the temporal bone; (2) into the posterior tubercles of the transverse processes of the three upper cervical vertebræ. The nerve-supply is from the posterior primary divisions of the cervical nerves.

Erector Spinæ and Branches.—Remove the vertebral aponeurosis (Fig. 282) and expose the fifth layer of muscles and its subdivisions. (Fig. 283.)

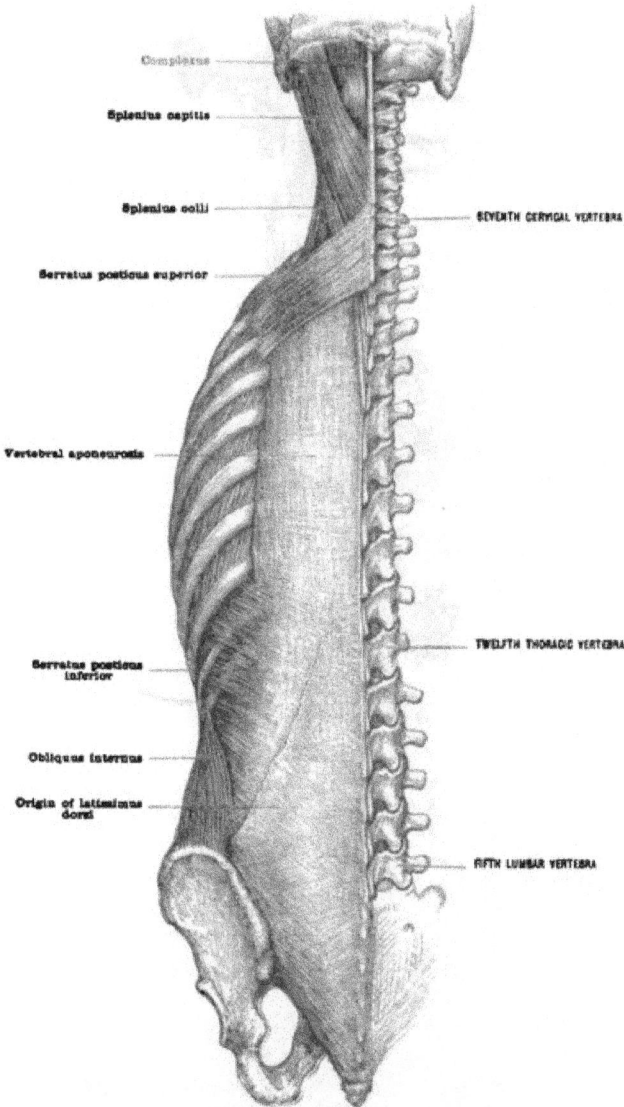

Complexus

Splenius capitis

Splenius colli

SEVENTH CERVICAL VERTEBRA

Serratus posticus superior

Vertebral aponeurosis

Serratus posticus inferior

TWELFTH THORACIC VERTEBRA

Obliquus internus

Origin of latissimus dorsi

FIFTH LUMBAR VERTEBRA

FIG. 282.—THE THIRD AND FOURTH LAYERS OF THE MUSCLES OF THE BACK.

Origin.—(1) The spines of the two last thoracic, all the lumbar, and the four upper sacral vertebræ; (2) the back of the side portion of the fourth sacral vertebra; (3) the posterior sacro-iliac ligament, a few of these fibres being continuous with the origin of the gluteus maximus; (4) the upper part of the

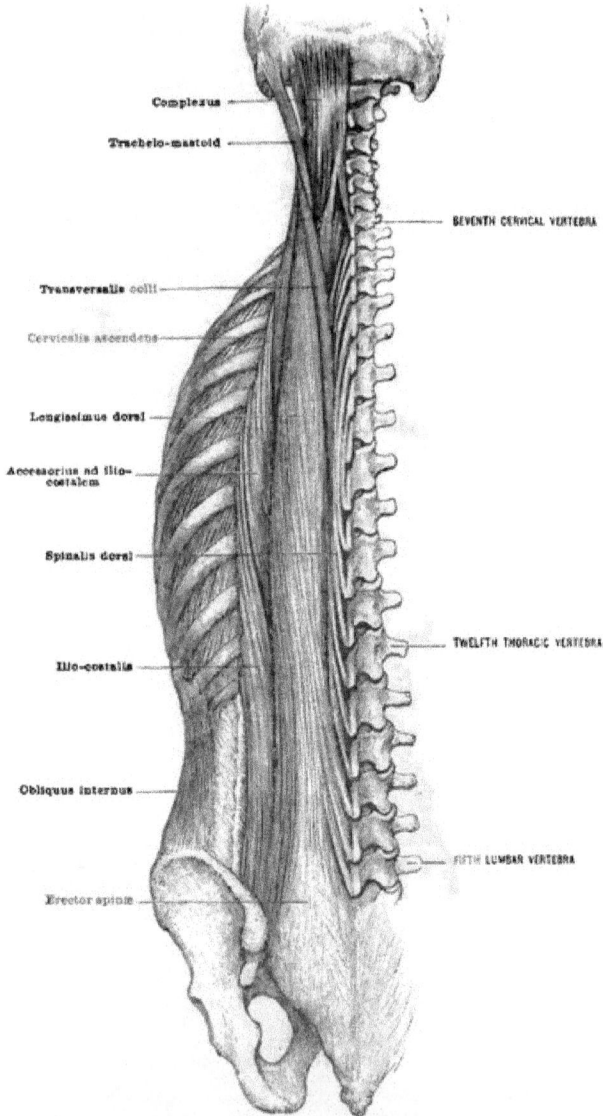

FIG. 283.—THE FIFTH LAYER OF THE MUSCLES OF THE BACK.

posterior superior spine of the ilium, and the posterior fifth of the iliac crest (Morris).

Insertion.—It is continuous above with the spinalis dorsi, longissimus dorsi, and ilio-costalis.

Locate the **spinalis dorsi** first, the inner insertion of the erector spinæ muscle. Remember it takes its origin from the inner part of the erector spinæ. Trace its tendons of insertion to the spines of the upper thoracic vertebræ.

Locate the groove between the outer and middle divisions of the erector spinæ. In this groove you will find the external branches of the posterior divisions of the spinal nerves, in company with the dorsal branches of the intercostal vessels. Turn the muscles apart, as in figure 284.

The outer division of the erector spinæ is called **ilio-costalis,** to its insertion into the angles of the lower ribs, from the sixth to the eleventh rib. It is continued upward through the back and neck as (1) the accessorius, (2) cervicalis ascendens.

Accessorius ad ilio-costalem arises from the angles of the ribs, seventh to eleventh, and is inserted into the angles of the ribs from the second to the fifth, and into the transverse process of the seventh cervical vertebra.

Cervicalis ascendens arises from the ribs—fourth to fifth upper—internal to the costal insertion of the accessorius, and is inserted into the posterior tubercles of the fourth, fifth, and sixth cervical transverse processes.

The *middle division* of the erector spinæ continues through the thorax under the name of longissimus dorsi; through the neck, to the head, as the transversalis colli and trachelo-mastoid. (Fig. 284.)

The longissimus dorsi arises: (1) From the middle part of the erector spinæ ; (2) from the transverse processes of some of the lower thoracic vertebræ. It is inserted into : (1) The ribs external to their tubercles ; (2) the transverse processes of the thoracic vertebræ ; (3) into the accessory tubercles of the upper lumbar and lower thoracic vertebræ ; (4) into the transverse processes of the upper lumbar vertebræ.

Transversalis colli arises from the transverse processes of the upper six thoracic vertebræ, internal to the insertion of the longissimus dorsi. It is inserted into the posterior tubercles of the transverse processes of the vertebræ from the second to the sixth cervical vertebræ inclusive.

The trachelo-mastoid (Fig. 284) is the inner part of the transversalis colli continued in the mastoid process. In some cases this muscle is unusually well developed.

The sixth layer of muscles comprises the following :

1. The complexus, or semispinalis capitis.
2. The semispinalis dorsi in the dorsal region.
3. The semispinalis colli in the cervical region.
4. The multifidus spinæ, found in all regions of the spine.
5. The rotatores spinæ, found in the thoracic region.

Describe the complexus muscle.

This muscle (Figs. 283 and 284) is covered by the splenius, with its two divisions, and the trapezius. In turn, this complexus covers the muscles that make the boundaries of the suboccipital triangle, thereby forming the roof of this important surgical area—the suboccipital triangle. The occipital artery lies on the muscle ; the suboccipital and great occipital nerves are under the muscle, being parts of the contents of the suboccipital triangle. The origin of the muscle is from : (1) The articular processes of the cervical vertebræ from the third to the sixth ; the transverse processes of the seventh cervical and the six upper thoracic vertebræ. The insertion is into the occipital bone, between the middle and inferior curved or nuchal lines. (Fig. 278.) Nerve-supply (1) The suboccipital or posterior primary division of the first cervical ; (2) the great occipital or posterior primary division of the second cervical ; (3) the posterior primary divisions of the third, fourth, and fifth cervical nerves.

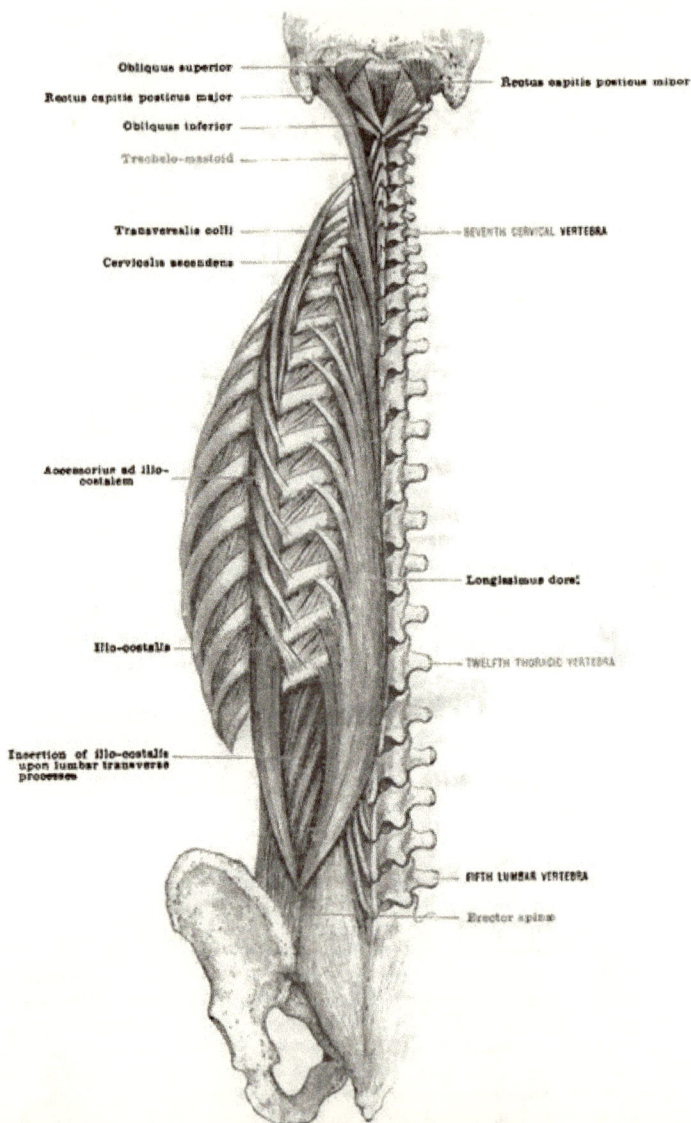

FIG. 284.—THE FIFTH LAYER OF THE MUSCLES OF THE BACK, AFTER SEPARATING THE
OUTER AND MIDDLE DIVISIONS.

Describe the semispinalis dorsi.

The idea in the name is founded on the facts (1) That one of the muscle is inserted into the spines, and (2) that the muscle is situated in the thoracic or dorsal region of the spine. (Fig. 285.) The muscles composing the series are small and tendinous. They extend obliquely inward and upward, from their origin, on the back of the transverse processes of the thoracic vertebræ, from the sixth to the tenth, to their insertion into the spines of the last two cervical and the first, second, third, and fourth thoracic. The nerve-supply is from the posterior primary divisions of the thoracic nerves. Remove this muscle and you expose the multifidus spinæ below. The muscle is covered by the spinalis dorsi and latissimus dorsi—the inner and middle parts of the erector spinæ, you will remember.

Describe the semispinalis colli.

The verbal idea is the same in this as in the preceding muscle. The muscle is located in the cervical region; it is inserted into the spines of the cervical vertebræ, from the second to the fifth inclusive. (Fig. 285.) The muscle originates from the transverse processes of the five or six upper thoracic vertebræ. The muscle is covered by the complexus; under it is the multifidus spinæ. Between the complexus and semispinalis colli you will find: (1) Branches of the posterior cervical nerves; (2) an anastomosis between the arteria princeps cervicis, a branch of the occipital, and the arteria profunda cervicis, a branch of the superior intercostal. In ligation of the common carotid and subclavian arteries blood may reach both the hand and brain by this channel.

Describe the multifidus spinæ.

The multifidus spinæ is found in every region of the spine. In the sacral and lumbar regions it is thick and fleshy; in the thoracic and cervical regions it is thin and aponeurotic. This muscle has the following origins, which must be carefully learned before this muscle can be dissected understandingly:

(1) From the deep surface of the erector spinæ. This fact of origin makes the dissection of the muscle a difficult task, except in cases of zinc or formaline prepared bodies where the cadavers have had a year to become very hard; (2) from the groove between the sacral spines and rudimentary articular processes of the sacrum; (3) from the mammillary processes of the lumbar vertebræ; (4) the transverse processes of all the thoracic vertebræ; (5) the articular processes of the cervical vertebræ from the fourth to the sixth, and from the transverse process of the seventh cervical. The segments making up the collective multifidus spinæ from these diverse regions are inserted as follows: into the lower borders of the vertebral spines, from the fifth lumbar to the second cervical. The nerve-supply of the multifidus spinæ is from the posterior primary divisions of the spinal nerves, from the second cervical to the third sacral nerve. This muscle covers the rotatores spinæ.

Describe the rotatores spinæ.

These are in the region of the thorax. They derive their name from the rotatory action they exert. Of these muscles there are eleven pairs. They originate from the back and upper part of the transverse processes, and are inserted into the lower border of the lamina of the next vertebra above. They are in relation above with the multifidus spinæ. Nerve-supply, the posterior primary divisions of the spinals.

Name and describe the seventh layer of muscles of the back.

(1) The interspinales arise from the upper surface of the spine of a lower, and are inserted into the lower surface of the spine of the vertebra immediately above. These muscles are very small. Nerve-supply, the posterior primary divisions of the spinals. (2) The intertransversales arise from the transverse process below and are inserted into the one above. They are small muscles.

They are principally in the cervical and lumbar regions. Nerve-supply, from the spinals as they emerge from the intervertebral foramina.

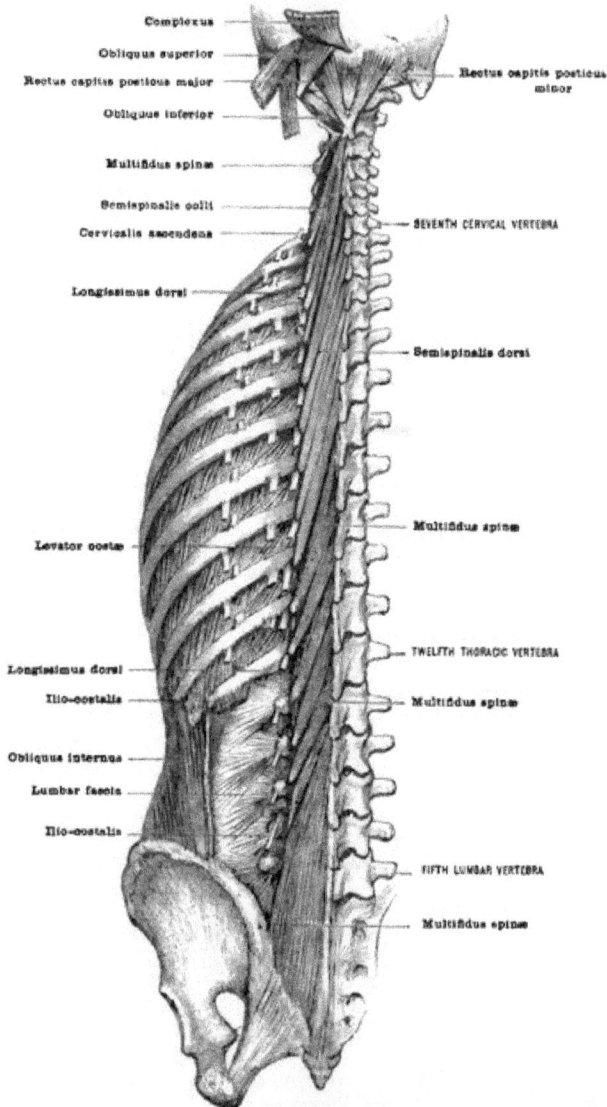

FIG. 285.—THE SIXTH LAYER OF THE MUSCLES OF THE BACK.

1. *The Complexus.*—Remove the longissimus dorsi and its continuation through the neck and to the head—the transversalis colli and the trachelo-

mastoid. The complexus will be seen arising from : (1) The back of the articular processes of the cervical vertebræ from the third to the sixth ; (2) the back of the transverse processes of the seventh cervical and the six upper thoracic vertebræ ; (3) generally also by an inner head from the spine of the seventh cervical vertebra. It is inserted into the occipital bone, between the superior and inferior nuchal lines. (Fig. 283.)

Carefully detach the insertion of this muscle and you will expose : (1) The *suboccipital triangle*, containing the vertebral artery and the suboccipital nerve ; (2) on the under part of the muscle you will find the *suboccipital nerve* and the *great occipital nerve ;* (3) the *recti and oblique muscles ;* (4) the *semispinalis colli*, coming to a large abrupt insertion into the spine of the axis.

The suboccipital triangle has (Fig. 285) :

A roof formed by the complexus muscle.

A floor formed by the arch of the atlas and atlo-occipital ligament.

An inner border formed by rectus capitis posticus major.

A lower border formed by inferior oblique muscle.

An upper border formed by superior oblique muscle.

The contents are the vertebral artery and suboccipital nerve.

It is traversed by the great occipital nerve.

The suboccipital muscles (Fig. 285) :

1. *The obliquus capitis inferior*, which extends from the spine of the axis to the transverse process of the atlas.

2. *The obliquus capitis superior*, which extends from the transverse process of the atlas to the inferior nuchal line of the occipital bone.

3. *The rectus capitis posticus major*, which extends from the spine of the axis to inferior nuchal line, middle one-third.

4. *The rectus capitis posticus minor* extends from the tubercle of the atlas to the inner one-third of the inferior nuchal line.

5. *The rectus capitis lateralis* extends from the lateral mass of the atlas to the jugular process of the occipital bone.

Now remove from its origin (1) the complexus, and as you turn the body of the muscle from the semispinalis colli, you will see between these two muscles some vessels—the anastomosis between the arteria profunda cervicalis, a branch of the first or superior intercostal, and the arteria princeps cervicis, a branch of the occipital. (Fig. 40.) You will find a large mass of veins here, too. In cases of ligation of the subclavian or common carotid artery, this is the principal channel by which the blood passes to form a collateral circulation. (2) Remove any remaining part of the spinalis dorsi or longissimus dorsi and make your work look like figure 285.

You have remaining the other members of the sixth and seventh groups (1) The semispinalis dorsi ; (2) the semispinalis colli ; (3) the multifidus spinæ. These were explained under the head of "Analysis of the Simple Movements of the Spinal Column," to which you are referred on page 395. Remove the three muscles just given, and dissect (4) the rotatores, (5) the intertransversales, (6) the interspinales, according to their previous description, just referred to.

1. *Give the cutaneous nerve-supply of the back.*

The scalp is supplied by (1) the great occipital ; (2) the suboccipital ; **(3) the** third occipital. These are all spinal nerves. The remainder is supplied **by the** posterior divisions of the spinal nerves. Next the spine you will see some **small** nerves coming through to the skin. (Fig. 279.) These are cutaneous twigs from the internal branches of the posterior divisions of the spinal nerves. A little distance from the spine you will see nerves coming through the trapezius and latissimus dorsi. These are the external cutaneous branches of the posterior divisions of the spinal nerves.

27

2. *Explain internal and external branches of the posterior divisions of the spinal nerves.*

The internal branches are small, and supply the sixth and seventh groups of muscles and the skin on each side of the spine for an inch or two inches. The external branches supply all the other proper muscles of the back and the remainder of the skin. All the thirty-one pairs of posterior divisions of spinal nerves divide as previously indicated, except the first. This is called the suboccipital nerve.

3. *Define what is meant by proper muscles of the back.*

All those muscles not acting on the shoulder girdle; those under the vertebral aponeurosis.

4. *Give a description of the superficial fascia of the back.*

It is very thick and dense, and contains granular fat. Abscesses often form here. A favorite place for carbuncles is in the superficial fascia of the neck.

5. *Name the muscles of the back not supplied principally by the posterior divisions of the spinal nerves.*

The trapezius, the latissimus dorsi, the levator anguli scapulæ, and the rhomboids, major and minor.

Short Summary of the Cranial Nerves.

How are the cranial nerves classified?

According to the new classification of cranial nerves there are twelve pairs, enumerated serially from before backward, from the olfactory, or first, to the hypoglossal, or twelfth, inclusive. This classification was proposed by Soemmering in 1778. Prior to this time the classification of Willis had been in use for more than a century, by which classification only ten pairs were recognized.

Make a further distinction between the classifications of the cranial nerves by Soemmering and Willis.

In the seventh pair Willis included both the *facial*, a nerve of motion, and the *auditory*, a nerve of the special sense of hearing. The eighth pair of Willis included the ninth, tenth, and eleventh nerves of our classification. The ninth pair of Willis included the hypoglossal, and the tenth pair of Willis included the first cervical, called the suboccipital nerve.

What is understood by superficial and deep origins of cranial nerves?

By the former is meant the place where the nerve is seen attached to the surface of the brain (Fig. 92); by the latter is meant the place deeply located, called a nucleus of gray matter, to which the motor part of nerves can be traced.

Does not the sensory part of nerves also have a deep origin?

Properly speaking, the sensory parts of nerves have their origin in the cells of the posterior root ganglia of the nerve-trunks, to which ganglia special attention must be given in the dissection of both cranial and spinal nerves. The sensory fibres originating in the ganglia grow inward to this nuclei, and, as far as is known, form there no direct connection with the nerve-cells.

Is there any correspondence between the origins of the cranial nerves and the origins of the spinal nerves?

Yes; each has a superficial and a deep origin. The superficial origin of the spinal nerves is readily seen on removing the spinal cord from its neural canal. The deep origin of the spinal nerves is from the anterior and posterior horns of the gray crescents of the interior horns of the spinal cord.

What further resemblance between cranial and spinal nerves may be mentioned?

(1) Fibres from the deep origin of each pass upward and are connected with the cortex of the cerebrum; (2) the pneumogastric nerve arises by motor and sensory roots from nuclei which are derivatives of the anterior and posterior

horns respectively of the spinal cord; (3) the ganglion on the sensory root of the fifth cranial nerve and the ganglion on the root of the vagus are homologous to the ganglia on the posterior roots of the spinal nerves; (4) cases are recorded of absence of the ganglion on the posterior root of the first spinal or suboccipital nerve; (5) cases are recorded of the hypoglossal—the twelfth cranial—nerve having a posterior root with a ganglion. Possibly anomalous cases like those cited in 4 and 5 may account for and even justify the tenth cranial nerve—the hypoglossal—in the classification by Willis.

In what respects do spinal nerves differ from cranial nerves?

In the following respects (1) Spinals arise from the spinal cord; (2) spinals have an anterior motor and a posterior sensory root; (3) spinals have a ganglion on each posterior root; (4) spinals have no special sense nerves.

State where the ganglia of the posterior or sensory roots of the **spinal nerves** *may be found in dissecting.*

The rule is that these ganglia occupy the posterior root of **the** spinal nerve, just behind the union of the anterior and posterior roots, in the intervertebral foramen.

Are **there any** *exceptions* **to the rule** *governing the location* **of the spinal** *ganglion?*

Yes; the ganglia of the first and second cervical nerves lie **on** the **neural** arches of the atlas and axis. The ganglia of **the** sacral and coccygeal nerves are in the spinal canal.

Describe the vagus or pneumogastric nerve.

This nerve is the tenth cranial, according to the classification of Soemmering; it is also called the nervus vagus and the nervus par vagam. The nerve leaves the base of the cranium by the central part of the jugular foramen with the spinal accessory and glosso-pharyngeal nerves.

Where is the ganglion of the root of the vagus located, and to what does it correspond morphologically?

It is located in the jugular foramen, and corresponds to the spinal **ganglion on** the posterior roots of the spinal nerves.

What is the importance of the ganglion of the root of the **vagus** **nerve** *?*

It has connections with the sympathetic, spinal accessory, glosso-pharyngeal, and facial nerves. (Fig. 293.)

Where is the ganglion of the trunk of the **vagus** **nerve** *and what its importance?*

It is a little below the ganglion of the root and below the base of the skull. Its importance is due to the fact that branches are given off from it **to** the sympathetic, to the cervical nerves, and to the hypoglossal nerve; **also the accessory** part of **the spinal** accessory nerve joins the vagus at this point. (**Fig. 293.**)

Describe **the olfactory nerve.**

The **olfactory nerve, as generally understood, is in reality a** dismembered part of the fore-brain **of the fœtus.** **It consists of four parts, as may be** appreciated by consulting figure **286.**

1. The olfactory roots, three in number.
2. The olfactory tract, a slender bundle.
3. The olfactory bulb, resting on the cribrosa.
4. The olfactory filaments, about twenty in number.

Name and give the source of the olfactory roots.

The *middle* root is attached to the under surface **of** the frontal lobe; the *internal root* to the gyrus fornicatus; the *external root* to the temporo-sphenoidal lobe. The roots meet to form the olfactory tract. The tract becomes bulbous and occupies a fissure on the under surface of the frontal lobe of the cerebrum, called the olfactory **sulcus.** The olfactory nerves are given off from the under part of the bulb. The nerves are about twenty in number on each side. They pass through the olfactory foramina in the cribriform plate of the ethmoid bone. In their course through the foramina, the nerves are invested by a sheath of dura mater. (Fig. 87.) The olfactory nerve filaments are distributed to the mucous membrane of the upper and middle turbinated bones, and to a corresponding part of the septum nasi. (Fig. 75.)

Describe the optic nerve.

This nerve is the special nerve of sight, being distributed solely to the eyeball. The commissure is the place on the sphenoid bone where the two sides

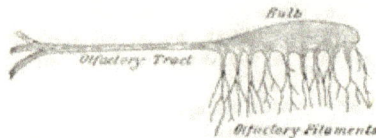

FIG. 286.—FIRST CRANIAL NERVE, OLFACTORY.

are connected. The optic nerve proper is between the commissure and the eyeball ; the optic tract is between the commissure and the brain.

Describe the optic tract.

It arises (Fig. 92) : (1) From the pulvinar of the optic thalamus ; (2) from the geniculate body ; (3) from the upper quadrigeminal body. The tract crosses the crus cerebri (Fig. 92), being closely attached thereto, and terminates in the optic chiasma or commissure. The student will recall the fact that the optic chiasma and the optic tracts assist the crura cerebri and pons Varolii in forming the boundaries of the interpeduncular space. (Fig. 92.)

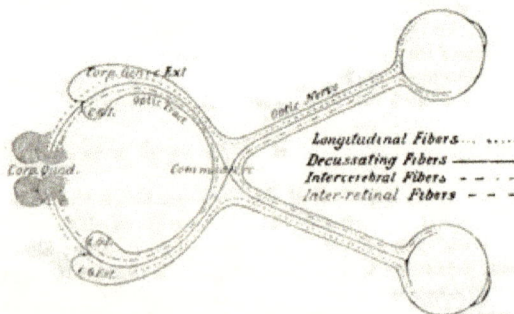

FIG. 287.—SECOND CRANIAL NERVE, OPTIC.

Describe the optic nerve proper.

This nerve extends from the chiasma to the eyeball. It leaves the cranium by the optic foramen, in the sphenoid bone, in company with the ophthalmic artery. The nerve has a cranial part and an orbital part ; the latter is flexuous, to accommodate the movements of the eyeball, and is surrounded by ciliary vessels and nerves.

In practical anatomy where do you find the arteria centralis retinæ ?

This must be seen in a dissection of the contents of the orbit. (Fig. 103.) You will find this artery piercing the under part of the nerve, and about midway between the optic foramen and the globe.

The sheath of the optic nerve is derived from what source?

It is derived from the dura mater coming through the optic foramen and delaminating; **one layer** ensheathes the optic nerve, the other forms the orbital periosteum. (Fig. 95.)

The third cranial nerve—the motor oculi—supplies: (1) The ciliary muscle **of** the eyeball; (2) the sphincter muscle of the iris; (3) all the muscles of the eyeball except the superior oblique and the external rectus.

This nerve arises beneath the floor of the aqueduct of Sylvius. On removing the brain from the cranium, you see the nerve between the crura cerebri. (Fig. 92.) The nerve pierces the dura midway between the anterior and posterior

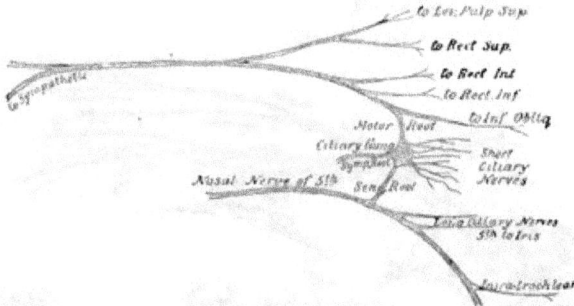

FIG. 288.—THIRD CRANIAL NERVE, MOTOR OCULI.

clinoid processes, and enters the cavernous sinus; **in the** sinus it lies above and internal to the fourth nerve. (Figs. 86–88.) The nerve divides behind the sphenoidal fissure into a superior and an inferior division. (Fig. 86.) These two divisions enter the orbit by the sphenoidal fissure, passing between the **two heads of the** external rectus **muscle.** (Fig. 101.) The superior division supplies **the** superior **rectus muscle** and the levator palpebræ; the inferior division **has the** remaining part of the distribution of the third nerve.

What can you say of the fourth cranial, or patheticus?

It is **also** called the trochlear nerve, from the fact that it supplies the superior oblique muscle of the eyeball only. This nerve has a long course in the cranial cavity. In dissection you see the nerve just under the margin of the anterior

FIG. 289.—FOURTH CRANIAL NERVE, TROCHLEAR.

free border of the tentorium cerebelli, and soon piercing the dura to the outer side of the posterior clinoid process, to gain the cavernous sinus. (Fig. 98.) This nerve is the smallest of the cranial nerves; it enters the orbit by the sphenoidal fissure. The nerve has its deep origin in the floor of the aqueduct of Sylvius.

Describe briefly the fifth cranial or trifacial nerve.

This nerve is specialized almost exclusively for the prehension and mastication of food directly or indirectly. It will be noted that the relation between this and other nerves depends upon the near or remote association these latter bear to the functions of the fifth nerve. Take notice, then, of the function of the trigeminus, as follows : (1) The fifth nerve supplies with sensation all the teeth. (2) The fifth nerve supplies with sensation the gums, by little nerves called the nervi gingivales. (3) The fifth nerve supplies the sides of the tongue, and **the anterior** one-half of the same through its lingual **or gustatory** branch. (4) The fifth nerve supplies the muscles of mastication with motion—muscles whose function is to move the mandible. (5) The fifth nerve supplies the temporo-mandibular articulation, because the muscles it supplies move this articulation. The same branch that supplies the articulation, the auriculo-temporal, also supplies the drum of the ear and the auditory canal, because these parts are secondary in the acquisition of food. (6) The fifth nerve supplies the skin covering **the fullest** region of the

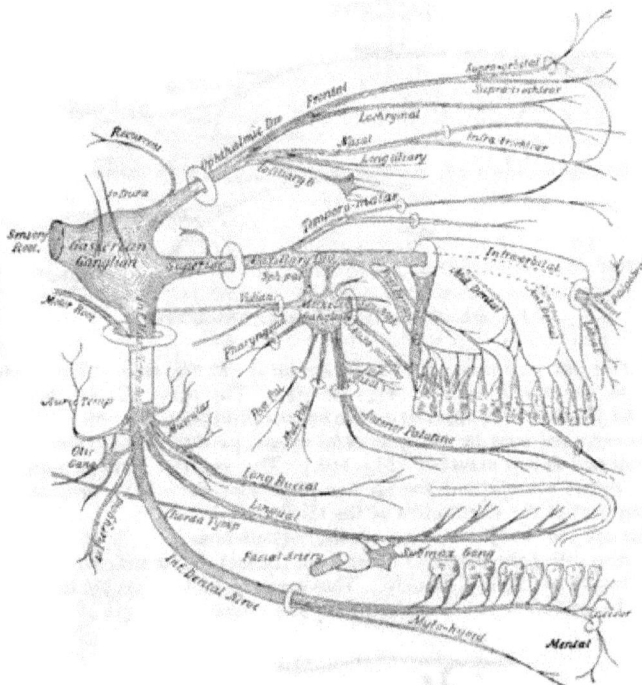

FIG. 290.—FIFTH CRANIAL NERVE.

muscles of mastication ; in round numbers, **the cutaneous region** in front of a line passing from side to side through the auditory meatuses is supplied by the auriculo-temporal, supraorbital, infraorbital, and mental branches of this nerve. (7) The fifth nerve supplies the anterior belly of the digastric muscle and the mylohyoid muscle, but these muscles are depressors of the mandible. (8) The fifth nerve has numerous connections with the facial nerve. Study the facial and temporal relations between these two nerves ; the chorda tympani ; the otic, submaxillary, and Meckel's ganglia. The trifacial nerve supplies the muscles of mastication with motion, and the skin covering these muscles.

What can you say of the sixth cranial nerve?

This nerve is distributed solely to the external rectus muscle. It arises from the floor of the fourth ventricle. It may be seen in dissection piercing the dura below and internal to the fifth nerve. (Fig. 88.) In the cavernous sinus the nerve lies on the outer side of and is attached to the internal carotid artery. (Fig. 88.) In the sinus the sixth nerve is joined by sympathetic filaments from the cavernous plexus, and receives a twig from the ophthalmic branch of the fifth nerve. (Fig. 98.)

The Seventh or Facial Nerve.—(1) Leaves the cranium by the internal auditory meatus in company with the auditory or eighth cranial nerve and the auditory artery, a branch of the basilar artery. (2) In the internal auditory canal it is connected to the auditory nerve by the pars intermedia, and enters a separate bony canal, called the **aqueductus** Fallopii, one-quarter of an inch from the internal auditory **meatus.** (3) On nearing the tympanum the nerve turns sharply backward and presents **an enlargement,** called the intumescentia gangliformis (gen-

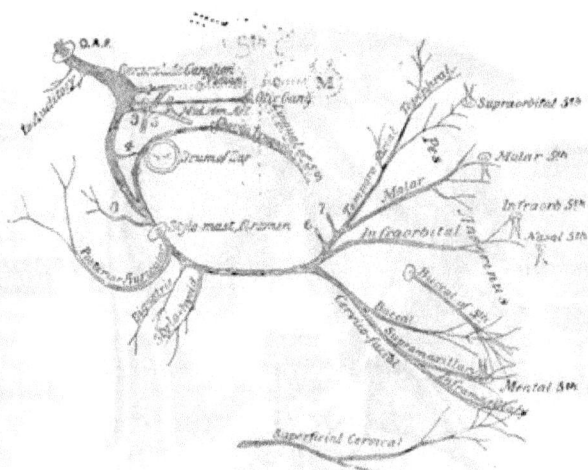

FIG. 291.—SEVENTH CRANIAL NERVE, FACIAL, OR PORTIA DURA.
1. Great petrosal, **to form Vidian** with No. 5. 2. Small **petrosal, to otic** ganglion. 3. External petrosal, to plexus on mid-meningeal artery. 4. Tympanic branch to stapedius, etc. 5. Branch from carotid plexus making Vidian, with No. 1. 6, 7. Branches to auriculo-temporal of fifth. 8. Branch to auricular of **vagus.** M. The ganglion **of** Meckel. O. A. F. Orifice of aqueductus fallopii.

iculate ganglion). (4) From the geniculate ganglion are the following branches : (a) The large superficial petrosal to Meckel's ganglion. This branch leaves the cranium by the foramen lacerum medium, and joins a branch from the carotid sympathetic plexus, called the large deep petrosal, to form the Vidian nerve. (b) The small superficial petrosal to the otic ganglion. This nerve leaves the cranium by the canalis innominatus, a small opening between the foramen ovale and the foramen spinosum, in the greater wing of the sphenoid bone. This branch receives a communicating twig from the tympanic branch of the glosso-pharyngeal nerve. (c) The external superficial petrosal nerve, to join the sympathetic on the middle meningeal artery. (5) The facial nerve traverses the middle ear in its bony canal, traversing between the roof and inner wall, and between the inner wall and the posterior, successively, emerging at the stylo-mastoid foramen. (6) In the middle ear the nerve gives off (a) a small branch to the stapedius muscle ; (b) the chorda tympani, which supplies the submaxillary and sublingual glands with secretory and vaso-dilator fibres, and the anterior part of the tongue with taste fibres. The facial supplies the stylo-hyoid, the posterior belly of the digastric, all the dermal muscles about the ear, the posterior belly of the occipito-frontalis, and all the muscles of facial expression. (7) The facial nerve communicates with the three divisions of the fifth on the face ; with the spheno-palatine, otic, and submaxillary ganglia ; with the auditory, vagus, sympathetic, and glosso-pharyngeal nerves.

Describe briefly the auditory nerve.

This is the eighth cranial nerve. It leaves the cranium by the internal auditory meatus with the auditory artery—a branch of the basilar—and the seventh or facial

OPHTHALMIC   SUPERIOR MAXILLARY   INFERIOR MAXILLARY

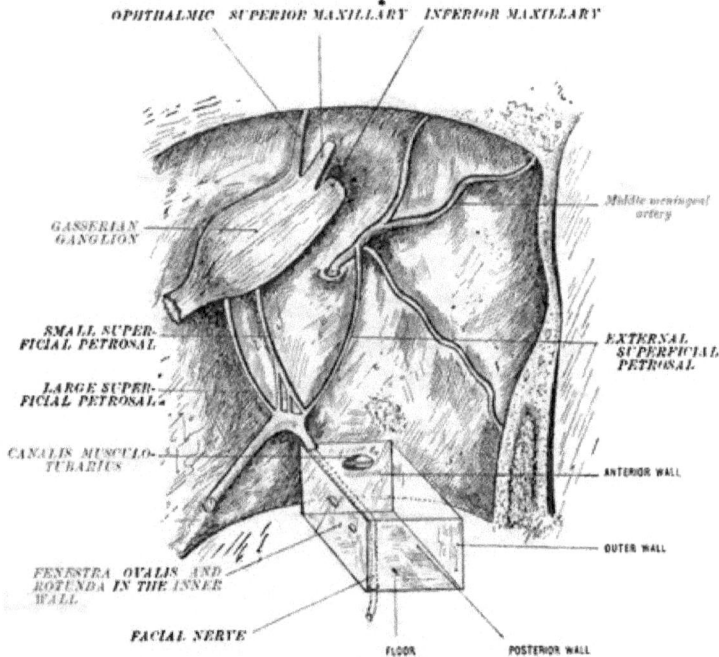

GASSERIAN GANGLION

SMALL SUPER-FICIAL PETROSAL

LARGE SUPER-FICIAL PETROSAL

CANALIS MUSCULO-TUBARIUS

FENESTRA OVALIS AND ROTUNDA IN THE INNER WALL

FACIAL NERVE

Middle meningeal artery

EXTERNAL SUPERFICIAL PETROSAL

ANTERIOR WALL

OUTER WALL

FLOOR   POSTERIOR WALL

FIG. 292.—BOX ILLUSTRATION OF SEVENTH CRANIAL NERVE IN RELATION TO THE MIDDLE EAR OR TYMPANUM.

For purposes of aiding the memory, the tympanum is compared to a box. The reader is referred to page 135 for a full description of the figure.

nerve. In the auditory canal it communicates with the seventh nerve by the pars intermedia. The eighth is the nerve of the special sense of hearing. (For the distribution of this nerve see Morris.)

Describe the glosso-pharyngeal nerve.

This is the ninth cranial nerve. It escapes from the cranium by the jugular foramen with the tenth and eleventh nerves and the jugular vein. The nerve has two ganglia : (1) The jugular and (2) the petrosal ganglion. The ninth nerve distributes branches to the mucous membrane of the tongue, pharynx, tympanum, and also to the stylo-pharyngeus muscle. It supplies the otic ganglion *via* its communications with the small superficial petrosal branch of the seventh nerve. This nerve communicates with the third division of the fifth nerve; with the facial, sympathetic, and vagus.

What can you say of the function of the vagus?

In animals less highly specialized than man the term pneumogastric expresses practically all that can be said of this nerve, since in its wandering distribution it supplies the organs of voice and respiration with motion and sensation; the organs of circulation and digestion with motion only. The vagus, then, has to do with the two greatest sources of metabolism : the ingestion of food and air. Man is not all stomach and lung. By synecdoche, however, stomach may represent a complicated digestive apparatus, and lung an equally differentiated apparatus, in which the blood throws off CO_2, and loads up with O, according to the law of the diffusion of gases. Now, the student must remember this : The names of the branches of the vagus nerve will be governed and determined by the following conditions :

1. The arbitrary subdivision of the alimentary canal, to which the vagus is distributed.

2. The arbitrary subdivisions of the respiratory system, to which the vagus is distributed.

3. The sympathetic nerve-branches to both the respiratory and digestive systems, since the sympathetic is the nerve of organic life.

4. The organs, or systems of organs, associated secondarily with the respiratory and digestive processes. The tongue, for example, is associated with both respiratory and digestive processes, and while this organ has its own specific nerve-supply, upon which its sensation and motion depend, still, the orderly adjustment of internal to external relations could not occur, were there no communication between the nerve-supply of the tongue, which forces food into the pharynx, and the organs that deliver this food to the stomach. In like manner there must be unanimity of action between the lungs, vocal cords, and tongue, for vocalizing and linguistic purposes; this unanimity must depend on an uninterrupted communication between the nerve-supply of the tongue and the vagus nerve. Observe the difference in facial expression of the two men, the one of whom is in the nausea period of a malarial attack, the other of whom is enjoying an after-dinner cigar. The facial nerve supplies the muscles of expression, and by its communication with the vagus both comfort and distress in the stomach are facially expressed.

Physiological Reasons for the Numerous Communications of the Vagus Nerve.

1. *With the spinal accessory nerve.* The spinal part of the spinal accessory nerve is distributed to the sterno-cleido-mastoid and trapezius muscles, and assists in forming the cervical plexus. It will be remembered that the mission of the cervical plexus is to supply the diaphragm and certain muscles of forced respiration, as well as the depressor muscles of the hyoid bone. (Page 73.) Now, the accessory part of the spinal accessory nerve supplies the muscles of the soft palate. Thus, by the communication between the vagus and the spinal accessory, harmony is established between the muscles that rhythmically enlarge and reduce the size of the thorax, and the lungs that occupy this thorax, and by their rhythmic movements carry on vocal and respiratory processes.

2. *With the petrous ganglion of the glosso-pharyngeal or ninth cranial nerve.* This nerve has a distribution as follows : (1) To the tongue ; (2) to the pharynx *via* the pharyngeal plexus ; (3) to the middle ear, Eustachian tube, and mastoid cells *via* the tympanic plexus ; (4) to the parotid gland *via* its small superficial petrosal branch and the otic ganglion ; (5) to the facial nerve ; (6) to the sympa-

thetic nerve. In the communications, then, between the pneumogastric and the
glosso-pharyngeal nerve, we see the connecting link between a mechanism
pumping air, under the direction of the vagus, and a territory to be constantly

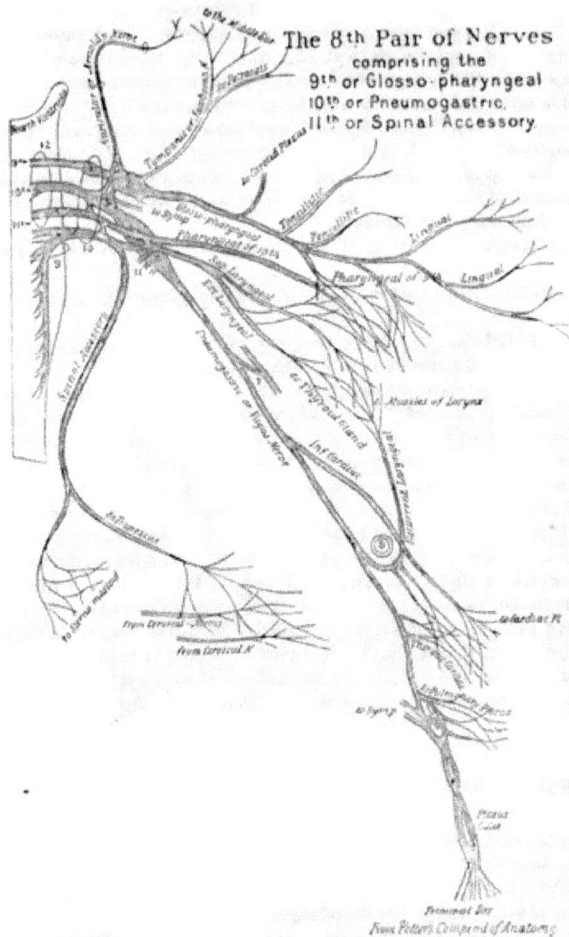

FIG. 293.—EIGHTH PAIR OF CRANIAL NERVES.

1. Jugular ganglion of ninth nerve.   2. Petrous ganglion of ninth nerve.   3. Ganglion of the vagus root.
4. Ganglion of the vagus trunk.   5. Medullary part of eleventh nerve.   6. Spinal part of eleventh
nerve.   7. Superior cardiac branch joining cardiac of sympathetic.   8. Subclavian artery, on right
side, arch of the aorta, on left side of the body.   9. Foramen magnum, receiving spinal part of the
spinal accessory.   10. Jugular foramen, transmitting all three nerves; also called the foramen lacerum
posterius.   11. Branches to hypoglossal, sympathetic, cervical nerves.   12. Olivary body (in broken
line).

ventilated—the middle ear and Eustachian tube, a territory whose delicate
structures might be irreparably damaged, were not the sensory sentinel, the
tympanic plexus, there to regulate the ingress and egress of air. And note, too,

the expression of anguish on the patient's face when the middle ear is the seat of pain; an expression determined by the auricular communicating branch of the ninth nerve with the seventh or facial nerve, and also with the pneumogastric nerve, as represented in figure 293.

3. *With the hypoglossal nerve*, by which the movements of the tongue are harmonized with both deglutition and vocal action.

4. *With the sympathetic nerve*, by which the action of the heart and lungs may be brought into harmony with their environment.

The physiological communications of the vagus nerve are :

1. With the spinal accessory nerve.
2. With the glosso-pharyngeal nerve.
3. With the hypoglossal nerve.
4. With the sympathetic nerve.
5. With the first and second cervicals.
6. With the facial or seventh cranial nerve.

This communication is for the purpose of harmonizing respiration and the ingestion of food, in their broadest terms, in consonance with a conservative and æsthetic adjustment of inner to outer relations.

Name the branches of the vagus nerve.

1. A recurrent branch to the dura mater.

2. The auricular branch, given off from the ganglion of the root, is joined by a branch of the petrous ganglion of the glosso-pharyngeal. (Fig. 293.) This branch divides into two : one communicates with the posterior auricular branch of the facial, the other joins the facial nerve in its canal. The auricular branch of the tenth cranial nerve is called Arnold's nerve.

3. The pharyngeal is the prime motor nerve of the pharynx. It unites with the glosso-pharyngeal, superior laryngeal, and sympathetic branches to form the pharyngeal plexus. The principal fibres of the pharyngeal branch of the pneumogastric are derived from the accessory part of the spinal accessory nerve.

4. The superior laryngeal nerve is the nerve of sensation to the mucous membrane of the larynx, and of motion to the crico-thyroid muscle.

5. The inferior laryngeal nerve is the motor nerve to all the intrinsic muscles of the larynx except the crico-thyroid. This is also called the recurrent laryngeal.

6. The cervical cardiac branches communicate with the sympathetic, and pass to the superficial and deep cardiac plexuses.

7. The thoracic cardiac branch has a double origin : (1) From the trunk of the vagus ; (2) from the recurrent laryngeal nerve. They are distributed to the cardiac plexus.

8. The anterior pulmonary branches unite with the sympathetic to form the anterior pulmonary plexus, which is found on the front of the root of the lung.

9. The posterior pulmonary branches unite with sympathetic nerves to form the posterior pulmonary plexus. This is larger than the preceding, and found on the posterior part of the root of the lung. (Fig. 151.)

10. The œsophageal plexus supplies the œsophagus and pericardium. It is formed by the union of branches from both the right and left pneumogastric nerves. The plexus is also called the plexus gulæ, because gula means gullet.

11. The terminal branches are the gastric. The branches from the left vagus supply the anterior, those from the right the posterior, surface of the stomach. They unite with the sympathetics and assist also in forming the splenic and hepatic plexuses.

Describe the spinal accessory nerve.

This is the eleventh cranial nerve. It leaves the cranium by the jugular foramen with the ninth and tenth nerves and the internal jugular vein. It has :

(1) A spinal part and (2) an accessory or internal part. The spinal part you will trace to the trapezius and sterno-mastoid muscles. The accessory part passes forward to the superior laryngeal nerve and pharyngeal plexus.

Describe the hypoglossal or twelfth nerve.

This is the motor nerve of the tongue. It leaves the cranium by the anterior condyloid foramen. It supplies the depressor muscles of the hyoid bone in association with the first three cervical nerves. It communicates with the vagus, lingual, sympathetic, and upper three cervical nerves.

ARTICULATIONS.—LIGAMENTS.

1. *What is an articulation?*

Union of bone to bone, bone to cartilage, or cartilage to cartilage, by means of modified periosteum called ligaments.

2. *What are some of the ends subserved by articulations?*

Stability, as in the joints of the cranium; motility, as in the extremities; semistability, as in the pelvis and vertebral column.

3. *Name and indicate the function of the structures found in a typical articulation for motility.*

(1) There is bone for strength and solidity; (2) a shell or covering of bone, free from vessels and nerves, adapted to bear pressure; (3) a highly polished articular cartilage to confer elasticity and reduce friction; (4) a capsular ligament, of periosteal derivation, attached above and below the joint to the bones party to the articulation; (5) a synovial membrane, with its articular vessels and nerves, closely investing the interior of the capsule, for the secretion of synovia.

4. *Name the structures found in an articulation intended for stability.*

Bones with variously shaped edges, whose union in the very young **is legal**ized by periosteum called sutural ligament.

5. *Name structures found in an articulation where semistability is to be attained.*

Bony surfaces firmly united by discs of cartilage, which admit of slight motion, and under certain physiological conditions—as pregnancy—develop partial synovial membrane.

6. *Are there any technical terms by which the articulations, as above classed,* **are** *designated?*

Synarthrosis includes all immovable articulations where bone is joined to bone by variously shaped borders, as toothed, grooved, scaled, sawlike, and seamlike.

Amphiarthrosis includes all articulations in which bone is firmly united to bone by cartilage, with slight movement,—as in the pelvis and vertebral column,—and a spasmodic production of synovia, as the case may require; as in pregnancy, this may be the case.

Diarthrosis includes all articulations with perfect capsules, free movement, and constant production of synovia.

7. *Does the degree of motion vary in articulations?*

Yes; and the following subdivisions of the class *diarthrosis* express the variety or kind of movement, in accordance principally with the degree of motion, as determined by the shape of the articular surfaces: (1) *Enarthrosis*, or ball-and-socket variety, in which there is movement in four modified angular directions. Examples of this are seen in the hip, shoulder, and carpo-metacarpal articulation of the thumb. (2) *Condylarthrosis*, in which free movement, as in the ball-and-socket articulation, is inhibited. Instance, the temporo-maxillary, occipito-atlantal, radio-carpal, metacarpo-phalangeal, and metatarso-phalangeal joints. In each, as you may demonstrate on yourselves, there is free movement—as flexion and extension—in two directions, but very limited movement from side to side. Observe the free movement of the fingers on the metacarpals forward and backward, and the limited movement from side to side. The movement of lower jaw is the same. We can see plainly that this variety of joint is a compromise between ball-and-socket and the hinge-joint. (3) *Ginglymus*, or *hinge-joint*, in which there is free movement in two directions, as in the elbow, knee, ankle, and interphalangeals of all the digits of both feet and hands. In these articulations note the impossibility of lateral movement without violence ; note the intermediate position between this variety and the ball-and-socket variety, occupied by the preceding condylarthrosis subdivision. (4) *Lateral ginglymus* is a term used to express a subdivision of movable joint which differs from the hinge movement in no essential, except in the direction of motion. The superior and inferior radio-ulnar articulations and the atlanto-odontoid belong to this subdivision. (5) *Arthrodia* includes the simplest subdivisions of all the movable articulations. In this kind the surfaces are plane, or nearly so. The movements are limited, by virtue of the strong and unyielding nature of the ligaments, which represent the lowest organized variety of capsular ligament. Examples of this subdivision are in the carpal and tarsal articulations ; between the metacarpals and metatarsals ; in the articular processes of the vertebræ ; in the costo-transverse and interchondral articulations.

I desire you to dissect and study the movable articulations—the diarthroses—according to the following outline :

1. Give the name of the class—diarthrosis, for example.
2. Give the name of the subdivisions of the class.
3. Give the technical name of the locality.
4. Give the osteological units in the joint.
5. Give subdivisional parts of osteological units.
6. Name the articular surfaces according to the rule.
7. Give the basis of a movable joint—a capsule.
8. Give the local subdivisions of the capsule.
9. Name the strengthening bands of the capsule, if any.
10. Name the incorporated tendons of obsolete muscles, if any.
11. Name the bony limitations of movements of the joint.
12. Name the nerve-supply and blood-supply of the joint.
13. Name the ligamentous muscles of the joint.

Let us review now the above *outline*, and understand its specific scope.

1. There are five subdivisions of the class diarthrosis. (Page 424.) By examining the articulation, you find movement in two directions. Your entry should read : Class, diarthrosis ; Subdivision, ginglymus.

2. Let us assume the particular joint is the elbow : then the technical name is humero-radio-ulnar articulation ; the common name, elbow-joint.

3. The osteological units in this joint are the humerus, radius, and ulna.

4. The subdivisional parts of the osteological units are the outer and inner humeral condyles, radial head of humerus, head of radius, olecranon process of ulna, coronoid process of ulna, greater sigmoid cavity of ulna.

5. Articular surfaces should take the name of the occupant. Then in this joint we find the radial and ulnar surfaces of the humerus ; the humeral surfaces of the radius and ulna. The ulnar surface of the humerus is called technically the trochlear or pulley surface ; the radial surface of the humerus is called the capitellum, or little head.

6. The basic principle in every movable joint is a capsule derived from periosteum, lined by a more or less extensive synovial membrane.

7. The subdivisions of the capsule are anterior, posterior, internal lateral, and external ligaments. A complete capsule is equal to the sum of its subdivisional parts. The usage by which capsules are thus subdivided is a useless (except

for locating lacerations in rare instances) as well as an arbitrary practice. There is no well-defined line of demarcation between a lateral and an anterior part of a capsule, as the student may expect to find, since transitions in anatomy are easy and gradual.

8. Strengthening or accessory bands are derived either from the deep fascia or from the aponeurosis of a muscle crossing a joint.

9. By incorporation of tendons in a capsule, the capsule becomes stronger where the incorporation occurs. Growth is the correlative of function, you will remember ; now, if from change in environment an animal acquire altered body movements, a muscle may migrate from its primitive insertion below a joint to an acquired insertion above a joint. Loss of function entails loss of specific character ; in this case loss of specific character—that is, loss of muscular element—leaves only the connective-tissue framework crossing the joint. The muscle migrated, the tendon became divorced, and by vigorous retrogression of muscle, a band of connective tissue remains which strengthens the capsule ; the original tendon became incorporated with the capsule. The divorced tendon of the adductor magnus obtains as the internal lateral ligament of the knee-joint ; that of the peroneus longus' as the external lateral ligament. The coraco-humeral ligament is the divorced tendon of the minor pectoral muscle. The greater sacro-sciatic ligament is the divorced tendon of the biceps femoris The ligamentum teres in the hip-joint is said by Morris to be in all probability the divorced tendon of the pectineus muscle.

10. The movements of joints are often limited by bone. Note should be taken of these physiological limitations on the cadaver when you dissect ligaments.

11. Ligamentous muscles to a joint are such muscles as cross a joint. The biceps, brachialis anticus, triceps, and all the muscles getting origin from the humeral condyles are ligamentous in their action to the elbow-joint. Muscles —as the rectus femoris and the sartorius—may be ligamentous to more than one joint. Ligamentous muscles are also called the elastic ligaments of a joint.

12. Note the entrance of nerves and vessels that supply the synovial membrane. . Such structures are designated articular. In this connection refer to Hilton's law governing articular nerves. This law is given in the introductory chapter of this book on page 19.

13. Finally, having found in the make-up of movable joints bone, articular lamella, cartilage, ligament, tendon, muscle, synovial membrane, nerve, and blood-vessels, be able to classify these structures histologically, according to the outline given in the beginning of this work under caption of anatomical tissues. The plan just outlined for your guidance will not only give you a thorough understanding of the ligaments, but will refresh your memory on those very practical points you will so much need in your subsequent medical studies, both as students and as practitioners of medicine.

Whenever you are in doubt as to the name of a particular subdivision of a class, consult the following table, taken from Morris ·

TABLE OF THE VARIOUS CLASSES OF JOINTS.

Class.	Examples.
I. Synarthrosis	
(*a*) **True sutures**	Lambdoid, sagital, coronal.
(*b*) **False sutures**	Internasal. Intermaxillary. Costo-chondral.
(*c*) **Grooved sutures**	Vomer and rostrum of sphenoid.
II. Amphiarthrosis	Bodies of vertebræ. Symphysis pubis, sacro-iliac, sacro-coccygeal.

TABLE OF THE VARIOUS CLASSES OF JOINTS.—(*Continued.*)

Class.	Examples.
III. Diarthrosis	
(*a*) Enarthrosis	Shoulder. Hip. Astragalo-scaphoid.
(*b*) Condylarthrosis	Temporo-mandibular. Occipito-atlantal. Radio-carpal. Metacarpo-phalangeal. Metatarso-phalangeal.
(*c*) Ginglymus or Trochlearthrosis .	Elbow. Ankle. Knee. Interphalangeal of fingers and toes.
(*d*) Trochoides or Lateral Ginglymus	. Atlanto-odontoid. Superior Radio-ulnar. Inferior Radio-ulnar.
(*e*) Arthrodia :	
(1) Simple	Lateral atlanto-axoidean. The joints between the articular processes of the vertebræ, costotransverse, and interchondral. Acromioclavicular. Carpal. Carpo-metacarpal of four fingers. Intermetacarpal. Tarsal. Tarso-metatarsal. Intermetatarsal. Calcaneo-astragaloid. Superior and inferior tibio-fibular.
(2) Saddle-shaped	Sterno-costo-clavicular. Carpo-metacarpal of thumb. Calcaneo-cuboid.

The following is a list of the principal articulations of the immovable class—synarthrosis. Let the student locate these joints on the skull :

COMPOUND.	BASE.	ADJUNCT.
1. Temporo-parietal.	Temporal.	Parietal.
2. Interparietal.	Parietal.	Parietal.
3. Fronto-parietal.	Frontal.	Parietal.
4. Occipito-parietal.	Occipital.	Parietal.
5. Masto-parietal.	Mastoid.	Parietal.
6. Squamo-parietal.	Squamosa.	Parietal.
7. Occipito-sphenoidal.	Occipital.	Sphenoid.
8. Occipito-temporal.	Occipital.	Temporal.
9. Spheno-temporal.	Sphenoid.	Temporal.
10. Spheno-frontal.	Sphenoid.	Frontal.
11. Spheno-ethmoidal.	Sphenoid.	Ethmoid.
12. Fronto-ethmoidal.	Frontal.	Ethmoid.
13. Fronto-maxillary.	Frontal.	Maxillary.
14. Fronto-nasal.	Frontal.	Nasal.
15. Fronto-lachrymal.	Frontal.	Lachrymal.
16. Fronto-malar.	Frontal.	Malar.
17. Lachrymo-ethmoidal.	Lachrymal.	Ethmoid.
18. Naso-maxillary.	Nasal.	Maxillary.
19. Malo-maxillary.	Malar.	Maxillary.
20. Costo-chondral.	Costal (Rib).	Chondrum.

The following are the members of the group of amphiarthrosis articulations. Their number is small. Remember, the opposed bony surfaces are united by discs of cartilage ; there **is** slight movement, due principally to the elastic nature of the discs of cartilage. A partial synovial membrane may be developed during pregnancy

1. The intervertebral or intercentral articulations.
2. The sacro-iliac synchrondrosis.
3. The sacro-coccygeal articulation.
4. The intercoccygeal articulations.
5. The interpubic articulation—symphysis pubis.

TEMPORO-MANDIBULAR ARTICULATION.

1. *Locate and give the synonym for temporo-mandibular articulation.*

This articulation is between the glenoid fossa of the temporal bone and the condyle of the mandible or lower jaw-bone. It is also called temporo-maxillary articulation.

2. *Name some of the conditions that make this articulation one of practical importance in medicine and surgery.*

Passing over the part this joint plays in the mastication of food, since this is too well understood by the student to need explanation, we may mention the following : (*a*) Temporo-mandibular fixity may occur in cases of exposed tooth-pulp. The nerve supplying the pulp, the trigeminus, also supplies the muscles of mastication. Here, then, we have a moto-sensory reflex circuit. The powerful contraction of the muscles will cease on the administration of an anæsthetic.

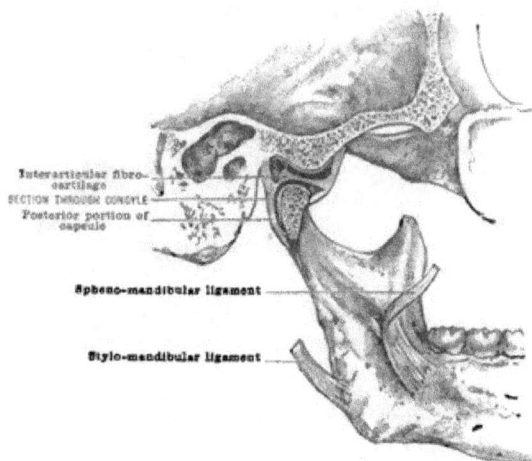

Interarticular fibro-cartilage
SECTION THROUGH CONDYLE
Posterior portion of capsule

Spheno-mandibular ligament

Stylo-mandibular ligament

FIG. 294.—VERTICAL SECTION THROUGH THE CONDYLE OF JAW TO SHOW THE TWO SYNOVIAL SACS AND THE INTERARTICULAR FIBRO-CARTILAGE.

I once saw a case of fixed jaw produced by a bug in the ear. Here was the same moto-sensory reflex circuit. The same nerve that supplies the muscles of mastication—the fifth cranial—also supplies the auditory canal through its auriculo-temporal branch. In this case no ordinary traction would open the mouth. An elderly lady suggested the curative property of warm salt-water in "lock-jaw." The ear was filled with water, the bug came out, and instantly the patient could open her mouth. (*b*) The condyle glides forward onto the eminentia articularis (Fig. 58) when the mouth is open. In this position a dislocation may occur. The condyle is deeply lodged in the glenoid cavity when the mouth is closed. In this position a blow on the chin may fracture the tympanic wall; while a blow on the angle of the jaw may break the thin dome of the glenoid, and produce injury to the brain.

3. *How is the glenoid cavity limited in front and behind ?*

Anteriorly, by the eminentia articularis ; posteriorly, by the Glaserian fissure and the post-glenoid tubercle.

28

4. How many synovial cavities has the temporo-mandibular articulation and how are they produced?

It has two; they are produced by interposition of the interarticular fibro-cartilage. The periphery of this disc is attached to the capsule. (Fig. 294.)

5. Explain the use of this interarticular cartilage.

When the mouth is opened, the condyle of the mandible slides forward onto the eminentia articularis, in the concave cup of the articular cartilage; when the mouth is shut, the condyle and cartilage recede into the glenoid. (Fig. 59.)

6. By what are the condyle and cartilage drawn forward?

By the external pterygoid muscle, a portion of the muscle being inserted into the cartilage as well as into the condyle. (Fig. 58.)

DISSECTION.

It is to be hoped that the foregoing has prepared the student to begin an intelligent dissection of the articulations.

TEMPORO-MANDIBULAR ARTICULATION.

1. *Class.*—Diarthrosis, because there is free motion.
2. *Subdivision.*—Condylarthrosis, because axial rotation is absent.
3. *Osteological Units.*—Mandible and temporal bone.
4. *Subdivisional Parts.*—Condylar process and glenoid cavity.
5. *Articular Surfaces.*—Condylar of temporal; temporal of condyle.
6. *Basis.*—A complete capsule lined by synovial membrane.
7. *Subdivisions.*—Anterior, posterior, internal, and external portions.
8. *Accessories.*—Spheno-mandibular, stylo-mandibular, fibro-cartilage.
9. *Incorporation of Tendons.*—None.
10. *Limitations.*—Articular eminence, post-glenoid tubercle.
11. *Ligamentous Muscles.*—Internal and external pterygoids, masseter.
12. *Nerve-supply.*—Masseteric and auriculo-temporal of the fifth.
13. *Blood-supply.*—Temporal, middle meningeal, ascending pharyngeal.

1. Name the osteological parts concerned in this articulation.

(1) The zygomatic arch; (2) the spine of the sphenoid; (3) the styloid process of the temporal bone; (4) the lingula of the mandible; (5) the eminentia articularis of the temporal bone; (6) the post-glenoid tubercle and tympanic plate of the temporal bone; (7) the Glaserian fissure of the temporal bone.

2. Give attachments of the spheno-mandibular ligament as you find them, and tell by what nerve this ligament is pierced.

The ligament extends from the spine of the greater wing of the sphenoid bone to the lingula or mandibular spine, and is perforated by the mylo-hyoid nerve.

3. Locate and give attachments of the stylo-mandibular ligament.

It is between the masseter and internal pterygoid muscles, it is attached to the styloid process and to the angle of the mandible. (Fig. 296.)

SHOULDER-JOINT.

1. *Class.*—Diarthrosis, because of free motion.
2. *Subdivision.*—Enarthrodia—angular movements and axial rotation.
3. *Technical Name.*—Humero-scapular; common name, shoulder-joint.
4. *Osteological Units.*—Humerus and scapula.

5. *Subdivisional Parts.*—Head of humerus, glenoid of scapula.
6. *Articular Surfaces.*—Humeral of scapula, scapular of humerus.
7. *Basis.*—A complete capsule lined by synovial membrane.
8. *Subdivisions.*—Not subdivided for description.
9. *Strengthening Bands.*—Three gleno-humeral ligaments.

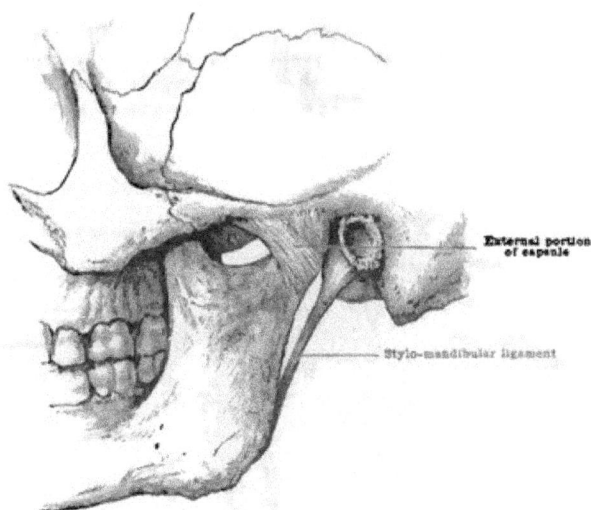

Fig. 295.—External View of Temporo-mandibular Joint.

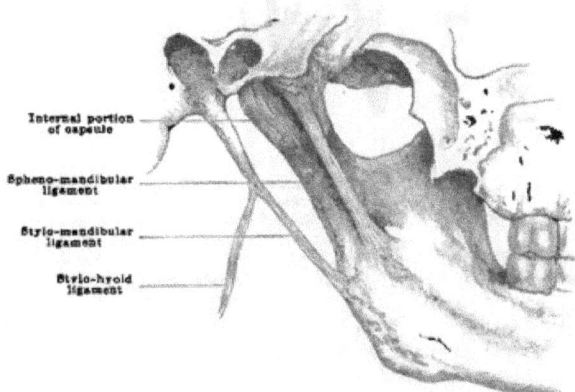

Fig. 296.—Internal View of Temporo-mandibular Joint.

10. *Incorporated tendon* of pectoralis minor in primitive man.
11. *Limitations.*—Clavicle, acromion, and coracoid process.
12. *Nerve-supply.*—Subscapular, circumflex, suprascapular.
13. *Blood-supply.*—Circumflex, subscapular, dorsalis scapulæ, axillary.
14. *Ligamentous Muscles.*—(1) Biceps ; (2) triceps ; (3) supraspinatus ; (4)

infraspinatus ; (5) subscapularis ; (6) teres minor ; (7) teres major ; (8) latissimus dorsi ; (9) coraco-brachialis ; (10) pectoralis major ; (11) deltoid.

1. *Name the osteological parts concerned in this articulation in any manner.*

(1) The articular surface of the humeral head ; (2) the glenoid cavity of the scapular head ; (3) the anatomical neck of the humerus ; (4) the anatomical neck of the scapula ; (5) the greater tuberosity with its three facets ; (6) the lesser tuberosity with one facet ; (7) the surgical neck of the humerus ; (8) the surgical neck of the scapula ; (9) the humeral bicipital groove and its lips ; (10) the supraglenoid tubercle—bicipital ; (11) the infraglenoid tubercle—tricipital ; (12) the scapular notch and foramen ; (13) the suprascapular notch and foramen ; (14) the fourth scapular angle ; (15) the acromion process of scapula ; (16) the coracoid process of the scapula.

2. *Give the morphology of the coraco-humeral ligament.*

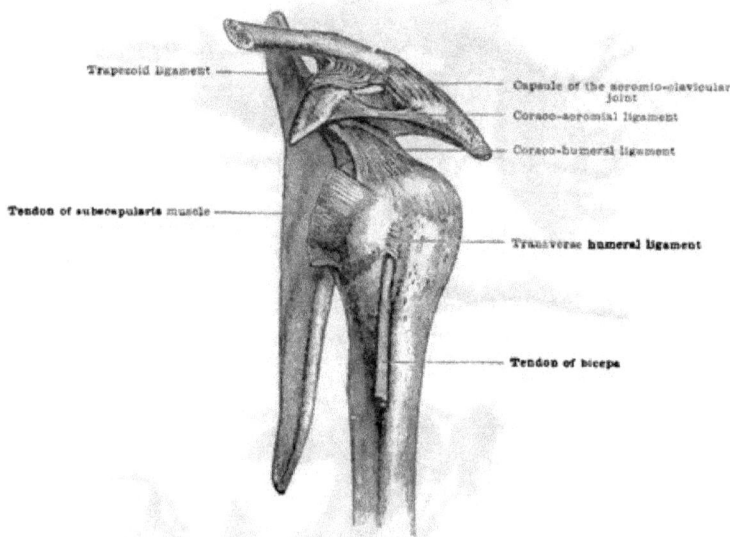

FIG. 297.—OUTER VIEW OF THE SHOULDER-JOINT, SHOWING THE CORACO-HUMERAL AND TRANSVERSE HUMERAL LIGAMENTS.

It is the divorced tendon of the pectoralis minor muscle, which in primitive man was inserted into the lesser tuberosity of the humerus.

3. *By what is the capsule of the shoulder-joint lined?*

By synovial membrane. This membrane is also reflected onto the long head of the biceps muscle in its transit of the cavity.

4. *Where is the transverse humeral ligament, and what is its function?*

It stretches from greater to lesser tuberosity, converting the bicipital groove into a canal for the lodgment of the long tendon of origin of the biceps muscle. The canal is lined by a vaginal synovial membrane.

5. *What is the glenoid ligament?*

A circumferential rim of cartilage whose function seems to be to deepen the cavity for the head of the humerus. It is attached to the capsule, and also to the margin of the glenoid cavity. It is also continuous with the long head of the biceps muscle.

6. *Through what osteological points would a fracture of the surgical neck of the scapula pass?*

(1) Through the suprascapular notch ; (2) through the scapular notch ; (3) through the fourth scapular angle or the deepest part of the subscapular fossa.

Tell where the ligaments are found by which the scapula and clavicle are held together.

THE SCAPULO-CLAVICULAR UNION.

1. Between the acromion process of the scapula and the clavicle.
2. Between the coracoid process and the clavicle.
3. Between certain given parts of the scapula alone. (Fig. 298.)

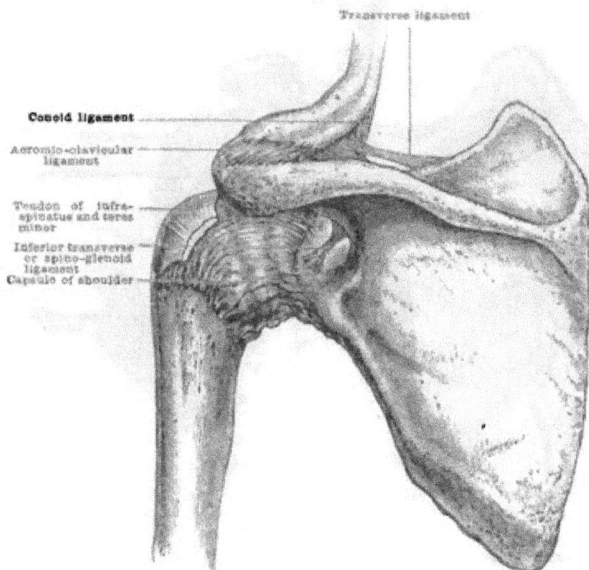

FIG. 298.—POSTERIOR VIEW OF THE SHOULDER-JOINT, SHOWING ALSO THE ACROMIO-CLAVICULAR JOINT AND THE PROPER LIGAMENTS OF THE SCAPULA.

THE ACROMIO-CLAVICULAR ARTICULATION.

1. *Class.*—Diarthrosis, because of capsule, synovia, and motion.
2. *Subdivision.*—Arthrodia, because of gliding, simple movement.
3. *Technical Name.*—Acromio-clavicular articulation.
4. *Osteological Units.*—Clavicle and scapula.
5. *Subdivisional Parts.*—Acromion and acromial end of clavicle.
6. *Articular Surfaces.*—Acromial of clavicle, clavicular of acromion.
7. *Basis.*—A capsule lined by synovial membrane.
8. *Subdivisions of Capsule.*—Superior and inferior, by some authors.
9. *Strengthening Bands.*—None. Purely periosteal.
10. *Incorporation of Tendons.*—None.
11. *Limitation of Motion by Bone.*—None.
12. *Ligamentous Muscles.*—Deltoid and major pectoral.
13. *Nerve-supply.*—The circumflex.
14. *Blood-supply.*—Circumflex, acromio-thoracic.

1. Describe the acromion process.

It is a projection of bone, a continuation of the scapular spine. It has an upper and an under surface, an inner and an outer border, and a tip. Its inner border articulates near the tip with the clavicle. Its outer border gives origin to the deltoid muscle in part. Its upper surface is occupied by the insertion of the trapezius muscle, by the origin of the deltoid, and by a subcutaneous area between the two. Its under surface is smooth and unoccupied. (Fig. 240.)

2. Describe the interarticular fibro-cartilage.

When present, it resembles others of the same class in function and attachments.

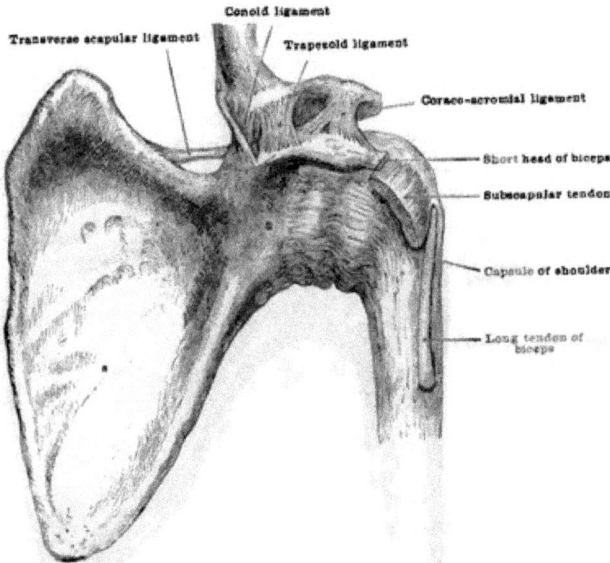

FIG. 299.—ANTERIOR VIEW OF SHOULDER, SHOWING ALSO CORACO-CLAVICULAR AND CORACO-ACROMIAL LIGAMENTS.

THE CORACO-CLAVICULAR LIGAMENTS.

Turn the clavicle upward, having cut the same through the middle third, and you will see a strong band of ligamentous tissue extending from the clavicle to the coracoid process. This consists of two parts : an anterior and outer part, called the trapezoid ligament ; a posterior and internal portion, called the conoid ligament. The action of these ligaments is to prevent upward dislocation of the acromial end of the clavicle in fracture of the bone.

The intrinsic ligaments of the scapula are : (1) The coraco-acromial ; (2) the transverse scapular (Fig. 299) ; (3) the inferior transverse scapular. (Fig. 298.)

The coraco-acromial extends from the outer border of the coracoid process by a broad base to the tip of the acromion process. Under it you will find the tendon of the supraspinatus muscle in a bed of fatty connective tissue.

The transverse scapular ligament bridges over the suprascapular notch, converting the same into a foramen, terminating externally in the coracoid process. On the ligament lie the suprascapular vessels ; beneath the ligament, in the foramen, you will find the suprascapular nerve. (Fig. 299.)

The **inferior scapular ligament** extends from the glenoid margin to the spine. . Under it pass the infraspinatous divisions of the suprascapular vessels and nerves. (Fig. 249.)

THE STERNO-CLAVICULAR ARTICULATION.

1. *Class.*—Diarthrosis, since capsule and motion are present.
2. *Subdivision.*—Arthrodia—simple gliding articular surfaces.
3. *Technical Name.*—Sterno-clavicular articulation.
4. *Osteological Units.*—Sternum and clavicle.
5. *Subdivisional Parts.*—Manubrium and sternal end of clavicle.
6. *Articular Surfaces.*—Sternal of clavicle, clavicular of sternum.
7. *Basis.*—A capsule lined by synovial membrane.
8. *Local Subdivisions.*—Should be none.
9. *Strengthening Bands.*—None.   Periosteal purely.
10. *Incorporated Tendons.*—None.
11. *Bony Limitations.*—Manubrium and first rib.
12. *Nerve-supply.*—Brachial plexus by nerve to subclavius.
13. *Ligamentous Muscles.*—Subclavius and sterno-mastoid.

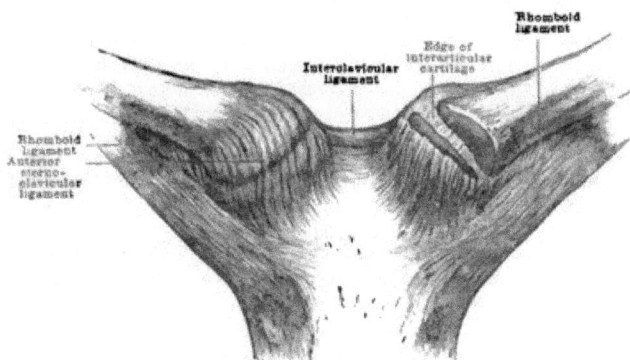

FIG. 300.—ANTERIOR VIEW OF STERNO-CLAVICULAR JOINT.
(The capsule is cut into on the left side to show the interarticular fibro-cartilage dividing the joint into two cavities.)

1. *Name all the ligaments of this articulation.*

The *capsular*, the basis of the joint.   The *interclavicular* binds the clavicles to the sternum.   The *costo-clavicular*, or rhomboid.   The *interarticular* fibro-cartilage.

2. *How many synovial cavities has the articulation?*

Two, separated from each other by the interarticular fibro-cartilage.   This cartilage is attached to the capsule like others of its class, as the temporo-mandibular and the inconstant acromio-clavicular interarticular fibro-cartilage.

3. *In practical dissection, when you wish to remove the clavicle, what is the most difficult structure to divide?*

The costo-clavicular ligament binding the clavicle very firmly to the first rib.

4. *What muscle do you find under the clavicle, and what is its surgical importance?*

The subclavius muscle, arising in front of the costo-clavicular ligament from the first rib.   It is inserted into the middle third of the clavicle on the under surface of the bone.   In fracture of the clavicle in the middle or outer third, the subclavius depresses the inner segment of the bone, while the weight of the arm elevates the outer fragment. These anatomical facts led up to the empiric practice of elevating the elbow and adjusting the inner to the outer clavicular fragments in the treatment of fracture of this bone.

ELBOW-JOINT.

1. *Class.*—Diarthrosis, because of free motion and a capsule.
2. *Subdivision.*—Ginglymus, because of motion in two directions.
3. *Technical Name.*—Humero-radio-ulnar articulation.
4. *Osteological Units.*—Humerus, radius, ulna.
5. *Subdivisional Parts.*—Two humeral condyles, a radial head, an olecranon process, a coronoid process, a greater sigmoid cavity, two condylar ridges, an olecranon fossa, a coronoid fossa, a radial fossa.
6. *Radial and Ulnar Articular Surfaces of Humerus.*—Humeral surfaces of radius and ulna. The ulnar surface of the humerus is called the trochlea. The radial head of the humerus is called the capitellum.
7. *Basis.*—A capsule lined by synovial membrane.
8. *Local Subdivisions.*—Anterior, posterior, internal, and external.
9. *Strengthening Bands.*—None.
10. *Incorporated Tendons.*—None.
11. *Bone Limitations.*—Olecranon and coronoid processes, humerus.
12. *Nerve-supply.*—Ulnar, median, musculo-cutaneous, musculo-spiral.
13. *Blood-supply.*—From the anastomosis about the joint.
14. *Ligamentous Muscles.*—Biceps, brachialis anticus, supinator longus or brachio-radialis ; triceps and anconeus ; pronator radii teres, flexor carpi radialis, palmaris longus, flexor sublimis digitorum, flexor carpi ulnaris ; supinator brevis, extensor carpi radialis longior, extensor carpi radialis brevior, extensor communis digitorum, extensor minimi digiti, extensor carpi ulnaris.

Through what would a fracture immediately above the condyles pass?

Through the external and internal condylar ridges, through the olecranon, radial, and coronoid fossæ of the humerus.

SUPERIOR RADIO-ULNAR ARTICULATION.

1. *Class.*—Diarthrosis, because of free motion.
2. *Subdivision.*—Lateral ginglymus, motion in two directions only.
3. *Technical Name.*—Superior radio-ulnar articulation.
4. *Osteological Units.*—Radius and ulna.
5. *Subdivisional Parts.*—Head of radius and lesser sigmoid of ulna.
6. *Articular Surfaces.*—Radial of ulna, ulnar surface of radius.
7. *Basis.*—An orbicular capsule lined by synovial membrane.
8. *Local Subdivisions of Capsule.*—None.
9. *Strengthening Bands of the Orbicular Capsule.*—None.
10. *Incorporated Tendons of Obsolete Muscles.*—None.
11. *Bony Limitations.*—The ulna and humerus.
12. *Nerve- and Blood-supply.*—Same as elbow-joint.
13. *Ligamentous Muscles.*—Supinator brevis and supinator longus.

Name the muscles inserted in the vicinity of this articulation.

The biceps into the bicipital tuberosity of the radius. The brachialis anticus into the coronoid process of the ulna. The triceps into the olecranon process of the ulna.

Give the origin and insertion of the orbicular ligament.

It describes three-fourths of a circle, and is attached to the anterior and posterior lips of the lesser sigmoid cavity of the ulna.

Locate the lesser sigmoid cavity.

It is on the outer part of the ulna, at the junction of the olecranon and coronoid processes ; it is limited in front and behind by the anterior and posterior lips of the cavity. It contains the head of the radius. It is covered by articular cartilage.

THE INTEROSSEOUS MEMBRANE.

This you will see extending from the interosseous ridge of the ulna to that of the radius. It is limited above by the ulnar origin of the supinator brevis ; below,

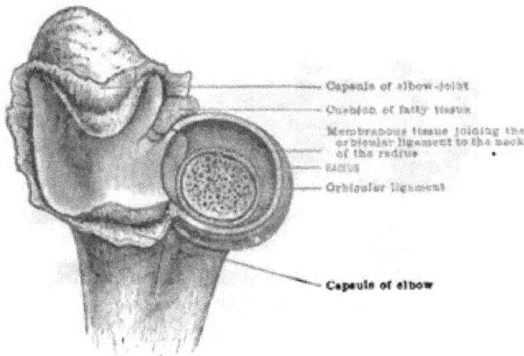

Capsule of elbow-joint

Cushion of fatty tissue

Membranous tissue joining the orbicular ligament to the neck of the radius

RADIUS

Orbicular ligament

Capsule of elbow

FIG. 301.—ORBICULAR LIGAMENT.
(The head of the radius removed to show the membranous connection of this ligament with the radius.)

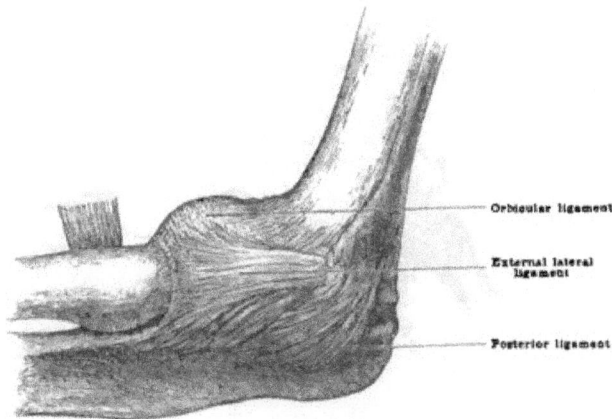

Orbicular ligament

External lateral ligament

Posterior ligament

FIG. 302.—EXTERNAL VIEW OF THE ELBOW-JOINT.

by the ulnar sigmoid of the radius. It is a modified form of radio-ulnar periosteum. It is an intermuscular septum between the flexor and extensor muscles on the forearm. It has a posterior and an anterior surface, a radial and an ulnar attached border. On its anterior surface you will find the origin of a part of the flexor profundus digitorum, and the flexor longus pollicis and pronator quadratus muscles. Between these you will see the anterior interosseous nerve and vessels.

On the posterior surface are the three extensors of the thumb, and the posterior interosseous nerve and vessels. An upper strong band of this membrane is called the oblique ligament. (Fig. 303.)

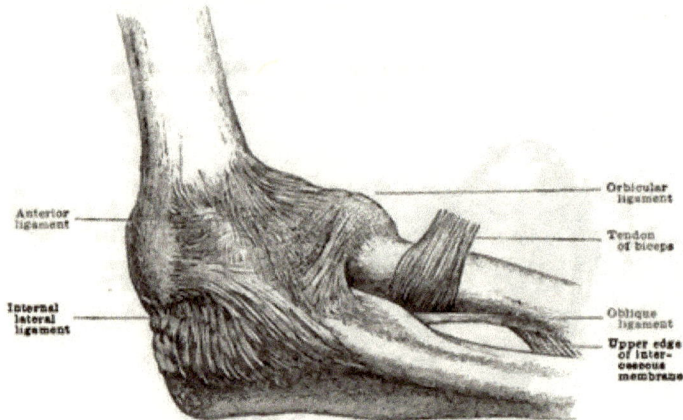

FIG. 303.—INTERNAL VIEW OF THE ELBOW-JOINT

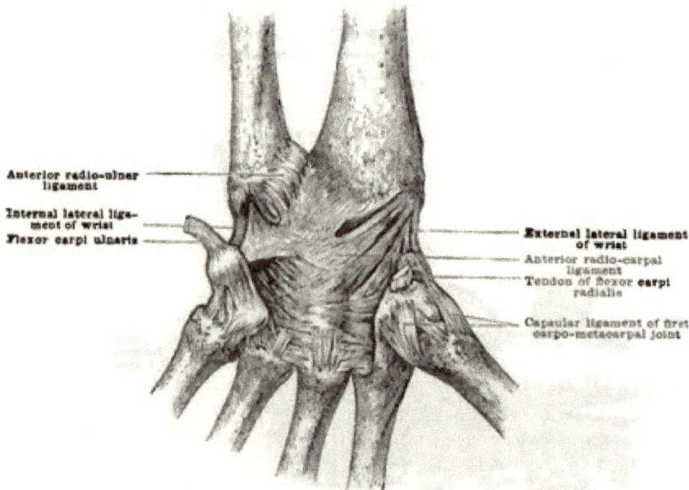

FIG. 304.—ANTERIOR VIEW OF WRIST.

THE INFERIOR RADIO-ULNAR ARTICULATION.

1. *Class.*—Diarthrosis, because of free motion and a capsule.
2. *Subdivision.*—Ginglymus, because of motion in two directions only.
3. *Technical Name.*—Inferior radio-ulnar articulation ; common, none.
4. *Osteological Units.*—Radius and ulna.
5. *Subdivisional Parts.*—Head of ulna, sigmoid of radius. (Fig. 246.)

6. *Articular Surfaces.*—Ulnar of radius, radial surface of ulna.

7. *Basis.*—A modified capsule, lined by synovial membrane.

8. *Local Parts of Capsule.*—An anterior radio-ulnar, posterior radio-ulnar, triangular fibro-cartilage.

9. *Strengthening Bands.*—None ; periosteal purely.

10. *Incorporated Tendons of Obsolete Muscles.*—None.

11. *Blood-supply.*—Anterior interosseous and anterior carpal.   (Fig. 247.)

12. *Nerve-supply.*—Anterior and posterior interosseous.

13. *Ligamentous Muscles.*—Pronator quadratus and supinator longus.

Locate and give the attachments of the anterior radio-ulnar ligament.

It is attached to the radio-ulnar triangular fibro-cartilage and styloid process. It is located on the anterior part of the bones.

Locate and give the attachments of the posterior radio-ulnar ligament.

It is attached to the radius, ulna, and triangular fibro-cartilage anteriorly and posteriorly.

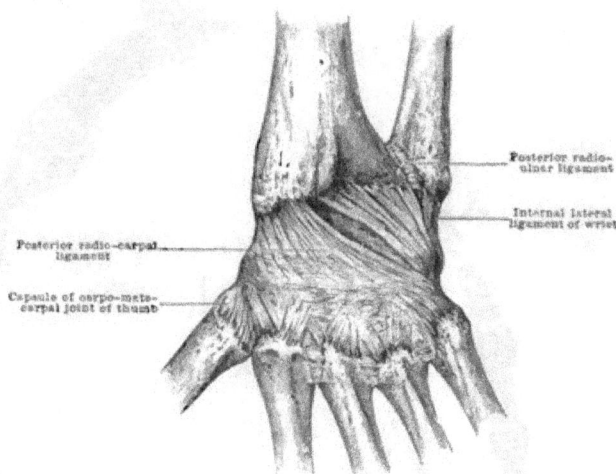

FIG. 305.—POSTERIOR VIEW OF WRIST.

THE RADIO-CARPAL ARTICULATION.

1. *Class.*—Diarthrosis, because of free motion and a capsule.

2. *Subdivision.*—Condylarthrosis, because rotation on a vertical axis is wanting ; *i. e.,* there is no axial rotation.

3. *Technical Name.*—Radio-carpal articulation.

4. *Osteological Units.*—Radius, scaphoid, semilunar, cuneiform, cartilage.

5. *Subdivisional Parts of Units.*—None.

6. *Articular Surfaces.*—According to the general rule.

7. *Basis.*—A modified and highly specialized capsule.

8. *Local Parts of Capsule.*—Anterior, posterior, internal, external.

9. *Strengthening Bands.*—None ; purely periosteal.

10. *Incorporated Tendons of Obsolete Muscles.*—None.

11. *Bony Limitations.*—Radial and ulnar styloids.

12. *Blood-supply.*—Anterior and posterior carpal arterial arches.

13. *Ligamentous Muscles.*—(1) The flexors of the carpus, on radius and ulna ; (2) the extensors of the carpus, on radius and ulna ; (3) the sublime and pro-

found digital flexors; (4) the flexor longus pollicis; (5) the extensor communis digitorum; (6) the special extensor of the little finger; (7) the special extensor of the index finger; (8) the three extensors of the thumb. Total, 14.

Explain fully the most important ligament in the radio-carpal articulation.

The most important is the triangular fibro-cartilage. It may be described as having: (1) A base, attached to the margin of the radius; (2) an apex, attached to the fossa at base of ulnar styloid; (3) an anterior border, attached to radio-

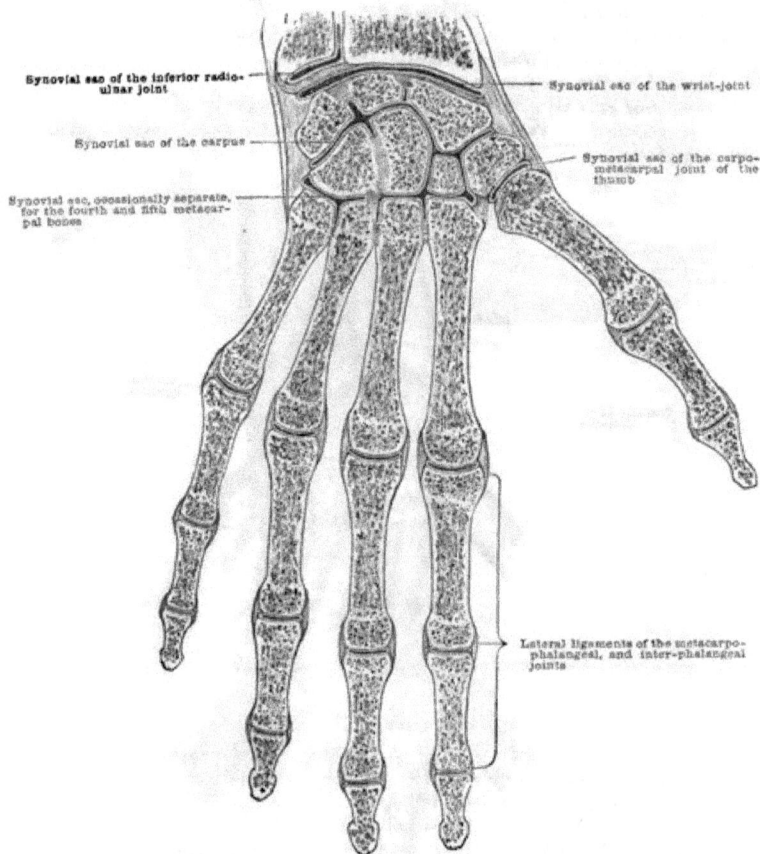

Synovial sac of the inferior radio-ulnar joint

Synovial sac of the wrist-joint

Synovial sac of the carpus

Synovial sac of the carpo-metacarpal joint of the thumb

Synovial sac, occasionally separate, for the fourth and fifth metacarpal bones

Lateral ligaments of the metacarpo-phalangeal, and inter-phalangeal joints

FIG. 306.—SYNOVIAL MEMBRANES OF WRIST, HAND, AND FINGERS.

carpal and radio-ulnar ligaments; (4) a posterior border, attached to the radio-carpal and radio-ulnar ligaments; (5) an upper or ulnar articular surface (concave); (6) a lower or cuneiform surface (concave).

How does this cartilage differ from all other fibro-cartilages.

It enters into the formation of two distinct articulations, the inferior radio-ulnar and radio-carpal, separating them completely.

With how many bones does the end of the radius articulate?

With the scaphoid and semilunar.

Give the attachments of the internal lateral ligament of the radio-carpal articulation.

Its apex is attached to the styloid process of the ulna; its base to the pisiform and cuneiform bones, and to the anterior annular ligament.

Give attachments of the external lateral ligament of the radio-carpal articulation.

Its apex is attached to the apex of the radial styloid process; its base to the scaphoid and trapezium.

Give the attachments of the anterior radio-carpal ligament.

It is attached above to the radius and anterior radio-ulnar ligament; below, to the two rows of carpal bones.

Give the attachments of the posterior radio-carpal ligament.

It is attached above to radius, styloid, and fibro-cartilage; below, to the first row of carpal bones.

Does the synovial cavity of the radio-carpal articulation communicate with the inferior radio-ulnar or carpal articulations?

No.

CARPAL ARTICULATIONS.

How may we consider the carpal articulations?

(1) The joints of the first row—consisting of four short bones; (2) the joints of the second row—consisting of four short bones; (3) the junction of the two rows—called the medio-carpal joint.

Name the short bones in the first and second rows.

The scaphoid, semilunar, cuneiform, and pisiform; trapezium, trapezoid, os magnum, and unciform.

Name the ligaments of the first row.

They are all connected by two dorsal and two palmar and two interosseous ligaments to each other, except the pisiform.

How may the pisiform bone be regarded?

It may be regarded as a sesamoid bone developed in the tendon of the flexor carpi ulnaris, and connected by a capsule to the cuneiform bone.

How are the bones of the second row united?

By three dorsal, three palmar, and two interosseous ligaments. The articulations of the first and second rows belong to the *diarthrodial class* and *arthrodial subdivision* of joints.

Name the ligaments of the medio-carpal articulation.

These are the anterior, posterior, medio-carpal, and transverse dorsal ligaments. To dissect this region cut through dorsal ligaments.

Give the attachments of the anterior annular ligament.

Internally it is attached to the unciform process of the unciform bone, and the pisiform bone; externally, to the scaphoid and trapezium.

How many insertions has the flexor carpi ulnaris?

Three: (1) Into the pisiform bone; (2) into the unciform process of the unciform bone; (3) into the base of the fifth metacarpal.

Trace the tendon of the flexor carpi radialis to its insertion.

It passes through a groove in the os trapezium, and is inserted into the base of the second metacarpal, sometimes also into the third as well.

How are the carpo-metacarpal articulations classified?

(1) Into an outer, corresponding to the thumb; (2) four inner, corresponding to the remaining four digits.

Why are they so classified?

Because they belong to different subdivisions of the class diarthrosis. The carpo-metacarpal of the thumb is a saddle-shaped arthrodia; the others are simply arthrodia.

Name the ligaments of the carpo-metacarpal articulation of the thumb. This is a capsular ligament.

In this joint are found all the movements except axial rotation.

Name the ligaments of the four inner carpo-metacarpal articulations.

They are dorsal, palmar, and interosseous.

How are these joints supplied with nerves?

By the ulnar median and posterior interosseous.

How many intermetacarpal articulations are there?

Four; bound together by dorsal, palmar, and interosseous ligaments.

How are the heads of the metacarpal bones held together?

By transverse ligaments. In front of these ligaments pass the lumbrical muscles, with the digital vessels and nerves; behind, pass the interossei muscles.

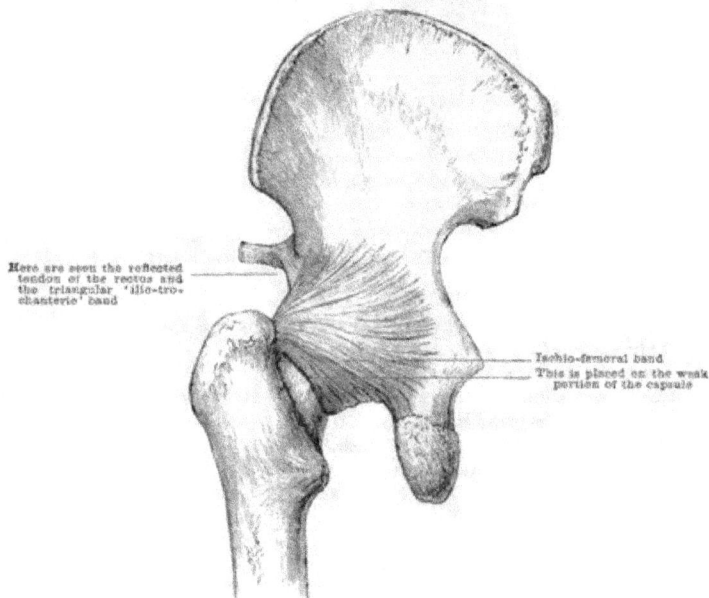

FIG. 307.—POSTERIOR VIEW OF THE CAPSULE OF THE HIP-JOINT.

The metacarpo-phalangeal articulations are five in number.

Class, diarthrosis; subdivision, condylarthrosis. The ligaments are lateral and glenoid. The nerves and arteries come from the digital branches.

Classify the interphalangeal articulations.

They belong to the class diarthrosis; subdivision, ginglymus. The proper ligaments are lateral and glenoid. Dorsally, the tendon of the extensor communis digitorum acts ligamentously; on the palmar surface, the tendons of the flexors. The articular arteries and nerves come, according to the rule, from the vessels and nerves that supply the muscles that move these joints.

THE HIP-JOINT.

1. *Class.*—Diarthrosis; axial rotation and four angular movements.
2. *Subdivision.*—Enarthrosis or ball-and-socket.

3. *Technical Name.*—Femoro-acetabular.
4. *Osteological Units.*—Femur **and** os innominatum.
5. *Subdivisional Parts.*—Ilium, ischium, pubes, **and femoral head.**
6. *Articular Surfaces.*—Named according **to the rule.**
7. *Basis.*—A capsule lined by synovial membrane.
8. *Local Subdivisions.*—Ilio-femoral, ischio-femoral, pectineo-femoral.
9. *Strengthening Bands.*—Tendino-trochanteric band.
10. *Incorporated Tendon of the Pectineus Muscle* (ligamentum teres).
11. *Bony Limitations.*—The brim and circumference of acetabulum.
12. *Nerve-supply.*—Anterior crural, obturator, great sciatic, sacral plexus.
13. *Blood-supply.*—Obturator, circumflex, gluteal and sciatic arteries.
14. *Ligamentous Muscles.*—(1) Sartorius ; (2) rectus femoris ; (3) ilio-psoas ;

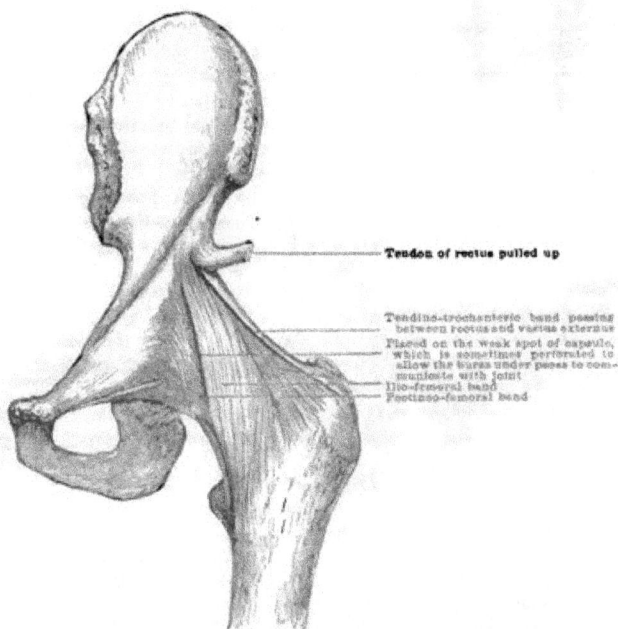

Tendon of rectus pulled up

Tendino-trochanteric band passing
between rectus and vastus externus
Placed on the weak spot of capsule,
which is sometimes perforated to
allow the bursa under psoas to communicate with joint
Ilio-femoral band
Pectineo-femoral band

FIG. 308.—ANTERIOR VIEW OF THE CAPSULE OF THE HIP-JOINT.

(4) pectineus ; (5) gracilis ; (6) adductors longus, brevis, and magnus ; (7) the glutei maximus, medius, and minimus ; (8) the obturators internus and externus ; (9) the gemelli, superior and inferior ; (10) the tensor vaginæ femoris ; (11) the semitendinosus, biceps, and semimembranosus.

Name the ligaments of the hip.
(1) The capsular ligament, the basis of the joint ; (2) the transverse ligament ; (3) the ligamentum teres, an interarticular structure ; (4) the cotyloid, a circumferential cartilage.

What can you say of the capsular ligament?
It is one of the strongest ligaments in the body. It is attached to the anterior intertrochanteric line in front ; posteriorly, it is attached to the back of the neck, one-half of an inch above the posterior intertrochanteric line.

What can you say of the ilio-femoral band?

This is the strongest band. Its apex is attached below the anterior inferior iliac spine; its base is attached to the anterior intertrochanteric line.

Describe the ligamentum teres.

Externally, it is attached to the lips of the cotyloid notch; between these extremes, fibres spring from the transverse ligament. Internally, the ligament is attached to the depression in the head of the femur called the tereal.

Name all the bony parts about the hip-joint. (Fig. 226.)

The greater and lesser trochanters, the neck of the femur, the anterior and posterior intertrochanteric lines, the anterior inferior spinous process of the ilium, the ilio-pubal eminence and line, the tuber of the ischium, the ischio-pubic ramus.

THE KNEE-JOINT.

1. *Class.—Diarthrosis,* because of free movement and constant synovia.
2. *Subdivision.—Ginglymus,* because of motion in two directions.
3. *Technical Name.—*Tibio-femoral articulation.
4. *Osteological Units.—*Femur, tibia, and patella.
5. *Subdivisional Parts.—*Femoral condyles, tibial tuberosities.
6. *Articular Surfaces.—*Named according to general rule.
7. *Basis.—*A capsule lined by synovial membrane.
8. *Local Subdivisions.—*Internal, external, anterior, posterior.
9. *Strengthening Bands.—*Biceps, sartorius, semimembranosus.
10. *Incorporated Tendons.—*Peroneus longus, adductor magnus.
11. *Bony Limitations.—*Patella, tibia, and femur.
12. *Nerve-supply.—*Great sciatic, anterior crural, obturator.
13. *Blood-supply.—*Articular arteries from the internal and external articular branches of the popliteal artery, both above and below the joint; from the azygos; from the recurrent branches of both the anterior and posterior tibial arteries; from the anastomotica femoris.
14. *Ligamentous Muscles.—*(1) Extensor quadriceps femoris; (2) gracilis and sartorius; (3) semitendinosus and semimembranosus and biceps; (4) gastrocnemius and plantaris.

Name all the ligaments of the knee-joint.

(1) The fibrous expansion of the extensors (strengthening bands); (2) capsular or anterior ligament; (3) posterior ligament or ligament of Winslow; (4) external lateral ligament; (5) internal lateral ligament; (6) ligamentum patellæ; (7) anterior crucial ligament; (8) posterior crucial ligament; (9) internal semilunar fibro-cartilages; (10) external semilunar fibro-cartilages; (11) the coronary ligament; (12) the transverse ligament.

What is the ligamentum patellæ?

It is a strong tendon by which the extensors of the leg are inserted into the tubercle of the tibia. Its borders form a guide to the surgeon in injecting and aspirating the cavity of the synovial membrane at the knee. Behind the tendon, between this and the true capsule, is a mass of fat and a small bursa, resting on the bursal segment of the tubercle of the tibia. In front of the ligamentum patellæ is a large prepatellar bursa, whose enlargement is known as "housemaid's knee." Laterally, the ligamentum patellæ is continuous with the fibrous expansion of the extensor muscles of the leg.

Describe and locate the ligament of Winslow.

It bridges the space between the internal and external lateral ligaments. It has perforations for vessels and nerves. It is strengthened by an aponeurosis of the semimembranosus muscle. It forms part of the floor of the popliteal space. On it rest the popliteal vessels.

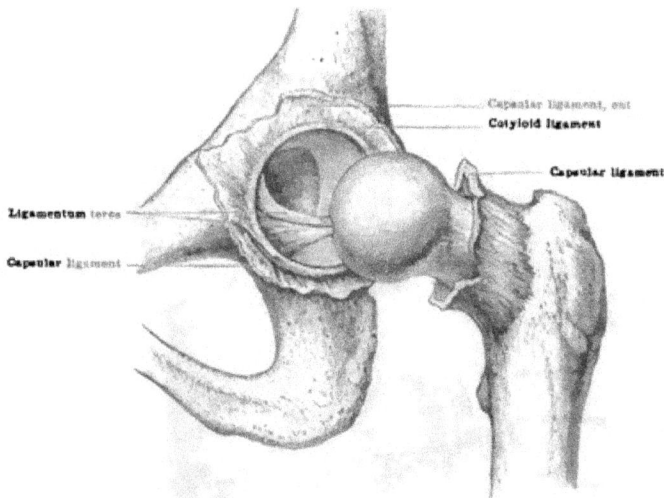

FIG. 309.—HIP-JOINT AFTER DIVIDING THE CAPSULAR LIGAMENT AND DISARTICULATING THE FEMUR.

Capsular ligament, cut
Cotyloid ligament
Capsular ligament
Ligamentum teres
Capsular ligament

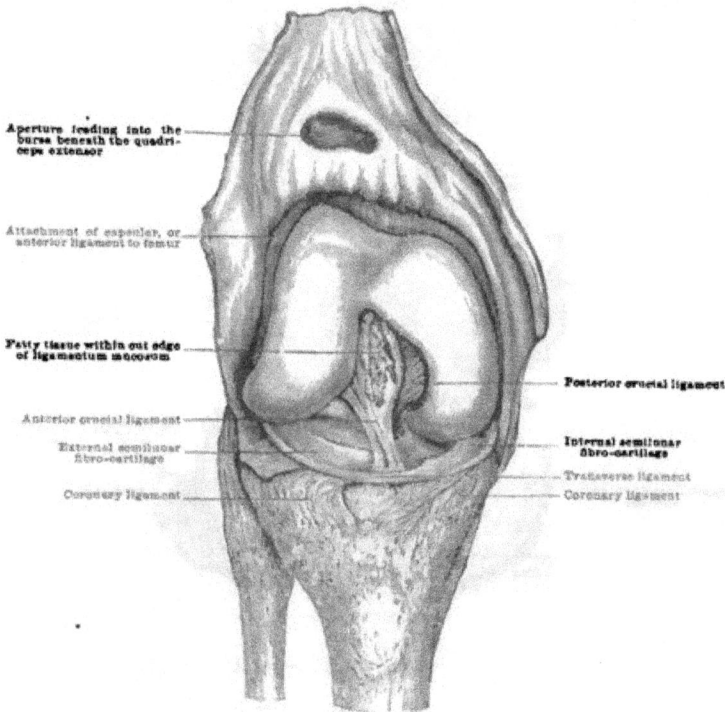

Aperture leading into the bursa beneath the quadriceps extensor
Attachment of capsular, or anterior ligament to femur
Fatty tissue within cut edge of ligamentum mucosum
Anterior crucial ligament
External semilunar fibro-cartilage
Coronary ligament
Posterior crucial ligament
Internal semilunar fibro-cartilage
Transverse ligament
Coronary ligament

FIG. 310.—ANTERIOR VIEW OF THE INTERNAL LIGAMENTS OF THE KNEE-JOINT.

Describe the location, length, and relations of the internal lateral ligament.

It extends from the inner condyle to the tibia, being about three inches in length. Its outer surface is related to the gracilis, the semimembranosus, and the sartorius muscles; its deep surface is in relation with the coronary ligament and

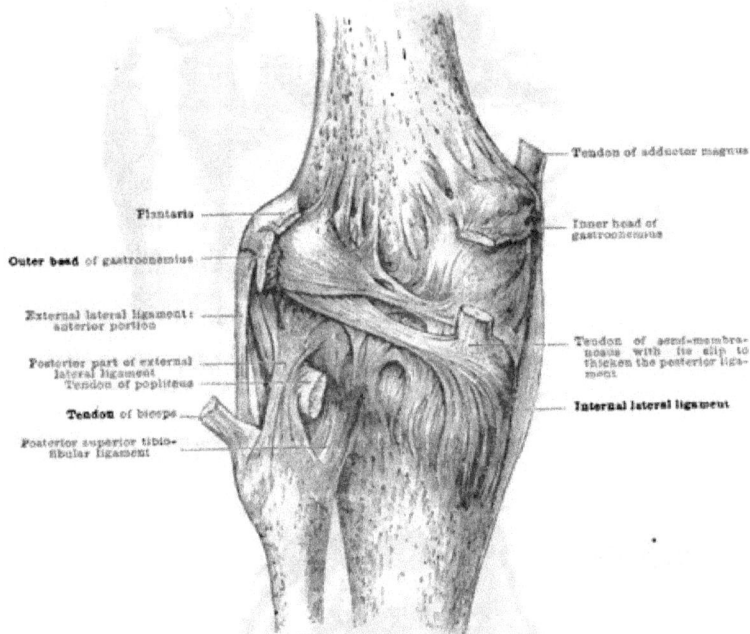

Plantaris

Outer head of gastrocnemius

External lateral ligament: anterior portion

Posterior part of external lateral ligament
Tendon of popliteus

Tendon of biceps

Posterior superior tibio-fibular ligament

Tendon of adductor magnus

Inner head of gastrocnemius

Tendon of semi-membranosus with its slip to thicken the posterior ligament

Internal lateral ligament

FIG. 311.—POSTERIOR VIEW OF THE KNEE-JOINT.

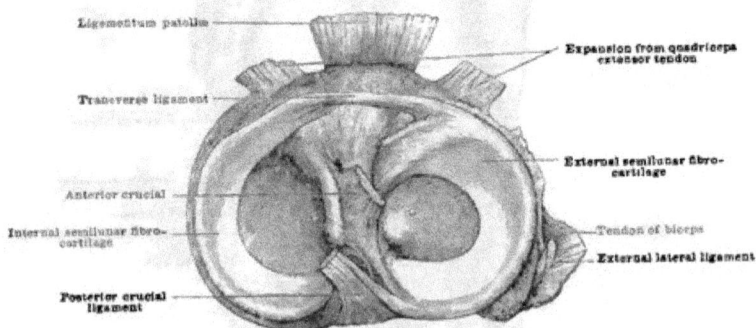

Ligamentum patellæ

Transverse ligament

Anterior crucial

Internal semilunar fibro-cartilage

Posterior crucial ligament

Expansion from quadriceps extensor tendon

External semilunar fibro-cartilage

Tendon of biceps

External lateral ligament

FIG. 312.—STRUCTURES LYING ON THE HEAD OF THE TIBIA.   (Right knee.)

semilunar cartilage. This ligament is the divorced tendon of the adductor magnus muscle.

The external lateral ligament consists of how many parts?

Of two—an anterior and a posterior. (Fig. 311.) The anterior, the longer, extends from the tubercle on the outer femoral condyle to the head of the fibula. Beneath this ligament is the popliteus tendon in its synovial sheath. The posterior portion is inserted into the styloid. It binds the popliteus to the outer tuberosity.

Describe the anterior or capsular ligament proper.

It is thin, and attached to the synovial membrane. It is attached to the femur, patella, popliteus muscle, and internal lateral ligament.

Locate and give function of the semilunar fibro-cartilages, and mention other members of the same physiological class of ligaments.

They are two circumferential cartilages resting on the articular facets of the tibia. In mechanics they correspond to a ball-bearing. They deepen the cavity which they surround. The cotyloid of the hip, the glenoids of the shoulder

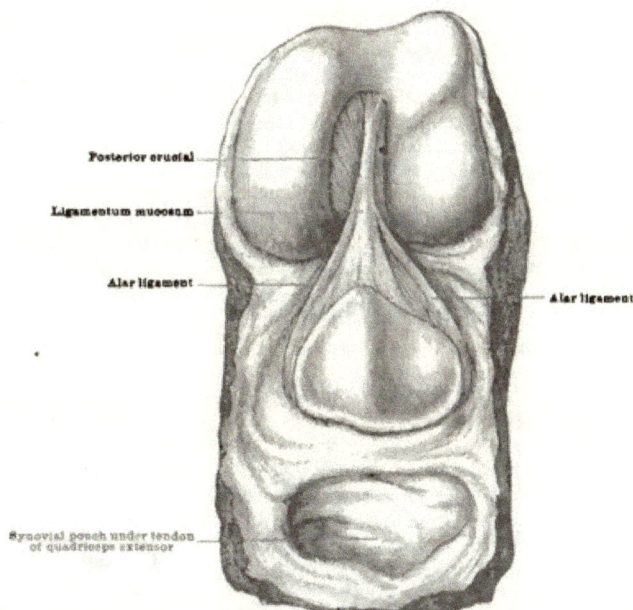

Fig. 313.—Anterior View of the Knee-joint, showing the Synovial Ligaments. (Anterior portion of capsule with the extensor tendon thrown downward.)

and of the interphalangeal and metacarpo-phalangeal articulations belong to the same class.

Name the attachments of the semilunars. (Fig. 312.)

They are attached loosely to the borders of the tuberosities by the coronary ligament ; to each other, in front, by the transverse ligament ; their extremities are attached to the depressions in front and behind the spine of the tibia. (Fig. 312.)

What can you say of the synovial membrane of the knee-joint?

It is the largest in the body. One part forms a sac beneath the patella ; another extends some distance above the attachment of the capsule above the joint ; it covers all the surface of the crucial ligaments except at two parts.

Give the origin and insertion of the crucial ligaments. (Fig. 312.)

The anterior crucial is attached to the tibia in front of the spine, and to the border of the inner facet, and to the inner and back part of the outer condyle. The posterior is attached to the popliteal notch and adjacent structures, and to the outer part of the inner condyle. (Fig. 227.)

Give synonyms for the crucial ligaments.

The anterior is called also external ; the posterior is called also internal.

Define ligamenta alaria and ligamentum mucosum.

The terms are misnomers. These structures are parts only of the synovial membrane. (See Fig. 313.)

NOTE.—The student should endeavor to make his dissection resemble the cuts, and then exercise his knowledge of osteology by minutely describing all the attachments and relations. Drill of this kind will make him familiar with the text, as well as with the cadaver, and be splendid discipline in cultivating descriptive powers—a faculty each one possesses to greater or lesser degree.

Name, locate, and define the varieties of synovial membrane.

1. The articular, which is found in all freely movable joints, as the knee and hip. In the foetus it is said this membrane covers the articular surfaces of the cartilage, as well as the interior of the capsule.

2. The bursal varieties are of two kinds. Mucous bursæ are found between the skin and bone, as the prepatellar. Synovial bursæ are found between tendons and bone, as about the knee between the ligamentum patellæ.

3. The vaginal kind. This variety will be found surrounding the flexor tendons of the digits, in the hand and foot ; surrounding the long tendon of the biceps in the bicipital canal of the humerus under the transverse ligament of this bone.

THE TIBIO-FIBULAR UNION.

1. **The superior tibio-fibular articulation** has a capsule, an anterior and a posterior tibio-fibular *ligament.*

It belongs to the *class* diarthrosis, because of its constant capsule and synovia, and to the *subdivision* arthrodia, because of its gliding movement.

The *synovial cavity* of this joint may communicate with the knee.

The *blood-supply* comes from the external articular and recurrent tibial.

The *nerve-supply* is from the recurrent branch of the external popliteal and the inferior external articular.

2. **The tibio-fibular interosseous membrane** is incomplete above for about an inch. The anterior tibial artery passes through here, between the two heads of the tibialis posticus muscle, to the anterior part of the leg. The function of membrane is principally for muscular origin. It has two borders—the tibial and the fibular ; two surfaces—the anterior and the posterior. The anterior is in relation with the anterior tibial nerve and vessels and with the muscles of the anterior region of the leg. The posterior surface is in relation with the tibialis posticus and the flexor longus hallucis.

3. **The inferior tibio-fibular articulation.** (Fig. 317.)

Class.—Diarthrosis, because of free movement and synovia.

Subdivision.—Arthrodia, because of a gliding movement.

The *ligaments* are anterior, posterior, transverse, and inferior interosseous, all of which are easily found. The synovial membrane is continuous with that of the ankle-joint. The nerve-supply is from the internal or long saphenous, anterior and posterior tibial. The blood-supply is from the peroneal and its anterior branch.

THE ANKLE-JOINT.

1. *Class.*—Diarthrosis, because of a capsule and free movement.
2. *Subdivision.*—Ginglymus, because of movement in two directions.

3. *Technical Name.*—Tibio-fibulo-astragaloid articulation.
4. *Osteological Units.*—Tibia, fibula, and astragalus.
5. *Subdivisional Parts.*—Inner and outer malleoli, astragalus.
6. *Articular Surfaces.*—Named according to the general rule.
7. *Basis.*—A capsule lined by a secreting synovial membrane.
8. *Local Subdivisions.*—Anterior, posterior, internal, external.
9. *Strengthening Bands.*—None.
10. *Incorporated Tendons of Obsolete Muscles.*—None.
11. *Bony Limitations.*—Malleoli of tibia and fibula, and the os calcis.
12. *Nerve-supply.*—Anterior and posterior tibial, and internal saphenous.
13. *Blood-supply.*—Anterior and posterior tibial and peroneal arteries.
14. *Ligamentous Muscles.*—Tibialis anticus, extensor proprius hallucis, extensor communis digitorum, peroneus tertius, peroneus longus, peroneus brevis, tibialis posticus, flexor longus digitorum, flexor longus hallucis, and the three muscles inserted by the tendo Achillis.

Give the attachments of the anterior ligament of the ankle-joint. (Fig. 317.)

It is attached above to the malleoli of the tibia and fibula, to the tibia, and to the anterior inferior tibio-fibular ligament ; below, it is attached to the neck of the astragalus.

Give the relations of the anterior ligament of the ankle-joint.

Anteriorly, it is in relation with the anterior tibial nerve and vessels, and all the muscles of the front of the leg ; posteriorly, with the synovial membrane and a fatty mass of connective tissue.

Give the attachments of the posterior ligament of the ankle-joint. (Fig. 317.)

It is attached above to the external malleolus, the tibia, and the posterior inferior tibio-fibular ligament ; below, to the posterior surface of the astragalus.

Describe the internal lateral ligament. (Fig. 318.)

It is called the deltoid ligament. It is attached above to the lower border of the internal malleolus ; below, to the astragalus, the sustentaculum tali, and the calcaneo-astragaloid ligament. Externally, this ligament is in relation with the tibialis posticus and flexor longus digitorum muscles ; internally, with the synovial membrane.

Describe the external ligament of the ankle-joint. (Fig. 317.)

It has three fasciculi, or bundles, two of which are horizontal and one vertical. The anterior bundle extends from the fibular malleolus to the astragalus in front ; the posterior bundle from the malleolus to the posterior surface of the astragalus ; the middle bundle from the malleolus to the outer surface of the os calcis.

Give the relations of the external ligament of the ankle-joint.

Externally, with the tendons of the peroneus longus and peroneus brevis muscles ; internally, with the synovial membrane.

Describe the movements in the ankle-joint.

This joint is a hinge ; hence it has movement in two directions, as in ordinary walking. If, however, you extend the foot fully, you will find considerable lateral motion. Flex the foot to the fullest extent, and no lateral motion is possible. Study the diameters of the articular part of the astragalus, and also the diameters of the intermalleolar slot, in which the astragalus works, and you will see reason for free lateral mobility of the foot in extreme extension, in the difference in these diameters.

What can you say of the synovial membrane of the ankle-joint?

It lines the ligaments, and is said by Morris to secrete more synovia than any other synovial membrane.

THE TARSUS—TARSAL ARTICULATIONS.

Define the word tarsus, and name its homologue in the upper extremity.
The word means instep; its homologue is the carpus, which means wrist.

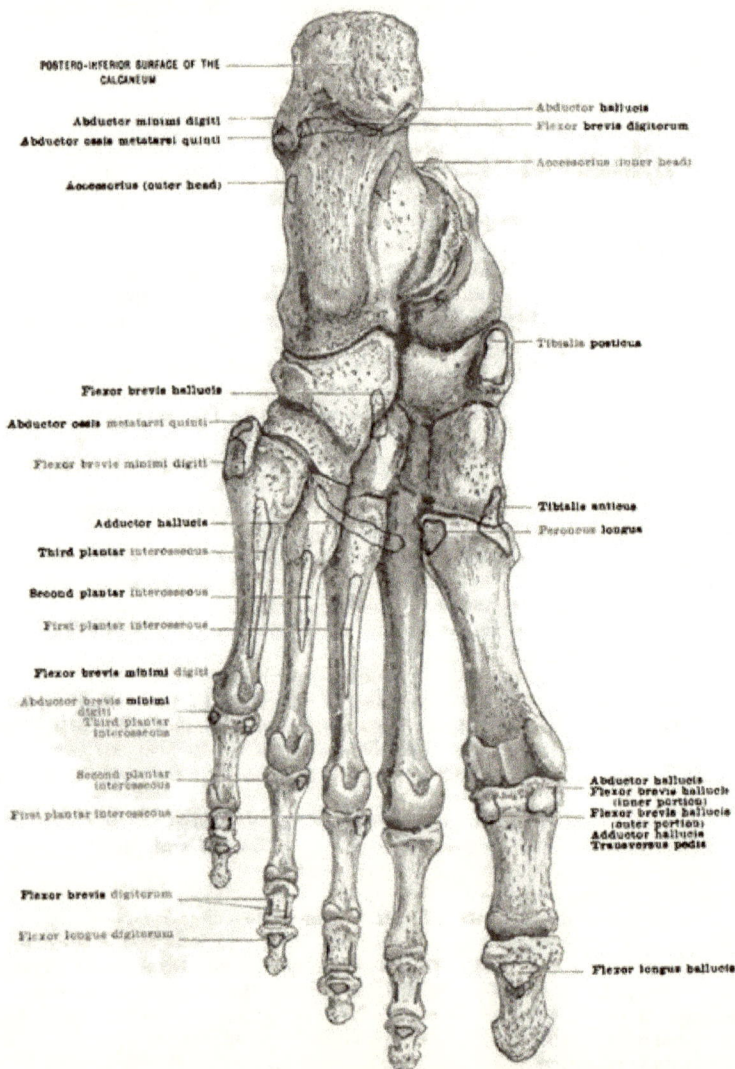

POSTERO-INFERIOR SURFACE OF THE CALCANEUM

Abductor minimi digiti

Abductor ossis metatarsi quinti

Accessorius (outer head)

Flexor brevis hallucis

Abductor ossis metatarsi quinti

Flexor brevis minimi digiti

Adductor hallucis

Third plantar interosseous

Second plantar interosseous

First plantar interosseous

Flexor brevis minimi digiti

Abductor brevis minimi digiti

Third plantar interosseous

Second plantar interosseous

First plantar interosseous

Flexor brevis digitorum

Flexor longus digitorum

Abductor hallucis

Flexor brevis digitorum

Accessorius (inner head)

Tibialis postica

Tibialis antica

Peroneus longus

Abductor hallucis
Flexor brevis hallucis (inner portion)
Flexor brevis hallucis (outer portion)
Adductor hallucis
Transversus pedis

Flexor longus hallucis

FIG. 314.—THE LEFT FOOT. (Planter surface.)
(Study the insertion and location of tendons on this and compare your dissection therewith.)

To what class do the tarsal bones belong? To what is their shape adapted?
They are classified as short bones; their function is to combine great strength with slight motion.

Name the bones of the tarsus and indicate their fanciful derivation.
(1) Astragalus, the dice bone : (2) calcaneum, the heel bone ; (3) scaphoid,

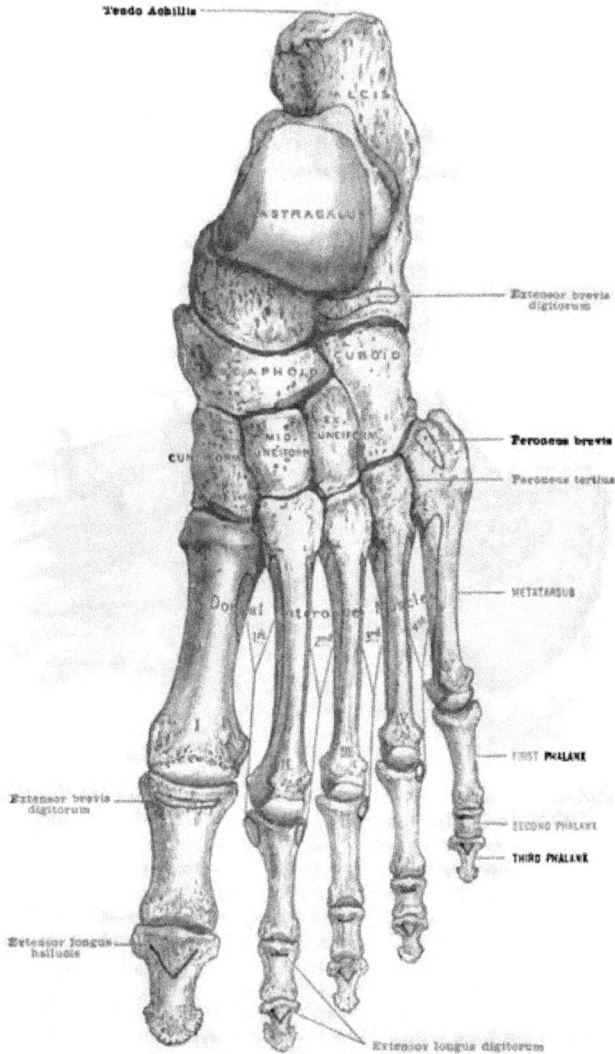

FIG. 315.—THE LEFT FOOT (Dorsal surface.)
(Study origin and insertion of muscles on this figure and compare with your dissection.)

resembling a boat ; (4) cuboid, like a cube ; (5) cuneiform, like a wedge. The
ancient shepherds threw dice. The bones they used were from the sheep, hence
our name, astragalus, or dice bone.

Describe the astragalus. (Fig. 315.)

This bone has (1) a head, which articulates with the scaphoid bone; (2) a neck, by which the head is joined to the body of the bone; (3) a body, which has superior, inferior, internal, external, and posterior surfaces.

Give the importance of each surface of the astragalus.

(1) The superior surface articulates with the tibia; (2) the external surface articulates with the fibular malleolus; (3) the internal surface articulates with the tibial malleolus; (4) the inferior surface articulates with the calcaneum; (5) the posterior surface has a groove for the flexor longus hallucis.

Describe the calcaneum or heel-bone. (Fig. 315.)

This bone has six surfaces—anterior, posterior, superior, inferior, internal, and external.

Give the importance of each surface.

(1) The anterior surface articulates with the cuboid bone; (2) the posterior surface has the tendo Achillis attached to it; (3) the external surface has two peroneal grooves and a tubercle between them; (4) the internal surface has the

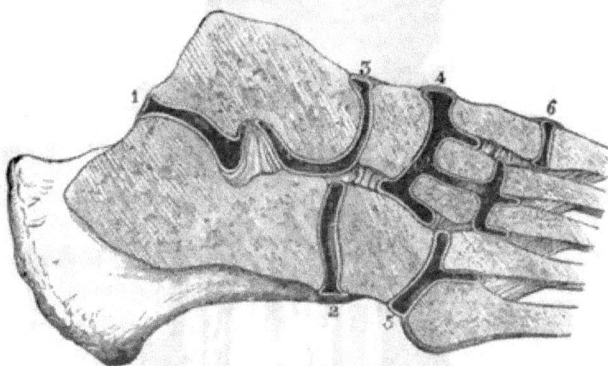

FIG. 316.—SECTION TO SHOW THE SYNOVIAL CAVITIES OF THE FOOT.

1. Posterior calcaneo-astragaloid.   2. Calcaneo-cuboid.   3. Anterior calcaneo-astragalo-scaphoid.
4. Tarsal.   5. Cubo-metatarsal.   6. First metatarso-cuneiform.

sustentaculum tali and a groove; (5) the inferior surface has the three important tubercles; (6) the superior surface has an astragaloid articular surface.

Name and give importance of the tubercles of the calcaneum. (Fig. 314.)

(1) The inner tubercle gives origin to the abductor hallucis, the flexor brevis digitorum, and the abductor minimi digiti; (2) the outer tubercle gives origin to the abductor minimi digiti, and to the inconstant abductor ossis metatarsi quinti; (3) the anterior tubercle gives attachment to the short plantar or calcaneo-cuboid ligament.

Describe the scaphoid bone. (Fig. 316.)

This bone articulates posteriorly with the head of the astragalus; anteriorly with the three cuneiform bones. The superior, inferior, and external surfaces are rough, for the attachment of ligaments. The internal surface has the tuberosity, into which is inserted the tibialis posticus muscle.

Describe the cuboid bone. (Fig. 316.)

It has six surfaces: (1) The anterior articulates with the bases of the fourth and fifth metatarsals; (2) the posterior articulates with the calcaneum; (3) the internal articulates with the scaphoid (sometimes) and external cuneiform; (4) the superior is rough, for attachment of ligaments; (5) the inferior has a groove for the peroneus longus ; (6) the outer is almost negative, and is continuous with the inferior surface.

Can you say anything more about the inferior surface of the cuboid bone?

Yes ; it has a ridge behind the peroneal groove, for the attachment of the long calcaneo-cuboid ligament ; this ridge terminates externally in the sesamoid articular surface, which articulates with the peroneal sesamoid bone, in the tendon of the peroneus longus. The flexor brevis hallucis also derives its origin on this surface of the bone. (Fig. 314.)

Describe the surfaces of the internal cuneiform.

(1) The internal surface has the insertion of the tibialis anticus ; (2) the external articulates with the second metatarsal and middle cuneiform ; (3) the anterior

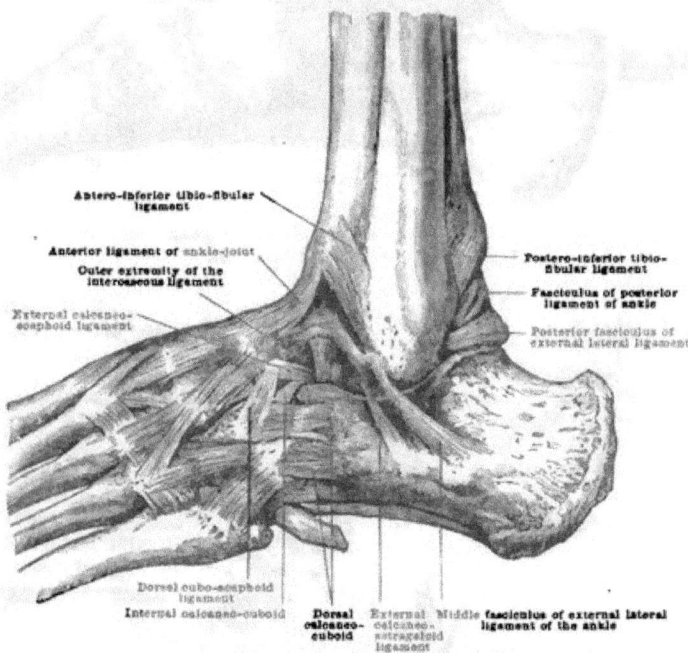

Antero-inferior tibio-fibular ligament

Anterior ligament of ankle-joint
Outer extremity of the interosseous ligament

External calcaneo-scaphoid ligament

Postero-inferior tibio-fibular ligament
Fasciculus of posterior ligament of ankle

Posterior fasciculus of external lateral ligament

Dorsal cubo-scaphoid ligament
Internal calcaneo-cuboid

Dorsal calcaneo-cuboid   External calcaneo-astragaloid ligament   Middle fasciculus of external lateral ligament of the ankle

FIG. 317.—EXTERNAL VIEW OF THE LIGAMENTS OF THE FOOT AND ANKLE.

articulates with the first metatarsal bone ; (4) the posterior articulates with the scaphoid bone ; (5) the inferior has the insertion of the tibialis posticus.

Describe the surfaces of the middle cuneiform.

(1) The anterior articulates with the second metatarsal bone ; (2) the posterior articulates with the scaphoid bone ; (3) the internal articulates with the internal cuneiform bone ; (4) the external articulates with the external cuneiform bone ; (5) the superior surface is rough, for ligamentous attachment ; (6) into the inferior surface is inserted the tibialis posticus muscle and ligaments.

Describe the surfaces of the external cuneiform.

(1) The anterior articulates with the third metatarsal ; (2) the posterior with the scaphoid bone ; (3) the internal with the middle cuneiform bone ; (4) the external with the fourth metatarsal and cuboid bone ; (5) the inferior is for tibialis posticus and ligaments ; (6) the superior is for ligamentous attachment.

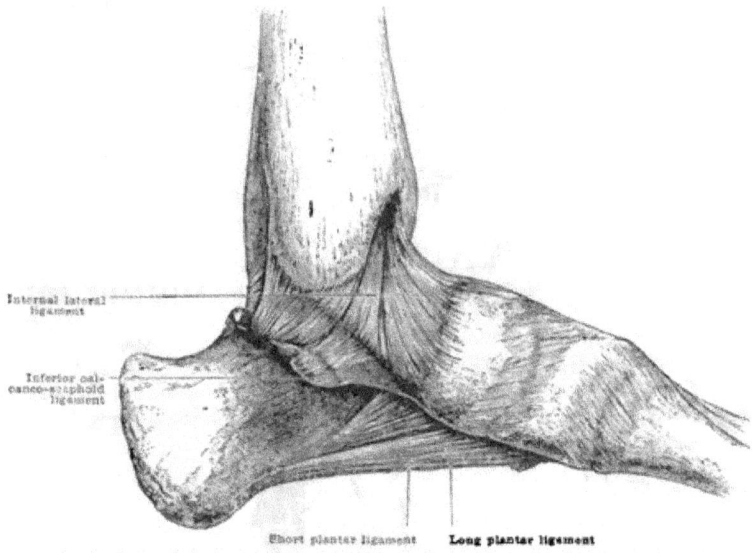

Internal lateral
ligament

Inferior cal-
caneo-scaphoid
ligament

Short plantar ligament   **Long plantar ligament**

FIG. 318.—INNER VIEW OF THE ANKLE AND THE TARSUS, SHOWING THE GROOVE FOR THE
TENDON OF THE TIBIALIS POSTICUS.

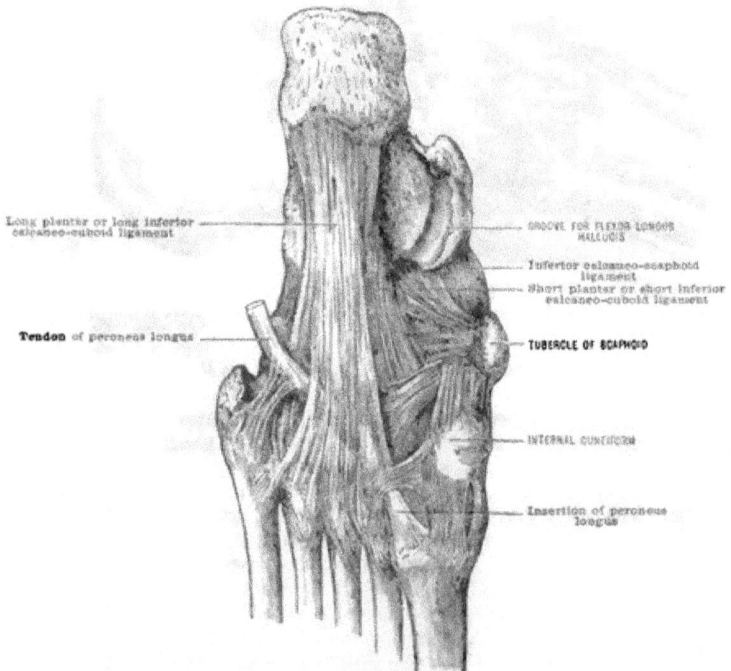

Long plantar or long inferior
calcaneo-cuboid ligament

GROOVE FOR FLEXOR LONGUS
HALLUCIS

Inferior calcaneo-scaphoid
ligament
Short plantar or short inferior
calcaneo-cuboid ligament

Tendon of peroneus longus

TUBERCLE OF SCAPHOID

INTERNAL CUNEIFORM

Insertion of peroneus
longus

FIG. 319.—LIGAMENTS OF THE SOLE OF THE LEFT FOOT.

Give comparative size of the cuneiform bones.

The internal cuneiform is the largest bone; the middle cuneiform is the smallest bone; the external cuneiform is intermediate in size.

Name the principal muscular-traction **points** *in and about the tarsus, and indicate the muscle attached thereto.*

(1) Calcaneum for the tendo Achillis; (2) calcaneum for the abductor hallucis and **flexor** brevis digitorum; (3) calcaneum for the abductor minimi digiti and **accessorius**; (4) calcaneum for the extensor brevis digitorum; (5) sustentaculum tali groove for the tendon of flexor longus hallucis; (6) the tuberosity of the scaphoid bone for the tibialis posticus; (7) the internal cuneiform for the tibialis **anticus**; (8) the **first** metacarpal **for** the peroneus tertius and tibialis anticus; (9) the **fifth** metatarsal for the **peroneus** brevis; (10) cuboidal groove for the tendon of peroneus longus.

Locate the medio-tarsal articulation by limitation, and name its ligaments.

This joint is limited posteriorly by the calcaneum and astragalus; anteriorly, by the scaphoid and cuboid. (Fig. 316.) The ligaments are: (1) The external calcaneo-scaphoid; (2) the inferior calcaneo-scaphoid; (3) the astragalo-scaphoid ligament; (4) the internal or interosseous calcaneo-cuboid ligament; (5) the long inferior calcaneo-cuboid ligament; (6) the short inferior calcaneo-cuboid ligament; (7) the dorsal calcaneo-cuboid ligament.

Give the attachments of the long inferior calcaneo-cuboid ligament and the common name for the same. (Fig. 319.)

It is also called the long plantar ligament. It is attached to the whole inferior surface of the calcaneum behind; to the cuboid and bases of the metatarsals in front.

Can you mention any other important fact in regard to the long plantar ligament?

Yes; it completes the canal for the peroneus longus muscle, and also gives origin to the adductor hallucis and the flexor brevis minimi digiti. (Fig. 319.)

Give the attachments of the short inferior calcaneo-cuboid ligament and its common name.

It is called the short plantar ligament. It is attached to the under surfaces of the calcaneum and cuboid. (Fig. 319.)

From what source does the medio-tarsal joint receive its nerve-supply?

From the outer division of the anterior tibial, the musculo-cutaneous, and the external plantar nerves.

From what source does the medio-tarsal joint receive its blood-supply?

From the anterior tibial, the dorsalis pedis, and the internal and external plantar arteries.

What class and subdivision does the medio-tarsal articulation belong in?

(1) The astragalo-scaphoid joint is diarthrosis by class, and enarthrodia by subdivision; (2) the calcaneo-cuboid is diarthrosis by class, and saddle arthrodia by subdivision.

Name the ligaments of the calcaneo-astragaloid articulation.

(1) The external calcaneo-astragaloid ligament; (2) the internal calcaneo-astragaloid ligament; (3) the posterior calcaneo-astragaloid ligament; (4) the interosseous calcaneo-astragaloid ligament.

Locate the interosseous ligament.

It extends from the calcaneum to the astragalus, connecting the parallel oblique grooves of these two bones.

Describe the **scapho-cuneiform articulation.**

Class.—Diarthrosis, because of free movement and a capsule.

Subdivision.—Arthrodia, because of a gliding movement.

Ligaments.—Dorsal, plantar, and internal.

Describe the **intercuneiform articulations.**
Class.—Diarthrosis, because of free movement and a capsule.
Subdivision.—Arthrodia, because of a gliding movement.
Ligaments.—Dorsal, plantar, and interosseous.
Blood-supply.—Metatarsal and plantar arteries.
Nerve-supply.—Anterior tibial and plantar nerves.
Describe the **tarsometatarsal articulations.**
Class.—Diarthrosis, because of free movement and a capsule.
Subdivision.—Arthrodia, because of a gliding movement.
Ligaments.—Dorsal, palmar, and interosseous.
Nerve-supply.—Anterior tibial and plantar nerves.

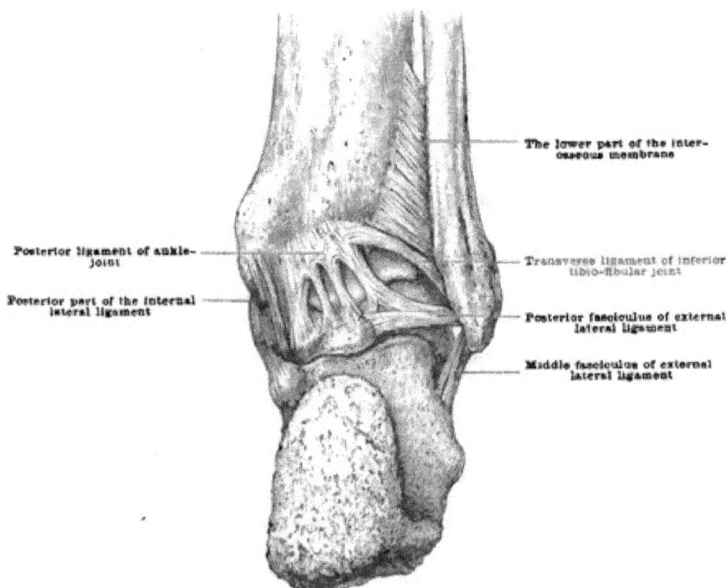

FIG. 320.—LIGAMENTS SEEN FROM THE BACK OF THE ANKLE-JOINT.

Describe the **intermetatarsal articulations.**
Class.—Diarthrosis, because of free movement and a capsule.
Subdivision.—Arthrodia, because of a gliding movement.
Ligaments.—Dorsal, plantar, and interosseous.
How are the heads of the metatarsal bones connected?
By rather strong, short, transverse ligaments.
Describe the **metatarso-phalangeal articulations.**
Class.—Diarthrosis, because of free movement and a capsule.
Subdivision.—Condylarthrosis, because of no axial rotation.
Ligaments.—Lateral, dorsal, plantar, sesamoid.
Ligamentous Muscles.—Flexors and extensors of the digits.
Describe the **interphalangeal articulations.**
Class.—Diarthrosis, because of free movement.
Subdivision.—Ginglymus, because of motion in two directions.

Ligaments.—Laterals, dorsals, and glenoids.
Ligamentous Muscles.—Flexors and extensors of the digits.
Nerve-supply.—Internal and external plantar nerves.

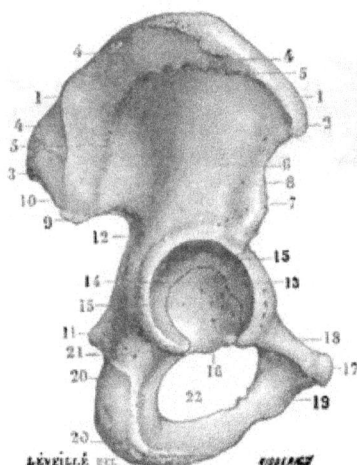

FIG. 321.—RIGHT INNOMINATE BONE, EXTERNAL ASPECT.

1. Superior border, or crest. 2. Anterior superior iliac spine. 3. Posterior superior iliac spine. 4, 4, 4.
Superior curved line. 5, 5. Inferior curved line. 6. Surface between inferior curved line and
acetabulum. 7. Anterior inferior iliac spine. 8. Anterior interspinous notch. 9. Posterior inferior
iliac spine. 10. Posterior interspinous notch. 11. Spine of ischium. 12. Great sacro-sciatic notch.
13. Acetabulum. 14. Fundus of acetabulum. 15, 15. Circumference of acetabulum. 16. Cotyloid
notch. 17. Spine of pubes. 18. Horizontal branch of pubes. 19. Descending branch of pubes.
20, 20. Ischium. 21. Groove for tendon of obturator externus muscle. 22. Obturator foramen.

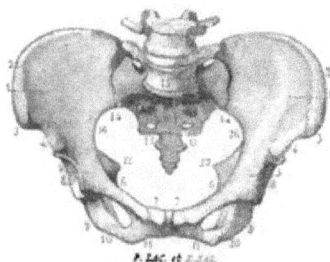

FIG. 322.—PELVIS, ANTERO-SUPERIOR VIEW, SUPERIOR STRAIT.

1, 1. Internal iliac fossæ. 2, 2. Iliac crests. 3, 3. Anterior superior iliac spines. 4, 4. Anterior inferior
iliac spines. 5, 5. Ilio-pectineal eminences. 6, 6. Horizontal branches of pubes. 7, 7. Bodies and
symphysis of pubes. 8, 8. Acetabula. 9, 9. Tuberosities of ischia. 10, 10. Ascending rami of
ischium. 11, 11. Descending rami of pubes. 12, 12. Spines of ischia. 13, 13. Posterior wall of
pelvic cavity. 14, 14. Sacro-iliac symphyses. 15. Sacro-vertebral angle. 16, 16. Superior strait.

THE PELVIS—ITS ARTICULATIONS AND LIGAMENTS.

The bones of the pelvis are the sacrum, coccyx, and ossa innominata. The
pelvis articulates with the lumbar part of the vertebral column, and with the
head of the femur. We may then classify the articulations, for purposes of convenience, as (1) intrinsic, where the articulation is between pelvic bones ; (2) extrinsic, where the articulation is between pelvic bones and bones not pelvic.

The intrinsic pelvic articulations are :

(1) The sacro-iliac, between the sacrum and ilium ; (2) the sacro-sciatic, between the sacrum and ischium ; (3) the interpubic, between the pubic bones ; (4) the sacro-coccygeal, between the sacrum and coccyx.

The extrinsic ligaments are :

(1) The sacro-lumbar, between the sacrum and the last vertebra of the lumbar region ; (2) the femoro-innominate group of ligaments.

Name the osteological points of interest in and about the pelvis.

(1) Symphysis pubis, between the pubic bones ; (2) pubic crest, limited by the angle and spine of pubis ; (3) ilio-pubic line, limited by the spine and ilio-pubic eminence ; (4) the ilio-pubic eminence, at the junction of pubis and ilium ;

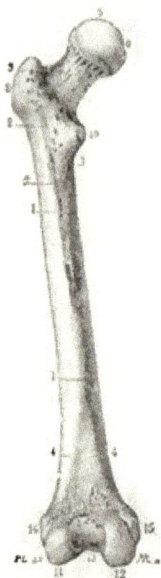

FIG. 323.—FEMUR, POSTERIOR ASPECT.

1, 1. Linea aspera.   2, 2. External division.   3. Internal division.   4, 4. Inferior divisions.   5. Head.
   6. Depression for attachment of round ligament.   7. Neck.   8. Great trochanter.   9. Digital or
   trochanteric fossa.   10. Lesser trochanter   11 Outer condyle.   12. Inner condyle.   13. Intercondyloid notch.   14. Outer tuberosity.   15. Inner tuberosity.

(5) the iliac crest, limited by the superior iliac spines, anterior and posterior ; (6) the acetabulum, at the junction of the ilium, ischium, and pubes ; (7) the fossa acetabuli, the non-articular part of the acetabulum ; (8) the cotyloid notch, an interruption in the acetabular brim ; (9) the lesser sacro-sciatic notch, between the ischial tuber and spine ; (10) the ischial spine, between the greater and lesser sciatic notches ; (11) the tuber of the ischium, one of the grand divisions of the ischium ; (12) the ischio-pubic ramus, between the pubes and ischium ; (13) the obturator foramen, bounded by the pubes and ischium ; (14) the pelvic inlet, between the true and false pelvis ; (15) the subpubic arcade, between the ischio-pubic rami ; (16) the anterior superior iliac spine, for the insertion of Poupart's ligament ; (17) the anterior inferior iliac spine, for the origin of the straight head of the rectus ; (18) the anterior interspinous notch, between the anterior iliac

spines; (19) the posterior iliac spines, superior and inferior; (20) the promontory
of the sacrum; (21) the alæ of the sacrum, consisting of transverse and costal
elemental parts; (22) the anterior sacral foramina, for the anterior sacral nerves;
(23) the sacro-iliac synchondrosis, right and left; (24) the greater and lesser

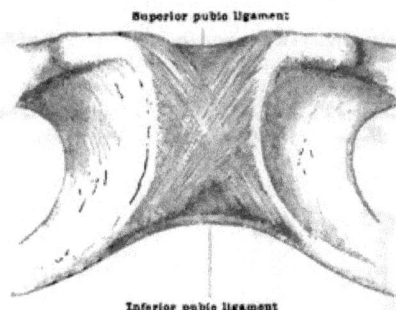

FIG. 324.—ANTERIOR VIEW OF THE SYMPHYSIS PUBIS (FEMALE, SHOWING GREATER WIDTH
BETWEEN THE BONES).

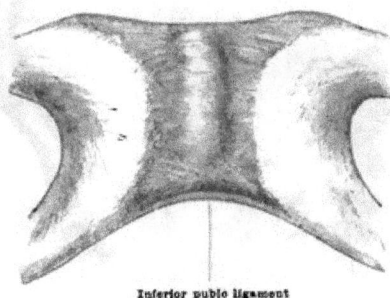

FIG. 325.—POSTERIOR VIEW OF THE SYMPHYSIS PUBIS, SHOWING THE BACKWARD PROJECTION OF
THE SYMPHYSIAL SUBSTANCE AND THE DECUSSATION OF THE FIBRES FROM THE INFERIOR
PUBIC LIGAMENT.

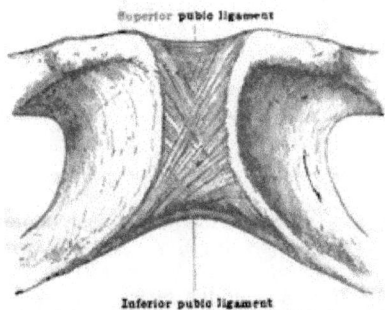

FIG. 326.—ANTERIOR VIEW OF THE SYMPHYSIS PUBIS (MALE) SHOWING DECUSSATION OF THE
FIBRES OF THE ANTERIOR LIGAMENT.

trochanters of the femur; (25) the anterior and posterior intertrochanteric lines;
(26) the digital or trochanteric fossa; (27) the superior and inferior cervical
tubercles of the femur.

Symphysis Pubis:

Class.—Amphiarthrosis, because of union by cartilage and slight motion.

Osteological Elements.—The bodies of the two pubic bones.

Ligaments.—Anterior, posterior, superior, inferior, and interosseous cartilage.

Is there any difference between this joint in the two sexes?

The joint is shorter and broader in the female than in the male. (Fig. 324.)

Is the nerve-supply of the symphysis pubis well understood?

No; but it probably is derived from the ilio-inguinal, ilio-hypogastric, and internal pudic nerves.

From what source does this articulation derive its blood-supply?

From the internal pudic, the obturator, deep epigastric, internal circumflex, iliac, and external pudic arteries.

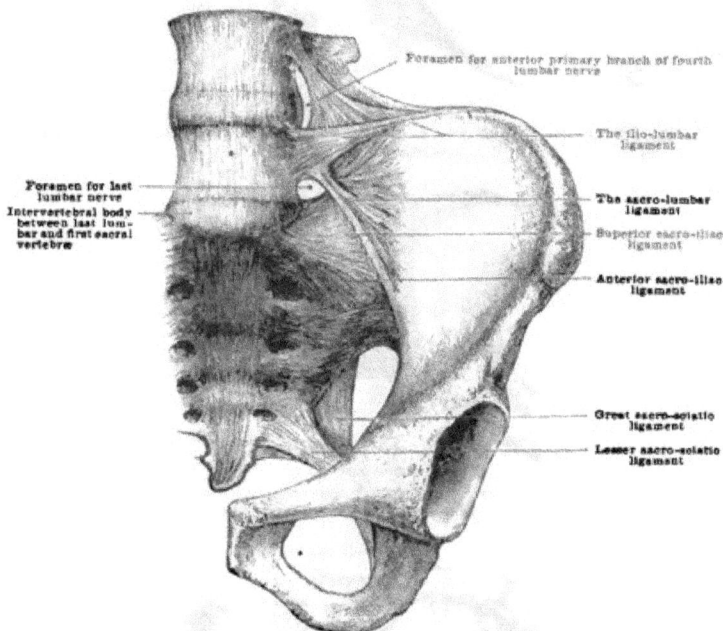

Foramen for anterior primary branch of fourth lumbar nerve

The ilio-lumbar ligament

Foramen for last lumbar nerve

Intervertebral body between last lumbar and first sacral vertebræ

The sacro-lumbar ligament

Superior sacro-iliac ligament

Anterior sacro-iliac ligament

Great sacro-sciatic ligament

Lesser sacro-sciatic ligament

FIG. 327.—ANTERIOR VIEW OF THE PELVIS.

What influence has pregnancy on the interosseous cartilage between the pubic bones?

The cartilage becomes softer and the blood-supply more abundant.

Is there any special difference in the interosseous cartilage in the male and female?

It is thicker in the female than in the male.

What is symphysotomy or symphysiotomy?

Cutting through the pubic symphysis to increase the conjugate diameters of the pelvic canal. In this obstetric operation all the ligaments of the symphysis pubis are cut.

Sacro-iliac Articulation and Its Ligaments:

Class.—Amphiarthrosis, bones united by cartilage and slight motion.

Osteological Units.—The sacrum and ilium.

Subdivisional Parts.—Auricular surfaces of the sacrum and ilium.

Nerve-supply.—Superior gluteal, sacral plexus, first and second sacral nerves.

Blood-supply.—Gluteal, ilio-lumbar, and lateral sacral arteries.

Morphologically, what is that part of the sacrum that articulates with the auricular part of the ilium?

It is the result of a peculiar modification of the costal elements of the first three sacral vertebræ.

Name and locate the ligaments of this articulation.

(1) Anterior sacro-iliac, from pelvic brim to greater sacro-sciatic foramen; (2) posterior sacro-iliac, from back of sacrum to iliac crest; (3) superior sacro-iliac, from sacral base to iliac fossa; (4) inferior sacro-iliac, from sacrum to posterior iliac spine; (5) interosseous ligament is between the two spines.

The function of the sacro-iliac ligaments is to bind the sacrum to the ilium.

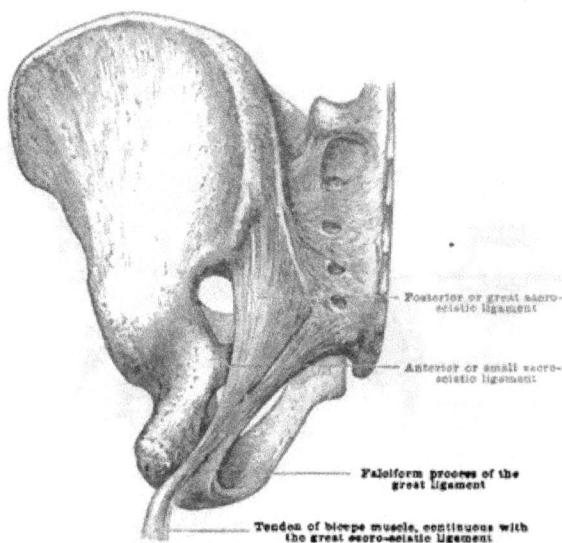

FIG. 328.—SACRO-SCIATIC LIGAMENTS. (Posterior view.)

They are purely of periosteal derivation. The plate of cartilage between the bones may become ossified, just as the ligaments of the vertebral column sometimes do. Occasionally, a distinct synovial cavity is found in the center of this cartilage.

Sacro-ischiatic Ligaments :

How many ligaments unite the sacrum and ischium? (Fig. 328.)

Two; the common name for these is sacro-sciatic, which is a contraction of the true word for purposes of euphony. The common names are: (1) Greater or posterior sacro-sciatic ligament; (2) lesser or anterior sacro-sciatic ligament.

Locate the greater ligament.

The greater or posterior sacro-sciatic ligament is attached above to the margin of the sacrum and coccyx and to the posterior inferior iliac spine; below to the inner margin of the tuber of the ischium and ascending ramus of the ischium for about two inches.

30

What is the special name for the two inches of the greater sacro-sciatic ligament that is prolonged on to the inner margin of the ascending ramus of the ischium?

It is called the falciform ligament. (Fig. 328.) This ligament unites with the obturator fascia to strengthen Alcock's canal, in which canal are located the internal pudic nerve and vessels.

What are the functions of the greater sacro-sciatic ligament, exclusive of its special part—the falciform ligament?

It gives partial origin to the gluteus maximus muscle, and converts the lesser sacro-sciatic notch into a foramen.

Name the structures that pass through the lesser sacro-sciatic foramen.

(1) The tendon of the obturator muscle and its nerve; (2) the internal pudic nerve and vessels.

Locate the lesser sacro-sciatic ligament.

The lesser or anterior sacro-sciatic ligament is attached above to the margin

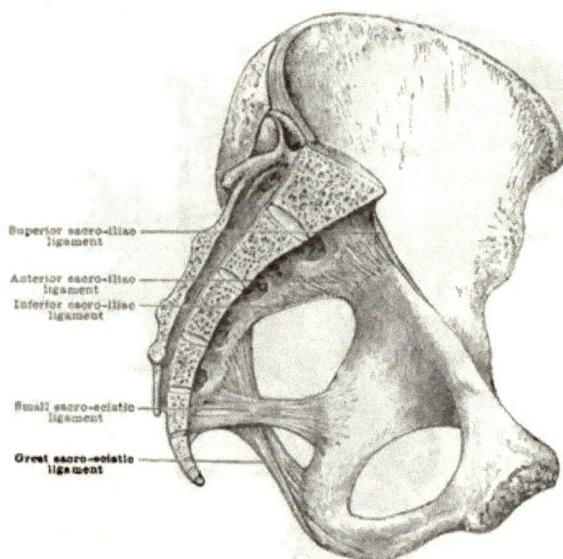

FIG. 329.—VERTICAL ANTERO-POSTERIOR SECTION OF THE PELVIS.

of the sacrum and coccyx; below, to the spine of the ischium. This ligament converts the greater sacro-sciatic notch into a foramen.

Name structures transmitted by the greater sacro-sciatic foramen.

(1) The pyriformis muscle, above which pass out the gluteal nerve and vessels, and below which muscle pass out the sciatic vessels and nerves, the internal pudic nerve and vessels, and the branch of the sacral plexus that supplies the obturator muscle.

Name all the structures that leave the pelvis through the greater sacro-sciatic foramen, cross the ischial spine, and reenter the pelvis through the lesser sacro-sciatic foramen.

(1) The branch of the sacral plexus to the obturator internus; (2) the internal pudic nerve; (3) the internal pudic vessels.

What becomes of these structures on reentering the pelvis?

They enter Alcock's canal, which is a delamination of the obturator fascia strengthened by the falciform ligament, a special part of the greater sacro-sciatic ligament.

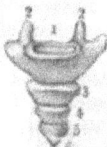

FIG. 330.—COCCYX, ANTERIOR ASPECT.

1. Base.   2, 2. Cornua.   3. Second coccygeal vertebra.   4. Third coccygeal vertebra.   5. Fourth coccygeal vertebra.   6. Fifth coccygeal vertebra.

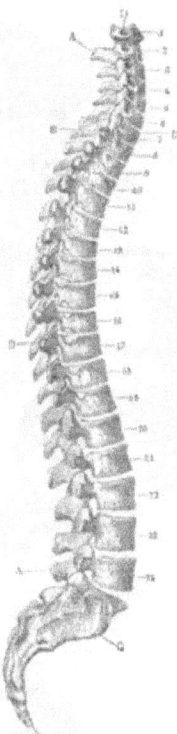

FIG. 331.—VERTEBRAL COLUMN, LATERAL ASPECT.

1-7. Cervical vertebræ.   8-19. Dorsal vertebræ.   20-24. Lumbar vertebræ.   A, A. Spinous processes. B, B. Articular facets of transverse processes of first ten dorsal vertebræ.   C. Auricular surface of sacrum.   D. Foramina in transverse processes of cervical vertebræ.

The Sacro-coccygeal Articulation and Ligaments :

1. *Class.*—Amphiarthrosis, bones united by cartilage.
2. *Osteological Units.*—The sacrum and coccyx.
3. *Subdivisional Parts.*—Sacral and coccygeal cornua.
4. *Nerve-supply.*—The fourth, fifth sacral, and coccygeal nerves.

5. *Blood-supply.*—Middle sacral and lateral sacral artery.

6. *Ligaments.*—Anterior sacro-coccygeal ligament, on the front of bodies; posterior sacro-coccygeal ligament, on posterior part of bodies; supracornual ligament, a prolongation of the supraspinous ligament; intertransverse ligament, analogous to the intertransverse ligaments of the true vertebræ; intervertebral substance of jelly-like cartilage.

The Sacro-vertebral Articulation and its Ligaments :

1. *Class.*—Amphiarthrosis between the bodies of the vertebræ. .
2. *Class.*—Diarthrosis between the articular processes.
3. *Subdivision.*—Arthrodia, because of a gliding movement.
4. *Osteological Units.*—Same as between two vertebræ.
5. *Subdivisional Parts.*—Same as in juxtaposition of vertebræ.

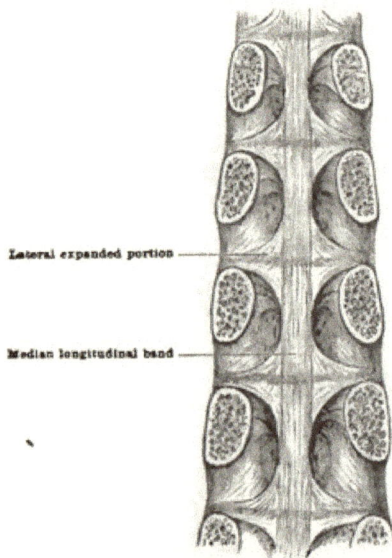

Lateral expanded portion

Median longitudinal band

FIG. 332.—POSTERIOR COMMON LIGAMENT OF THE SPINE. (Thoracic region.)
(Pedicles cut through, and posterior arches of vertebræ removed.)

6. *Nerve-supply.*—Fourth and fifth lumbar and sympathetic.
7. *Blood-supply.*—Lateral sacral, last lumbar, ilio-lumbar.
8. *Special Name.*—Sacro-vertebral angle.
9. *Ligaments of Sacro-vertebral Articulation.*—(1) Prolongation downward of the ligaments of the vertebral column under the same name. They may be modified to some extent to meet the particular demands of a special region; for this modification you can account philosophically: growth is the correlative of function. (2) Sacro-lumbar ligament, triangular in shape. (3) Ilio-lumbar ligament, triangular in shape.

THE VERTEBRAL COLUMN AND ITS LIGAMENTS.

Of what is the vertebral column composed?

Of thirty-three irregular bones, called vertebræ. They are classified as true and false vertebræ. The false vertebræ, nine in number, five sacral and four coccygeal, make by their fusion the sacrum and coccyx.

Name the regions of the vertebral column and give some reasons for a regional subdivision of the column.

The vertebral regions and the number of bones in each are: (1) Cervical region, containing seven bones; (2) dorsal or thoracic region, containing twelve bones; (3) lumbar region, containing five bones; (4) sacro-coccygeal region, containing nine bones. Subdivision of the vertebral column into regions depends on a difference in form of the bones, while this difference in the form depends on the function of the region in which the bones are located; growth predetermines form, and is the correlative of function.

What is the natural arrangement of adjacent vertebræ to each other?

They are in juxtaposition—that is, one next to another, with parallelism of

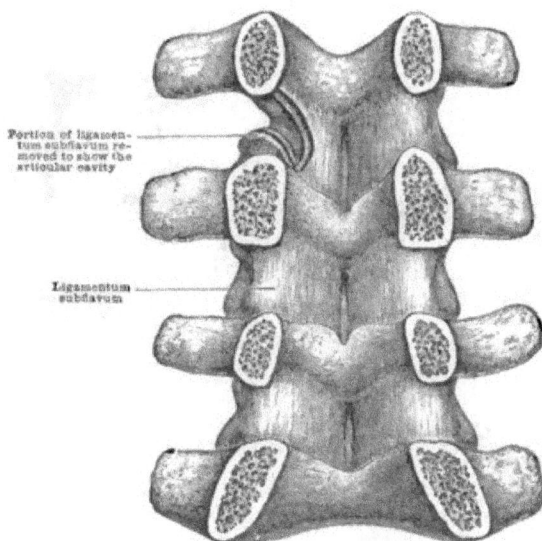

FIG. 333.—LIGAMENTA SUBFLAVA IN THE LUMBAR REGION.

homologous parts. Thus, this arrangement among the bones of the vertebral column gives a logical basis for studying the ligaments of the column, since we find: Juxtaposition of bodies; juxtaposition of pedicles; juxtaposition of laminæ; juxtaposition of articular processes; juxtaposition of spinous processes; juxtaposition of transverse processes.

What is the function of the bodies or centra, and by what ligaments are they held together?

They give solidity and strength to the column. The ligaments of the bodies or centra are the anterior common, posterior common, and intercentral discs of fibro-cartilage. Their union forms an amphiarthrodial articulation, since bone is united to bone by cartilage with limited motion.

Give the function and relations of the pedicles.

The pedicles support the neural arch. Above and below them are found the intervertebral notches, which are converted into intervertebral foramina when the vertebræ are in juxtaposition. (Fig. 335.) These intervertebral foramina transmit the spinal nerves and blood-vessels.

Give the function of the laminæ, name their proper ligaments, and tell wherein these ligaments differ histologically from all others of the vertebral column.

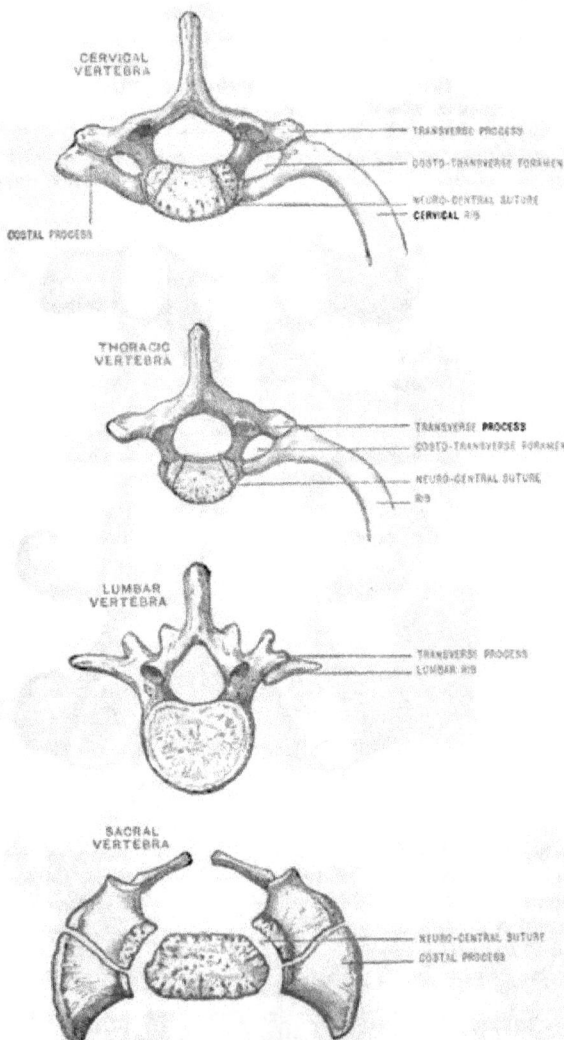

FIG. 334.—MORPHOLOGY OF THE TRANSVERSE AND ARTICULAR PROCESSES.

The laminæ unite posteriorly to enclose the neural canal. They are bound together by the interlaminar ligaments. These are composed of yellow elastic connective tissue, hence they are called the ligamenta subflava. The term subflava means yellowish.

Give the function of the articular processes and tell what kind of joints they form when in juxtaposition.

The articular processes, by their union, make the vertebral column a concrete whole, the separate parts of which column move or turn on each other. They form diarthrodial articulations of the arthrodial subdivision. The basis of the joint is a capsule lined by synovial membrane.

Give the function of the transverse processes and explain their serial morphology in the four regions of the column.

The transverse processes are for the attachment of ligaments, muscles, and either fully developed ribs, or rudimentary ones, called costal elements. The simplest series is in the thoracic region. Here we find a transverse process articulating with a rib in such a manner as to leave a space bounded by a rib, transverse process, and pedicle. This space is a vascular opening, and is called a costo-

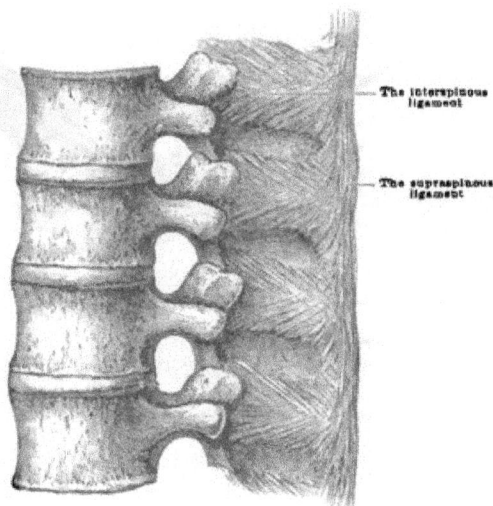

The interspinous ligament

The supraspinous ligament

FIG. 335.—THE INTERSPINOUS AND SUPRASPINOUS LIGAMENTS IN THE LUMBAR REGION.

transverse space. The extremes of the column show this costo-transverse space : a well-formed vertebral canal in the cervical region ; a mere collection of small foramina in the lumbar region ; a depression transmitting a few small vessels in the sacrum, between the costal and the transverse element of the ala of this bone. (Fig. 334.) In the cervical region the rudimentary rib obtains as the anterior tubercle ; in the thoracic region the costal portion is maximal—a rib ; in the lumbar region, the transverse process is suppressed and the costal element is very large ; in the sacral region the costal elements of the first three sacral vertebræ are modified to form the auricular part of the sacrum, for articulation with the ilium.

In view of the great importance of the subject of the serial morphology of the vertebræ in your dissections, I here introduce figure 334, which will illustrate the foregoing, leaving the subject of morphology for the student to read up in Morris.

Give the function of the spinous processes and name their ligaments.

They are for the attachment of muscles and ligaments.   Their ligaments are periosteal, and are named interspinous and supraspinous.   (Fig. 335.)

What is the ligamentum nuchæ?

Literally, the ligament of the nape of the neck.   This is a septum of connective tissue between the muscles of the back of the neck.   Its attachments are the external occipital protuberance (Fig. 338) and the spines of the cervical vertebræ.   It is rudimentary in man—impossible of demonstration in practical anatomy, except as a very feeble intermuscular connective tissue ; not being demonstrable even, in my experience, as an intermuscular septum, such as we find between the peronei muscles and those on the front part of the leg.

Describe the atlas.

This is the first of the cervical series of vertebræ.   It consists of an anterior

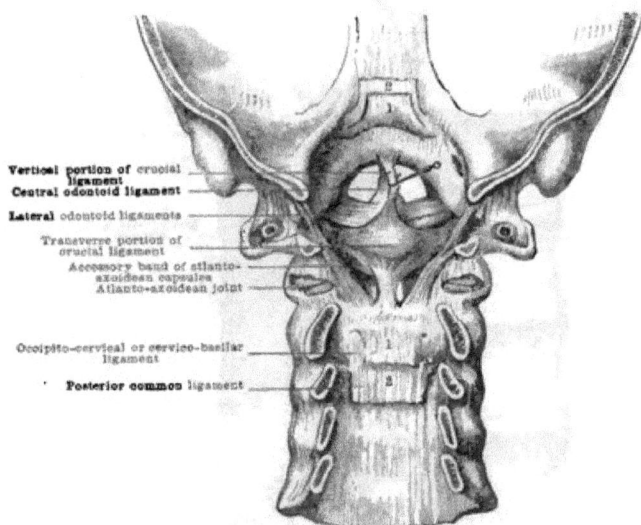

Vertical portion of crucial ligament
Central odontoid ligament
Lateral odontoid ligaments
Transverse portion of crucial ligament
Accessory band of atlanto-axoidean capsules
Atlanto-axoidean joint
Occipito-cervical or cervico-basilar ligament
Posterior common ligament

FIG. 336.—VERTICAL TRANSVERSE SECTION OF THE SPINAL COLUMN AND THE OCCIPITAL BONE TO SHOW LIGAMENTS.

(The cervico-basilar (1), though shown as a distinct stratum, is really the deeper part of the posterior common ligament (2).)

arch, a posterior arch, and two lateral masses.   Morphologically, its divorced or dismembered body obtains as the odontoid process of the axis.

Give composition of the lateral mass.

The lateral mass consists of (1) a superior articular process that articulates with the occipital condyle ; (2) an inferior articular process that articulates with the axis ; (3) costal process—a rudimentary rib ; (4) a transverse process for muscular attachment ; (5) a costotransverse foramen for the vertebral artery ; (6) tubercles for attachment of the transverse ligament.

What do you find at the junction of the lateral mass and the posterior arch?

A groove for the lodgement of the vertebral artery.

What is peculiar about the relation of the articular processes of the axis and atlas to the spinal nerves?

In these vertebræ the nerves issue behind the articular processes ; the remaining spinal nerves issue in front of the articular facets or processes.

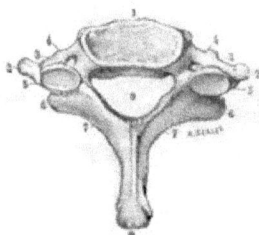

SEVENTH CERVICAL VERTEBRA, POSTERO-SUPERIOR VIEW.

1. Body. 2, 2. Transverse processes. 3, 3. Anterior or costal roots of transverse processes. 4, 4. Foramina for vertebral arteries. 5, 5. Superior articular processes. 6, 6. Inferior articular processes. 7, 7. Laminæ. 8. Spinous process. 9. Spinal foramen.

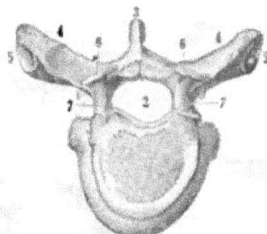

DORSAL VERTEBRA, ANTEROSUPERIOR VIEW.

1. Anterior surface. 2. Vertebral foramen. 3. Spinous process. 4, 4. Transverse processes. 5, 5. Articular surfaces for tubercles of ribs. 6, 6. Superior articular processes. 7, 7. Pedicles.

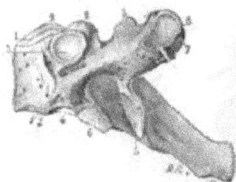

FIRST DORSAL VERTEBRA, LATERAL VIEW.

1. Superior surface of body. 2, 2. Semilunar processes. 3. Articular facet for head of first rib. 4. Demi-facet for head of second rib. 5. Superior articular process. 6, 6. Inferior articular processes. 7. Transverse process. 8. Articular facet for tubercle of first rib. 9. Spinous process.

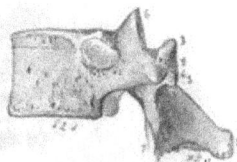

ELEVENTH DORSAL VERTEBRA, LATERAL VIEW.

1. Articular facet for head of eleventh rib. 2. Transverse process. 3. Superior tubercle of transverse process. 4. Inferior and anterior tubercle. 5. Inferior and posterior tubercle. 6. Superior articular process. 7. Inferior articular process. 8. Spinous process.

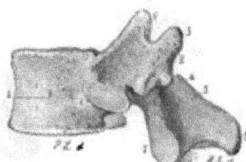

TWELFTH DORSAL VERTEBRA, LATERAL VIEW.

1. Articular facet for head of twelfth rib. 2. Transverse process. 3. Superior and posterior tubercle of transverse process. 4. Inferior and posterior tubercle. 5. Inferior and anterior tubercle. 6. Superior articular process. 7, 7. Inferior articular processes. 8. Spinous process.

FIG. 337.

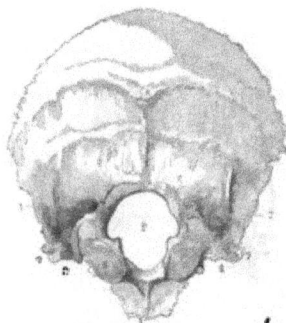

FIG. 338.—OCCIPITAL BONE, POSTERO-INFERIOR VIEW.

1. Basilar process. 2. Foramen magnum. 3, 3. Posterior condyloid foramina. 4. Crest. 5. External occipital protuberance. 6, 6. Condyles. 7, 7. Jugular processes. 8, 8. Jugular fossæ.

PRACTICAL ANATOMY.

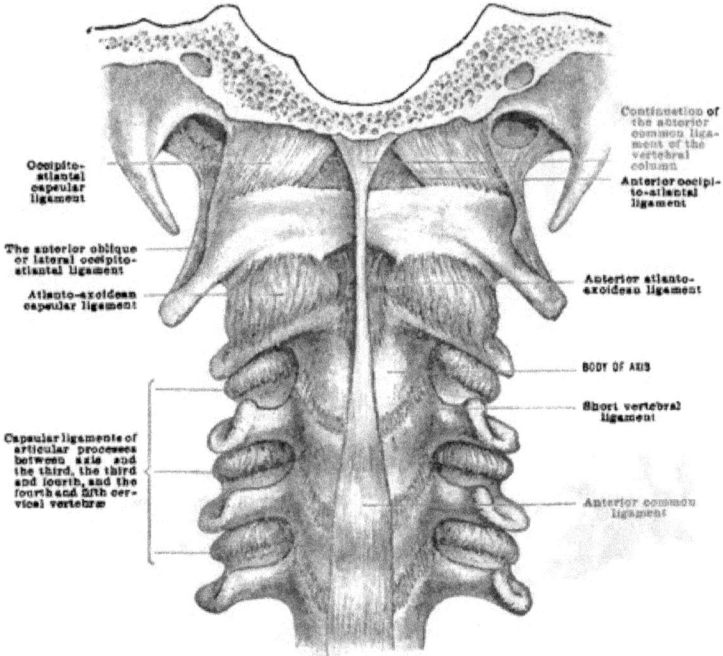

Fig. 339.—Anterior View of the Upper End of the Spine.

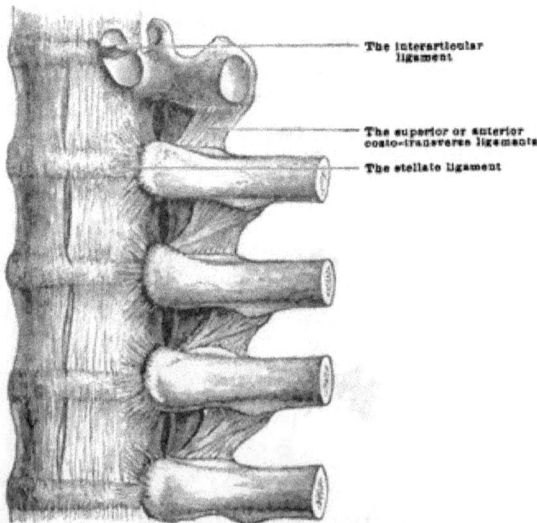

Fig. 340.—Showing the Anterior Common Ligament of the Spine, and the Connection of the Ribs with the Vertebræ.

What is peculiar about the axis?

The odontoid process, the dissociated body of the atlas, is fused to the axis.
Name the ligaments that bind the axis to the atlas.

These ligaments have to do with two distinct classifications :

1. The lateral atlanto-axoidean   { *Class.*—Diarthrosis.
                                  { *Subdivision.*—Arthrodia.

2. The central atlanto-axoidean { *Class.*—Diarthrosis.
                                  { *Subdivision.*—Trochoides.

The ligaments are of periosteal derivation, called: (1) Anterior atlanto-axoidean ; (2) posterior atlanto-axoidean ; (3) two capsular ligaments lined by

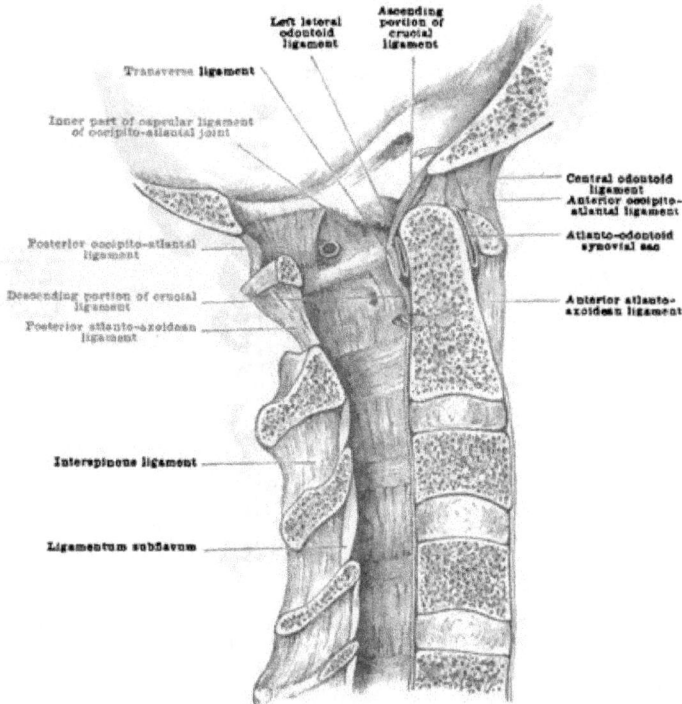

FIG. 341.—VERTICAL ANTERO-POSTERIOR SECTION OF SPINAL COLUMN THROUGH MEDIAN LINE, SHOWING LIGAMENTS.

synovial membrane ; (4) the transverse ligament extending between the tubercles of the lateral masses ; (5) the atlanto-odontoid capsular ligament.

THE ATLANTO-OCCIPITAL ARTICULATION AND LIGAMENTS.

1. *Class.*—Diarthrosis, because of free movement and synovia.
2. *Subdivision.*—Double condylarthrosis.
3. *Technical Name.*—Occipito-atlantal, or atlanto-occipital.
4. *Osteological Units.*—Occipital bone and atlas.
5. *Subdivisions.*—Condyles of occipital bone and lateral masses of atlas.

6. *Nerve-supply.*—Anterior division of the suboccipital nerve.

7. *Blood-supply.*—The vertebral and ascending pharyngeal arteries.

8. *Ligaments of Atlanto-occipital Articulation.*—(1) Anterior occipito-atlantal; (2) posterior occipito-atlantal; (3) two capsular; (4) two anterior oblique.

THE OCCIPITO-AXOID ARTICULATION AND LIGAMENTS.

1. The occipito-cervical.
2. The crucial.
3. Two lateral odontoid or check.
4. The central odontoid or suspensory.

For detailed description of these ligaments the student is referred to "Morris' Anatomy."

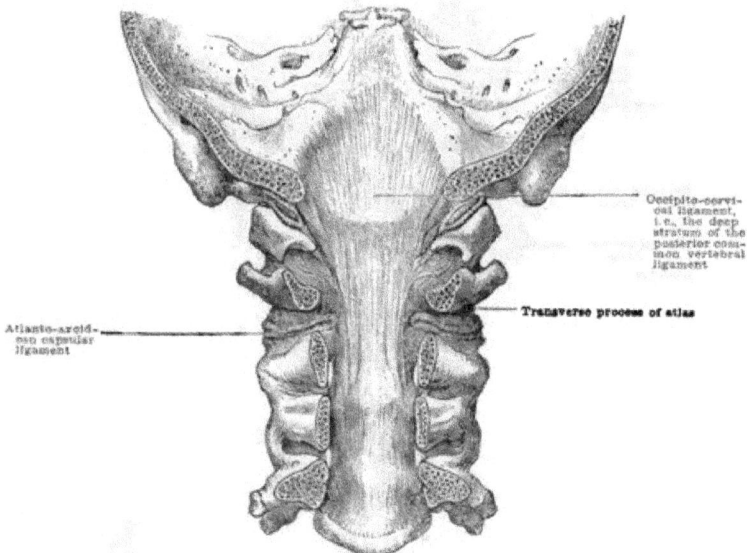

Occipito-cervical ligament, i.e., the deep stratum of the posterior common vertebral ligament

Transverse process of atlas

Atlanto-axoidman capsular ligament

FIG. 342.—THE SUPERFICIAL LAYER OF THE POSTERIOR COMMON VERTEBRAL LIGAMENT HAS BEEN REMOVED TO SHOW ITS DEEP OR SHORT FIBRES. THESE DEEP FIBRES FORM THE OCCIPITO-CERVICAL LIGAMENT.

ARTICULATION OF THE RIBS WITH THE VERTEBRÆ.

1. *Class.*—Diarthrosis, because of free movement, and a capsule.
2. *Subdivision.*—Condylarthrosis, because of no axial rotation.
3. *Osteological Units.*—Vertebræ and ribs.
4. *Subdivisions.*—Head and tubercle of rib; body and transverse process.
5. *Technical Names.*—Costocentral joints; costotransverse joints.
6. *Basis of Joint.*—A capsule lined by a synovial membrane.

THE BONY THORAX AND ITS INTRINSIC LIGAMENTS.

The bony thorax, plus certain soft parts,—as ligaments, fasciæ, muscles, and skin,—is a cavity for the protection of the major organs of respiration and circulation. Geometrical analysis of the thorax shows it to have:

1. *An apex*, formed by the manubrium, first rib, first thoracic vertebra.

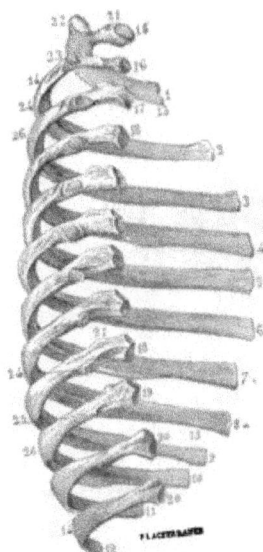

FIG. 343.—RIBS OF LEFT SIDE, POSTERIOR ASPECT.

1–12. Anterior extremities of twelve ribs of left side.   13, 13. Internal surface.   14, 14. External surface.
15. Head of first rib.   16. Head of second rib.   17. Head of third rib.   18, 18. Heads of ribs from
fourth to ninth.   19. Head of tenth rib.   20, 20. Heads of eleventh and twelfth ribs.   21, 21. Necks
of ribs.   22. Tubercle of first rib.   23. Articular facet of tubercle of second rib.   24, 24. Articular
facets of tubercles of ribs from third to ninth.   25. Articular facet for tubercle of tenth rib.   26, 26.
Angles of ribs.

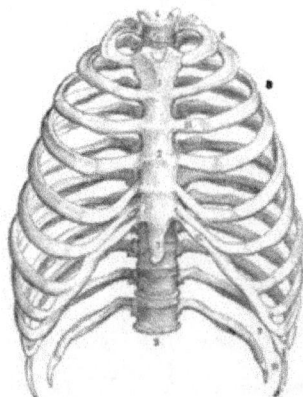

FIG. 344.—THORAX, ANTERIOR VIEW.

1. Manubrium sterni.   2. Gladiolus.   3. Ensiform cartilage or xiphoid appendix.   4. Circumference of
apex of thorax.   5. Circumference of base.   6. First rib.   7. Second rib.   8, 8. Third, fourth, fifth,
sixth, and seventh ribs.   9. Eighth, ninth, and tenth ribs.   10. Eleventh and twelfth ribs.   11, 11.
Costal cartilages.

2. *The base*, formed by the musculo-aponeurotic diaphragm.
3. *An anterior wall*, formed by the sternum and costal chondra or cartilages.
4. *A posterior wall*, formed by the thoracic vertebræ and the ribs to the angle.
5. *Lateral walls*, formed by the rib from angle to costo-chondral joint.

There are twelve ribs on each side—twenty-four ribs in the typical human thorax. Each rib articulates with the vertebral column posteriorly; the seven upper ribs articulate in front with the sternum, being called, on this account, true ribs. Of the five lower ribs, three have their ventral end secured in this manner: the eighth is attached to the cartilage of the seventh, the ninth to the cartilage of the eighth, the tenth to the cartilage of the ninth rib. The ventral ends of the eleventh and twelfth ribs are free,—*i. e.*, unattached,—and are called floating.

A typical rib, the seventh, should be studied systematically as to:

1. *A head*, which has two facets and a horizontal crest.
2. *A neck*, which intervenes between the head and the tubercle.
3. *Tubercle*, which consists of an articular and a non-articular part.
4. *An angle*, the place where the rib bends in two directions.
5. *The Shaft.*—This has two surfaces and two borders.
6. *The outer surface* is for the attachment of muscles.

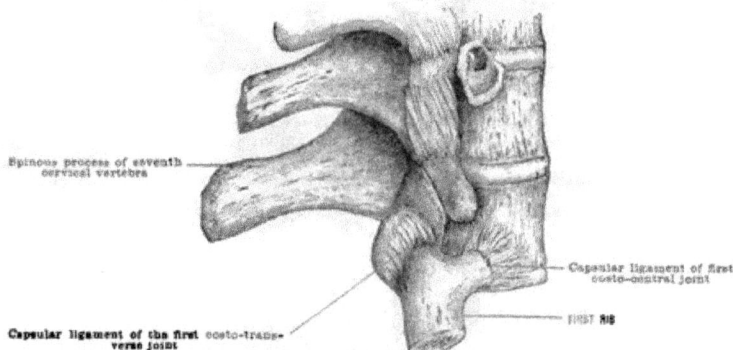

Spinous process of seventh cervical vertebra

Capsular ligament of first costo-central joint

FIRST RIB

Capsular ligament of the first costo-transverse joint

FIG. 345.—THE CAPSULAR LIGAMENTS OF THE COSTO-VERTEBRAL JOINTS.

7. *The inner surface*, has the subcostal groove for nerve and vessels.
8. *The superior border*, which is thick and rounded.
9. *An inferior border*, which is sharp and thin.
10. *The subcostal groove* lodges the intercostal vessels and nerves.

Give the rule for the articular part of the head of a rib.

There are two facets, separated by a horizontal ridge. The ridge is for the attachment of the interosseous ligament between the rib and the intercentral disc of fibro-cartilage.

What practical observation can be made regarding the lower articular facet on the head of a rib?

This one is the larger of the two, as a rule, and articulates with the thoracic vertebra which corresponds to it in number from above down. The upper is the smaller, and articulates with the vertebra next above.

Are there any exceptions to the rule of costal facets on the heads of ribs?

Yes. As exceptions, may be mentioned the first, tenth, eleventh, and twelfth ribs, each of which has one facet only on its head, as a rule.

Does the rib articulate with the vertebral column at any other point? If so, explain fully the rule.

The rule is, the articular part of the costal tubercle articulates with the transverse process of the lower of the two vertebræ with which the head of the rib articulates; thus, the first articulates with the first, the second with the second, and the third with the third.

Are there any exceptions to the rule governing the articulation of transverse processes and articular parts of costal tubercles?

Yes; the eleventh and twelfth ribs do not articulate with the transverse processes.

Enumerate the articulations of the ribs with the vertebræ.

(1) Costocentral, between heads of ribs and bodies of vertebræ; (2) costotransverse, between tubercles of ribs and transverse processes.

COSTOCENTRAL ARTICULATION—LIGAMENTS.

1. *Class.*—Diarthrosis, because of capsule and free movement.
2. *Subdivision.*—Condylarthrosis, there being absent axial rotation.
3. *Osteological Units.*—A rib and a vertebra.
4. *Subdivisional Parts.*—Head of a rib, centrum, and cartilage.

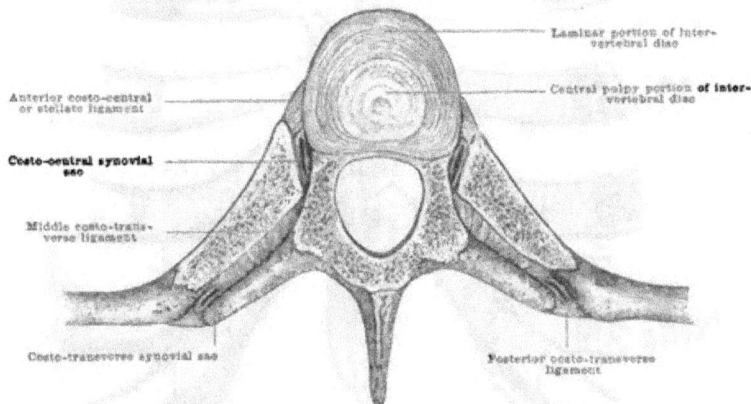

FIG. 346.—HORIZONTAL SECTION THROUGH THE INTERVERTEBRAL DISC AND RIBS.

5. *Basis.*—A capsule lined by synovial membrane.
6. *Subdivisions.*—Interarticular and costocentral or stellate ligaments.

From what point to what point does the interarticular ligament extend?

From the interarticular cartilage to the horizontal ridge on the head of the rib, between the articular surfaces of the same.

From what point to what point does the stellate ligament extend?

From the front of the head of a rib, externally, to the bodies and cartilage with which the head articulates.

Of how many parts does the articulation between the head of the rib and the centrum consist?

This cavity is divided into two distinct parts by the interarticular ligament.

COSTOTRANSVERSE ARTICULATION—LIGAMENTS.

Class.—Diarthrosis, because of free motion and a capsule.
Subdivision.—Arthrodia, because motion is simple and gliding.
Osteological Units.—A rib and a vertebra.

Subdivisions.—Vertebral transverse process and tubercle of a rib.
Basis.—A capsule lined by synovial membrane.
Accessory Ligaments.—Middle, superior, posterior, costotransverse.
Locate the middle costotransverse ligament.

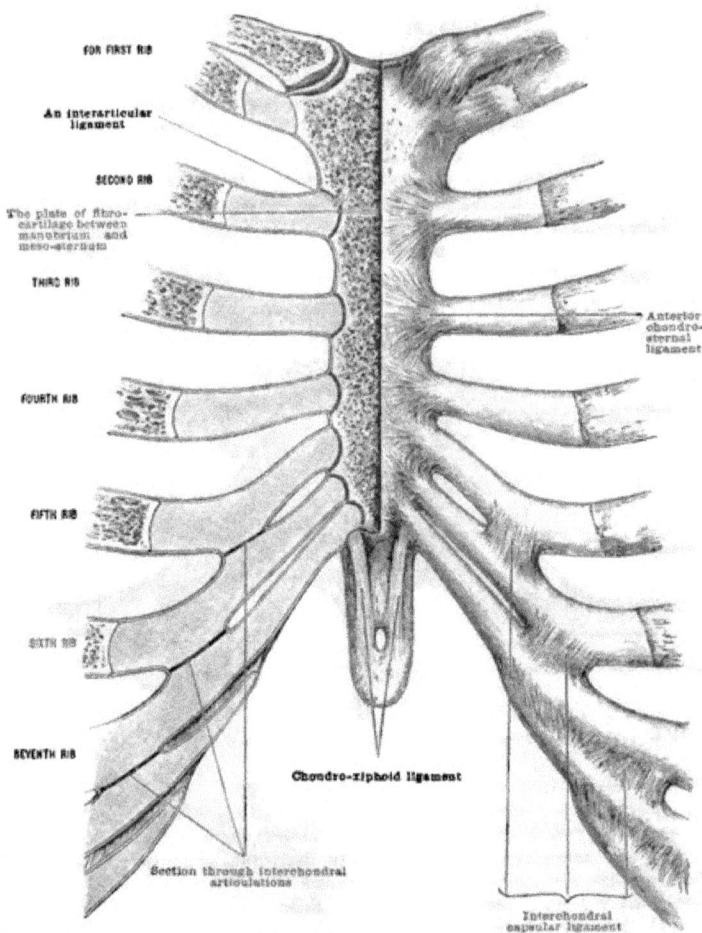

FOR FIRST RIB

An interarticular
ligament

SECOND RIB

The plate of fibro-
cartilage between
manubrium and
meso-sternum

THIRD RIB

FOURTH RIB

FIFTH RIB

SIXTH RIB

SEVENTH RIB

Anterior
chondro-
sternal
ligament

Chondro-xiphoid ligament

Section through interchondral
articulations

Interchondral
capsular ligament

FIG. 347.—THE STERNUM.
(Left side, showing ligaments; right side, the synovial cavities.)

It is between the back of the neck of the rib and the front of the transverse
process. It is called the costotransverse interosseous ligament.

Locate the posterior costotransverse ligament.

It extends from the non-articular part of the tubercle of the rib to the tip of
the transverse process.

Locate the superior costotransverse ligament.

It extends from the crest of the neck of the rib to the lower border of the transverse process of the vertebra next above.

Have all the costal heads capsules?

No; the eleventh and twelfth have none, their union being fibrous.

The anterior part of the thorax has the following articulations:

1. *Intersternal*, binding sternal ligaments together.
2. *Costo-chondral*, binding the ribs to their cartilages.
3. *Chondro-sternal*, binding the costal cartilages to the sternum.
4. *Interchondral*, binding the sixth, seventh, eighth, ninth, and tenth costal cartilages together.

The superior intersternal articulation is an amphiarthrodial joint.

The manubrium and gladiolus are bound together by an anterior and a posterior common sternal ligament and by an interosseus fibro-cartilage. This ridge or junction may be considered valuable in physical diagnosis as a guide to counting the ribs, since it corresponds to the cartilage of the second rib.

The inferior intersternal articulation is between the gladiolus and ensiform parts of the sternum.

By **this** union these two pieces form a synarthrodial joint.

THE CHONDRO-STERNAL ARTICULATION.

Wherein does the articulation of the first rib differ from the others in its articulation with the sternum?

It is synarthrodial; the others are diarthrodial, the hinge or ginglymoid variety.

Name the ligaments of the chondro-sternal joints.

1. Anterior chondro-sternal.
2. Posterior chondro-sternal.
3. Superior chondro-sternal.
4. Inferior chondro-sternal.

Explain the nerve-supply of the intrinsic thoracic joints.

(1) The costocentral is supplied by the intercostal nerves; (2) the costotransverse is supplied by intercostal nerves; (3) the chondro-sternal is supplied by the intercostals.

SOME INTERESTING FACTS ABOUT VEINS.

Veins have more extensive anastomoses than arteries.

Veins carry blood toward the heart; arteries from **the heart.**

Pulmonary veins return aerated blood from the lungs.

Systemic veins return CO_2 laden blood from the body general.

Veins are arbitrarily classed as superficial and deep.

Systemic veins terminate in the superior and inferior venæ cavæ.

Superficial veins are located in the superficial fascia.

Superficial veins communicate with deep veins through deep fascia.

Deep veins accompany arteries in one and the same sheath.

The portal vein takes blood from abdominal digestive organs to liver.

Visceral veins usually take the name of the organ they drain.

Deep veins below the elbow and knee are called venæ comites.

The larger arteries are attended by one vein only.

Veins collapse when empty; arteries remain open.

Superficial veins have thicker walls than deep veins.

31

Veins are **stronger** than arteries proportionate to their thickness.

The walls of **veins** are thinner than those of corresponding arteries.

The following **veins do** not accompany **arteries :** The large veins from bone ; veins from the skull and spinal canal ; the **hepatic veins**—these latter take all blood from the liver to the ascending **vena cava.**

The veins in the lower are thicker than those in the upper extremity.

Some **veins have** semilunar valves formed by the inner coat.

The following veins have no valves : The facial, angular, Vesalian, and ophthalmic ; the intracranial veins and dural sinuses ; the hepatic, **renal**, spermatic, ovarian, and uterine ; the superior and inferior venæ cavæ **and veins** of bone ; the veins of the whole portal system.

All veins have nutrient or trophic vessels called vasa vasorum.

The rule governing the relation of veins to arteries is as follows : Above the diaphragm veins are anterior to their arteries ; below the diaphragm, veins **are** behind their arteries, where they **are** not on the same plane. Exception to the rule, the renal vessels.

Name the veins of the head and neck and indicate any points of special or practical importance in connection therewith.

The occipital vein (Fig. 18) **is** the principal vein of the posterior region of **the scalp.** It perforates the trapezius muscle and **is** tributary **to the** deep cervical vein. It communicates by an emissary with the lateral **sinus** by the mastoid foramen. This may explain the rationale of the empiric practice of counterirritation in **the** region of the mastoid.

The common temporal vein (Fig. 18) communicates with the deep temporal **plexus of veins** by the deep temporal vein. Infection of the region of the scalp, **drained by tributaries of** this vein, may reach the cavernous sinus *via* the deep **temporal vein, the** pterygoid plexus, and the Vesalian vein. In the parotid gland **this vein unites** with the internal maxillary to form the temporo-maxillary vein.

The facial vein (Fig. 18) corresponds in distribution to its companion artery. Near the angle of the jaw it will **be** seen to communicate with the external jugular vein. The vein is tributary to the internal jugular below the jaw ; the submaxillary gland, the digastric and stylo-hyoid muscles, and the hypoglossal nerve (Fig. 31) intervening between the vein and its artery in this locality The most important tributary of this vein **is the** angular, which communicates with the ophthalmic vein. Infection **may** reach the cavernous sinus *via* the angular vein. (See p. 151.)

The external jugular vein (Fig. 18) is the **cervical** continuation of the temporo-maxillary vein. It perforates the deep cervical **fascia** behind the middle of the clavicle, and opens into the subclavian or into the **internal** jugular vein. The special feature of this vein is its inability to collapse in the region of the clavicle. The deep cervical fascia prevents collapse. This inability to collapse renders puncture of the external jugular vein liable to the admission of air.

The internal maxillary vein is the companion of the artery bearing **the same name.** This vein **returns** blood from the deep parts of the face, the muscles **of mastication,** the teeth, and the greater part of the dura mater. The veins from **these regions form a** plexus called the pterygoid. This plexus is between the **temporal** and external pterygoid muscles, around the latter muscle, and on the inner surface of the internal pterygoid muscle. The internal maxillary vein proper begins at this plexus. It unites with the temporal vein in the parotid gland to form the temporo-maxillary vein. (See p. 151.)

The pterygoid plexus communicates with the cavernous sinus *via* the vein of Vesalius ; hence infection anywhere in the region of the radicals of the internal maxillary vein may reach the dural sinuses in the same way.

The internal jugular vein (Fig. 31) begins at the jugular foramen. It is in direct continuation with the **lateral** sinus. It conveys all the blood from the brain that is distributed to this organ by the circle of Willis. This internal jugular vein, by its confluence with the subclavian vein, forms the innominate—on each side. In the neck the vein accompanies the internal carotid and common carotid arteries, in the upper and lower parts respectively. At the jugular foramen the vein is in relation with the ninth, tenth, and eleventh cranial nerves, which also emerge from this jugular foramen. This vein has a pair of valves two inches above its terminus.

The Innominate Veins.

Explain fully the innominate or brachio-cephalic veins on the right and left sides.

These veins are formed by the confluence of the subclavian and internal jugular veins. (Fig. 150.) The left innominate is much the longer of the two. It crosses all the branches of the second part of the arch of the aorta, both pneumogastric nerves, the trachea, and the œsophagus. Tributary to it are both inferior thyroid veins. This vein meets the innominate of the right side to form, with **the** vena azygos major, the descending vena cava.

Veins of the Upper Extremities.

These are both deep and superficial. The deep veins in the forearm are the radial and ulnar, two for each artery, called radial and ulnar venæ comites respectively. They are connected by transverse branches, one to the other. The deep veins of the forearm unite in the retiring angle of the elbow (cubital fossa) to form **the** brachial vein or veins.

The Brachial Vein or Veins.—This vein extends from the bend of the elbow **to** the lower border of the latissimus dorsi tendon—*i. e.,* from the cubital fossa **to** the base of the axillary space. Where there are two veins on each side **of** the artery they communicate freely with each other and with the superficial veins opposite to them in the superficial fascia. Sometimes the brachial receives the basilic vein.

The axillary vein is the continuation of the brachial through the axillary space. This vein has a valve in its lower one-third. It is in very close relation with two axillary glands, on account of which this vein is of great surgical importance in the operation for the removal of infiltrated axillary lymphatic glands. This vein conveys nearly all the blood of the upper extremity. The axillary vein is noted for the following large and important tributaries which it receives

1. The circumflex (anterior and posterior).
2. The three subscapular veins.
3. The long thoracic vein.
4. **Small** and numerous **alar** thoracics from **the axillary glands.**
5. Short thoracic veins, the highest in the **axillary space.**
6. The long thoracic or external mammary **vein.**
7. The acromio-thoracic or thoracic axis vein **of** Morris

The subclavian vein is the upward continuation of the axillary vein. It unites with the internal jugular vein behind the sterno-clavicular articulation to form the innominate or brachio-cephalic vein. On the outer surface of the first rib it lies in a depression, and is separated from the subclavian artery by the scalenus anticus muscle. In front of the **vein** are the clavicle and the subclavius muscle. Tributaries are the cephalic and the external and anterior jugular veins. On the left side, the thoracic duct opens into the beginning of the innominate, between the subclavian **and** internal **jugular** ; on the right side the right lymphatic duct opens **into a corresponding** point of the right innominate vein.

The superior vena cava is formed by the confluence of the two innominate **veins.** It begins near **the junction of the sternum** and first rib on the right side.

It is about three inches long. About one-half of the vessel is in the pericardium. It returns practically all the blood to the heart from above the diaphragm. Its minor tributaries are the vena azygos major and minor and the pericardiac, mediastinal, and œsophageal veins. The vena cava superior has no valves, but its tributaries are provided therewith. The vessel opens into the right auricle of the heart.

THE VEINS OF THE LOWER EXTREMITIES

Are but superficial and deep. The superficial veins are very numerous, anastomose quite freely, but only two are of sufficient import to be entitled to special names. The general name by which structures in the superficial fascia are designated is cutaneous or superficial. The special names for the two important veins of the lower extremity are: (1) Long saphenous; (2) short saphenous.

The long saphenous vein begins at the inner end of the dorsal arch of the foot, passes in front of the inner malleolus, behind the inner tuberosity of the tibia and the inner condyle of the femur, and passes through the saphenous opening in the fascia lata to become tributary to the common femoral vein. Below the knee this vein is attended by the long saphenous nerve. This vein has valves. It communicates with venæ comites in its course. Above the knee its tributaries are the superficial external pudic, the superficial epigastric, the superficial circumflex iliac, and the internal and external femoral cutaneous veins.

The short saphenous vein begins at the outer end of the dorsal arch of the foot, passes behind the outer malleolus, gains the mid-line of the leg posteriorly, and pierces the popliteal fascia to become tributary to the popliteal vein.

The deep veins of the lower extremity accompany the arteries, and take the same name, where there is but one vein to attend the artery; where, however, there are two, as is the case below the knee, then the term venæ comites is used. About one inch below Poupart's ligament the common femoral vein is formed by the confluence of the superficial and deep femoral veins. The common femoral vein is in the femoral sheath, between the common femoral artery and the femoral canal. Above Poupart's ligament this vessel is continued upward, under the name of external iliac vein. Opposite the junction between the sacrum and ilium the external iliac becomes confluent with the internal iliac vein, to form the common iliac vein.

The ascending vena cava is formed by the confluence of the common iliac veins. It begins opposite the fourth or fifth lumbar vertebra. It lies to the right of the abdominal aorta. It passes through a groove on the posterior surface of the liver. It passes through the caval opening in the diaphragm and opens into the right auricle of the heart. The sacra media vein is the first tributary, the hepatic vein the last tributary, that the vena cava receives in its course. Between these two extremes the tributaries are numerous and important: the lumbar, right spermatic or ovarian, renal, suprarenal, and phrenic. The left spermatic vein is tributary to the left renal, as is also its homologue, the ovarian.

The portal vein is formed by the gastric, splenic, superior mesenteric, and inferior mesenteric arteries. These veins have no valves. Blood coming from the abdominal organs is laden with bile and sugar, both of which are removed from the portal blood by the liver. The bile is stored up in the gall-bladder; the sugar is stored up as potential energy in the hepatic cells. The portal vein enters the root of the liver, between the hepatic artery and the common bile-duct.

INDEX.

477

www.ingramcontent.com/pod-product-compliance
Lightning Source LLC
Chambersburg PA
CBHW031932220326
41598CB00062BA/1710